AF843045

Mathematical Psychology

Mathematical Psychology

The World of Thoughts as a Quantum Space-Time with a Gravitational Core

Norbert Schwarzer

JENNY STANFORD PUBLISHING

Published by

Jenny Stanford Publishing Pte. Ltd.
101 Thomson Road
#06-01, United Square
Singapore 307591

Email: editorial@jennystanford.com
Web: www.jennystanford.com

British Library Cataloguing-in-Publication Data
A catalogue record for this book is available from the British Library.

Mathematical Psychology: The World of Thoughts as a Quantum Space-Time with a Gravitational Core

Copyright © 2025 by Jenny Stanford Publishing Pte. Ltd.
All rights reserved. This book, or parts thereof, may not be reproduced in any form or by any means, electronic or mechanical, including photocopying, recording or any information storage and retrieval system now known or to be invented, without written permission from the publisher.

For photocopying of material in this volume, please pay a copying fee through the Copyright Clearance Center, Inc., 222 Rosewood Drive, Danvers, MA 01923, USA. In this case permission to photocopy is not required from the publisher.

Cover Picture: The Path of Desire: "Knowing Desire—Thoughts Are Freer than the Thinker," by Livia Schwarzer.

ISBN 978-981-5129-27-4 (Hardcover)
ISBN 978-1-003-58203-8 (eBook)

"For the right mind it always holds that thoughts are freer than the thinker!"

Contents

Abstract

In this book we are going to discuss the usage of mathematical tools for a fundamental description of the human psyche. Thereby we walk the graph every clever clinical psychoanalyst would take when performing the anamnesis of a patient. We use a set of attributes, allocate them to the human being and try to adjust the attributes' (or properties') scale to our observations. This set of attributes, then describing our individual, mathematically forms a space-time, and by assuming that this space-time (a) is of Riemann character and (b) has to be subjected to a Hamilton extremal principle in order to allow the object (or system), described by, respectively residing within the space-time, to exist in a universe governed by extremal principles, we simply require a mathematical apparatus, which can handle arbitrary—and rather general—ensembles of attributes. This apparatus already is at hand. It was introduced by David Hilbert in 1915 [1] and it directly leads to the so-called Einstein–Hilbert action. Solving the corresponding variational problem gives generalized Einstein field equations [1, 2], which—most interestingly—also contain all major quantum equations [3–5]. We will see that it seems to be that what classically in quantum physics has been known as the wave-function makes out what we see as individuality.

We will also see that it seems to be possible for "certain minds" to bypass condition A (the Riemann condition) and produce so-called g=0-spaces of thought. We will cover this unconventional "physics of the mind" in appendix R, sub-section "5.18.7 The Conventional and the Unconventional Physics Realm" and we suspect that it is the conscious mind which not only makes use of this option but actually needs the ability to produce such mind states in order to be conscious. Yes, we are going to work out the quantum gravity physics of the mind and along the way come across a set of options, which seem to be necessary for consciousness to actually be. These options are:

(a) The ability of the mind to set itself, respectively, parts of it (ensembles of ideas or systems of thought) into states of non-Riemann character;
(b) The ability to steer the individuality or subjectivity via the uncertainty within the variation as
  (i) Volume uncertainty, which is to say, a scalar factor of the metric to the system, we see as the mind-system or the sub-mind-system in case we only consider parts of the mind;
  (ii) Ricci-uncertainty, which is to say, a scalar factor of the scalar curvature presented in the Einstein–Hilbert action as Ricci-scalar;
(c) The ability of a conscious mind to actively switch sub-systems of thoughts from fermionic to bosonic behavior and vice versa;
(d) The ability to apply a local cosmological constant to achieve a greater flexibility in dealing with the Einstein–Hilbert action task;
(e) The ability to actively adjust the mind's dimensionality via symmetry-breaking and/or x-polarization (x denotes the number of different charges) thereby thwarting the covariance principle [6].

Further below in this abstract, we will come back to this "other physics realm."

Taking on this basic conjecture or rather ensemble of conjectures leads to a variety of insights regarding the inescapable escape of "intellectual property," the inoculation of ideas, the tunneling of alien "ideologies" through an individual's mind guard, the storage of thought, the emittance of confidence, the impossibility of thought prisons (Thoughts are always freer than the thinker!) and many other things.

The subsequent recipe one can extract from the equivalence of quantum gravity field concepts and individuality, regarding one individuum, does seem to lead to some kind of "molecular dynamics" (multi-dimensional and complex, though) when considering ensembles of individuals. This concept will be discussed in connection with the problem of mass formation psychoses.

The most important outcome of the book may be seen in the proof of the following sentence:

"For the right mind it always holds that
thoughts are freer than the thinker!"

Thereby we will only need a quantum gravity derivation of the Hawking radiation out of a black hole and connect it to our newly found knowledge about the nature of thoughts, ideas and the conscious (awake, not woke) mind in general.

Further, as already said above, it will be worked out in this book that not only the general recipe brings us the conventional physics in form of a quantum gravity for everything, but that there is even more. Above we simply called it the "unconventional physics," but especially in connection with the aspect of "love" and everything having to do with it we are looking for a somewhat more illustrative expression. We will therefore start to consider also solutions to a physics realm which more or less "correctly" has been described throughout the history of mankind as "chaos." As said, the corresponding math will be presented in appendix R, sub-section "5.18.7 The Conventional and the Unconventional Physics Realm." This author sees this as an essential aspect of the attempt to holistically try and give tools for the human mind for all its possible states of existence. There may be purists who will claim that this has nothing to do with science and thus should not be presented in a book like this, but as it will be shown that the math for such an unconventional physics exists and that it even also follows from a Hamilton extremal principle, we see the presentation of such possibilities almost as a must in order to justify the claim of a true holistic approach. The same holds for questions about "life after death" and states of the mind where the thinker claims "to have left reality"—or at least had the feeling of having done so—and entered "another world" or a "different form of existence." This author would be a bad caricature of a scientist if he just dismissed such claims as esoteric nonsense only because they did not fit into his classical physical understanding. Instead, he prefers to look for the uncharted land in the fundaments of physics so far in order to make sure that nothing is overlooked. Yes, there may be many false priests and lying shamans who only pretend to "have access to other realities," but it would be very unscientific to sentence them all ... at least this should not be done without a fair trial.

Attention

Please note that the author does not consider this book a completely independent work. Instead, it has to be considered a rigorous theoretical contribution (if not to say mathematical tool box) to a project this author is currently undertaking with many partners from a great variety of fields. In here, we will mainly concentrate on "mathematical psychology," but it should be pointed out that due to the general applicability of the math being used, it may well happen—here or there—that other fields are being mentioned or even discussed. Usually, however, there will be a significant bearing on the field of psychology.

For convenience, we will—partially—use local references throughout the book.

After a brief and purely verbal introduction about the mathematical approach, we proceed with a motivator section, followed by applications and the appendices, which contain most of the derivations and fundamental mathematics necessary for a quantum-gravity-based mathematical psychology.

References

1. D. Hilbert, Die Grundlagen der Physik, Teil 1, *Göttinger Nachrichten*, 1915, 27, 395–407.
2. A. Einstein, Grundlage der allgemeinen Relativitätstheorie, *Ann. der Phys.*, 49, ser. 4, 769–822.
3. N. Schwarzer, *The Theory of Everything—Quantum and Relativity Is Everywhere—A Fermat Universe*, Jenny Stanford Publishing, 2020, ISBN-10: 9814774472.
4. N. Schwarzer, *The World Formula: A Late Recognition of David Hilbert's Stroke of Genius*, 2022, Jenny Stanford Publishing, ISBN: 9789814877206.
5. N. Schwarzer, *The Math of Body, Soul, and the Universe*, Jenny Stanford Publishing, 2023, ISBN 9789814968249.
6. N. Schwarzer, *The Covariance Principle Is Dead, Long Live the Covariance Principle: How Quantum Gravity Kills a Cornerstone of Physics*, self-published, Amazon Digital Services, 2023, Kindle, ASIN: B0C96L41LS.

About the Mathematical Approach Applied in This Book

Many people are of the opinion that it is impossible to mathematically describe human behavior in a satisfactory manner. Thereby they accept millions of artificial intelligence solutions doing exactly this in every second and with respect to almost every aspect of their life. Even more people are of the opinion that such impossibility regarding the mathematical description, modelling or simulation holds even stronger for the conscious human mind. Thereby—again—they overlook that advanced artificial intelligence systems are already extremely close in exactly doing that and that in some cases such systems appear to be even better than the original thing.

It is widely agreed that human beings can be characterized by a list of attributes, which partially entangle with each other ... thereby we mean that the attributes entangle, not the humans. Even though this list may be extremely long and the interaction (or entanglement) of the various attributes might be very complex, in essence we are characterizable by such a list. After all, nothing else is it when we try to—more or less comprehensively and fairly—describe another person or—usually with much less vigor, enthusiasm and rigorousness—ourselves.

Accepting the fact that humans can be described by a list of entangled attributes, however, still does leave us with the question how to mathematical handle this list in order to describe the processes of action, human interaction, individuality and other characteristics. Most surprisingly, this work had been already done. Over 100 years ago it was the great mathematician David Hilbert who found a way to put a set of properties, attributes, degrees of freedom or—as all these other terms are just coding—**dimensions** into a mathematical formalism, which was already in his time known for years as the one fundamental principle for the determination of the laws of our universe, the Hamilton extremal principle. All truly fundamental natural laws originate from this principle. The only problem is to find the right ingredients to "feed the monster of the Hamilton principle." The essential ingredient namely, the Lagrange density function, seems to differ from problem to problem and this cannot be when assuming that the Hamilton mechanism should be truly fundamental. Here now came the stroke of genius of Hilbert [1], who realized that a given set of attributes is nothing else but a set of dimensions and, thus, forms a mathematical space or space-time. The corresponding Lagrange density function is then always the same, namely the so-called Ricci scalar of the very system or space-time. Hilbert applied his calculus on our ordinary 4-dimensional space-time and derived the Einstein field equations and thus, obtained in just one fluid and conclusive derivation the complete mathematical basis for the General Theory of Relativity of Albert Einstein [2]. Admittedly, Hilbert never extended his calculus to the space or space-time of attributes characterizing human beings, but the fact that the title of his work [1], in which he derived Einstein's great theory, translates "The Fundaments of Physics" clearly shows us that he saw his approach in a much more general way than just the minimum-principle-origin of Einstein's field equations. Thus, no matter the system we are considering, the moment we have its basic attributes and know their entanglement (various interactions), we also immediately know how to place all this knowledge into Hilbert's calculus to come up with a most fundamental and unbiased model for this very system.

Hilbert's creation, de facto, was—no, is—a Theory of Everything [3] or World Formula [4, 5], even though he himself had little chance of fully realizing this. Even in physics, where we now can show that Hilbert's fundamental equation covers both great theories, the General Theory of Relativity and the Quantum Theory, the time was not ripe for such a discovery, simply because the mathematical apparatus of the Quantum Theory was not fully developed then. While Hilbert brought out his great work in 1915 and knew about the Einstein field equations at the time, the basic quantum equations like the Schrödinger, the Klein–Gordon and the Dirac equation would not follow before the second half of the 1920s.

For convenience, falsifiability and rigorousness, we not only present the Hilbert approach adjusted to our arbitrarily dimensional space-times in the appendices of this book but also derive the Klein–Gordon, the Schrödinger and the Dirac equation (the main relativistic quantum equations), because it will be our mathematical fundament for the consideration of a variety of psychological processes like the falling in love, like decision making and like mass formation.

References

1. D. Hilbert, Die Grundlagen der Physik, Teil 1, *Göttinger Nachrichten*, 1915, 27, 395–407.
2. A. Einstein, Grundlage der allgemeinen Relativitätstheorie, *Ann. der Phys.*, 49, ser. 4, 769–822.
3. N. Schwarzer, *The Theory of Everything—Quantum and Relativity Is Everywhere—A Fermat Universe*, Jenny Stanford Publishing, 2020, ISBN-10: 9814774472.
4. N. Schwarzer, *The World Formula: A Late Recognition of David Hilbert's Stroke of Genius*, 2022, Jenny Stanford Publishing, ISBN: 9789814877206.
5. N. Schwarzer, *The Math of Body, Soul, and the Universe*, Jenny Stanford Publishing, 2023, ISBN 9789814968249.

1

The Motivator

Dedicated to all children who don't let themselves be forbidden to think for themselves—no matter how difficult it may be or how much it may hurt ...

But remember:

"For the right mind it always holds that thoughts are freer than the thinker!"

"Common Field Mouse, Comrade Major!"

My Battle of Seelingrad and the True Ending of the GDR
Dr. rer. nat. habil. T. Bodan

Mathematical Psychology: The World of Thoughts as a Quantum Space-Time with a Gravitational Core
Norbert Schwarzer
Copyright © 2025 Jenny Stanford Publishing Pte. Ltd.
ISBN 978-981-5129-27-4 (Hardcover), 978-1-003-58203-8 (eBook)
www.jennystanford.com

1.1 Information about the Motivator Book

1st illustrated edition, April 2022

With illustrations by Livia Schwarzer (14)

ISBN: 9783756202218

Production and Publishing: BoD—Books on Demand, Norderstedt, Germany

Text and Cover Copyright © 2015 Dr. T. Bodan

Translated into English by Peggy Heuer-Schwarzer, ESAE

Cover picture & design: Peggy Heuer-Schwarzer, ESAE

Second picture (including title picture) with many thanks to E. Gucinski, 2021

All rights reserved

1.2 Preface

This story really happened this way, but it may well be that we do not mention one or the other name or have to present one or the other episode with a little bit modification. This only partly has to do with our memory, but rather with the fact that the heroes of that time do not necessarily want to walk around with a sign "I was a hero of Seelingrad" and the much more numerous twerps and sneaks of the regime of the German Democratic Republic (GDR) hopefully feel stupid enough as it is. Meaning, even without naming them with rank, name and day number.

The leadership of the GDR thought it was a great idea to call up all of its male students to the military reserve force at the beginning of their second academic year in order to, on the one hand, put them on the right track with the omnipresent military-political line and, on the other hand, to give them the tools for a career as an officer in the reserve force of the National People's Army (NVA) which was loyal to the Socialist Unity Party of Germany (SED) and the GDR regime.

In the fall of 1987, however, this went massively wrong. Two years later, the GDR fell. The culprit—as only insiders know—was a small rodent, a "common field mouse."

1.3 Seelingrad

The village is not really called that. Therefore, it cannot be found on any map, and since we do not know how the people who live there today will react to the fact that now, after more than 25 years, the true role of this former bulwark of the GDR regime is finally being revealed, we do not want to name the location. In this regard, the fighting power of the place never lie in the strength of the garrison stationed there but always in the psychological depth of the ways of thinking and behaving that were indoctrinated there. The fact that the deliberately preserved desolation of the area, which could hardly be surpassed in depressive capitalism, especially in the fall, was used mercilessly to get the minds of the young students as small and at the same time as nasty and inhuman as possible, was in the deep logic of the GDR leadership. The region had a lot to offer, but it took unbelievable effort and some resistance to even raise one's head out of the mud enough to see more than the mud- and clay-smeared tips of one's own boots or the heavy ammunition belt of the man in front. However, those who managed to pay attention to more than the shouts of superiors during the daily forced marches and assault drills could marvel at impressive sunrises and sunsets. For those who managed to elicit more from the area than the next best dry spot to throw themselves at the command "take cover!" without getting too wet, they saw an impressive abundance of rare plants and creatures and heard a wonderful variety of bird calls ... even in the fall. Inside the bulwark, however, also called the training base, there were no such moments of relaxation and acoustic-visual refuges. Here, psychologically perfect and ideologically deliberate attention was paid to maximum emptiness and desolation in order to keep instructors and trainees trapped in maximum depression and thus, politically focused.

Nevertheless, as a small geographical hint, one may be told that a frustrated guard named Martin Konrad F. carved the slogan "Better sperm in the bulky waste than uranium in the urine!" into one of the nationally owned watchtowers on October 7, 1987, Republic Day.

1.4 The Private

He was different. He stood out. Yet it would never have become clear to an outsider without help what the reason was for this conspicuousness, the obvious standing out from the crowd. Admittedly, compared with many of his fellow students, he was a handsome figure at a good 2.04 meters tall and with an enormous torso, but by no means so unusual that one could have explained the constant turning of heads and the respectful murmuring in his vicinity. Anyway, the outsider, the uninitiated probably saw the special only in the behavior of his environment, rather than in the appearance of the Private himself. Even if one could then definitely see at a closer look that the guy was somehow different. He walked differently, he carried his things, but especially his heavy weapon, as if they were made for him. Above all, however, he held his head differently, that is high up. While all his fellow students went colorfully and individually into the "clothing and equipment storage" and then came out uniformly green and oppressively pressed, one somehow noticed no real difference in this Private. It seemed as if one could have put a bag over him and he would have remained just him ... also in appearance. The only thing that was really visible and "tangible" if you will, apart from the fact that only this Private carried a heavy RPG6 and not, like all his fellow students, the lightweight Kalashnikov machine gun, was the fact that the Private had the lowest rank among all the thousands of students. He was a simple soldier, a private[1], while everyone else had "made it" to at least non-commissioned officer in the obligatory and regular service of the GDR's National People's Army before going to university. That was actually impossible, because as a private in the GDR, you could not actually get a decent place at university. At least not without "proving oneself" in the NVA at the same time, so that one was also promoted. It should be noted here that in the GDR, and even more so in the NVA, proving oneself simply meant not stepping out of line. In general, the military system provided for automatic promotion to at least lance corporal even for the most recalcitrant "comrades," as all NVA members had to call each other. According to this, something very special must have happened that allowed this soldier here to remain a simple Private and yet, and this was the totally crazy thing about it, to wear the identification of the paratroopers and, on top of that, to study at university. It was just a tiny, narrow band of orange paint around his otherwise blank and, especially, stripe-less shoulder pieces, and yet this was the epitome of the unbelievable, the impossible, the "this can't be," even the rebellion.

But it wasn't just their own fellow students who were amazed. The training officers, who were actually standing here to drive the "gang of effeminate academics," stopped their shouting and stared at the newcomer with the heavy weapon dangling casually over his shoulder. Actually, they should have intervened now and admonished to the "suit and marching order," even shouted, but they were just silent and quite obviously did not understand the world anymore. Only the "elderlies," meaning the real soldiers, who had to serve here on the training base, behaved differently. With them, there was no

[1]In the National People's Army (NVA), the ordinary soldiers were subdivided threefold: the "private" had the lowest rank and was a nobody. He was followed by the "lance corporal," who in turn was followed by the "staff corporal."

stiff astonishment, but a kind of casual grin, a welcoming nod that one gives to a highly respected comrade-in-arms. They all wore combat gear and the good old AK47 at the ready. The leadership of the facility had put the elderlies here for intimidation and apparently normally it worked, because none of them liked the students. For them, these were all arrogant idiots who could go home again after six weeks, while one still had at least 190 days and more to go in this disgusting hell. The leadership wanted and promoted this hatred of the student recruits and each of the elderlies here had been carefully selected—without ever realizing it, of course—exactly for this purpose. It was not uncommon for them to be penal transferees or real criminals, and this also applied to a large extent to the officers. However, when the elderlies saw the Private, they forgot their hatred for a moment. They grinned and—this was already bordering on mutiny—saluted as the "Academic Soldier of the Reserve" walked past them. The Private simply nodded at them, while paying no attention to the rest of the staring and gesturing. At the end of the line of instructors, without paying attention to the officer there, he sheered out of the line of pressed men and addressed one of the long-serving elderlies. Nobody did hear what he said, but the bystanders saw the elderly jump in attendance and give a brief reply. The Private nodded, turned and headed purposefully to his billet, which the elderly had directed him to. Behind him, the elderly saluted with clicking heels.

1.5 The Election

Thanks to the shortcut the elderly had shown him, the Private was the first in his room. He looked for a bed near the window—there were six double bunk beds—and a locker nearby. Needless to say, he slept upstairs. On the table, he discovered the magazine *Armeerundschau* and the journal *Neues Deutschland*[2]. Disgusted, he pushed the newspapers aside and threw his duffel bag on the table. By the time his roommates moved in, he had already put away his locker, was lying on his bed, and was scribbling formulas from a book on nuclear physics on a small piece of paper.

"Hi Sven," a voice well known to the Private called out, "Looks like they didn't completely tear us apart after all."

The Private looked up. Sure enough, there stood one of his fellow students, someone everyone just called Schmitti, the top of the class, and next to him six more from his study program. All of them were physicists or students who wanted to become physicists. The Private grinned and replied, "No, they must have thought you were going to enforce some etiquette and discipline on me."

Schmitti just laughed hoarsely and made a dismissive hand gesture. The man next to him, a very unmilitary-looking, delicate non-commissioned officer named Jan, was about to open his mouth when a resounding "ATTENTION!" rang out. One of the elderlies had entered, immediately followed by a lieutenant, and loudly announced the superior officer. Even in here, the elderlies were running around with their assault rifles at the ready. Everything jumped to a standstill. Only the Private remained lying on his bed, engrossed in his book again. The lieutenant looked over at him briefly and then called out rather disinterestedly, "At ease, men!" He was one of the few upright men who had started an officer's career in the NVA out of conviction and then realized much too late what a lying and inhuman association this army was. "Let's not make life difficult for each other here, people. You want to get your six weeks over with and I basically want my peace and quiet, got it?"

"Yes, Sir!" came a general murmur, but the officer repeated the question. When no one responded this time, the Private looked up and realized that all eyes were on him. He eyed the officer and, after what felt like an eternity, gave him a nod that was not necessarily unfriendly, almost encouraging. The officer seemed relieved and explained the day's tasks.

---

[2]*Armeerundschau* (ar) = *Army Review*, magazine published by the Military publishing house of the GDR; *Neues Deutschland* (nd) = *New Germany*, journal published by the German Party SED.

"You have to appoint a group leader. I should give you a few instructions, but I see that you are all at least non-commissioned officers and you don't need such instructions. Besides, you need a cashier to collect the money for the newspapers."

"What newspapers?" asked one of the four, whom the student did not know from his course.

"These!" the lieutenant pointed to *Armeerundschau* and *Neues Deutschland* on the table.

"But we didn't order those!" protested Schmitti.

"Come on people, don't make trouble and deliver the money by the end of the week," the lieutenant already seemed annoyed. Still, his tone was rather conciliatory, intent on persuasion and with just a tiny hint of "You guys don't stand a chance anyway!"

"We all know ...," but he didn't get any further, because at that moment another officer entered, a first lieutenant, one rank above the lieutenant.

"What's going on here?!" he roared, "Why is that soldier lying there on the bed?"

The Private scanned the newcomer with a glance and knew that this jerk was one of the "old farts" who had only gotten their study places because they had committed themselves to the NVA for eternity and three days. Regularly the worst students, they could now show the "young academics," who were otherwise far ahead of them, that they were the masters and were in charge. The little busybody crossed the room at a furious storming pace and stood up at the head of the bed, where the Private was still lying calmly, eyeing him. Apparently assuming that his fellow officer, who was firmly serving here, needed support, the "chief nut of the reserve," as these guys were secretly called, now yelled directly at the Private.

"Why are you lying here in bed? And then with boots on! There's an officer in the room. Get the hell down and report!" like machine-gun fire, he screeched each of his sentences. He almost spat; he was so excited.

"AN OFFICER IS PRESENT!

How dare you, you ..."

Slowly, the Private sat up. His impressive upper body and bald head did their work. The first lieutenant fell silent with his mouth half open. The Private leaned forward, grasped the bed rail and slowly but fluidly, he came to stand directly in front of the reserve officer with an inverted upward circle forward. To execute such a movement so slowly cost normal people, if they managed it at all, an incredible amount of strength and energy. With Bodan, it looked as if he would never climb beds or leave them any other way. He stood casually, his eyes fixed on the first lieutenant, slowly pulling his headgear out of his pants pocket. The angry officer's jaw dropped when his obviously very strong and skilled counterpart, without taking his eyes even a whit from his opponent, put on the NVA paratrooper's beret and pointedly calmly adjusted it on his bald head. When he was done, he suddenly pulled his hands down, slammed his heels together so hard, and yelled in the face of the man standing directly in front of him:

"Correct, C O M R A D E!

ONE

officer is present."

Then he turned away, skillfully and briskly turned to the lieutenant, whose rank, formally speaking, is actually below that of the first lieutenant, and to the elderly who had accompanied the lieutenant, slammed his heels again, shouted loudly and surely, "Comrades, attention!" and saluted in a military manner so accurate that at that moment no one had any doubts that someone here must have "sufficient practical experience." Involuntarily, as if they had received an electric shock, all those present jumped to a standstill, even the mingy first lieutenant. Who, realizing his mistake, embarrassedly tried to conceal the movement, which made him look even sillier. The so-honored lieutenant grinned.

The elderly was the only one who did not seem to have any problems with the situation. He made a brisk turn towards the Private, shouldered the rifle in parade style, and saluted back with a straight face. An awkward silence fell, but only for a brief moment. Then Schmitti also turned emphatically accurate, although not as perfect as the elderly in the direction of Bodan, also slammed his heels loudly and raised his hand likewise. Immediately the other fellow students followed. The lieutenant seemed to be thinking. Finally, he too raised his hand to his cap and saluted in the direction of the Private.

"What is your name, Comrade Private?" he asked.

"Bodan, Comrade Lieutenant."

"Thank you, Comrade Private ... Comrade Bodan, you may make yourself comfortable again!"

With that, the Private fixed his gaze again rigidly on the reserve officer in front of him, calmly tucked his beret into his pants pocket, and then made an upward circle forward back onto his top bed just as he had come down. To promptly drown out the awkward silence, the lieutenant called out:

"Well Comrades, that's it then. In two hours at the latest I want to know your group leader, including deputy and treasurer. Thank you! Carry on!"

With that, he left and the elderly with the AK47 followed him. His grin had almost taken on grotesque features. The first lieutenant stood around awkwardly in the room for a moment, but in the end disappeared without another word.

"Oh dear!" Schmitti just said, scratching his head thoughtfully.

For quite a while, there was silence and everyone looked at each other, more or less embarrassed, all except the bald Private, who was again engrossed in his book and scribbles. Finally, Jan, one of the non-commissioned officers, went to the door, closed it and started laughing unrestrainedly and uproariously—just as if he had had an incredibly funny flash of inspiration. His fellow students looked at him questioningly at first, but at some point, they joined in as well. Even the Private, lying on the bed, laughed into his book without really knowing what Jan or the others found so funny. When silence finally returned, the lanky Jan walked up to the Private, put his delicate hand on his massive upper arm, and said in a deadpan tone, "I'm going to propose you as our squad leader now, and I'm sure the others will go along with that." He paused and looked around. Schmitti and some others nodded immediately, but two guys, whom nobody else in the group knew, dropped their jaws, and another guy obviously looked as if he thought he had misheard. Jan did not let himself be distracted, however, and said quite calmly and well-formulated, as if he were quoting from a manual, "Group leaders shall always be those outstanding soldiers who have distinguished themselves by combat experience and willingness to serve and who have earned the respect of their comrades. Rank and grade in previous periods of service play a subordinate role. On the contrary, attention is to be given to maximizing the combat effectiveness of the group."

He, the lanky little fellow, looked provocatively into the faces of the three obvious doubters and continued, "Is anyone here seriously suggesting that he could demonstrate more fighting prowess than Sven here, never mind in what field ...," he nodded in Bodan's direction, "Or even that he would have our respect?" No one reacted. You could see that the brains of the two strangers were working in panic, but they just couldn't think of anything so quickly. Meanwhile, the one with the hearing problem shook his head vigorously. "Good!" Jan now continued, "With that we have a proposal—a good one, I think, not to say the best—and proceed to the vote." Again, he paused and looked at Schmitti. He grinned and nodded his thumb down at the two strangers. Jan understood immediately and called out, "Ok, whoever is against our experienced paratrooper, please speak up now!"

You could literally feel the shock waves emanating from the features of the two strangers, but they remained silent and Jan continued, "So no votes against. Any abstentions?"

This time Jan barely paused but announced, "With that, Comrade Bodan is unanimously elected group leader. Very good then!"

Jan turned to the elected one, "Do you, Comrade Private, accept this election?" he asked.

Sven Bodan pulled his lower lip over his upper lip and pretended to think hard. Then he grinned broadly and said, "But with me at the helm, we won't pay that shitty newspaper money, that's for sure!" All but the two strangers cheered and the one who had just been hard of hearing shouted at the top of his voice a 100% correct military "Hooray!" in such a piercing and crystal-clear voice that everyone froze. When the others looked at him questioningly, he just laughed and said, "Sorry, classical vocal training and then three years of music corps Leipzig."

"Good Lord!" wailed Schmitti, "a singer ... We really need all the help we can get."

But the Private rubbed his chin with a smile and asked the singer for his name.

"Non-commissioned officer Kramer, Comrade Private!" he shouted and saluted briskly.

"The training at the music corps certainly included military and working class and battle songs, didn't it?" asked Bodan with interest.

"Yep, the full program. Plus, a huge range of soldier songs in Russian."

"Perfect!" said Bodan, and no one in the room understood what excited him about it ... no one except Jan and Schmitti, perhaps.

"Great!" said Jan, turning back to the Private, pointing to his comrades, "Your squad!"

The Private just nodded, came down from his bed in a proven manner by a reverse upward circle forward and stood in front of "his" people.

"Good!" he said after quite a while, "There is work to be done. Jan, I know it's not requested, but I have to ask you to play the recorder and protocol the election so far. **Now!**"

Without hesitation, the man grabbed the necessary writing utensils, took a seat at the table and began to write. The Private gave him some time and then continued, "As my deputy, I propose non-commissioned officer Schmidtbauer.

Opposed votes ... none.

Abstentions ... none.

Thus, unanimously elected.

Thank you!

Any questions?"

At first there was silence. Finally, however, one of the strangers raised his hand, yes, raised his hand (!). Almost amused, the Private gave him the word with a casual wave of his hand, "Yes, Comrade Ensign, you wanted to say something?!"

"Um ...," began the one so addressed, "What about the treasurer?"

The Private strode towards the ensign until he was only a hand's breadth in front of him, then said, "I expect attention in my squad. Only then will we be able to perform at our best. Attention and focus on the essentials, you got that, Ensign?"

"Yes, but ...,"

"From the context of my previous statements, it was clear that we do not need a treasurer. Whoever would still like to make a donation for cultural purposes please contact our recorder. Jan, did you understand that?

However, you will not accept money for the newspapers we did not order, understand?"

"Crystal clear!" replied Jan as briskly as he could and continued writing his protocol.

"Splendid, guys! With that, we can now get on with our real task."

"And what is that?" asked the second stranger, a simple non-commissioned officer.

"Strengthening our combat power, of course!" replied the Private, assuming the typical posture of a military trainer with fists pressed together behind his back. Without waiting for further questions, he continued, "We're about to learn some interesting alternatives for getting on and off our bunks. In a week on the assault course at the latest, you'll know what this is good for—damn good ...!"

And then they began to practice. They practiced, first incredulously and doubtfully, finally doggedly and with commitment, until even the well-trained Schmitti's abdominal muscles ached. Then they pushed the beds aside, threw the mattresses on the floor and the Private showed them some close combat tricks and then let them practice them. When in the meantime an elderly of the rank of a sergeant looked into the room, he only shook his head in disbelief and quickly closed the door again without saying a word. In all the other rooms, they were either sitting at the tables drinking coffee or tea or lying on their beds. Certainly, no one used the time for the "agitational and propagandistic" purposes that were actually scheduled, but certainly no one voluntarily trained in close combat—only total lunatics or the convinced did such things.

1.6 Muster

"Is there anything we can do?" the company commander asked the reserve officer. As he did so, he waved a small bundle of notes in front of the officer's face.

"Do something about what?" asked the addressed man through clenched teeth. He knew exactly what the captain meant, but it was repugnant to him how this small mind treated him, a major after all. And the guy was only a captain and had been punitively transferred here.

"Oh man come on, in the election of Squad Eleven/Two of course!"

The major groaned, "The election was conducted cleanly and according to the regulations. To be honest, it's the only cleanly documented and therefore probably the only cleanly conducted election, and what are you going to argue against the rationale their recorder wrote down there?" A short pause followed. "... and where all of the squad signed. ALL of them!?" the captain looked inquiringly into the major's face, but when he did not respond, he continued, "I guess something didn't quite work out with your political distribution, huh?!"

The major took a deep breath. "I wasn't responsible for that. That's done by completely different agencies and you should know that."

"Now I know, thank you!" the captain nodded to the major, then handed him the bundle of notes and added, "Write up a brief report, then just hand the whole package up to the battalion commander and ...," he paused briefly, "... for heaven's sake mention that we found the rationale given 'in the interest of maximum defense capability' to be extraordinarily valuable, sincere, and thus incontestable ... and that was after we had talked intensively with the comrades."

"But we haven't!" retorted the major.

"Oh yes, I have ... right now." with that, the captain rose and left the room.

"But it's night rest!" the major called after him.

"Not for the Second and certainly not for Squad Eleven!" he got in reply and then the door slammed into the lock.

Shortly thereafter there was a murderous roar in the corridor.

"Alarm! Step out, everyone! Alarm!"

"Shit!" the major only muttered and angrily pushed the typewriter away from him. As he scowled and stepped out into the long hallway, he saw, as expected, the young officers running back and forth yelling, opening the doors to the squad rooms, and yelling again:

"Alarm! Step out, everyone! Alarm!"

The company commander stood at the front staircase with the stopwatch, seemingly amused, watching the goings-on. His features froze, however, when he looked up and glanced behind. That was completely impossible! There, lined up and ready to march off, stood Squad Eleven/Two with the squad leader in front of his comrades, checking the order of the suits and correcting the fit of the steel helmet here and there. Then he turned jaggedly towards the company command, raised his right hand to his forehead, and roared in an almost unbelievable voice down the two hundred meters of the seemingly endless corridor:

"Comrade Company Commander, I report Squad Eleven/ Two ready to move out!"

The captain saluted him rather mechanically, then thought about it and finally beckoned the Private to him. When he stood in front of him, the captain eyed him carefully. 'Damn, that guy looks tough,' he thought and said, "Comrade Private," he paused emphatically, then raised his voice, "Comrade Squad Leader, that was the best time I've ever stopped, and I'm certainly not going to punish your squad for the sleepy headedness of the other twerps here and make them wait for the rest. You are hereby ordered to receive your weapons downstairs on the first floor, walk a lap around the block, then you will return your weapons, wash up and go to bed. In doing so, you will lead your squad alone and receive no support from higher command. Do you think you can do that, Comrade Squad Leader ... *Comrade Private*?"

"Yes, Comrade Company Commander ... Comrade Captain! My men and I will manage that," the Private replied without making a face. But the captain grinned briefly before his face turned to stone again.

"Then march off, Private!" he roared.

The Private turned and yelled as before down the length of the hallway: "Squad Eleven/Two, right turn!

On the double ...!"

Then the Private turned back to the captain, saluted briefly, then turned back to his squad, which was only a few feet from the stairway, and called out to them, "To the armory and grab weapons, quick march!"

When the last one had run past, he followed.

The captain looked after them and scratched his head. "Wouldn't it be better for an officer to control the execution of the task, Comrade Captain?"

Almost unnoticed, the first lieutenant of the reserve had appeared next to the captain. The captain gave him a questioning look. "Well, I guess ...," the first lieutenant hurried to explain, "uh, after all, this is just a Private." The captain laughed to himself. Of course, he knew about the incident of this morning. 'Another eager one,' he thought, and replied, "Well, Reserve Lieutenant, if you think you can keep up with them, you're welcome to run along and check on the **execution of the task**, but the squad leader calls the shots. How else are we going to get good leaders if we don't let them lead?"

"But ...,"

"But what? Hmmm? Man, damn it, Lieutenant, I need you here with the ones who can't get out of their feathers and not where everything goes by itself and officer work is easiest. **Now get the rest of the company on their feet, go!**"

"Yes, Comrade Captain!" it came back quite meekly and somewhat miffed.

By the time Squad Eleven/Two reappeared, still not everyone else was in the hallway. The Private stopped just before the major of the reserve—the captain was nowhere to be seen—and reported while his men headed directly for their room. At a casual run they tramped along the corridor, past more or less puzzled looking fellow soldiers. The Private had let them rest at the armory after a very fast run. "I want you to look relaxed when you run past them up there, all right?!"

"Sure!" it came as if from one mouth.

"Running in those clothes, boots, and then with weapons is quite different, isn't it?" the Private put his hand on the back of one of the strangers. The ensign had his hands on his knees, bent over to the front, and was breathing heavily. You could see he was completely exhausted. "And of course, that's even more true when you sprint the eight hundred meters, isn't it Ensign?" The Private didn't wait for an answer but continued immediately, "But it takes nothing but will and some practice; then soon you won't mind it any more than a little 'run' on the beach ... past all those pretty chicks lying there."

The ensign straightened up and managed a grin.

"Well folks, and right now I'm going to show you a few techniques on how to recover faster in situations like this. Even if it seems silly or weird to you, proper breathing makes all the difference." After barely five minutes, not only had everyone from Eleven/Two surrendered their weapons, but they already looked really fresh again. Thus, they could stomp easily through the corridor upstairs past the company and thus convey exactly the impression that the Private wanted to leave.

"Geez Schmitti, you're totally screwing up the norm, that's exactly what they want!" one of Squad Twelve/Two whispered to them.

"On the contrary, that's not what they want right now, but you'll understand."

When they finally emerged from the washroom, looking forward to their beds, the rest of the company was just about ready to leave. The captain stepped out into the hallway and roared a proper curtain sermon at everyone, all but Eleven/Two of course.

"Almost 20 minutes ... you must be out of your mind! Where do you think you are ...!"

When he was finished, he turned to the eager first lieutenant of the reserve, "Lieutenant, lead the company to drill, that means grab weapons, five laps, turn in weapons, and then wash and bed rest, got it?!"

"Yes!"

"Yes what?"

"Yes, Comrade Captain!"

"That's better, and please see to it that ALL the officers run along and that there is silence up here at the end ... after all, we don't want to rob the people of Eleven/Two of their well-deserved sleep, do we?"

"Certainly not, Comrade Captain, Sir!"

1.7 Extra Training

Tough days of combat training followed. This part of the six weeks' reserve was deliberately put at the beginning by the leadership to exhaust the people. They were to be brought physically to their limits in order to have no more mental resistance when it came to the subsequent agitation and propaganda. This part was also blandly referred to as "theoretical training." The Private and his squad, except for the two strangers, all future physicists, could only laugh about this kind of "theory" later; for the most part, they were even exempted. But before that, the hard days of "breaking bones" had to be endured. After consulting with his entire squad, the Private had decided to take the direct way.

"Guys!" he said, "they will try to break us physically, that is, they will try to break you physically. But they haven't considered one thing: in order to break us, they have to keep up with us, otherwise we'll break them."

"What?" the ensign shouted in disbelief, "We are wearing equipment and weapons and they wear nothing. How are we supposed to take them down and not them us?"

"Easy: by being better. Better and smarter. Got that, Comrade?!"

The ensign shrugged meekly.

"And besides, we have Kramer, our singer," Bodan added, and the entire squad stared at him as if he had gone mad.

Bodan, however, just grinned, "Non-Commissioned Officer Kramer: You are responsible for ensuring that by tomorrow the entire squad has sufficiently mastered the following treasury of songs:

Sabota u nas prostaja
Katjuscha
Kalinka. The first two verses are sufficient in each case. Furthermore:
Spaniens Himmel breitet seine Sterne
Bella Chiao
..."

"*Überlegung, Du hast ja ein Ziel vor den Augen, Sag mir wo stehst, Die Moorsoldaten, Und weil der Mensch ein Mensch ist*?" Kramer continued the list.

"Excellent suggestions, they should be halfway familiar by now too, but we need more than just the first verse, all right? So, practice!" Some in the squad groaned, but Bodan wasn't done yet, "And then you make sure that every day a new and more difficult song is added. And definitely one or two songs for several voices. First of all, the Soviet national anthem, understood?"

"Sure," came a rather muffled response, and the ensign nagged, "And when should we do it? In our spare time, perhaps?" Bodan grinned. "Comrade Ensign," he began, "in case you haven't noticed, there is no spare time in our schedule here."

"Yes, there is!" protested the person addressed loudly, "before and after dinner."

Bodan rubbed his chin, just as if he were thinking. A few of the others were already laughing inwardly. Then Bodan took the ensign by the arm and led him to the timetable. "What are you reading in those time frames you mentioned, Comrade Ensign?"

The latter didn't have to look at the schedule to know what it said. "'ind. agi-prop', but everyone knows that's actually ..."

"... spare time?" Bodan finished the sentence as a question. "I hardly think so, because with us it's fully pronounced 'individual agitation and propaganda' and what could be better than a cheerful workers' struggle song sung in a tight march, hmmm?"

The ensign's jaw dropped, but he could not answer. "You see," Bodan slapped the speechless ensign on the drooping shoulder, "you can't think of anything better either. Let's get going then, we've got something to do after all."

With that, Bodan turned to his secretary.

"Jan!"

"Yes chief?"

Bodan had to laugh at the salutation, but did not correct Jan. "Go to the political officer and tell him that we need 'Leben Singen Kämpfen'—one copy for each of our man for the entire duration of the reserve! You know, the little blue FDJ songbook."

Jan nodded.

"The rest of you: get ready to march! Full gear, no weapons. To make up for the weight, you fill your backpacks with gravel downstairs, got it?"

"Yes!"

"And go!"

1.8 Poor Officers

Less than a quarter of an hour later, the company commander could not believe his eyes. Eleven/Two marched at attention around the parade ground, warbling one song after another. At first, they marched with the hymn books in their hands and when everyone had memorized the song sufficiently, the books were put aside and the march became faster. At the end, they practiced one verse at a time, running, then the next song and so on. It sounded, one simply could not say it differently, excellent. The squad obviously had not only one good singer, but besides Kramer, all of them could sing reasonably well and did so gladly and often. In some songs, they even managed to do it in several voices.

"What the heck are you up to, Private?" the captain asked, half muttering to himself. The major of the reserve, behind him at the desk, looked up and although he knew he hadn't even been asked, he answered: "Maybe they're just in the mood for some singing."

But at that moment, the books were just being packed away again downstairs, but the squad did not finish its exercises, but went to the lawn behind the muster ground and practiced there, still singing and running, throwing themselves on the ground on command. The captain was unable to believe it and, as if the major's hint had never existed, repeated his question, "What the hell are you up to ...?"

When the weather got worse two days later, raining for hours cold and damp, he began to understand. In the meantime, the harder part of the combat training had begun. They were constantly on the move in the boggy field and, as every year in the fall, the captain had instructed his officers to keep pace with the men and "make some pressure." But none of his officers were able to keep up with Eleven/Two. If one made it for a while, it was never long before the huge squad leader, who was responsible for his men, this private named Bodan, appeared dominating and threatening right next to the officer. Like a huge dark shadow, Bodan then simply strode or rather stomped along beside the officer. Sometimes he would also suddenly appear behind the officer and blow his warm breath down his neck. For some reason, the officers' observations never lasted long, and they always reported to the company commander that everything was going perfectly at Eleven/Two, and therefore there was no need to look too closely there. To let the company march as a whole and thus squeeze Eleven/Two into a mass, the captain gave up very quickly, because it was clearly evident that this was like a vacation for them. The little tricks of Eleven/Two, namely that they helped each other, cleverly swapped the marching order so that everyone had to take the first steps into the mire once, that the stronger ones regularly carried the weapons of the weaker ones or even parts of the luggage, these went unnoticed or no one took them seriously any further. But these little things made all the difference. Above all, Bodan made sure that they spared their singer and when, after agonizing hours, they were approaching the garrison and everyone, from the last private to the highest officer, could only tiredly put foot in front of foot, Eleven/Two belted out one battle song after another through the little town. The accompanying officers did not even have the strength to stop it. And what should they have had against it? They were after all straightforward workers' struggle and military songs from the socialist school and the singing was impeccable on top of that. In addition, Bodan had deliberately chosen the busy village for the beginning of the performance and how would it look if officers of the NVA would forbid a private to sing battle songs.

However, words could not describe the devastating effect this final show of willpower had on the leadership qualities of the officers thus demonstrated.

Soon, Eleven/Two was left completely to its own devices when it came to combat training. The squad was simply given a battle order, a strategic objective, start and end times and meeting points, and then Bodan and his people were left to it. This individual freedom made the squad even stronger, because now Bodan could train them the way he liked and, most importantly, the way they, his people, could keep up. When a fair amount of this time, so difficult for everyone else, was over, it happened that a high general visited the battalion. Because they didn't have much else to show him, they took him to the combat training area to demonstrate the technical capabilities. To the surprise of the battalion commander, an officer with the rank of colonel, however, the site was not empty. A single squad, although dispatched just an hour ago at the main base, had unbelievably already arrived and was about to launch a strategic attack against an imaginary enemy. The colonel, who had no idea who he was dealing with, thought it would be a good idea to show his distinguished guest the training possibilities with the help of these soldiers. He had his driver drive closer to the training area and ordered the squad leader to join the two senior officers. The colonel and his guest were astonished when suddenly a huge, obviously well-trained private stood in front of them and reported briskly, but unimpressed by the ranks. However, they did not let on and asked the otherwise obviously not incapable group leader on the training ground to demonstrate a bunker defense together with antitank defense and then an infantry attack on such a "well-fortified" position.

What then was offered to the senior officers was found by both to be uplifting beyond measure. They had the squad leader called to them once again and praised just about everything they had seen. They believed they were dealing with a regular squad, but by no means with reservists, and Bodan, who had quickly ascertained this, saw no reason to enlighten them about the actual facts. After all, he was not asked directly about it.

"Comrade Private," said the celebrity guest after finding nothing more to talk about the combat exercise, "if I am to be honest, I have rarely seen such a dynamic attack technique and defense. How on earth did you manage to motivate your people like that?"

"I don't have to motivate them, Comrade Lieutenant General," Bodan replied calmly and modestly, "they always do their best because they are convinced that this is the only way to fight at all."

The two high-ranking gentlemen looked at each other. "I think," the guest now interjected, "these outstanding soldiers deserve an award." He cast a questioning glance at his host, who hastened to nod.

"Very well, Comrade Private," the general finally continued with a patronizing smile, "you are now in the unique position of making the most sensible proposal possible."

Bodan was aware that they were expecting something like special leave or furlough or something like that, after all they considered his people and him to be just normal soldiers. But he replied, "In order to be able to improve our training even further, Comrade Lieutenant General and Comrade Colonel, I would be so free as to wish for running cards for my people."

"Are you serious, Comrade Private?" the two men could not hide their surprise. A "running card" meant that, although one could independently leave and enter the army object in combat or sports clothing at will during one's free time, this was always only within such a narrowly defined set of rules and connected with a hard sports or concrete training program that no soldier would want to do such a thing voluntarily. Usually there was someone in charge who was entrusted with the control of the sporting performances, but Bodan had already included this in his considerations.

"Well, I must say," the senior guest confessed and his comrade in arms nodded resolutely, "I haven't come across such an enthusiastic bunch yet. I would say, then, that your people may see themselves distinguished with running cards until the end of the next term of service. And as for you, Comrade Private," the general paused theatrically and without taking his eyes off Bodan ordered, "Comrade Colonel, have them stand at attention!"

"Aaaaaaaattention!" came as agreed, and everyone assumed even more posture than they already did, which Bodan thought was a bit silly, but didn't let on. "Comrade Squad Leader, because of your outstanding performance in leading your squad in various combat training situations, I hereby award you with a leadership running card. From now on, you are free to decide when you and your squad may and may not leave the base."

And this was exactly what Bodan had speculated on. Whom else should they have assigned to control the "normal running card carriers" in this object, if not the squad leader of the decorated ones?

With that, he extended his hand to Bodan and said, "Congratulations, Comrade Private!"

"I serve the German Democratic Republic!" replied Bodan prescriptively. After that the colonel also congratulated and Bodan repeated his sentence.

"Comrade Colonel, make sure that the men receive their well-deserved award this very day!" the colonel then called for his driver and ordered him to take all the necessary personnel data. "When you're done with that, come to the vehicle immediately! We want to be back inside by noon ... so make it quick!" The two officers then saluted Bodan once again and the colonel called out, "And for today, you and your people are off duty, understand!?"

"Yes Sir, Comrade Colonel!" Bodan replied, slamming his heels together again so hard that it rattled loudly.

"Oh, shit, man, you guys really fought them hard," the driver, a staff corporal, said to Bodan when the higher-ups were out of earshot. "How come a guy like you is just a simple Private, man?"

Bodan just grinned and said, "It's GKdoS: highest level of secrecy and stands for: secret command thing—Mate, sorry. And believe me, you really don't want to know."

The staff corporal looked a little irritated, but then began to take down Bodan's data. When he realized that Bodan and his men were only reservists from one of the academy companies, he almost had a laughing fit, "Now I see why the tickets are worth more to you guys than a day of special leave like that. Man, that's sneaky."

"They don't really need to know that, do they?" whispered Bodan to him quietly, nodding in the direction of the officers, who were by now pacing along near the bunker complex, apparently talking animatedly.

"What for?" the driver asked hypocritically, laughing again. "Don't worry, you'll get your tickets today! It's going to be fun at the battalion. I'll use the car phone in a minute and get everything going."

When Bodan returned to his squad, his people looked at him expectantly. But he didn't let on and said, "You've been really good, guys. So good, in fact, that now we're going to go back together on the double ...," he paused for a while and finally added with a grin "... because we're off for the day because of the excellent performance!" His people cheered. At first, each one for himself and somewhat disorderly, but then all together roared a resounding "Hooray! Hooray! Hooray!" into the wilderness and set off without any grumbling.

"A song!" roared Bodan and Kramer intoned, "*Wann wir schreiten Seit an Seit*." The mood could not have been better. All carried their weapons crosswise in front of their chests, which saved strength and, on top of that, looked far more casual than the prescribed slinging over one shoulder. When they met the rest of the company on the way, which was only now arriving, and the puny first lieutenant demanded an explanation for the early march back, Bodan only replied coolly, "Special mission! By order of Colonel Breitschuh and Lieutenant General Kirner. Please inquire more directly there, Comrade Lieutenant! We now have to obey our withdrawal order from higher authority."

The lieutenant's face froze with his mouth slightly open, and he stood motionless as the extremely fast, but also extremely easy and casual bunch with the identification Eleven/Two passed him by. He had actually still intended to criticize the dress code and in general this whole disorder, but at the moment his head was completely empty.

A few kilometers before the object, Bodan stopped and arranged a campfire. He himself quickly ran to the butcher and bought some bratwurst and bockwurst for his men. It was probably the most comfortable lunch break these NVA privates had had so far. Strengthened in this way, the remaining three kilometers were a piece of cake and, as usual, the men sang their way through the town, whereby Kramer for some reason had come up with the crazy idea of singing "*Fröhliche Weihnacht*," a Christmas carol, in early November. Apparently, however, precisely because it was so crazy, everyone joined in and sang so enthusiastically that, despite all the shaking of heads, the people in the village also had a fair amount of appreciation for the young soldiers.

In the afternoon, the running cards arrived. They were valid until May of next year, about seven times as long as it would have taken to finish the reserve in the first place. The squad laughed and all slapped Bodan appreciatively on the shoulder.

"And what are we going to do with them?" asked the ensign.

"We're going to the assault course," Bodan replied without batting an eye, and the ensign's grin turned into a grotesque mask.

In fact, Bodan led his people past the baffled guards—one simply ran past and held up the cards—directly to the assault course and trained with them for about an hour some tricks on the escalade wall, the climbing and inclined ropes and at the so-called hand grenade long target throw. Afterwards, however, they played volleyball for the rest of the afternoon, which was great fun for everyone and led the ensign to remark to Schmitti, "And I really thought he was going to let us run around the assault course all the time."

He was somewhat surprised when the latter answered him dryly and completely seriously, "That's what I thought, too."

The ensign's face threatened to freeze again for a moment, but then Schmitti began to laugh out loud.

1.9 Newspaper Money

In the evening, unfortunately, socialist forced reality caught up with the squad around their charismatic group leader again in the form of newspaper money. Bodan was called to the political officer in his office.

"Why don't you stop fussing around," the political officer said to the group after some back and forth, "everyone has already paid, except your group. That's twelve men out of five thousand."

"Fourteen!" replied Bodan.

"I beg your pardon?" asked the political officer.

"That's fourteen out of five thousand, not twelve!" repeated Bodan, "Two non-commissioned officers from Twelve/Two have asked me to join our refusal. They wish to be represented by me in this matter. They have not paid either and do not wish to pay what they have not ordered."

"Why don't I know about this?" the political officer snapped at the puny lieutenant sitting next to him. The latter started to feel uncomfortable and after a while admitted that he had "persuaded" the others in the group to advance the missing money in order not to have more troublemakers in the squad or to have to make report to the top.

"Indeed," Bodan confirmed bitingly, "he has suggested to the others of the group to get the money from the two troublemakers with code red methods."

"What?!" the political officer almost hit the roof, "Code red methods?" Addressing the lieutenant, he shouted: "Have you gone completely crazy, man? You know very well that this kind of harassment is strictly forbidden. Fix this right now! Go, man, go!"

The puny lieutenant stood up half-heartedly. He was furious at being shown up like this by the political officer in the presence of Bodan. This one was, after all, only a private. Halfway to the door, he stopped. Apparently, his anger had gotten the upper hand. He pointed his index finger at Bodan and hissed, "Because of him, it had no effect anyway."

"Excuse me?" asked the political officer, "Would you mind explaining that?"

"I can explain," Bodan now interjected, "if I may, Comrade Major."

"Really?!" the major raised his eyebrows, "Well, get on with it, Private!"

"Quite simple," Bodan replied, "We told the others from Twelve/Two that code red methods are forbidden and that they must account to us for the safety and physical as well as mental integrity of the two non-payers in their group."

"You threatened them, admit it!" the lieutenant had stepped back up to the table and his voice was accusatory and incredibly angry.

"Since when is a reference to service regulations a threat?" asked Bodan. "And besides, shouldn't it be the duty of a good officer to protect his subordinates against forbidden assaults? Shouldn't he have the courage to stand up to injustice, even if it requires his own sacrifices instead of using injustice to make his own job comfortable?"

Silence fell in the room. The political officer slowly stroked his head. "You may go, Comrade Private!" And when Bodan was already standing in the open doorway, he called after him, "I understand your displeasure, Private, but at the same time I would like to ask you once again to influence your squad to make these people pay."

"Comrade Major! Please excuse me, but I will not do such a thing against my inner conviction."

"Then I will soon no longer be able to help you and your people, and you have no idea what kind of forces you then will have to deal with."

1.10 Chicanery

Bodan had expected everything: extra toilet cleaning, punitive marches, night alert, withdrawal of running cards, cleverly organized deprivation of food, things like these. But nothing of the sort happened. Instead, the system had developed such a perfidious strategy for "softening them up" that even Bodan's now very self-confident squad, slowly but steadily became more and more insecure. The high lords granted Bodan's group all privileges. After all, there would hardly have been any reason to take them away. His squad was the best everywhere and if there had been no way to break them physically, at least not without getting dirty and that had to be avoided at all costs with the academics, then nothing would be found now. No, they even put a few privileges on top. The respective best performances of Bodan's group were read out at every muster from one day to the next. Even worse, there were extra roll calls. This meant that the company, several companies or even the entire reserve battalion had to line up to witness nothing other than once again an award for someone of Bodan's group or even just Bodan, the simple private and yet best group leader ever. After all, the whole point of training academics was to get reserve officers who could really lead. And so, it obviously didn't do the prospective reserve officers much good to have a private paraded before them every day and to have to listen to: "You of all people, Comrade Officers of the Reserve or Future Reserve Officers, you should take this Private as an example."

In the beginning, Bodan and his men still thought this was a joke and secretly laughed about it. Schmitti had even created the term "ossiciers" and explained that in this system no people with leadership qualities could emerge at all, let alone rise up.

"They only want twerps, so they only get twerps. The fact that we have the best one and he's only a private says it all." Everyone laughed about that, too.

However, they were not laughing for long. The system was never about fighting power or defense; it was always about control, control over its own people.

But Bodan and his men only understood this when at some point during the roll calls the subject of newspaper money was raised. Almost casually, it was pointed out that still not all comrades had paid. Later it was specified that it was only a tiny group of subversive individuals who did not want to read and pay for either the "beautiful, colorful *Armeerundschau*" or—and that beats everything—the *Neues Deutschland*, which is the "central organ" of the Socialist Unity Party of Germany. Of course, no one was interested in that, but the fact that they had to stand here because of this alone, and that they didn't get to eat, or that their free time was stolen from them, that was what people were interested in. And of course, everyone knew who these subversive individuals were and that they were the reason for the constant musters, because they broke all the norms.

Bodan would never have thought how hostile and devoured by the most primitive hatred his own fellow students from the other groups could become. People, fellow students, with whom he had already experienced the greatest things in the first year of study, with whom he had let the socialist harvest mission ("students to the fields"—also just a means to keep the future academics small and to discipline them) wither away to a fun. In just a few days, they had gone from friends and supporters, even admirers, to the worst and vilest enemies. One could literally grasp and knead the hatred when walking through the corridors of the dreary company building and past the doors of the other groups. Of course, Bodan knew that these flabby, wimpy weenies were secretly just annoyed that they had paid without any resistance. That they had allowed themselves to be pushed around, and that they found neither the means nor the courage to fight back. It was ultimately their own wretchedness that made them enemies of freedom, at least of spiritual freedom, and beasts to dissenters and deviants. Bodan knew this, and he could almost have felt something like pity, but he could not afford it for the sake of his men. Because in the meantime there had even been assaults ... or let's call it better attempts of assaults. Three particularly pathetically frustrated, but ultimately physically quite strong fellow students from Nine/Two had tried to grab Kramer in the restroom to "make him eat shit," which was meant quite literally. Cowardly they had waylaid him, grabbed him from behind and pushed him towards the toilet cubicle. It was probably his cheeky singing that apparently did not suit them at all. In addition, they had noticed very well that the "voice of Eleven/Two" always had less to carry than they themselves and was then naturally more rested when an exercise was done. Everyone in Eleven/Two took great care to keep their "precentor" at the height of his vocal form, for they knew only too well how important this was to them all. For this they all made sacrifices and carried his weapon, his luggage or even sometimes himself. But it was precisely these sacrifices that the others envied Kramer for; above all the particularly "flabby weenies." When they were exhausted from the exercises, bent by their own inner weakness and depression, Kramer stood there with his people, fit and erect, and sang a song or two or three. Cheekily they sometimes changed the lyrics or the melodies and mercilessly they used the so-called "socialist songs," so that no officer could really say anything against it ... especially not if they sang in Russian, which is in the language of the first class and brother in arms.

Unfortunately for the three of them, Kramer was not only a damn good singer but also a tough dog. Besides, it was, of course, really stupid to ambush in bathing slippers someone who was still wearing his combat boots. So, Kramer might well have been a little smaller than any of the three, but the screams did not come from him, and no one may certainly like to imagine how a toe on which the heel of a good stiff military boot had just landed with power looks. Lured by the fuss, a comrade of Kramer's Eleven/Two finally joined in. After all, Eleven/Two and Twelve/Two were accommodated at the very back of the corridor that is directly in front of the toilets and the washroom. The constant close combat training with Bodan paid off immediately. Although some audience had appeared in the meantime, the three attackers from Nine/Two were now eating shit—their own, by the way, because that was what they had intended and "prepared" for Kramer in the cubicle. No one made an effort to help them, although many were full of inexplicable hatred for each of Eleven/Two. They simply didn't dare. Kramer had insisted on turning the tables and explained to both, the three guys and the other frustrated spectators, in his loud, carrying voice, "It's like with young dogs. They only learn it this way, that you can't poop in the parlor."

Bodan was anything but pleased about this development, "I think it's good that you fought back, and really now Ensign, I wouldn't have thought you had that much guts, but whether the dunking was such a good idea ..." and he was to be proved right. Less than a quarter of an hour later, Kramer was ordered to the company commander. When he stood up, everyone just looked at him sadly and Kramer said, "Take care guys. It was great to be with you ... Thanks Bodan, you tyrant!" Then in the corridor, he shouted so loud that it echoed through the hallway, "And don't let them get you down!" As the pounding footsteps moved away, the ensign suddenly jumped to his feet and, without uttering a sound, hurried after Kramer and the non-commissioned officer who had picked him up.

"Get lost, Ensign, it was my idea!" Kramer was heard scolding in the hallway, "Don't let them punish you for helping me!"

"Don't worry about it, Kramer!" countered the ensign, unusually confident and calm by his standards. "That was the best thing of my life, and I'll answer for it myself. And who knows ...," but he didn't finish the sentence.

~

"What?" exclaimed Jan in disbelief, "Three to ten days of after duty for the three of Nine/Two and you two ..."

"NOTHING!" added Kramer, grinning from ear to ear.

"How did that go?" Schmitti wanted to know.

"You'll have to ask the ensign," replied Kramer, "he was first at the interrogation and when he came out after only 5 minutes, we were allowed to leave again and everything was settled."

"What did you tell them, Ensign?" asked Schmitti.

But the ensign just shrugged his shoulders and replied, "I don't know, actually it was just the truth." Then he turned away and it almost seemed as if he wanted to get something else off his chest, but finally he just raised his shoulders once more and turned to his bed, onto which he then threw himself immediately.

Fortunately, the next muster was not about the good performance of Eleven/Two, but about the punishment for the three wrongdoers of Nine/Two. Two limped forward when their names were called. The third walked normally, but they all hung their heads.

"I didn't quite manage to get him, he was lucky," Kramer muttered and received nasty glances from the squad next to him. But Kramer just gave them the finger and hissed, "You're welcome to taste it next time, if you feel like it, you twerps!" and because he was in such a good mood at the time, he added angrily, "Well, is there a lot of interesting stuff in *Neues Deutschland* and your *Armeerundschau*?"

"Shut up, Kramer!" warned Bodan, who was standing directly in front of him. The admonition, however, came in a clearly amused tone. He added, facing the squad next to Kramer, "Even if it's hard for you, but as in battle it holds: the enemy who knows nothing is to be ignored as **harmless**!" Kramer grinned, as did the rest of Eleven/Two.

Blatant threats hissed from the surrounding squads, "We'll get you eventually, count on it!" But Bodan and his men now stubbornly adhered to the commandment of battle over the unimportant enemy and looked stubbornly ahead.

When finally, after a long sermon about the seriousness of the offense and the impossibility of such behavior, the sentence was announced, there was a real outcry.

"We will not tolerate such so-called code red methods in our National People's Army and even the slightest attempts in this direction are therefore to be punished most severely. The three perpetrators, and may this serve as a warning to all, will therefore receive 10 days of detention, which are to be served here afterwards. In addition, their university institution will be informed and the staff will request immediate ex-matriculation. The decision to do so is, of course, up to the respective educational institutions themselves."

The latter was an outright lie, because if a military superior wanted someone to be removed from the university, then he got his wish in 99.9% of cases.

"Aw shit, that's mean!" even Kramer said.

"Captain, dismiss!" roared the battalion political officer.

The final straw, however, was, when at one point the deputy battalion commander took charge of the roll calls and announced at the very first muster, "There are still some among you who do not show the necessary respect for the central organ of our party. Until that changes, we as a command staff can't pretend that everything is fine either. For this reason, ..." he paused for a substantial moment, licked his lips, and continued yelling, "Comrade Bodan?!"

"Yes, Comrade Lieutenant Colonel!"

"You will lead your squad to lunch on your own, then free time, or what you and your running cards understand it to mean. Squad Eleven/Two, dismissed!

The rest, right turn!

Comrade Major, you take over! 20 rounds, if there is sloppiness, 5 will be added on top. Understood!?"

"Yes, Comrade Lieutenant Colonel!"

Bodan had his men disengage and led them thoughtfully to the canteen. Something was fishy here, very fishy.

When Bodan and his men were finished after a long, drawn-out lunchtime, the others finally came to the canteen.

"Pay up, you idiots!" they said from all sides, but no one jostled them or even came near them. They had become lepers.

"Aren't collective punishments forbidden in our country, too?" asked Jan to the group when they were back in the room and everyone was putting on their sports clothes.

"Yes, but in the context of marching, have you heard the word 'punishment' even once?"

"Those sneaky bastards!" the ensign scolded, "I can never venture among my fellow students like that again."

"Do you want to give up?" asked Schmitti calmly.

The ensign gave him a shocked look, "Did it just sound like that?" But he didn't even wait for an answer, but continued right away, "No damn it," the ensign shook his head vigorously, "I've never felt so good. No way I'm giving in now! In general, I never again want to be another one of those ...!"

He did not finish the sentence, but only made a deprecating hand gesture. He didn't need to say more, because the others understood.

"Me neither!" came from all sides.

Bodan was proud of his men, and as he nodded inwardly with satisfaction, he saw Jan eyeing him from the side. Finally, Jan put his rather delicate hand on his shoulder and said, "You lead dude, we'll follow."

Bodan laughed and called out, "Ok guys, now follow me to the gym. There we want to hit some smashes today that the world hasn't seen yet, because we're playing volleyball!"

1.11 Big Combat Exercise

The special marches for the battalion, always with the ostentatious exception of Bodan's group, now took place daily at least once, usually more often, and the hatred and vituperation against Eleven/Two became almost unbearable. But Bodan's people withstood the pressure. Thanks to the running cards and their own motivation, they had enough opportunities to distract and strengthen themselves at the same time.

But Bodan knew that the final act had to come soon, and so he was only moderately surprised when Jan took him aside one evening and whispered, "Tomorrow at the big combat exercise!"

Bodan only nodded and looked Jan inquiringly in the eyes. But the latter looked straight back and just shook his head, "Please don't ask me how I know, but I think they'll try to take you away from us."

And so, it happened. But before that, there was another roll call. The deputy battalion commander announced that today the ringleaders of this blatant provocation, of this counter-revolution would be seized and mercilessly brought to justice. "All of them will be hanged!" he shouted at the assembled reservists, and there were more than a few who nodded with enthusiasm and promptly sent gloating looks in the direction of Bodan's men. There were thousands of sardonic looks, from thousands of poor, downtrodden, puny weenies.

"Move out!" the lieutenant colonel bellowed, and now himself, for the first time, cast a direct glance in the direction of Eleven/Two. He grinned, and his grin was the most spiteful thing Bodan had ever seen. Shortly after that, it was clear how they were going to do it.

"Comrade Private!" roared the puny first lieutenant. Before Bodan had even turned to face him, the little busybody continued shouting, and it was clear that he was speaking more to Bodan's men and the other reservists still nearby than to Bodan himself, "You will be given special assignments with the staff today. Your SPECIAL skills are indispensable there. Your squad will be commanded by me today. You will report immediately to the adjutant of the battalion commander. Do you understand?!"

Bodan did not answer him but turned to his men instead. "Hang in there, no matter what they do to you! I'll try everything to rejoin you somehow." Then he turned and headed for the staff building.

Once there, he was greeted by the commander's driver, who still remembered him well, "Gee dude, still a private?" he joked, adding, "I guess you couldn't get a lazier job than on the staff, huh?"

"That's just a nasty trick to keep me away from my people. I've got to get back there somehow," growled Bodan, and the lance corporal didn't understand how in God's name anyone could wish himself back to combat training when there was nothing there but stress and backbreaking work. "And that today, when this high-ranking visitor will come who wants to see something. No, you'd better stay where you are now and ..."

"What sort of high-ranking visitor?" asked Bodan immediately with interest.

"Huh ... well that lieutenant general from last time. Forgot the effin' name, but you guys must have made a big impression on him, and that's why he's coming back today as a surprise, and even wants to bring another big shot with him ..."

"Who knows about this?" interrupted Bodan the lance corporal again.

"I don't know, just came in ... well, in with me. According to this, so far, the colonel, myself, and now you, of course, know."

1.12 False Reports

"Damn shit!" the deputy commander cursed and ran to his car. He yanked the phone up and barked at the duty officer, "Get me the commanding officer, now!"

"Major Gerber?"

"Yeah man, is there any other commanding officer? And hurry the hell up!"

Normally the driver understood what was going on, but what the deputy had been saying since this morning hardly made sense to him.

"... Yes exactly, somehow keep him busy and keep the man at a distance, that is, absolutely far away from his people."

"How'd mean it ..., away from the staff and his people, how do you picture it, Comrade Lieutenant Colonel?" The driver recognized the major's voice and he sounded quite perplexed.

"I don't care how you do it. If the General asks for him, he is sick. Do you understand? SICK!

... How, what illness does he have? How am I supposed to know, you'd better make something up!"

1.13 Occupation

"Here dude, this is supposed to go to the Kastenwald[3] somehow and you're to bring it over there!" The lance corporal pointed to two large boxes with rather dangerous-looking warnings.

---

[3]Military slang for a small forest in a rectangular setup.

"What's that?" asked Bodan.

"Combat stuff, like practice hand grenades, barrage simulation and imitation grenade fire ... all that kind of stuff. There's more of it back there, but just these two boxes need to be brought over there."

"Oh!" Bodan said, continuing to ask, "and what's the exact order?"

"To the Kastenwald with it, and quickly." He leaned over to Bodan and whispered, "I think the visit is coming and they just kind of want you out of here."

"So, no more orders as far as the junk here is concerned, right?"

"No, Sir!"

Bodan could not hide a grin. 'So,' he thought, 'they want me gone and quiet, and in return they give me the loudest noisemakers the NVA has to offer. If that's not a hint of fate. Who do they think they're dealing with here?'

"Well then, I guess the order is 'use exercise elements at your discretion,' isn't it, Corporal?"

The addressed understood immediately, jumped to a standstill, saluted overly formally and correctly, and roared, "Yes, Comrade Private! Then probably at your own discretion and just don't ask too many questions!"

"What are you doing, Lance Corporal?" someone murmured from the staff car. "Will you stop saluting the private? And you, PRIVATE, grab those boxes and get the hell out of here!"

"At your command!" roared Bodan, grabbed the two boxes and was out of sight and earshot before anyone could shout "Halt!" anymore.

Bodan knew he had to time things well and most importantly, the visitors had to arrive before he could start his show. First of all, therefore, it was necessary to gain time. So, he dawdled along the field that led from the makeshift command post to the Kastenwald. According to the battle plan, which Bodan had studied intensively as soon as he arrived at the base, the "enemy aggressors" of the first wave were "lurking" there. They were three squads of "prospective reserve officers" with a reserve lieutenant in charge. Since they had only been assigned to be used as cannon fodder anyway and had to "fall" quickly in the first attack wave, the motivation of this troop was certainly very low, and Bodan would take advantage of that. In general, he, the troublemaker, had certainly been sent to exactly this department for this reason. No one would pay attention to them, because they were just the pawns to be sacrificed and this in the first few minutes of the battle exercise that lasted several hours. The average body mass index of the entire troop was several points in the adipose range, and the somewhat leaner colleagues were consistently, if at all, just barely off sick leave. So, nothing exciting was to be expected there and no guest would develop interest in this squad ... definitely not, or would they?

1.14 Backing

A box of firecrackers on top of each of his two shoulders, Bodan marched leisurely along and looked back at the staff from time to time to see if the visitors had already arrived. In doing so, he almost stepped on a small mouse that was sitting in front of him between the plants in the field. It sat there just like that, apparently without fear, and nibbled on the last juicy leaves that the otherwise bare harvested field had to offer.

"Hey, you're funny!" Bodan observed in amazement. He bent down and lowered the boxes to the floor. The mouse remained seated, calm as a cucumber. Slowly, very slowly, Bodan dug some of the cheese out of his pocket, which he had taken from breakfast as a backup, and held out a tiny piece to the mouse. The mouse sniffed the air with interest and then slowly came closer. When it spotted the source of the scent, it actually put a paw on Bodan's huge fingertips and gleefully ate the offered cheese.

"Wow!" said Bodan, "you are a brave little mouse. A battle mouse!" and then he had to laugh, and an idea came to his mind. He had no clue what it was good for, but he took off his winter cap, put a piece of the cheese on it, and held both together to the mouse. It didn't take long and the mouse climbed onto the cap. Bodan laughed again and put the cap back on his head, together with the mouse and the cheese. Then he continued on his way with the two boxes. At first, he walked very cautiously, but soon realized that the mouse could hold on to the woolly winter cap very well. In addition, the cap had a depression in the middle, so that even in the run not even the cheese fell off. When the Kastenwald was only about a stone's throw away, the special vehicle Bodan had been expecting appeared at the staff. He quickened his pace and soon spotted the "first enemy attack wave." There they lay, the written-off, the worthless, the fat and lazy, the ones that could only be sent into battle to be used up and were otherwise useless. The fact that all these reserve officers possessed abilities in "normal life" that would have seemed like magic to any of the local officers was of no interest here. Nothing was expected of them and they expected nothing from being here. Accordingly, the entire troop was bored, including their commander. Bodan quickly recognized him. He was the only one standing and constantly peering with his binoculars in the direction of the staff, so as not to miss his mission signal, which had to come at any moment. For him at least, because he could not yet know that everything would be delayed a little because of the high visit.

Bodan approached him directly and quite silently and cleared his throat. The guy was startled and when he wheeled around and saw this giant man in front of him, barely reaching his chest, he almost panicked. Probably he had recognized the private, which was no wonder, after all, he had been decorated a couple of times before the battalion. The lieutenant knew that some of his fellow officers did not like the "super trooper" because he and his "damn squad screwed up all the standards" and there had been some trouble about the newspapers, but what in God's name was this guy doing here of all places?

"W ... w ... what's going on?" he stammered.

"Change of plans, Comrade Lieutenant. You must have seen the new arrivals at the staff, right?"

"Yes, looks like a gene ..."

"Lieutenant General Kirner and company, Comrade Lieutenant."

"Blimey! What are they doing here?" gasped the astonished lieutenant.

"Good question, Comrade Lieutenant, but that's exactly why I'm here. The distinguished visitors want a good show, of course, and that's where you and your people come in."

"Oh no!" that was real desperation, that much was certain. This man really had no desire to be observed in any way with his loser troop even by the last backwoodsman. But there were one or even two generals waiting there, that was pure horror. Immediately, small beads of sweat appeared on the lieutenant's forehead and he put his hands in front of his face.

"Don't worry, Comrade, I have the right solution to your problem, but to do that, I would need to say a few words to your people. May I?"

A moment ago, the little lieutenant would have agreed with each of his fellow officers that such a norm-breaker, such a wretched super-trooper was nothing but a hassle, but now, at this moment, he could just hug him.

"Sure man, of course!" the lieutenant was so full of grateful relief that it almost made Bodan uncomfortable. The lieutenant eagerly turned to his three squads and shouted as loudly as his somewhat delicate voice would allow:

"Hey guys, listen up, this private here has something to say to you."

1.15 The Disabled Company

Bodan couldn't believe it, what was this bunch? Where, and especially how, had these twerps, these fat-asses, gotten their NCO ranks in the first place, or even become ensigns? That was simply grotesque. In the direction of the lieutenant, he hissed half aloud, "But Comrade, if we want to get through this demonstration in one piece, then we need a little bit of military order and discipline now."

The lieutenant immediately understood what Bodan meant, and more sweat began to appear on his forehead. Meanwhile, those lying closer lolled and looked towards Bodan and the lieutenant, half interested but certainly unwilling to be told anything. To Bodan's complete astonishment, the latter whispered back, "I can't do it, really man. Wouldn't you rather do it yourself?"

Bodan swallowed. He'd been through a lot, after all, but this exceeded, or rather undercut, all expectations. "Oh dear, Lieutenant, I can only do that if you appoint me as your deputy first. What is this anyway, I mean, does this count as a company or what?"

"Yes, the three squads are grouped together as a company ... the disabled company, so to speak," the lieutenant raised his shoulders apologetically, "Oh, and uh ... I hereby appoint you my second-in-command and place you in charge!"

"Louder, man!" hissed Bodan, "and add 'because of the special circumstances'!"

"Okay!" The poor man actually tried to yell now, and though his voice was rather delicate in nature, it carried far and echoed from the edge of the forest, "Comrade Private, because of the special circumstances, I hereby appoint you my second-in-command and place you in charge of the company!"

The surrounding people's jaws dropped, but before they had even half-understood what was going on, they heard the deafening, lion-like roar of Bodan yelling at them:

"Coooooompany ATTENTION!"

Bodan didn't even wait for the shock to settle. He had already targeted his victim in advance, rushed at the poor midshipman, the highest rank Bodan could locate in the immediate vicinity, in one leap and, grabbing him by the upper arms, yanked him upwards in one mighty jerk, holding him there just above the ground despite what must have been a good 95 kilograms of him, yelling, "What's the matter with you, Middy, don't you know the command ATTENTION?" and then put him down. The midshipman remained standing stiffly in a perfect still position as if he had been pounded into the ground, which in a way he was. The effect was instantaneous and fatal. Everyone immediately jumped to his feet. Even the lieutenant stood at attention.

"There you go, there you go!" exclaimed Bodan now a little more calmly. "Form up for roll call in threes, chop, chop!"

Bodan probably noticed that the guys had hardly ever had a proper marching training, but they tried and that's all he needed now. When they were all standing in front of him, more or less in line, and he was about to start including them in his plans, that is to give Bodan's very personal order of the day, he remembered the mouse on his head or winter cap and thought, 'Oh shit, I totally forgot about it ... the guys are already looking quite questioning ... Anyway, close your eyes and go for it.'

"Comrades Officer Candidates," now Bodan roared again, pacing back and forth in front of those gathered in front of him, "Special circumstances require special solutions. Your commanding officer has placed me, a supposedly ordinary soldier with a mouse on his hat, in command because special circumstances make it imperative." Bodan paused and looked into the crowd. Quite as expected, he saw here and there the hint of a laugh or at least an amused expression. He picked the one with the highest rank, strode heavily towards him, and yelled directly into his face, "And what on earth do you think is so funny about that?"

Silence, all laughter died in a split second.

"I asked you a question, Comrade Officer Candidate!"

As expected, there was no response, except for an uncertain stammer.

"As a prospective officer of the National People's Army you always have to know what you are doing and why you are doing it, do you understand?"

"Yes, Comrade ... uh ..."

"Private!" added Bodan sharply, "simply PRIVATE."

"Yessir, Comrade Private!"

Without saying a word, Bodan pointed at the twisted strap of the Kalashnikov, the trouser legs sticking out of the boots, and the jacket, which was completely wrinkled, at the man standing in front of him. Then he turned around, walked back to his line about 4 meters in front of the company and yelled, "At ease! Establish suit order!"

Then he turned to the actual company commander, the lieutenant, and asked, "Comrade Lieutenant, aren't you going to help your comrades?"

"Oh yes!" the latter hastened to say, glad that he had been assigned such a simple task.

When, after about 5 minutes, the squads finally looked halfway decent, at least as far as the suit order was concerned, Bodan called out, "Get ready! Attention!" This time everything happened much more briskly and Bodan began his pacing again.

"Comrades, as I'm sure your company commander on duty will be happy to confirm, high-ranking visitors have just arrived at the battle staff. It is General Kirner and another high member whose name I am not at liberty to mention here ..." (In fact, Bodan did not know the name, but did that matter? In a system where one never knew what one was allowed to say, or even to know, and where there were ears everywhere, an ominous unknown was always better than even the army general Kessler would have been.) "... the high executives are present today because they were recently able to observe a small combat exercise right here, which convinced them very much. In fact, this exercise has convinced the top brass to such an extent that they expect much more today, in this much larger and more real combat simulation. They also expect more from you, Comrades Officer Candidates. That's why they're here ..."

Bodan paused, and as expected, he looked into completely disbelieving faces. This much was certain: these people here had been completely disillusioned, they didn't trust themselves with anything. So, he continued, "... and that is why I am here with you. Because it was I who commanded the squad a few days ago, which was able to show this excellent performance. This performance, which demanded the highest respect even from a general, and of course you now know which squad we are talking about, don't you?"

Bodan paused, turned to the company, and looked closely at the midshipman, whom he had earlier brought to his feet. "Now, Comrade Midshipman, what squad are we talking about here?"

"Eleven/Two, Comrade Private!" came as if shot from a pistol.

"Yes, Comrade Midshipman ... and who is leading this squad?"

"You, Comrade Private!"

"Ten out of ten points, Comrade Midshipman. I confess to being seriously impressed."

Quiet chuckle.

"However, as you may not know, until a few weeks ago, even the soldiers in Squad Eleven/Two had been just ordinary students. Students of physics to be more precise ... if you disregard the three teachers and the two mathematicians plus the musician as well as the many other professionals in the squad."

Another quiet laugh.

"None of them had ever had any special military training or anything like that. So, Comrade Midshipman, what do you think the high command was thinking when they sent me to you?"

"So that you would lead us, Comrade Private!" the midshipman hastened to reply.

"Lead you?!" roared Bodan again now, "To what? And answer clearly and loudly!"

"To do our best and deliver a good combat exercise, Comrade Private!" now roared the midshipman as well.

"Almost correct, Comrade Midshipman!" and Bodan turned to the entire company again, "not a GOOD, but an EXCELLENT exercise you will perform, do you understand?"

From the midshipman and a few others came an audible, "Yes!"

"Did you really ALL understand that?" repeated Bodan his question, yelling even louder and looking at them all.

"Yes, Comrade Private!"

"Good, that's better, because then you'll already know where you're going today." Bodan's voice rose again, becoming admonishing once more, "But we all know that this is not going to be easy, because we also all know that you are not the people you would call the first selection for a platoon of paratroopers or combat divers, are you?"

"Yessir, Comrade Private!" shouted three dozen mouths.

"But the fact that you are aware of this, your weakness, is also your greatest strength. That is, we will make it to your greatest strength today and you will do your best, do you understand me?"

"Yessir, Comrade Private!" shouted three dozen mouths again, louder than before.

"We all know that this is supposed to be nothing more than an enemy commando point that will be spied out by our reconnaissance, stormed by artillery, and eventually taken by infantry. A formality and with only one task for you, a seemingly pointless task, namely, to die as convincingly as possible. You may think this is unimportant, but just the opposite is true. Because it depends on your performance, **it depends quite decisively** on how demanding the exercise will be for everyone else. Because you make the beginning, you determine the level and you have it in your hand to make this whole battle simulation in front of the high spectators a success or a total failure. So now I ask you again: are you willing to give your best for a brief moment of your otherwise miserable military existence?"

"Yessir, Comrade Private!" it now roared from 3 dozen and one mouths, for the actual company commander on duty had also joined in the shouting.

Bodan turned to the lieutenant and saw that he had clenched his fists and was positively beaming with excitement. Bodan could have laughed out loud, but instead he jumped to a standstill himself, slamming his heels together audibly, saluted, and yelled out to the stunned officer, **"I congratulate you on this great company, Comrade Lieutenant!"**

The person addressed was utterly speechless and stood for a moment completely motionless with a rather stupid expression on his face, but Bodan did not pay any further attention to him, but had already turned back to the company.

The praise had obviously done them a world of good, for they too were beaming and looking determined.

But Bodan was not finished yet, "By the way, the comrade with the laugh on his face was absolutely right. Having a mouse on your head is, of course, a good reason to laugh. Consider it my spleen, Comrades Officer Candidates. You will be allowed to have one later, when you will be officers of the National People's Army and have a command ... And now I'll show you what I have in these boxes and how it will help you make your performance even more impressive."

1.16 Resistant Losers

When finally, the signal was given to start the "reconnaissance offensive," which meant nothing else than that the enemy commando point was to be located and "broken up" by the own artillery, something happened that nobody had expected. The "disabled" and "adipose" of the "scrap company," as they were called within the staff, acted extremely skillfully, were well spread out and fought off the reconnaissance forces to the best of their ability.

Company A4, which consisted of 100% prospective reserve officers, was entrusted with the "reconnaissance of the commando point and its later deployment." These were one hundred and twenty fit and quite well-motivated young men with twelve eager squad leaders, a bored company commander who left everything to his deputy, and a deputy who was bursting with ambition and wanted to show what he was capable of. When the latter found the commando point defense, classified as extremely weak, surprisingly so strong, he thought of a trick by the combat training command. A trick to put him and his company of prospective reserve officers to the test.

So, he decided on a tactical retreat, thoroughly screwing up the command's schedule and, after briefly consulting with his squad leaders, who also all believed it was a serious test, called for reinforcements.

The battle command could not believe their ears when reinforcements were requested by field telephone from Company A4 to "overcome the completely misjudged enemy." This had never happened before in Seelingrad. The staff was immediately completely out of their minds and everyone was just yelling in confusion ... everyone except the major who had to coordinate everything. He had a hunch why, in God's name, the "scrap company" was suddenly anything but scrap and he just couldn't help it, he smiled. It simply could not get any worse: the plan completely messed up, utterly idiotic suggestions from A4, a lazy squad that was no longer lazy, a staff that was panicking entirely and right next to him, staring fascinated into the binoculars, two high-rank generals on surprise inspection. He, who had already seen and experienced really a lot, had arrived hardly thirty minutes after the beginning of the exercise already in the stage "To hell with it" and would have liked to simply leave. It's just the way it is sometimes when you think that you can't save anything anymore anyway, and that it doesn't matter what you do ...

At that moment, Lieutenant General Kirner turned to him and whispered, just loud enough for only him and the guest he had brought along, also a general, to hear, "Very good, Comrade Major, you've given the people a hard nut to crack ... Putting in something unforeseen, that's finally an exercise we need. Very good indeed, Comrade!" The general patted the major on the shoulder. He wondered a bit about the strangely sour, almost somewhat dim-witted grin on the commended man's face but thought, 'He sure is pretty jumpy today.'

"And how do you intend to resolve this hopeless muddle now, Comrade Major?" asked Kirner's companion, still looking fascinated through his binoculars and observing the region around the Kastenwald.

The major swallowed. He had to think of something. He would ignore the order of the deputy battalion commander concerning this crazy Bodan and his people for the time being. Here it was a matter of his own head. The lieutenant colonel should see for himself how he got Eleven/Two down.

While the major was still brooding, the first practice hand grenades went off in the Kastenwald and Kirner's guest called out in fascination, "Look Wolfgang, you must have seen that: he's about to throw the hand grenade one hundred meters ... damn, that's more than one hundred ... from a standing start!"

Kirner immediately took the binoculars up again and looked in the same direction as his companion, "Golly, I know that one. It's that private, the squad leader of ... of ..., Major?"

"Eleven/Two," the major answered in a long sigh, immediately throwing all the solutions he had thought up in the meantime out the window.

"Yes, that's right, Bodan is his name. What's he doing over there, that's not his squad at all?" Kirner asked emphatically.

The major was about to reply that the leader 'over there' was Lieutenant so-and-so, when the lieutenant general's companion interrupted him. "Not his squad? My ass! That guy, Bodan, as you said, is leading the company, I've been watching that all along ... and it's a sight for sore eyes."

"And you have to know, Fritz, he's only a private!" Kirner informed his companion.

"No way!"

"Yes, he is, but he's an airborne soldier ... from one of those young special units where they take specially selected one-and-a-half-year-olds."

"And why only private?" asked the attendant.

"I don't know, it's top-secret stuff, and I haven't had time to call that in and look it up."

"You should definitely do that Wolfgang, because the man is incredible."

Kirner set down the field glasses again and turned to the major.

"Damn man, you actually shit on rank and send a common soldier to our officer candidates?"

"Uh ...," the major just made uncertainly.

"You're using the best leaders to challenge the others. That's what I call smart. Where's the one who leads the company otherwise?"

'Finally, a simple question,' the major thought, and replied, "Well, he's over there, too, of course."

"So, he can learn something. Boy is that good, Comrade Major!"

At that moment, the major had an idea. At least this way he could save his butt. As for the plans of the deputy battalion commander, he didn't care about them at the moment. "Yes!" he confirmed, "the battle choreography doesn't require Eleven/Two until later, so I figured weaker leaders could definitely use some help ... well, you know."

Both generals nodded. "Then when and where is Eleven/Two scheduled?"

The major thought quickly and strained. He knew that Eleven/Two had been completely planned out at the insistence of the deputy. The squad was not to be given any further opportunity to present itself, whether with Bodan or without him. This plan had now, at this moment, become completely obsolete.

"The men of Eleven/Two will end up, surprisingly and also unplanned for all other officers, reinforcing the defenses of the enemy bunker complex, that is, after all other forces there have already been eliminated ... or as good as, because of the fluid transition. No one but us knows this yet, and that's the way it should stay. After all, we need to know to what extent our future leaders will also be able to react to surprises."

"Excellent, Comrade Major!" the companion of Kirner was enthusiastic, "and how do you make sure in the end that our troops will still win?"

Following a completely insane idea, the major replied, "Not at all, Comrades Generals!"

Both generals' jaws dropped, but now there was no turning back and the major declared, "If our officer candidates don't make it against this excellent squad, then they will just have to take a beating and hopefully learn something in the process. I think it's better they learn that in an exercise than in a real battle. What do you think, Comrades ... Comrades Generals?"

Lieutenant General Kirner nodded, but his companion looked thoughtful, "That sort of thing can be pretty darn demotivating, though, Comrade Major," he said.

"Yes, and that's exactly why our future officers lose, if at all, to the best squad this battalion has ever seen, and everyone knows it ... Please excuse me now Comrades, I must attend to my exercise," he said politely.

Then he strode briskly to his staff and bellowed, "Comrades, what kind of chaos is this? If this were a real combat situation, you would all be dead by now. Kindly concentrate on your tasks, keep calm and stick to the battle plan!"

"But Comrade Major, the commando point ..." a captain timidly spoke up.

"What?" the major roared at him, and roaring he could. "What? ... like the commando point is resisting, right? Is that what you're trying to tell me? I know the commando point is resisting; why do you think I sent Bodan from Eleven/Two there?"

The other officers on staff were left open-mouthed, but now there was no turning back and hopefully the other losers realized the wind had changed. "Remember this, Comrades, the problem is not a good enemy commando point, it's a bad own attack." He turned to the communications officer, "Give me A4!"

Immediately the major had the receiver in his hand and no sooner had the company commander there answered in his typical indolent manner than the commanding officer was already yelling at him, "What do you and your people think you are here for? You are outnumbered 4 to 1 against a miserable to non-fortified position of disabled. And there you request WHAT? Can you please repeat that!? ...

Louder man, because there are two generals here watching and listening to you, so kindly show us why the National People's Army has given you an officer rank! ...

Yea, that's better, and now march off!"

The major hung up. "Captain Spitz, you're coming with me! We have a battle to assess. Captain Kassner, you take over here and take care of our guests.

Driver, take us to the Kastenwald!"

When the captain had squeezed into the back of the Trabant Kübel and the major had taken his seat next to the driver, he quickly called out to the guests, "I'll be right back!" then the major saluted from his seat and the Trabant howled.

The two generals looked at each other and nodded in approval. This was leadership at its best!

Meanwhile, in the distance of the Kastenwald, there was banging, shooting and hooting as if the battle of the century had actually broken out there. However, it raged only briefly, because since no one really fell down dead and the exercise participants had the instruction to lie down on the ground as "hit" if the situation was "clear," it was indeed quickly clear with this superiority that the enemy commando point would be captured after all. The major and the captain got out at some distance from the Kastenwald and approached the "battlefield" on foot. There were still bangs here and there, and fog and gunpowder smoke were slowly drifting away to the east. All the battle participants, whether "hit men," "enemies" or "friends," were sitting on the ground breathing heavily. Only a few were walking around, such as the lieutenant of the commando point and Bodan, who walked from man to man, patting their shoulders.

"Aaaaaaaattention!" the captain yelled when they got close enough to the largest group. The shout echoed across the level field and soldiers everywhere jumped to their feet and immediately stood at attention. For a stretch of a good one hundred and fifty meters they stood, spread out in a winding line across the field.

"You were great, Comrades! All of you!" roared the major as he slowly walked down that line. "Our spectators are very pleased. The defenders of the enemy site did not simply give up despite being vastly outnumbered, and the attackers did not let themselves be demotivated by the first setbacks. That's what I call courage, Comrades."

1.17 "Common Field Mouse, Comrade Major!"

Now the major reached the lieutenant and Bodan and looked at them. What was that? He had actually intended to say something on the subject of "good leadership in the face of quantitative inferiority," but then he noticed a movement on Bodan's winter cap. No, that couldn't be. The guy was even weirder than everyone already thought.

"Comrade Private, what is that on your head?" roared the major. He roared as loud as he could, even though Bodan and he were standing just eight or nine meters apart.

But Bodan could also roar, and this was no less impressive thanks to his enormous chest.

"Common Field Mouse, Comrade Major!"

It took the major a good twenty seconds before he digested the answer. The soldiers all around were about to burst, but no one dared to laugh out loud or even grin. Which partly led to terrible grimaces. But the major kept his temper and roared:

"Field mouse? What? Put that down, Comrade Private!"

Bodan, too, remained serious, seemingly completely ignorant of the members of the glorious People's National Army now giggling around him, who were almost torn apart inside ... as if suffering from terrible but somehow funny flatulence.

"I'm afraid that's not possible, Comrade Major."

"And WHY the hell not, COMRADE PRIVATE?"

By now, the mutual roaring had reached unbelievable proportions and the major was almost certain that the two generals could also hear well ... even though they were standing about a kilometer away.

"Because we are standing at attention, Comrade Major!"

All around Bodan there could be no more talk of standing still, because the surrounding soldiers were shaking with suppressed laughter. In fact, only the major and Bodan were still standing in a correct attention, but one in which they yelled at each other, anything but still.

Unimpressed by the absurdity of the situation, however, the major turned to the captain and called out:

"Comrade Captain, make the men stand at ease!"

It took a while until the person addressed could speak or give a command, but finally, after what felt like an eternity, came the redeeming signal:

"At ease!"

In the laughter that now began, thighs slapping and falling down in a fit of laughter, the major did not even try to bring back order to the company. He strode towards Bodan and said almost calmly, "Good, Private, and now please take down this mouse!"

"Yes, Comrade Major!" replied Bodan and did as he was told.

Almost amused, the major watched as the mouse sniffed its way around the ground and finally scurried off in the direction of the Kastenwald under some turnip leaves. Then he looked at the hooting bunch of men, thought of his two high guests at the staff and considered how he should get order back into this company. Then he had an idea. The person who had caused the trouble should deal with it, and he would do it somehow. How, the major was completely indifferent. And even if Bodan had resorted to the Roman method of decimation, it would not have mattered to him, because he, Major Gerber, had two generals breathing down his neck. He turned on his heel, strode vigorously back to his position, a dozen paces ahead of the soldiers, and yelled:

"Comrade Captain, give the order to attention!"

Not that there was any need for the captain's "Attention!" at all, because everyone was already standing at attention again when he gave the command. Now the major turned back to the long line of soldiers, yelling. That is, he looked and roared at everyone equally, but addressed Bodan exclusively in his speech. "Comrade Private, you will take over both companies, establish order and lead them to the emplacement in the very best marching order! Do you understand?"

"Yes, Comrade Major!"

"The senior officers are subordinate to you until then and have to support you. You are in charge and in command. Is that clear?"

"Yes, Comrade Major!"

"When you arrive at the emplacement, you will take over your regular squad again, which will be waiting for you there. The current squad leader of Eleven/Two, lieutenant ... what was his name again?" Bodan just raised his shoulders. "... Oh, never mind," the major made a dismissive gesture with his hand, "you're sending him to me, anyway!"

"Send lieutenant whatsit to you. Understood, Comrade Major?!"

The major did not make a face. "After that, you turn these squads here back over to their officers, understand?"

"Yes, Comrade Major!"

The major thought again. He had already turned away when something else struck him after all. "But you will only hand over these squads when they are fit to be led again, is that clear?" he roared.

"Yes, Comrade Major!"

Then he stomped off, speaking to his captain as he passed, "Have them stand at ease and moved off!"

1.18 Spaniens Himmel and Freedom

The small ambitious first lieutenant had received the task from the deputy battalion commander to "mop up Eleven/Two" by means of dull marching. He had imagined it to be quite easy. Just let the guy's march, always around the huge emplacement. When he himself got tired, he would simply order them to run, without running himself, of course, and that would wear down the squad for sure. However, he had reckoned without his host, that is, Bodan's men. Because they had so much practice by now that they knew very well how to save energy even when marching with heavy luggage or how to move at a run. The disadvantage of the double quick, at least from the lieutenant's point of view, was that the men became extremely slow as soon as the lieutenant did not walk beside them, and he soon gave up on that. So only the normal marching remained. Here, however, Bodan's men had a perfidious defense at the ready, the so-called tank step. This involved dragging every fourth step across the ground. When a dozen boots performed this perfectly at the same time on the loose chippings with a cement base, there was a wonderful sound:

Boom, Boom, Boom, SHRUMMMMM ...

Boom, Boom, Boom, SHRUMMMMM ... and so on.

It was enough to make you jump out of your skin. The lieutenant raged. He tried punishments, ordered push-ups, but the people just didn't listen to him and did jumping jacks or squats instead. In the end, when he started threatening jail days and extra duty, they ignored him completely and marched across the square at tank pace without him.

Boom, Boom, Boom, SHRUMMMMM ...

Boom, Boom, Boom, SHRUMMMMM ...

Suddenly, singing was heard in the distance. It seemed to be getting closer, louder and louder.

"That's a whole company," Schmitti noted, and Kramer added, "At least they are allowed to sing."

"We can, too, or do you think that Wally will stop us?" the man next to the ensign, whom everyone just called Ulli, although his real name was Bernhard, had taken his weapon from his shoulder and carried it casually in front of his chest, as he was used to doing during the marches with Bodan.

> *"Spaniens Himmel breitet seine Sterne*
> *über unsren Schützengräben aus ..."*

The singing could now be clearly heard. Still, however, the choir was behind the gentle hill that led up from the emplacement and to the sprawling battle and practice field. Bodan's men could not see it, but it had to be a mighty choir.

"And they seem to be in a good mood," said Jan, who did not even try to hide his displeasure.

"Carry your weapon as required by the NVA marching order!" suddenly shouted a shrieking voice behind them. The lieutenant hurried up. He yelled, "Squad, haaalt!"

"Just keep marching!" whispered the ensign, and so absolutely nothing changed about the:

Boom, Boom, Boom, SHRUMMMMM ...

Boom, Boom, Boom, SHRUMMMMM ...

"Stop where you are. We're in a combat exercise!"

"Thanks to you not anymore, you moron," Jan shouted back and on it went with

Boom, Boom, Boom, SHRUMMMMM ...

Boom, Boom, Boom, SHRUMMMMM ...

Meanwhile, in the distance, the singers had once again reached the chorus. They had to be close behind the hill now and one could hear:

> *"Und der Morgen grüßt schon aus der Ferne ..."*

"Come on, let's match the marching beat and sing along," Kramer suggested, and almost instantaneously the

Boom, Boom, Boom, SHRUMMMMM ...

Boom, Boom, Boom, SHRUMMMMM ...

perfectly matched the beat of the now not-so-distant song.

> *"... bald geht es zu neuem Kampf hinaus."*

Boom, Boom, Boom, SHRUMMMMM ...

Boom, Boom, Boom, SHRUMMMMM ...

They finally joined in on the chorus:

> *"Die Heimat ist weit.*
> *Doch wir sind bereit.*
> *Wir kämpfen und siegen für dich:*
> ***F R E E D O M***"

Boom, Boom, Boom, SHRUMMMMM ...

Boom, Boom, Boom, SHRUMMMMM ...

The little puny lieutenant had turned bright red and was shouting and yelling. But no one understood what he was trying to say anymore.

On the other side of the hill, Bodan had the double company march up, aligned lengthwise to the hill, chanting. Simply again and again "*Spaniens Himmel*," the first verse including refrain. In between, his commands, followed by the corresponding repetitions of the senior officers now subordinate to him. From behind the hill, they heard a

Boom, Boom, Boom, SHRUMMMMM ...

Boom, Boom, Boom, SHRUMMMMM ...

That quite excellently set the rhythm, a rhythm that they could never have stomped in such a way on the soft meadow.

So finally, one hundred and sixty men stood in rows of three with a high-ranking squad leader every five meters or every dozen men, along the whole long line. Bodan knew who was there on the other side of the hill celebrating such a wonderful tank pace and he knew that he owed his faithful followers something.

Boom, Boom, Boom, SHRUMMMMM ...

Boom, Boom, Boom, SHRUMMMMM ...

His people, on the other hand, had no idea what was coming, although at one point, Schmitti thought he heard Bodan's voice.

Boom, Boom, Boom, SHRUMMMMM ...

Boom, Boom, Boom, SHRUMMMMM ...

Exactly at a renewed beginning of the song finally came the command:

"Right turn! In broad front: forward, march!" and the double company started to move up the hill singing.

Boom, Boom, Boom, SHRUMMMMM ...

Boom, Boom, Boom, SHRUMMMMM ...

In addition, the stomping of the more than one hundred and seventy pairs of boots in the field. The singing carried so far from the hill that once again the generals took notice and observed the imposing marching movements.

"By golly, Comrade Major, they sure can do some marching, too!"

And Kirner added, "It's no wonder Fritz, look who's leading them again!"

The major, who had just returned from a short coffee break in the supply shack, took a pair of binoculars and could not believe his eyes. Not that he had a problem with this show, the more impressive the performance the better, but he would never have thought such a thing possible with the "undisciplined giggling troop" of a moment ago.

"Well, they're just our best people you see there," he said dryly, and, practicing modesty himself, he added, "Anyone could do good combat choreography with them."

General Kirner nodded knowingly. "Yes, but not everyone is able to see the potential and use people properly, are they, Comrade Major?"

"Yes, I suppose that's true," the major agreed, secretly wondering if the two of them didn't know more than they were letting on. Maybe their visit wasn't as coincidental as everyone here thought.

'All the more reason to be careful and drop the lieutenant colonel and his little war against Bodan,' the major thought to himself and concentrated again on his binoculars and the interesting deployment in the distance.

Boom, Boom, Boom, SHRUMMMMM ...

Boom, Boom, Boom, SHRUMMMMM ...

When the chorus was on, the men of Eleven/Two could see the first heads of the company marching up the hill in a front almost a hundred meters wide, and they almost lost their rhythm when they realized who was leading this march. Exactly at "*Freedom*" the wide front reached the small summit ridge, and it became probably the loudest and most resounding

"F R E E D O M"

that this landscape had ever heard.

Continuing to chant, Bodan had the large company march down the hill and line up in a U. Then he yelled

"Song off!"

Momentarily there was silence. Bodan turned to the lieutenant, who had been watching the action at a fair distance, apparently not daring to come closer.

"Comrade Lieutenant, by order of major Gerber, I am now taking over my squad again. You yourself are to report to the major immediately."

Without waiting for a reaction, Bodan now turned back to the company and called out, "It has been an honor, Comrades, and I thank you for your service.

Senior Officers, please now resume your units and command as you see fit!

Thank you!"

In a perfect fluid motion, Bodan clicked his heels and saluted. All squad leaders returned the salute in unison, and if that had not been enough, Kramer's operatic voice now bellowed a perfect:

"Hooray!" into which immediately fell his comrades, "Hooray! Hooray!"

When the sound of this call reached the battle staff, the major thought, 'Now don't you disappoint me Bodan, don't you disappoint me!'

1.19 The Enemy Wins and Turns Out to Be a Brother in Arms

But Bodan and his men did not disappoint anyone. The superior force, which arrived with several tanks and two full companies of infantry, should have had an easy time of it, despite the fortified defenses. But when Eleven/Two was sent into the field, shortly before the total collapse of the "enemy defense," something happened that no one had expected. Instead of rigidly taking over the task of defense and entrenching himself with his squad in the trenches and bunkers, Bodan ordered an attack. He had taken over half a dozen "not yet fallen" but already quite exhausted defenders and ordered them to "just make one more real noise" and to detonate all smoke grenades at once on a certain signal.

"What signal?" asked one of the remaining.

"A whistle from me, you always hear it everywhere, don't worry," and he gave a taste.

"Yes, you hear it everywhere," said the "survivor," rubbing his ears.

"You're taking care of everyone who comes running across the middle in your direction in a panic. You've got a perfect open field there so no one should be able to get through, understand?"

The seven survivors nodded, but it was clear they didn't understand why the enemy would be so stupid as to run across the open field into their direct paths of fire.

"And when we're done with them," Bodan continued, "Kramer will call out 'Victory' to let you know it's all done and you can come out .. or at least stop shooting so you don't end up hitting us. All right?"

Again, nodding of heads, but rather disturbed looks instead of real understanding.

Then Bodan split his squad and both groups crept to the opposite edge of the trench. When after the whistle the shots and practice hand grenades banged, and at the same time the smoke grenades took the view, the two groups of Eleven/Two streamed in a wide pincer movement from the side behind on the enemy, who was entirely confused. They were completely forward oriented and had nothing to oppose to this paratrooper tactics, nothing at all.

"That's almost a textbook example," Kirner enthused, impressed.

"Yes, it's almost like Hannibal's strategy at Cannae ... only with modern weapons," agreed his companion. "What a dynamic!"

"Yes, and look Fritz, both attacking groups mercilessly stick to the slightly backward angle of attack so they can't hit each other. That's real perfection."

When Bodan's groups had completely roughed up the enemy line, Kramer's tremendous voice rang out, shouting:

"Victory!"

And then two more times:

"Victory! Victory!"

Behind the cheering that followed, one might have guessed over a hundred men, but it was only Bodan's eleven companions and the seven survivors, as well as Bodan himself, who made this infernal noise. Then Bodan's voice was heard:

"A song!" and Kramer sang "*Sabota u nas prostaja*," the "*Song of the restless youth*" in Russian. Singing, the three groups thus united somewhere in the open field, where swaths of artificial fog and battle smoke were still wafting around. The fact that two of the survivors from the previous defensive exercise still had simulated bloody bandages around their heads could hardly have fit better into the picture. When the groups had found each other, the song was over and Kramer sang the Soviet national anthem, which at least the men of Eleven/Two had learned from their "choirmaster," as they occasionally called Kramer, and could sing perfectly by now.

The major, who observed the entire action standing next to the two generals, put his hands in front of his face and groaned. Now those lunatics from Eleven/Two had really gone completely over the top, he thought, and could have cried.

But suddenly he felt the heavy hand of Lieutenant General Kirner on his shoulder and heard him say, "Don't take it so hard, Comrade Gerber! We may have lost, but it seems we lost to our brother in arms from the glorious Soviet Army, and that wouldn't be the first time in history, would it?"

Since the major didn't know whether he should cry about it or laugh, he decided on the latter, and because he couldn't stop laughing once he started, his laughter became increasingly free and infectious. So infectious, in fact, that soon the two generals joined in and didn't stop until all three of them had tears in their eyes and could only gasp. It must have been the first and last time that such high-ranking members of the National People's Army laughed out loud while the Soviet Union's anthem was sung with fervor nearby.

1.20 The Interrogation

Neither Major Gerber, nor the little mingy lieutenant colonel, nor anyone else from the entire staff had the guts to enlighten the deputy battalion commander about the "situation in the case of Eleven/Two or Bodan." That is, the lieutenant colonel had no idea that Bodan and his men had not only somehow gotten back together after all, but that they had taken the cake again.

So, it was not surprising that the deputy battalion commander, assuming that he would have to deal with a humiliated bunch of pathetic, now submissive and insecure, meaning ultimately typical, NVA members, had prepared a show tribunal. He summoned not only high-ranking political officers and all the more important personalities from the battalion staff who were subordinate to him but also all the battalion's senior officers at once. He promised nothing less than the demonstration of an "example to rebellious elements." Major Gerber and the company captain made one last attempt, but without sticking out his neck too far.

"But Comrade Lieutenant Colonel, won't you at least let the soldiers shower first? After all, they have just come from a combat exercise," the major tried for the last time and the captain agreed with him, "Yes, or at least let them hand in their weapons first."

But the lieutenant colonel remained obstinate. He wanted to make these puffed up "heroes" even smaller than they would be today anyway and he wanted to see them in their dirty, damp clothes with their heavy weapons draped in front of them. He wanted to see how they would look around in vain for a chance to sit down, to rest, and how he, the master of their damned well-being, would not grant them just that ... and that until they would be ready to cave in and publicly apologize. It irritated him somewhat that a man in civilian clothes had been sent to the interrogation "from higher up." He had not been told anything more about this person, but it was not unusual for other state forces besides the military to have a say in the matter, because after all it was about the academic elite that they had to "bring in line" here. The guy in civilian clothes, however, was friendly, seemed completely harmless, and gave the impression that his participation in the interrogation of these "rebels and insurgents" was nothing more than a formality. Finally, the sergeant announced the arrival of the company and, indeed, a heavy stomping on the stairs announced that the moment had come.

"Have the company take a seat in the aisle for weapon cleaning. Then, call Bodan to me! We'll start with him, he's the key."

The captain just nodded. He knew very well that the lieutenant colonel wanted to get the squad leader of Eleven/Two as publicly as he possibly could away from his fellow soldiers, past the other soldiers of the company as if running the gauntlet, to the front of the staff rooms. He knew well that he was counting on Bodan being finished and depressed today, because in all the haste with which the deputy battalion commander had wanted to conclude with this personnel matter, he had not taken a single message from today's exercise. Yes, no one had dared to inform him specifically about what had happened either.

All right, then, that's the way it should be. The soldiers sat on their stools in two endless rows on both sides of the wide corridor and began to clean their weapons.

The whole company knew what was going on, when suddenly, in the silence of this process, the call, "Bodan forward to the staff! Interrogation by the deputy battalion commander!" resounded.

Bodan, still in full gear and with his heavy RPG6 and, as usual, Kramer's AK47 in front of his chest, nodded to his men. They looked uncertain, but also determined. Then Bodan turned towards the staff and roared so that the hallway trembled:

"Yessir, Comrade Captain!"

Then he stomped forward.

At the latest when this mighty warrior stood before the so well-draped audience, the higher officers seated and the lower ones all standing on one side of the room, with Bodan on the other, it dawned on the lieutenant colonel that it had not been such a good idea to let Bodan appear in full combat gear. They should have at least let him take off his weapons first. To make matters worse, the lieutenant colonel had also instructed the captain to order Bodan to stand at attention, but that only seemed to make things worse. Now, however, it was too late for a correction. This giant stood in the room, rigid and tough, filthy from battle, but still a colossus of energy and willpower.

'I don't care ...,' the lieutenant colonel thought to himself and was sure he knew how to tackle this guy. He would grab him by his arrogance and either he would buckle or get himself into hot water. The lieutenant colonel, who had his experience in such things, did not see any other way and he would not let this soldier have any other way either. 'And God have mercy on you, young man, if you don't buckle,' he thought, and immediately got down to business.

"You know why you are here, Comrade Private?"

"Yes, Comrade Lieutenant Colonel! I think so."

"Well, then you can probably guess what we, all of us here, want to hear from you now—EXACTLY NOW—can't you?"

"Yes, Comrade Lieutenant Colonel! I can guess that, too," Bodan answered calmly and almost modestly, at least without the slightest sign of rebellion or revolt. That was exactly what the lieutenant colonel had expected. He could not have known that Bodan was already manipulating him and steering him exactly where he wanted him. Having to play the modest one for a short time was a comparatively small and therefore justifiable sacrifice.

"Excellent, Private," the lieutenant colonel's voice clearly betrayed his satisfaction at this easy victory, "then we're all ears now."

The lieutenant colonel turned to a young lieutenant at the desk behind him. The latter was apparently taking the minutes, and the lieutenant colonel asked, "Do you have everything, Lieutenant ...? And are you ready for the rest?" Broadly, he then turned back to Bodan and in a grand gesture patronizingly allowed him, "Stand at ease, Comrade Private!"

He had hardly uttered the words when the lieutenant colonel was already regretting them again. The two weapons, one of them a huge bazooka, casually in front of his chest, this private in front of him now assumed the broad-legged stance of the paratroopers, brought his fists together behind his back and thus made such an imposing picture that a murmur went through the ranks of the young officers. Taking alone the precisions of movements made everybody in the room hold his breath for a moment or two.

One could literally feel the crackling atmosphere when Bodan began to speak. His voice was calm, deep and serene ... far too calm for one who was about to grovel.

"Correct me if I am wrong, but we are here because we have a task as future academics of our socialist homeland. A task that cannot consist solely of studying well and later doing excellent work. Then there would be no need for all this here. No, we are here because this is also where the future leaders are produced ... are to be produced. As these future leaders, character traits such as sincerity and righteousness must always be inherent in us. That's the only thing that distinguishes good leaders, and that's why that's one of the goals that our training strives for ... isn't it, Comrade Lieutenant Colonel?"

The lieutenant colonel felt himself nod, he saw his surroundings nod, and he heard himself, to his own horror, say an audible "Yes!" This went quite differently than the lieutenant colonel had imagined, and even as his surprised brain searched for a solution to turn things around, Bodan struck the final blow:

"Then it must also not be the goal of the training here to give the prospective leadership cadres of the GDR the idea, or even the impression, that you can simply pass on mistakes that you have made yourself to others ... even more so if the others are subordinates."

Nodding in the room again. But now Bodan's voice became threatening.

"Unfortunately, however, I have a feeling that this is exactly what is happening here. Someone made a mistake. Something was ordered without ever asking those who are supposed to pay for it. Now those who didn't even want what was ordered are supposed to pay for the mistake. This is injustice!"

"But now look, Comrade Private!" the lieutenant colonel attempted a defensive blow, "If we were talking about a bar of chocolate or an ice cream here, you might be right, but we are talking about the central organ of the Socialist Unity Party of Germany. This is something the soldiers simply have to get AND READ, and if they have to get it and read it, they have to pay for it. It's as simple as that." He tried to sound confident and intimidating, but that didn't catch on with Bodan, who already had his answer ready.

"You think so?" he parried. "So, we have ONE *Armeerundschau* and ONE *Neues Deutschland* for TWELVE TO FOURTEEN men with only thirteen minutes a day to inform. How on earth is that supposed to work?"

Complete silence in the room was the only audible response.

"That's why I let my people watch the television newscast *'Aktuelle Kamera'* every day and then we have fifteen minutes to discuss it. With the newspaper, not even one person could read everything, let alone inform everyone. So, this is a completely nonsensical solution as far as the necessary political information needs of the future cadres of this country are concerned. I don't even want to know who came up with this nonsense."

The lieutenant colonel had astonishment written all over his face, but Bodan wasn't finished yet, "What I want to know, however, is who cannot see this mistake and instead forces subordinates, who have to get by on a minimum scholarship and pay, to cover the damage for the person or persons responsible. As I said before, this is injustice!"

Bodan took a deep breath and since no one could or would say anything he continued, "I have a well-working typewriter that has certainly written letters, which have been read, to the Central Committee of our Socialist Unity Party and I will use it to make sure that this story does not end like this. Because I cannot, in good conscience, support that an entire cohort of male cadre is dismissed with the understanding that you can pass off your own mistakes on subordinates."

The lieutenant colonel, whose mouth was still open, had turned white as a sheet. Everyone looked at him, but he did not move.

Finally, and completely unexpectedly, there was a sudden applause from the last row behind the wide desk at which the minute writer was sitting as well. The inconspicuous looking man in civilian clothes stood up. He clapped so naturally and looked so sincere, so radiant, that gradually the junior officers and finally the senior officers, except for the lieutenant colonel, also applauded. Then the man in civilian clothes stopped clapping and there was an instant silence. He spoke:

"This is the spirit we need, Comrades!" Then he came directly to the point, "Of course, you and your squad, as well as the two of Twelve/Two who wisely joined you, Comrade Private, Comrade SQUAD LEADER, do not have to pay the newspaper money ... and you may still keep the newspapers."

Bodan nodded.

"However," the civilian ran his right hand through his hair, apparently brooding heavily, "I would like you to suggest what we should do with the funds already collected from all the other reservists in the battalion. Do you have any ideas about that?"

Bodan didn't have to think long. "Leaders also learn from mistakes," he replied. "I think redistributing the funds is pointless, but you should use the process to educate people about leadership. Inform your comrades of the circumstances and thank them for their support, even if it is not entirely voluntary. I think it's instructive when they realize that timely questioning would have led to quicker clarification and sincerity ... and also might have saved them cash."

"That's how it's going to be!" spoke the man in civilian clothes and Bodan had great trouble not to look into the shocked, stupid face of the lieutenant colonel and laugh out loud.

"We're done here?!" the civilian now turned to the lieutenant colonel, and it was clear that he wasn't really expecting an answer or clearance. When the lieutenant colonel showed no reaction, he said, "Good!" Then he pushed his way forward past the desk and motioned Bodan to follow him.

"Come with me for a moment, please!" he said in a whisper, almost conspiratorially. He led Bodan out of the staff meeting room into the small anteroom, called over his shoulder, "Two minutes, please!" and then closed the door behind him and Bodan, which, incidentally, was the only exit to the corridor.

"Have you ever thought about a career as an officer ... alongside your career as a physicist, which is what you want to become, I mean? Possibly in a special unit or organ of the GDR?"

Bodan now looked surprised for the first time, and he almost seemed shocked. Finally, he replied, "Oh, you know, I'm not so good at belittling people."

"We know that" the plainclothes man replied, "that's exactly why I'm asking. Because there will be days when we need the other leaders, the ones like you, who build people up and push them to do their best, not the ones who keep them down," and he gave an unmistakable wave in the direction of the door they had just closed behind them.

Bodan did not answer. He would not have known what to answer either. Finally, he said, "Give me time until after this training. I already have a command now and my people need me."

The civilian smiled and said in a very gentle voice, "No problem, we'll get what we want anyway." These words seemed so incongruous with his thoroughly friendly expression. He paused for a moment and seemed to be brooding again, just as if he were at a crossroad, and Bodan sensed very well that this was all an act. He would have liked to know where this was going, but he didn't have long to wait.

"By the way," the civilian finally continued, "we know, of course, about your little side-study in Leipzig. Child psychology, physics, and hands that can kill if they want to ... **or have to**.

What great potential."

The man greeted Bodan by casually putting his hand to an imaginary headgear.

"You may leave now ... **and celebrate**!"

Bodan saluted back just as casually, and as he strode to the door leading to the hallway, he heard the civilian call into the meeting room, "We're done, you may come out now!"

1.21 Triumph

'No one at home is going to believe this,' Bodan thought, and looked down the long corridor to his people at the very back. He stood in the middle of the hallway and saw how they fixed him inquiringly, anxiously, excitedly, partly also fatalistically resigned. He must have looked somehow defeated, or at least exhausted, because now he noticed the stares of the other soldiers. Unconcealed malice and hatred spoke from their faces.

'Poor bastards,' thought Bodan, looking at them expressionlessly. There they sat on their stools and were pleased that one of their own, one who had rebelled and not gone along with every nonsense from those up there, had obviously taken a beating. And of all things, this now gave these pathetic sneaks satisfaction. Bodan had nothing but contempt for these people at that moment. But then he thought about how lucky he had been with his squad, how they had stuck together and gone through thick and thin. What if he had ended up on one of the other platoons? How would he be standing here then? Or would he even be sitting in the long aisle, looking spitefully at whoever had dared to rebel in his place ... another 'one of five thousand', another 'colorful bird among gray sparrows'?

'Never!' thought Bodan, 'Never would I have sunk so low!'

Slowly, infinitely slowly, Bodan took the bazooka from his shoulder. He grasped it thoughtfully with his right hand, stretching it above his head as if to a typical Red Army war memorial. Then he roared so that the hallway just shook:

"VICTORY! Comrades, VICTORY!"

In the farthest part of the aisle, cheers instantly erupted. It didn't take Kramer long to turn the uncoordinated yelling into a wonderful "*For he's a jolly good fellow. For he's a jolly good fellow* ..." that boomingly filled the aisle and gave something thoroughly mystical to the seemingly overman-sized war memorial standing in the front section with the huge bazooka weapon on its outstretched arm.

The hatred in the faces almost immediately gave way to pure astonishment, even horror. There it was again this stupid expression of "This can't be true!" There they sat on their stools, polishing their weapons and some already cleaning their boots, or more precisely, holding these things in their hands, because at that moment no one was cleaning anything anymore, and looked as if they had now finally been robbed of any illusion. And gradually the hatred returned, became anger, and focused itself in Bodan's direction.

'If looks could kill ...,' he thought.

But when Bodan looked more closely, he noticed a tiny change. The glances were no longer directed at him. They went past him or through him and were directed at those who now stepped out of the anteroom of the staff meeting room.

And then it happened. In the back, they were done with "*For he's a jolly good fellow*," Bodan had just lowered his heavy weapon and begun his seemingly interminable walk down the hall to his squad, while behind him the officers streamed out of the briefing room. As Bodan took the third step, he heard a soft word in front of him:

"Victory"

He took the next step and again he heard:

"Victory!" this time already a little louder. After three more steps, the stomping of his feet on the stony floor could already no longer be heard and when Bodan passed the staircase, reached the first stools, there was the

"Victory!"

"Victory!"

"Victory!"

and it swelled to a thunderous roar that punctuated each of his steps, as if he himself would bring forth such a natural event with his movements.

The eyes Bodan now looked into had something thoroughly defiant, unruly and, as far as he was concerned, appreciative.

1.22 The Last Posse

Bodan had reckoned that the lieutenant colonel would at least be smart enough to have somehow quietly cleared up the announcement of the "affair with the newspapers," but the man was so full of hatred and thirst for revenge that he simply could no longer read the signs, let alone construe and interpret them.

The stories of the "Common Field Mouse" and the "Victory!," as well as the renewed, incredible performance of Eleven/Two, had long since made the rounds throughout the battalion. And when roll call was held the other morning, it was as if all of Eleven/Two were not a squad of simple reservists, but a staff of the highest generals the Warsaw Pact could muster. Wherever they went, they clicked their heels and saluted; the senior officers of the garrison as well as the junior officers. More and more often the shouts "Victory!" or "Common Field Mouse, Comrade Private!" were incorporated into the salute. Sometimes they just shouted "Field Mouse!," later "Major Field Mouse" or "Colonel Field Mouse" and in the end even "General Field Mouse."

On some of the wall newspapers, which were usually full of political propaganda and thus, widely ignored, funny pictures of "field mice" appeared.

The political officers only realized this when suddenly there was quite some gathering around these sections of the corridors, because, as said, usually the political wall newspapers attracted no attention at all.

When they started to take these cartoons down, more realistic drawings appeared, showing things like "real field mice in their natural habitat."

The greatest attractor, however, was a rather simple picture of a soldier being entitled "field mouse with shoulder-fired anti-tank missile."

Bodan and his men could hardly stop laughing. They graciously returned the compliments and Kramer got used to regularly giving the papal blessing or at least casually waving his backhand like the Pope.

At the roll call, however, Eleven/Two lost their laughter. In his stubbornness, the lieutenant colonel announced that at the "suggestion of the soldier Bodan," the newspaper money would not be given back. He went even further and stated that the battalion leadership was very indebted to Bodan and his squad. He said that this was not only for the help in clearing up the "oversight," but above all for this excellent suggestion.

"How many times does he want to pick on it, and they would never have given the money back anyway? What a nasty pig!" grumbled Jan.

But the lieutenant colonel continued his speech. It was precisely this suggestion of private Bodan, he explained, that would make the solution easy, because it would spread the damage over many broad shoulders.

"Well say it you bastard: just not on ours!" Jan had now become loud, and the other men in Bodan's squad were also grumbling and voicing their displeasure almost unmuffled.

But the lieutenant colonel didn't seem to care. On the contrary, he seemed very pleased and now looked directly at Eleven/Two. "As a token of gratitude," the lieutenant colonel continued, "the battalion command ..."

"So, that would be you, you ass!" shouted the ensign.

"... decided to give the entire squad Eleven/Two two days off. These may also be used for egress."

"That sneaky prick," Schmitti scolded, "now we're the bad guys again for everyone."

"Squad Eleven/Two, you are dismissed!"

Neither of Eleven/Two moved.

It seemed the lieutenant colonel had expected just that. With a few long strides, he stepped in front of Bodan and roared, "Have your squad step away, Private, otherwise absolutely nothing will move here, for anyone. No one will get anything to eat here either, until you are gone. Not the whole battalion. Now march off and have fun on your special leave!"

Bodan was about to grab this disgusting guy in front of him and break his neck with one quick grip, but he knew that then not he, but the lieutenant colonel would have won. He was just about to decide to stay put when shouts rang out from everywhere, "Come on, just go, we'll be fine!"

Or, "Have fun out there, Private, and give my regards to the 'Common Field Mouse'!"

Or: "Just have a beer for us!"

Or: "No, even better, bring something in, because the seniors don't control you anyway!"

When Bodan looked into the broad crowd, he saw nods of agreement everywhere and eyes that winked encouragingly at them.

'Well, you didn't expect that, did you, Lieutenant Colonel?' he thought.

Then he turned on his heel and roared, "Squad Eleven/Two: Attention! To our loyal comrades of the entire glorious Seelingrad Reserve Battalion, three cheers!"

His men grinned and they shouted in unison into the 1987 Advent sky:

"Hurrah! Hurrah! Hurrah!"

With a skillful sweep, Bodan led his men out of roll call order, and no sooner had they taken two steps than he commanded, "A song!" and Kramer, who as always had the right nose, hit the famous nail on head with *"Sag mir wo du stehst!"*

1.23 Five Thousand Men at Tank Pace

As he led his men in song to the canteen, Bodan glanced once more at the lieutenant colonel and saw him seething inside. But the lieutenant tried not to let his displeasure and disappointment show. He had not yet completely spent his powder. He had stridden back to his speaking place and his voice was no longer bellowing. On the contrary, it was almost quiet, hissing, and yet it carried and seemed all the more threatening.

"Unfortunately, I have recently observed several times that higher ranks greet lower ranks first. However, this is only permissible if the lower rank occupies a higher position, otherwise ...," he paused and then roared again, "**... it is strictly forbidden!**"

"In this facility," he continued in his threatening hissing voice, "which even some SPECIAL SOLDIERS say is meant to produce leadership cadres, such silly behavior is anything but acceptable. I therefore order five rounds of penalty march before noon. Move out!"

The lieutenant colonel had probably reckoned that no particularly orderly marching would now follow. In God's name no, they should mutter and grumble. At some point, the grumbling would turn back to the right direction, and then Bodan and his men would have nothing to laugh about. They would be outcasts again and would also be treated like that by their fellow students. And that is exactly how it should be. The lieutenant colonel knew all about this kind of psychology. He had already solved quite different problems in this way. This time it would not be any different. Eventually, they would dump Bodan and hate him and his squad for making them suffer because of him. He, the deputy battalion commander, just had to make them suffer enough and that was easy, very easy in this wonderful National People's Army.

But then something completely unexpected happened again. The battalion had completed the fifth penalty lap and should now have turned off to the canteen. Instead, however, they simply continued marching on the round ground. Even the senior officers marched along. Couldn't they count anymore? The lieutenant colonel was about to roar off and tell those in front that they were effin' done, when a noise rang out, a tremendous scraping rumble that would have made any attempt to communicate absolutely impossible. The lieutenant colonel recognized the sound immediately and he panicked. But there it was again, and again, and then there was this wretched rhythm:

Boom, Boom, Boom, SHRUMMMMM ...

Boom, Boom, Boom, SHRUMMMMM ...

Boom, Boom, Boom, SHRUMMMMM ...

Boom, Boom, Boom, SHRUMMMMM ...

Boom, Boom, Boom, SHRUMMMMM ...

Boom, Boom, Boom, SHRUMMMMM ...

...

If a hundred men do that, that's already impressive, but with five thousand it is simply a force of nature. In the canteen, Jan's spoon almost fell out of his hand. The men around Bodan looked at each other. First shock, astonishment, and then laughter on their faces. They jumped up, knocked over the chairs and ran outside.

The battalion had almost finished the first round at tank pace and was now marching again like a mighty roaring storm towards the large canteen. Next to them, that is, next to this huge noisy roller, the lieutenant colonel was fidgeting, small and ignored by everyone.

Boom, Boom, Boom, SHRUMMMMM ...

Boom, Boom, Boom, SHRUMMMMM ...

It was obvious to Bodan then that these people would not stop now, not for a long time.

Boom, Boom, Boom, SHRUMMMMM ...

Boom, Boom, Boom, SHRUMMMMM ...

He had his squad line up next to the march on a large loading platform and, as in the May Day parade where the bigwigs took in the military parade from a grandstand, Eleven/Two stood and saluted the battalion marching by at tank pace.

Boom, Boom, Boom, SHRUMMMMM ...

Boom, Boom, Boom, SHRUMMMMM ...

And just like in the real parade, the marchers jerked their heads to the right to pay homage to the bigwigs in the stands.

Boom, Boom, Boom, SHRUMMMMM ...

Boom, Boom, Boom, SHRUMMMMM ...

Boom, Boom, Boom, SHRUMMMMM ...

Boom, Boom, Boom, SHRUMMMMM ...

Boom, Boom, Boom, SHRUMMMMM ...

Boom, Boom, Boom, SHRUMMMMM ...

And because all this was not enough, suddenly another sound came from the mass, at first isolated, then louder and louder until it became one compact monstrous roar:

"Field mouse!"

"Field mouse!"

"Field mouse!"

"Field mouse!"

...

Perfectly in the rhythm of the tank pace went this call.

Bodan's men couldn't believe it and they could hardly do anything but laugh. But they made every effort to continue saluting properly and so it became probably the funniest standstill of all time.

Boom, Boom, Boom, SHRUMMMMM ...

Boom, Boom, Boom, SHRUMMMMM ...

"Field mouse!"

"Field mouse!"

...

When the battalion had passed, they went back inside to finish their meal, but the lieutenant colonel snorting with rage blocked their way.

"What do you think you were doing just now?" the lieutenant colonel roared in all the din.

"We saluted higher ranks, Comrade Lieutenant Colonel," Bodan replied, also drowning out the noise, and he and his men walked past the deputy battalion commander as if he was already gone.

Boom, Boom, Boom, SHRUMMMMM ...

Boom, Boom, Boom, SHRUMMMMM ...

"Field mouse!"

"Field mouse!"

...

The battalion marched united, for a full 10 hours, into the late night and in addition to

"Field mouse!"

"Field mouse!"

Boom, Boom, Boom, SHRUMMMMM ...

Boom, Boom, Boom, SHRUMMMMM ...

...

they called in between occasionally

"Victory!"

"Victory!"

...

and in the end even

"Bodan!"

"Bodan!"

...

1.24 Discharge

The remaining days were all fun for Bodan and his men. No superior wanted anything more from them and one was even glad when they came up with suggestions that somehow brought them, that is, squad Eleven/Two, out of the facility. Because ban or not, saluting Bodan and his people, "field mouse" and "victory" shouts were simply unstoppable. For example, when a volleyball tournament was held in the nearby town, the superiors were only too happy to give their OK for the entire squad to participate. They even allowed them to participate in a Christmas carol singing in the nearest larger church, but ordered them to wear civilian clothes. Bodan ignored this, however, and so it happened that on the third Advent of 1987 a festively uniformed "delegation" of the National People's Army made a quite wonderful choir in a service of the beautiful neo-Gothic St. Johannis Church in Gera.

On the day of discharge, Bodan used his connections with the "elderly" of the staff to obtain for himself and his men the early issue of their military service cards. Without these, they could not have really and above all permanently disappeared.

As agreed, he waited with the passes about a hundred meters in front of the gate, where his people were to join him with the luggage. When they finally arrived, the ensign was missing.

"Where is our highest rank?" Bodan asked his men jokingly, but they just shrugged.

"I think he had something to settle with the sergeant ..."

"We can't go out without him, or if we go out now, he won't be able to come with us, because your running cards only work together with mine, and ..."

"All right, we'll wait!" interrupted Kramer good-humoredly.

Bodan nodded contentedly, for he would have suggested the same.

Relaxed, he watched the hustle and bustle in front of the gate, or rather on its inner side. A considerable number of reservists had already gathered there and wanted to get out. Although it was known that the deputy battalion commander wanted to demonstrate his power once again and not let "this bunch of rotten academics" out "until the last train would have already left Seelingrad," people had nevertheless gone to the gate and were pushing and shoving to get out. There had to be several hundred of them, and like Bodan and his men, they were all already in civilian clothes.

"He's just writing his report, you know!" Unnoticed, Jan had stepped next to Bodan and whispered this sentence to him.

"What? Who's writing what?" asked Bodan, puzzled. But Jan put his finger to his mouth and explained, "Well, the ensign, he's writing his report ... his last one."

Now Bodan understood, and he said only, "Aha! ... The ensign, then."

Jan wasn't done with his information, however, so he continued whispering, "At first these reports were shit, but then they got better and better, that is, we were portrayed better and better in them." Suddenly he grinned, "Now we are true heroes ... even you Bodan! We are the true heroes of this land. Who would have thought it, right?"

Bodan just shook his head. Then he looked inquiringly at Jan, who immediately understood. "How do I know all that, you want to know? ... Well Bodan, but don't tell anyone else ..., because I'm the one who reads the reports." And he put his index finger to his lips again and turned around.

"Oh look, speaking of the devil ...," he exclaimed, and sure enough, at that moment the pleased ensign appeared.

"Let's go people, rally behind me and present your cards!" shouted Bodan and his men were immediately ready.

The tumult at the gate came to an immediate halt when Bodan's squad appeared at the gate. They were saluted and those jostling at the gate respectfully stepped aside. The guard commander, a lance corporal, came out of the guardhouse, unlocked, opened and saluted as the squad went outside. However, when he tried to close the gate again, there was a scuffle.

"Come on dude, let us out!" some of the close ones shouted and pushed at him. The man pushed against them with his rifle. "Jeez guys, cut the crap or I'll get in trouble!"

"**Hey**!" yelled Bodan at the pushers and came back to the steel grate. The jostling stopped immediately. "Leave the man alone, ok?! Climb over that barbed wire for all I care, if you've got it in you, but don't any of you touch this lance corporal here again!"

"C'mon, field mouse, we want out, too!" apologized one of the pushers.

"Then discharge yourself, but not at the expense of others! At least you should have learned that, right?"

The man looked at Bodan ashamedly and said nothing.

The guard commander adjusted his Kalashnikov and gave Bodan a grateful look before removing the key and returning to the guardhouse. On the way there, however, the lieutenant colonel approached him. He must have literally snuck up from behind the guardhouse, because no one had noticed him.

"Hold it, Lance Corporal! Not like that, first you maintain order here!"

"And how?" asked the lance corporal defiantly.

"Send these reservists to their quarters until I decide to let them out!"

"What?!" shouted his vis-à -vis, stunned.

The man who was yelled at turned red and lost his temper, "Then use your weapon, Lance Corporal!" he yelled.

There was a moment of almost perfect silence. Then the lance corporal tapped his forehead, muttered loudly and audibly, "Do it yourself, if you dare," and strode past the lieutenant colonel to the guardhouse, shaking his head. When he arrived at the door, it was opened from the inside and three more lance corporals, and a sergeant stepped out. All of them had their AK47s ready to fire and the sergeant had his right hand resting on his Makarov pistol, holster open. "Is there a problem here?" he asked, his gaze fixed solely on the lieutenant colonel. Without another word, the latter stomped off. The guards then disappeared completely into the guardhouse and the sergeant locked the door behind him.

"I'll be jiggered!" it came out of Schmitti, and it was once again up to Kramer to bundle the mood in a song. It only took one line and the masses sang along, 12 men in front of the gate in freedom and by now almost a thousand on the other side:

> *"Und weil der Soldat ein Soldat ist,*
> *drum braucht er seine Freiheit bitte sehr,*
> *Es macht ihn ein Befehl nicht satt,*
> *das schafft keine Freiheit nicht her.*
> *Drum links zwei drei, drum links zwei drei,*
> *wo dein Platz Genosse ist,*
> *reih dich ein in unsere Freiheitsfront,*
> *weil du auch nur ein armes Schwein bist!"*

1.25 Gift Giving

Then Bodan and his men moved off. Behind them the song was sung again and again, and one could tell by the volume that more and more reservists were joining them, reservists who wanted to get out, out into freedom.

On the way to the station, Bodan's men bought a case of beer at the *Konsum*, the local small store, and a stollen at the baker just across the street. When the baker wouldn't take money for it, they thanked him with a perfectly polyphonically sung "*Es ist für uns eine Zeit angekommen*," whereupon the baker even gave them a jug of mulled wine as an encore. As a joke, Bodan then "ordered" Christmas, and so they sang their way through the town with all the Christmas Carols they knew and, arriving at the station, gave each other all sorts of nonsense as presents. Bodan received, for example, Schmitti's old toothbrush, Jan's worn-out shaving brush ("for the bald head") and, what he found particularly significant, a very worn-looking fountain pen from the ensign.

Later, one of the twelve would write in a book, "That was the best Christmas Eve we ever had on a December 21."

Together, they first drove to Gera. There they parted ways and continued on different trains to their respective hometowns. In Bodan's direction, three of his comrades still had to go with him. But they wanted to get some "travel beer" first, while Bodan was already stomping to the train. When he was alone on the train, he risked a look at his military service card. He flipped straight to the page with the promotions, and for the first time in over six weeks he was speechless. This had to be about the highest and—what is more—rather silent promotion a soldier had ever received, at least in the history of the

National People's Army. The countersignature confirming the promotion and new rank should also be anything but ordinary.

"Shit, ..." gasped Bodan and he stroked his bald head with his hand.

"Well, what have they turned you into?" asked Schmitti, who was just entering the compartment with the promised beers. Jan and Kramer followed him.

"Damn!" replied Bodan, "I'm not a common private any longer."

"Oh, you poor bastard!" Schmitti grinned and handed him a beer.

"Hooooooray!" roared Kramer, raising his bottle in cheers.

When they had drunk, Jan asked, "What have they made the best leader ever into?"

"Well to a lance corporal of course, but surely you know that!" replied Bodan and everyone laughed uproariously ... all except Jan, who looked thoughtfully outside and finally muttered half aloud, "Don't even want to know what this LANCE CORPORAL is up to when he's already moved so much as a private ..."

No one understood what he was trying to say, and no one really wanted to know. Finally, Kramer just yelled another perfect "Hooooooray!" and then sang a beautiful "*Süßer die Glocken nie klingen.*" Bodan sang along cautiously at first, but somewhere in the second line of the third verse he decided for himself that he would make the military service card with the number 76/0457883* disappear and always answer anyone who asked him that he, Bodan, was "only" a simple lance corporal of the reserve.

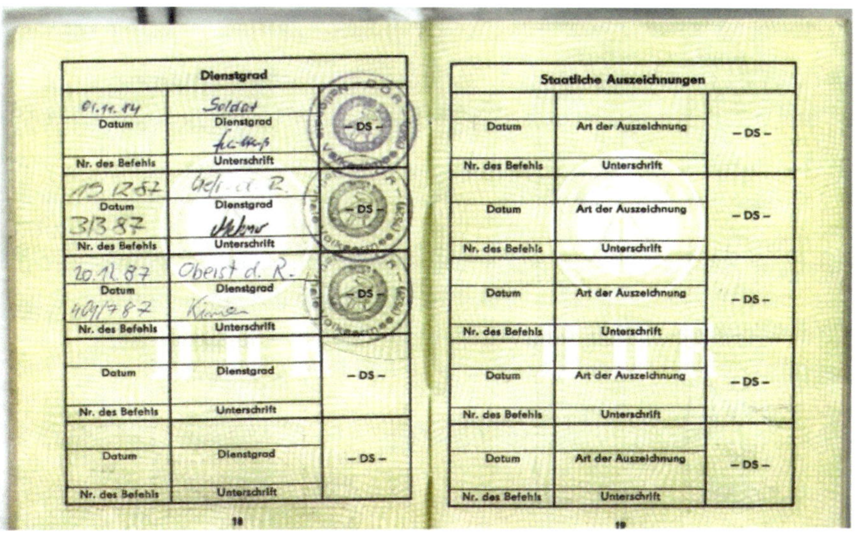

Dienstgrad

Datum	Dienstgrad	Nr. des Befehls	
01.11.84	Soldat		– DS –
13.12.87	Gefr. d. R.	3/3 87	– DS –
20.12.87	Oberst d. R.	401/7 87	– DS –
			– DS –
			– DS –

18

Staatliche Auszeichnungen

Datum	Art der Auszeichnung	Nr. des Befehls	Unterschrift	
				– DS –
				– DS –
				– DS –
				– DS –
				– DS –

19

Apparently the NVA military service card no. 76/0457883* was never disposed of after all, as you can see here. It shows that the simple private Bodan had become a colonel of the reserve in the end ...

1.26 The Call for Freedom

At some point, even the best steel will no longer hold. At some point, even the sturdiest gate gives way. And so, it happened on that day in December 1987, when thousands of reservists broke open the steel gate of the garrison and streamed out into freedom.

On that day, the GDR had finally achieved with its "military academic training" what, on paper, would have been the goal of this institution: It had produced independently thinking, strong-willed and imaginative personalities. The leadership of the GDR certainly did not really want this.

In December 1987, the GDR "discharged" an entire cohort of male students from Seelingrad. Students who had learned not only what freedom was but also how to obtain it.

Less than two years later, the GDR fell.

1.27 The End

A small field mouse sits on the edge of the gloomy Kastenwald near Seelingrad. It looks longingly into the distance, as if waiting for something.

But nothing comes. What the little field mouse is waiting for will never come again at all.

And the little mouse doesn't know that it itself changed history so that what it is looking for has disappeared forever ...

The common field mouse from my battle of Seelingrad.

Since the author is aware that there are still many people who idolize the GDR so much that they would like to see those who are responsible for their downfall dead, we have refrained from showing the real "common field mouse" here. Instead, we present a faithful replica in clay. Clay extracted, mind you, from the loamy soil of the place where this wondrous story took place. This is done exclusively for the protection of the mouse and of course not to provide a stage for the artist, for example.

"For the right mind it always holds that thoughts are freer than the thinker!"

2

Appetizers

2.1 How Do We Emit Confidence or Can the Dunning–Kruger Effect Be Mathematically Derived?

In the opinion of this author, there are two dominating processes in which we emit confidence. Of course, there are also other options, often only derivatives of the two dominant ones, but we leave these apparently recessive processes for later. Those two dominating ways can—briefly and dramatically simplified—be described as follows:

(a) We absorb information and give it away (show or mirror it in our confidence level to the outside world) depending on how well we have digested it.

(b) We internally reflect upon a certain topic and radiate to the outside world the pieces of our inside "confidence" depending on our inner state of "excitement" and/or self-assurance.

Thereby, the outside observer can only detect the sum of the confidence emittance of the two sources (a) and (b) and does not truly see the inner confidence level. This is hidden inside the individual (one may call it individual censorship or confidence horizon).

We therefore will prefer to use the expression "confidence emittance" instead of just the word "confidence" as the two might be severely different.

Mathematical Psychology: The World of Thoughts as a Quantum Space-Time with a Gravitational Core
Norbert Schwarzer
Copyright © 2025 Jenny Stanford Publishing Pte. Ltd.
ISBN 978-981-5129-27-4 (Hardcover), 978-1-003-58203-8 (eBook)
www.jennystanford.com

While we can assume that the source (a) should be approximately a linear dependency, which is to say that our "confidence emittance" increases linearly with the properly absorbed and digested information, we have to perform quite some contemplation on point (b). This is our more or less innovative source, where new ideas could come from and where we mull over things, before making up our minds. It is only fair to consider this source as some kind of cavity, in which bits of information are being processed, meaning they are permanently internally exchanged, respectively, intrinsically emitted and absorbed. This, however, is nothing else but the description of a black body or cavity radiation system. As the derivation of the corresponding radiation law is textbook knowledge, we will not present it here, but only refer to corresponding literature [1]. When we now take into account that all information has to be discrete so that (b) results in a quantum system or quantum-dominated object, automatically producing a Planck-like radiation, we can evaluate the connection between information being absorbed and "confidence emittance" for an individual.

We immediately recognize the typical Dunning–Kruger dependency [2, 3] with the well-known Mount Stupid (local maximum of the yellow curve) at relatively low knowledge levels (see red ellipse in Fig. 2.1). As far as this author knows, this is the first time that this effect has been mathematically derived.

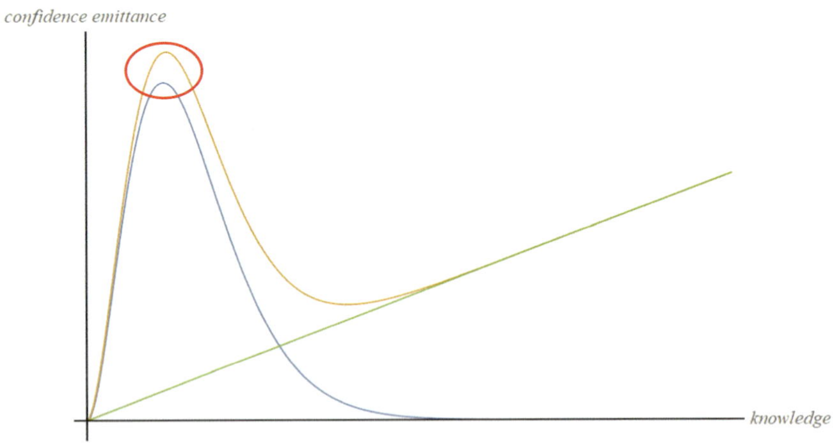

Figure 2.1 Individual confidence emittance to the outside world as a sum of source (a) (green), source (b) (blue) and the sum of the two sources (yellow). The red ellipse marks the so-called Mount Stupid (see text).

2.1.1 The Other Temperature

But having a mathematical description of this effect now at hand, we can try to find out more. For here and now, we are only interested in the influence of the inner state of "excitement" and or self-assurance. The dependency is shown in Fig. 2.2, where we clearly see how with growing self-assurance, the quantum-dominated source (b) becomes ever more dominant at lower amounts of absorbed information or knowledge and consequently the Mount Stupid effect increases.

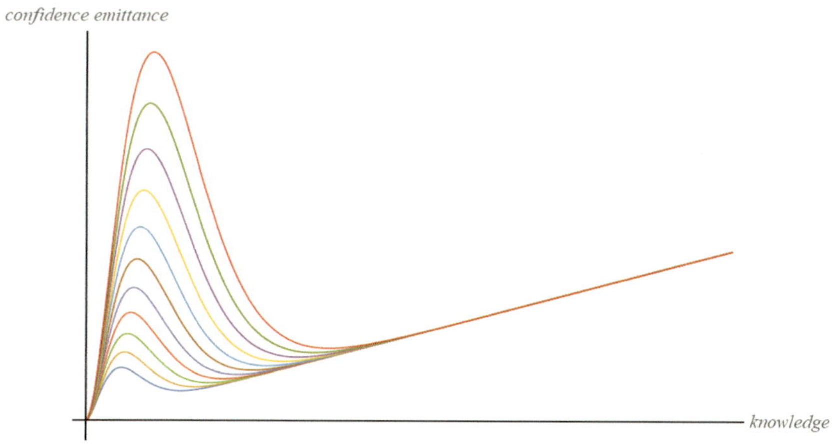

Figure 2.2 Individual confidence emittance to the outside world as a sum of sources (a) and (b) for a variety of inner states of "excitement" and/or self-assurance. Thereby, from blue to red, the Mount Stupid effect increases with growing self-assurance.

Most interestingly, we find that what in Planck's radiation law plays the role of the temperature, becomes the intrinsic self-assurance in connection with the confidence emittance. This makes intuitive sense, because the self-assurance could be seen as to be comparable with some kind of inner excitement. Seeing the latter as an average of the "kinetic" energy of our inner thoughts, we immediately recognize the classical temperature definition and hence, the congruence between self-assurance/self-excitement and physical temperature. Unfortunately, the inner (b)-confidence is not coupled to any real knowledge, but only the result of more or less isolated inner reflection. It can therefore—just as the Dunning–Kruger effect illustrates—lead to a confidence emittance which is more an embarrassment than reality-based wisdom. Hence, the expression "Mount Stupid."

2.1.2 How to Avoid Climbing Mount Stupid

After sending the calculations and Fig. 2.1 to a few friends, including even some experts on the field, one of them pointed out to me that he had hoped to obtain hints about how to get over the mountain of stupidity, i.e., avoid or overcome it, and asked what ways and means there are of not making this mistake that is obviously inherent in us.

Well, first of all, we should realize from which time in the evolutionary development of man our brain structure originates. It was constructed and formed when we roamed the countryside as hunters, gatherers and fishermen in at best medium-sized groups and the group already ensured—via its integral collection of experiences and knowledge—that the Mount Stupid did not become too high. Those who did not learn this quickly became victims of the omnipresent dangers, landed under the mighty hooves of the mammoth, in the clutches of a cave lion, poisoned themselves on plants or mushrooms, froze to death like Ötzi somewhere in the mountains or simply died of an avoidable disease or injury. It was simply the experience that one should be careful with the imaginary knowledge that made people quickly realize that they had to be careful in unknown areas of knowledge in order not to fall into the trap of being too stupid to realize that one is too stupid.

At the same time, however, the center of innovation was an advantage for those who had ideas over those who were not quite so imaginative. One could assume that this is why Neanderthals were inferior to *Homo sapiens sapiens*[1], even if they were physically the more robust species. So, despite the Mount Stupid disadvantage, the innovation center was an advantage and was therefore retained by evolution.

How could evolution have foreseen that there would one day be politicians, Tuna-Gretas and CO_2-smellers, media scribblers and fact-checkers who, in their protected bubbles, would never be exposed to the experience of the Dunning–Kruger effect and its negative outcomes? Evolution could not have known that there would one day be self-contained areas of gigantic groups of individuals in which not individual mountains of stupidity would be able to grow, but whole continents of total stultification would sprout from the ocean floor into the sky.

Real experts, i.e., those who have already experienced Mount Stupid at some point because they have worked their way beyond it in a certain field, therefore hardly ever show the Dunning–Kruger effect. The same holds true

[1]Something this author definitively intends to calculate through in one of his next little projects [6].

for sensibly educated people who have not been kept in a bubble of well-being or a protected area and who have therefore automatically learned the pitfalls of the competence conceit through the "corrective of external reason" ... who, yes, here the word is appropriate, have adopted nothing other than modesty. This also explains why there are so many Mount Stupid climbers today. These are all people (politicians, mainstream media flacks and fact-checkers above all) who have been and are being bred matrix-like into spheres and to whom one is not allowed to say anything, least of all the truth about their own stupidity. In climate or COVID-19 "debates" alone, it has happened to me all the time that I could not get through with facts and evidence (e.g., introduction sections of [4, 5]) only because my counterpart, sitting in his bubble, seriously considered the presentation of facts and evidence as "unfair and inappropriate." Such people (and only too often they are "academics," i.e., students who have been made stupid on purpose) thus refuse precisely to take in that very knowledge which could bring them to the realization that they do not know nearly enough to justify their bigheadedness, which is to say, to cavort so arrogantly and self-confidently in a certain field, as they are doing right now as actually stupid know-nothings, but not being aware of this fact.

So, back to our question:

How can we avoid becoming victims of the Dunning–Kruger effect?

Well, we just have to be aware of it and get as much supervisory authority and constructive criticism as we can get ... to each and every topic. A tiny bit of humility helps along the way.

2.1.3 Summary

It was found that by assuming that the internal thinking process of the human mind consists of discrete units (quantizable pieces), we can easily derive the Dunning–Kruger effect as a necessary characteristic of human behavior.

2.1.4 References for Section 2.1

1. W. Greiner, Quantenmechanik—Teil 1 Einführung, in: *Theoretische Physik*, 5th ed., in German, Verlag Harri Deutsch, Frankfurt am Main, vol. 4, 1992, ISBN 3-8171-1206-8.
2. J. Kruger, D. Dunning, Unskilled and Unaware of It. How Difficulties in Recognizing One's Own Incompetence Lead to Inflated Self-Assessments, *J. Personal. Soc. Psychol.*, 77(6), 1999, 1121–1134.

3. D. Dunning, Chapter five—The Dunning–Kruger Effect: On Being Ignorant of One's Own Ignorance, *Adv. Exp. Soc. Psychol.*, 44, 2011, 247–296, doi:10.1016/B978-0-12-385522-0.00005-6.
4. N. Schwarzer, *The World Formula: A Late Recognition of David Hilbert's Stroke of Genius*, 2021, ISBN: 9789814877206, T&F and Jenny Stanford Publishing.
5. N. Schwarzer, *The Math of Body, Soul, and the Universe*, 2023, Jenny Stanford Publishing, ISBN 9789814968249.
6. N. Schwarzer, *The Mathematical Darwin—How Quantum Gravity Explains The Miracle of Evolution*, upcoming book-project with Jenny Stanford Publishing (e.g. see: https://www.jennystanford.com/9789814968249/the-math-of-body-soul-and-the-universe/#tab-4).

2.2 The Mass Formation in the Quantum Gravity Model Picture

2.2.1 Mass Formation Psychosis: A Very Brief Overview

In this chapter, where we want to concentrate on the illustrative discussion of the application of our fundamental extremal-principle-based (quantum gravity founded) math [1–5] to potentially every psychological problem and effectively describe this psychological process of the masses, we are only going to give a very brief overview about the process of mass formation and mass formation psychosis. For more information, see Prof. Mattias Desmet's book *The Psychology of Totalitarianism* [6]. It may be very hard to believe or even imagine that the math of quantum gravity [4, 5], with its roots so deep in classical physics and abstract mathematical concepts [7–15], could provide more than an entertaining dingbat to the oh-so-complex field of psychology, the branch of science, dealing with the human mind, but the moment we accept that our consciousness is—perhaps—not more than a cleverly entangled and rather dynamic ensemble of attributes of a very general character, being governed by the most fundamental laws of the universe, we might come to the conclusion that our mind is just a very sensitive nonlinear calculation "machine." And that therefore, a mathematical description of the latter is possible. The success in our evaluation of the Dunning–Kruger effect in the previous section out of the simple assumption of discrete thoughts (quantized ideas) gives us quite some confidence that our approach has some merit. Thus, having found this connection, we now want to move on to some aspects in connection with the psychological behavior of greater groups of individuals. The example we are most interested in is the so-called mass formation process.

In difference to dictatorships, where obedience has been created from a basic fear of the dictator, totalitarianism produces obedience via mass hypnosis (Fig. 2.3). In psychology, this mass hypnosis is called mass formation.

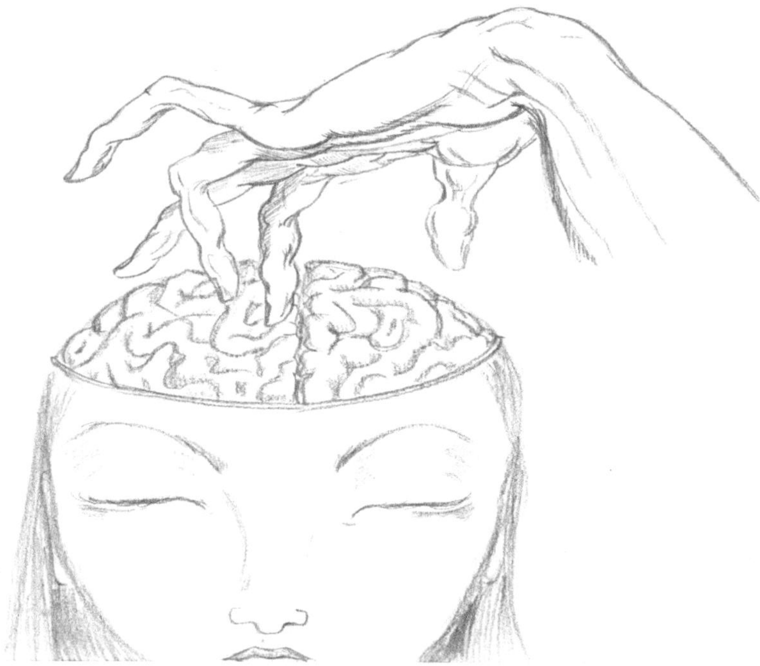

Figure 2.3 Hypnosis: Meddling with the brain in order to make a human being ready for a mass formation psychosis (artist Laura Kube).

Such a process requires 4 conditions:

(1) The masses must feel alone and isolated.

(2) Their lives must feel pointless and meaningless.

In our modern societies, these conditions have been growing for years with social media (i.e., narcissism, sociopathy, addiction and mental illness) and an education system which generally suppresses holistic analysis and broader, if not say lateral, thinking. In other and very simple words, people are mentally squeezed into boxes and these boxes are not just becoming smaller and smaller, they are also having lesser and lesser dimensions or attributes. It is therefore of little wonder that psychological studies [6] and the daily experience show us a greater susceptibility to totalitarian control mechanisms with higher grades of education. The education system creates box-thinkers and Pavlov dogs (Fig. 2.4) instead of lateral and holistic thinkers and independent true individuals.

(3) The masses then must experience constant free-floating anxiety.
(4) They must experience free-floating frustration and aggression.

Figure 2.4 The full and oh-so-empty mind (artist Laura Kube).

Free-floating means that there is no discernible source for the anxiety or aggression. And so, the person begins to irrationally crave a remedy, no matter how absurd or destructive it may be.

The moment these conditions are met, people are ripe for hypnosis. It suffices to present a "solution" to them.

No matter how stupid, senseless or even harmful the "solution" is, people are made to feel solidarity, to contribute in some free-floating, undiscernible way, when following the proposed "solution." This validates the whole thing for them and they are becoming blind towards the damage the remedy might cause to them, to the society and yes, to even their own children.

They are lesser than Pavlov's famous dogs, but even degenerated ones.

They are now changed, no longer rational.

People become intolerant and cruel.

By the way: When trying to understand this process, some experts try to apply game theory. However, the creation of irrationality as one goal of the process automatically excludes mathematical approaches like Nash's game theory, where all agents are assumed to act completely rationally.

2.2.2 Illustrative "Mathematical," Quantum Gravity Discussion of the Process of Mass Formation Psychosis

In this discussion, we will now aim for an effective mathematical quantum gravity description of the process of mass formation psychosis. We will do so without explicitly introducing the math at this point but only refer to the corresponding mathematical parts given in the appendix of this publication and the corresponding previous books and papers (mainly [4, 5, 16–32]).

The important aspect for our goal of a mathematical description of the process of mass formation requires the understanding and falsifiable presentation of dimensional mass creation and its entanglement with the individual fields of all agents or players in a given socio-economic space-time.

After introducing the Hamilton principle of psychology via a generalized Hilbert approach in appendix B, in appendices C and D we concentrate on the metric derivation of the generalized metric Klein–Gordon equation and mass as one of the essential parameters in this very equation. From there it is simple to also extract the metric appearance of mass for Dirac [7] (appendix A2 of chapter 11 in [5]) and appendices F and H in this book), Schrödinger [8, 9] (appendix B of chapter 11 in [5]) and any other quantum equation also for arbitrarily big systems. The reader finds an extensive introduction to the process of dimensional mass creation in chapter 6 of [5]. An example of the dimensional mass creation in a 10-dimensional space-time is given in section 11.5.5 of [5]. We will also discuss other physical equivalencies to the possible driving forces of a social or socio-economic mass formation. In this context, it is important to note that in this book we cannot present all the mathematical outcomes of the Einstein–Hilbert action and its derivatives and therefore refer to our other papers and books, especially with respect to Dirac-like scenarios and the third-order differential equations leading to three generations of masses (particles). Here, we will mainly base our simulations and illustrations on second-order differential quantum gravity equations as direct outcomes of non-classical variations of the Einstein–Hilbert action.

We could have a simple model, which does not aim for a "perfect" mirroring of a society of individuals, because this would not be possible even on the best super-computers, but only for the illustration of the method on a small scale. In such an approach, the individuality has been illustrated as wave functions (which actually it is anyway). The more dynamic, healthy and active the individual is, the more complex the multidimensional orbital appearance of the individual space-time wave density. It is also important to note that the individual aspect requires the wave function to be fermionic, which means it follows the Pauli exclusion principle and wants to keep its own individuality[2]. Suitable solutions can be obtained directly from our derivations in section 11.5 of [5]. There is nothing artificial about this picture, because, as explained in the introduction part, all our oh-so-human individuality summed up is just a huge set of attributes (partially entangled) and each attribute contributes to the complexity and appearance of the individual's very own halo (psi-field some call it) in the space-time of society. Of course, we cannot show the full dimensionality of the wave function, because we are unable to visualize more than 3 dimensions, but we can sum up a few dimensions and show the result in the 3 dimensions we can illustrate. This still gives us at least a feeling about the appearance of a person's individual field inside a "socio-economic space-time" (just another word for society). So, to give an example, the wave function figures given in chapter 11 of [5] or Fig. 2.5 here were actually calculated from individual fields of up to 10 dimensions of which only the 3 spatial ones are being shown. Nevertheless, the other dimensions influence the appearance of the spatial ones. In this chapter, we used some of the solutions given in [5] to evaluate examples of such wave fields (Fig. 2.5). For convenience, more solutions are given in the appendices of this book.

[2]Please note that this holds only for healthy individuals, while those ready for mass formation are behaving like bosons.

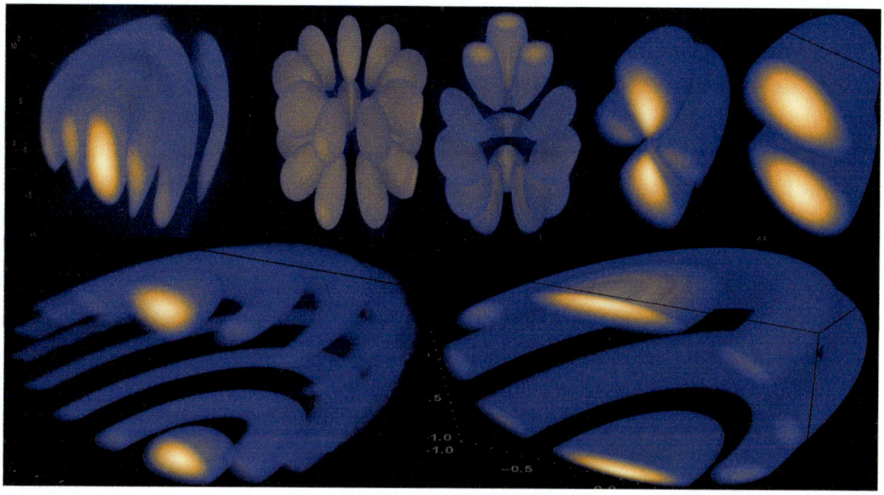

Figure 2.5 Example wave fields from half spin solutions given in chapter 11 of [5].

Please note that none of the fields has any sharp structures, but always comes as a slightly fuzzy density distribution. This is in perfect congruence with its natural equivalent, namely the field of individuality. This, too, can never be a sharp object but always only is a nebular something. In section "2.3.1 Is Developing the Ability of Tweaking the Rules of the Universal Game Some Kind of Consciousness, Respectively Its Awakening?" we are going to try and give an explanation for this fact. It even needs to be pointed out here that the sharper such an orbital object becomes, the lower its creative content, its innovative potential, its individuality. The sharpest fields or orbitals are to be found near the ground states[3]. Thus, when aiming for low individuality in order to herd the masses into a formation psychosis, the goal has to be to load the individual agents within a socio-economic system with so much mass that they are forced into their ground states, where they not only lose their mobility, but also their potential to individually excite anything or bind to other individuals. In other words, they stop thinking, communicating and their individual acting (seen from the perspective of the physical meaning of action) and thus, their true human being. Interestingly, as shown in appendix D, section "5.4.1 The Cosmological Constant as Balance Factor," a cosmological constant could compensate for the influence of the mass creation field and thus leads to some kind of immunity against any "herding" attempts. In a

[3]The exception is the individual's self-construction of mental boundaries by the means of superposed basic solutions, which we will not discuss here, but for which we have to refer to another project [33].

socio-economic space-time, this cosmological constant could simply be a critical number of holistic thinkers or a general climate of reason, rationality and a deep sitting desire for elucidation within the society. Such a state would also act as a field, counterbalancing or even completely deactivating the hypnotizing potentials. Just as in cosmology, where such fields act as inflation sources, the cosmological herding immunity would be equivalent to a permanent psychological desire of mind expansion, healthy curiosity and an unquenchable thirst for falsification and getting to the bottom of things ... in an integrated manner.

In a mass formation psychosis, the individual fields, wave functions or **"personality fields"** need to be squashed, distorted, made less complex, made less fermionic (→ bosonic then) in order to switch off the Pauli principle and make the individual compliant and ready for compactification ... in other words, we need a process of de-individualization. This has been done by external—free floating—fields and one already sees that anxiety is just one such field. In both socio-economic reality and the model, such a field requires the introduction or activation of an additional degree of freedom (could be a fear factor against a certain omnipresent entity like a virus). But this does not suffice, because there can always be threats around and nobody gives a damn about them. In order to make the process work, there also needs to be a strong entanglement with dimensions defining the individuals (again it is the same in reality and the model, which is to say the quantum gravity model already is quite realistic). And here the media, politicians and false experts come into play with their fear-mongering. This entanglement then automatically creates what in physics has been called inertia and as inertia always distorts the space-time they are residing in, we have the desired distortion and squashing of the individual wave fields. As a side note, it should be pointed out that the so-called central bank digital currency (CBDC) [34][4] would provide the perfect tool to contribute to all aspects necessary for a mass formation. Taking the four necessary steps from Desmet's theory (see the text above), it is almost frightening to realize how a CBDC could couple into all these aspects and almost freely (and cruelly) play with all scales and frequencies in the socio-economic system. Even systems of nowadays artificial intelligence could be coupled to such a thing and create all sorts of formations and at all scales in a completely automated manner.

So far—even though it appears kind of obvious now that we have the "Mathematical Psychology" and CBDC plans being laid out before us—

[4]This author suggests this book, because it does not only present a comprehensive explanation about CBDCs but also shows clever alternatives.

apparently nobody has seen the connection. Of course, the CBCD already "is" or rather provides a fear field on its own, but as said, there is much more.

And there is another important effect.

We know that the increase in inertia due to an additional (free floating) field of fear degenerates individuality, depending on the strength of entanglement (coupling of the fear field with the individual wave functions or metrics). More inertia, however, leads to a lower probability of excited states (ideas, innovation, evolutionary progressive freedom, readiness to assume a productive risk, seeking new opportunities, ...) and a higher probability of ground states of mainly bosonic character (compliance).

With respect to the recent or—yes—the current COVID-19 crisis, we think that statistical data of vaccine compliance, on the one hand, and the innovative potential, activity or even job-market developments, on the other, would directly fit into the model of inertia and the multitudes of states of the individual wave fields. Stagnation or even decline here (in reality) would be directly correlated to a lower number of excited states (or a shift of their local distribution), more bosonic wave fields and much more confined orbitals in the model. As we know that this reduction of individuality in the model comes from the inertia, produced by the fear field, we can extract some measurable/ quantifiable effect from the fear-mongering and fear-entangling, taking place in the real world.

In other words, we'd have a coupling constant and could make our model more robust (mathematically we mean) and thus, also falsifiable in a Karl-Popper-like sense.

All these smaller, distorted and bosonized (de-fermionized = de-individualized) residuals of those former healthy individuals are then ready for herding into whatever small barn and way of life they (whoever they is) want the masses of former healthy individuals in (Fig. 2.6).

Not just in the model, but also in reality the individual with such a degenerated individuality will accept very little for itself ("having nothing but being happy all the same," which in fact is rubbish as it cannot work for true human beings ... at least not for elongated time-spans).

But there are severe problems involved.

Problem A with all this: Because the add-on dimensions of fear (after all it is a free-floating and thus, non-graspable anxiety) produces entanglement not only with the target group but in general (per definition it cannot be contained), the whole will also affect the source. It is bound to backfire onto the group who created the fear field, and this makes it very likely that the whole system collapses in the end (like in a Black Hole formation as illustrated in Fig. 2.7).

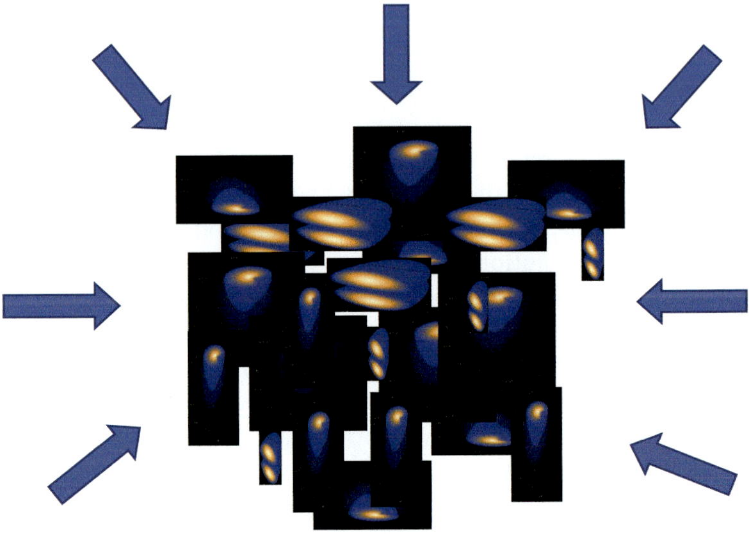

Figure 2.6 Herding: Distorted and deindividualized wave fields are ready for compactification (see text).

Figure 2.7 The COVID-19 Collapse.

Problem B with all this: Humans are humans because they have individuality. They are defined by it and they define themselves throughout their individuality. Striping them off it makes each and every individual degenerated wave function unstable and renders it ready for all sorts of phase transitions ... in other words, violence is one of the more likely outcomes. The latter then also contributes to the collapse scenario as illustrated in Fig. 2.7.

As we saw that the process of creation of inertia (mass) is the fundament for both, the practical process and the model of the mass formation psychosis, we again refer to chapter 6 and section 11.5.5 of the book [5], where the process of dimensional mass creation has been elaborated in quite some detail. After all, this aspect is essential for the understanding of the process of deindividualization (and thus, perhaps also for seeing possibilities to hold off this process or even reverse it). The math is not too complicated and we even think that the fact that we all have this inner picture of ourselves losing individuality and psychologically crumbling when being loaded with too much "mass" or "mess" (whatever type of mass or mess it is) helps in understanding why such a Hamilton–Einstein–Hilbert approach is anything but far off. A simple example of the dimensional mass creation is given in appendix D of this book. The interested reader will find more examples and details in [5].

As this dimensional creation of inertia is somehow the key, we probably have to endure the math in this book, because this is the tool to truly and fundamentally understand the whole process ... at least in the appendices and in one of its many incarnations.

One immediate counterargument against the mechanism described here could be that the masses may easily find a way to avoid the unpleasant mass creation fields. After all, "life always finds a way" and so—usually—also do humans. Thus, in addition to the mass creation fields and their intensive entanglement with the individual personality fields, there needs to be a potential ensuring a sufficient demobilization of the individuals. This can be done in three ways[5]:

(a) a significant increase in the viscosity of the environment in general
(b) the introduction of a polarization field allowing the confinement of the masses within suitably charged traps
(c) the creation of a truly omnipresent mass creation field

While it is mathematically clear how—from a variational or extremal principle—charges and subsequent polarizations (point (b)) can be derived

[5]Once again we point out the danger of CBDCs as it can easily be seen how such money systems could steer all the aspects contribution to mass formations.

(see Noether's theorem, e.g., [13], pp. 46), we struggle with such a derivation in the case of points (a) and (c). Here the so-called Higgs mechanism [14] can be applied, but as the corresponding field (the famous Higgs field) was only postulated without any idea about its fundamental origin, we do not simply want to introduce this concept without properly deriving it first.

This has been done in appendix F.

The action of the Higgs field can be demonstrated as the amount of attraction of an environment towards an individual moving within this very environment. With the attraction being small, the individual can move quite freely, like in thin air. With the attraction being strong, however, the individual "feels" like moving in honey. This demobilization of individuals and their personality fields is essential for the success of the mass formation activities, because without it, the personality fields would simply elude the mass-formation fields by using the degrees of freedom (dimensions) still being unaffected by the herding forces and potentials[6] (and there are usually almost infinitely many of such free dimensions). Isolation is an important aspect of mass formation [6] and too dynamic and flexible individuals will not comply, simply because they cannot be isolated. It should be pointed out that here we do not restrict the Higgs effect of demobilization of individuals to their real spatial mobility (even though this is also being meant), but see this much more generally also as a form of demobilization in thought, cognition and information processing (linearization of the latter).

It may be argued here that with the complexity of humans and societies, the resulting variational tasks are getting too complicated to ever be solved correctly or even remotely comprehensively. Yes, especially when taking into account the possibility of generalized Einstein–Hilbert actions [15] and the subsequent increase in complexity with the resulting field equations, one has to admit that such critique has a point. However, as shown in appendix I in this book, clever transformation of the metric tensor can render even the most complicated task treatable again.

2.2.3 References for Section 2.2

1. D. Hilbert, Die Grundlagen der Physik, Teil 1, *Göttinger Nachrichten*, 1915, 395–407.

[6]One may argue now that there would also be the option to apply the herding field to just ALL dimensions in a given socio-economic system, but as this would be equivalent to a Black Hole formation (see fig. 7 for illustration), we will not discuss this option in here.

2. A. Einstein, Grundlage der allgemeinen Relativitätstheorie, *Annalen der Physik*, 49, ser. 4, 769–822.
3. N. Schwarzer, *The Theory of Everything—Quantum and Relativity Is Everywhere—A Fermat Universe*, Jenny Stanford Publishing, 2020, ISBN-10: 9814774472.
4. N. Schwarzer, *The World Formula: A Late Recognition of David Hilbert's Stroke of Genius*, Jenny Stanford Publishing, 2022, ISBN: 9789814877206.
5. N. Schwarzer, *The Math of Body, Soul, and the Universe*, Jenny Stanford Publishing, 2023, ISBN 9789814968249.
6. M. Desmet, *The psychology of Totalitarianism*, Chelsea Green Publishing, 2022, ISBN 978-1645021728.
7. P. A. M. Dirac, *The Quantum Theory of the Electron*, Proceedings of the Royal Society A, Published 1 February 1928. DOI: 10.1098/rspa.1928.0023.
8. E. Schrödinger, Quantisierung als Eigenwertproblem (erste Mitteilung), *Ann. Phys.*, 384(4). 1926, 361–376.
9. E. Schrödinger, Quantisierung als Eigenwertproblem (zweite Mitteilung), *Ann. Phys.*, 384(6), 1926, 489–527.
10. N. Schwarzer, *Science Riddles—Riddle No. 11: What is Mass?*, self-published, Amazon Digital Services, 2019, Kindle.
11. N. Schwarzer, *Science Riddles—Riddle No. 12: Is There a Cosmological Spin?*, self-published, Amazon Digital Services, 2019, Kindle.
12. N. Schwarzer, *Science Riddles—Riddle No. 13: How to Solve the Flatness Problem?*, self-published, Amazon Digital Services, 2019, Kindle.
13. W. Greiner, J. Reinhardt, *Feldquantisierung, Theoretische Physik, Band 7A*, Harri Deutsch & Thun, Frankfurt am Main, 1993, ISBN 387144975X.
14. P. W. Higgs, Broken Symmetries and the Masses of Gauge Bosons, *Phys. Rev. Lett.*, 19 October 1964, 13, 508.
15. C. A. Sporea, *Notes on f(R) Theories of Gravity*, 2014, arxiv.org/pdf/1403.3852.pdf.
16. N. Schwarzer, *The Quantum Black Hole, Part 7B of Medical Socio-Economic Quantum Gravity*, self-published, Amazon Digital Services, March 2021, Kindle.
17. N. Schwarzer, *Quantum Cosmology—A Simple Answer to the Flatness Riddle, Part 7C of Medical Socio-Economic Quantum Gravity*, self-published, Amazon Digital Services, April 2021, Kindle.

18. N. Schwarzer, *How Systems Evolve—A Red Pill Course on Evolution*, self-published, Amazon Digital Services, 2022, Kindle.
19. N. Schwarzer, *The Metric Electron—Derivation of Spin-1/2-Objects from the Einstein-Hilbert Action*, self-published, ASIN: B0BTPQM26X, Amazon Digital Services, 2023, Kindle.
20. N. Schwarzer, *Towards the Metric Quark—Derivation of More Spin-1/2-Objects from the Einstein-Hilbert Action*, self-published, Amazon Digital Services, B0BWDQDBFW, 2023, Kindle.
21. N. Schwarzer, *Derivation Newton's Gravity Law from Quantum Einstein Field Equations*, self-published, Amazon Digital Services, ASIN: B0BXPS2ZJH, 2023, Kindle.
22. N. Schwarzer, *Derivation Hawking Radiation of Black Holes from Quantum Einstein Field Equations*, self-published, Amazon Digital Services, ASIN: B0BXQ5ZG4Z, 2023, Kindle.
23. N. Schwarzer, *Quantum Gravity—Brief Derivation of Quantum Einstein Field Equations*, self-published, Amazon Digital Services, ASIN: B0BWTWZHP5, 2023, Kindle.
24. N. Schwarzer, *Quantum Gravity and Einstein's Famous Moon Problem—The Theory of Perspectivity*, self-published, Amazon Digital Services, ASIN: B0BXPQQWZN, 2023, Kindle.
25. N. Schwarzer, *Shouldn't Quantum Gravity Solve the 3-Generation Problem*?, self-published, Amazon Digital Services, ASIN: B0BXPXP33L, 2023, Kindle.
26. N. Schwarzer, *About Apparent Asymmetries in the 3-Generations*, self-published, Amazon Digital Services, ASIN: B0BXPYGMNF, 2023, Kindle.
27. N. Schwarzer, *The Universe's Half-Spin Deception Game*, self-published, Amazon Digital Services, ASIN: B0BXPKWPQZ, 2023, Kindle.
28. N. Schwarzer, *Metric Interpretation of Quantum Theory—How Can Quantum Gravity Help Us to Understand Quantum Strangeness*, self-published, Amazon Digital Services, ASIN: B0BXQRKPV7, 2023, Kindle.
29. N. Schwarzer, *The Metric Potential—How Can Quantum Gravity Help Us to Understand the Origin of Forces*, self-published, Amazon Digital Services, ASIN: B0BXSHQZX1, 2023, Kindle.
30. N. Schwarzer, *Understanding Quantum Tunneling—How Can Quantum Gravity Help Us to Get the Gist about This Strange Phenomenon*, Amazon Digital Services, 2023, Kindle, ASIN: B0C241FQH3.

31. N. Schwarzer, *The Death of Schrödinger's Cat—How Quantum Gravity Solves the Problem of the Wave Function Collapse*, Amazon Digital Services, 2023, Kindle, ASIN: B0C2444JZX.
32. N. Schwarzer, *Dirac's Little Spinor Problem—CAN Quantum Gravity Explain the Occurrence of this Postulated Object?*, self-published, Amazon Digital Services, 2023, Kindle.
33. N. Schwarzer, *Love Fields—The Application of Mathematical Psychology for the Understanding and Use of the Supposedly Greatest Force in the Universe*, upcoming book-project with Jenny Stanford Publishing.
34. Thor Alexander, *Dezentrales Geld & DeFi—Geld regiert die Welt—Über die Zukunft von Geld, Finanzen und Dezentralität*, 2nd ed., Das Blockchain-Institut, 2021, ISBN 978-3-9821271-5-6.

2.3 Time and Uncertainty in Connection with Individuality and Subjectivity

In appendices A to H, we derive the main quantum equations just out of the Hamilton extremal principle via the Einstein–Hilbert action [1], thereby applying nothing more than a scaled metric tensor and already find—as shown above—not only quantum physics inside the General Theory of Relativity [2] but many other things, too [3–15]. In appendix I of this book, we also show that perturbations of the kernel in the Einstein–Hilbert action also lead to quantum equations. On the other hand, Desmet suggested a closeness of uncertainty and individuality or subjectivity [16] as the ability to realize one's own position in the world and perhaps even to influence one's own fate. Thereby subjectivity has to be understood in the meaning of "something being a subject, narrowly meaning an individual who possesses conscious experiences, such as perspectives, feelings, beliefs, and desires" [17].

In addition, in appendix J of this book we show how fundamental solutions to the governing laws of the universe lead to constellations, which not only explain the existence of what we note as time but also result in fundamental limits of accessibility. At least with respect to the aspect of certain scales, this can be easily understood. So, just repeating from appendix J, when assuming a flat space in Cartesian coordinates, we see (appendix J) that the Hamilton extremal principle gives solutions where at least one coordinate has to be time-like. The simplest of such solutions are oscillations and we may consider the possibility that our notion of space-time is just the result of the fact that we actually live on such a solution instead of "existing" on the underlaying space. What we then realize as time is just the result of the fact that in order to allow for the solution to actually be, it has to sport at least one coordinate, which is acting as time. Some critics may now say that there is no big difference from the concept of just defining a space-time and using this as starting point instead of only having a space and saying that quantum gravity solutions to this space make our notion of space and time, but we think that in fact there is a huge difference. Living on a solution to something is very much different from the direct existence on this very something. For one thing, not being a direct part of the space but only of one of its solutions does not allow us to directly access the space at all scales but restricts this very access to only those scales which are in correspondence with "our" solutions, meaning the ones on which our existence is manifest or "defined." Taking the oscillating solutions, for instance, we cannot access anything smaller than the wavelengths of the oscillations. We are kind of stuck in a scale trap which only allows us certain areas or ranges of the space to observe and these areas or ranges are determined by the solutions to the space-laws governing the true foundations of the universe. It is shown in appendix N that this conclusion results not only from the simple flat space Cartesian solutions, which we have derived in appendix J and all other appendices, but also from spherically symmetric solutions. In addition, we have to conclude that the restriction of our existence to the solutions (oscillating ones) automatically gives us a lower limit for the circuit time we realize as an upper limit of velocity. In our case this is the speed of light in vacuum.

Placing one's own existence on a space-time is something completely different than only existing on certain solutions to this space. Time and scalar restrictions are the direct consequences of a "way of life" based on solutions to a space rather than the space itself.

And just similar to the scalar and the cycle time trap we are caught in due to the fact that our whole existence is based upon solutions to the laws of space instead of a life on the space itself, we have to realize our individual existence and our conscious mirroring of the reality being only a composite of solutions ... solutions to the quantum gravity equations governing everything. As these solutions are—in fact—only solutions and thus, not the "thing" (space) itself, they can only give us a limited accuracy of what we are trying to comprehend by using these solutions. In other words, our thoughts are automatically scale and cycle time restricted by their own wavelengths or frequencies, because they are nothing but waves and oscillations. The almost daily experience of not being able to grasp a certain thought and hold it still for at least a nanosecond is just the manifest realization of our permanently jittering wave-like cognitive apparatus.

In connection with the mass formation psychosis from the previous chapter, this is a very important aspect, because on one side it is exactly this fuzzy wave character of our cognition which allows the creation of degenerated, deformed and completely misplaced mind complexes in the first place, but on the other side the wave-consciousness can also allow for interference and tunnel effects and entanglements, potentially leading to an unexpected and quick resolution of the distorted states. In other words, spontaneous self-healing of a psychosis or mass psychosis is a principal possibility and it follows from the same quantum gravity laws which allow the existence of the most strange and funny effects in the world of elementary particles. In fact, taking the aspect of a mass formation psychosis requiring a great deal of coherence and knowing how difficult it is to keep these states stable even for rather small ensembles of quantum dots over longer periods of time in quantum computing processes immediately gives us one important counter measure against all attempts to form a mass psychosis. One simply needs to destroy the coherence field, which deindividualizes the personal (individual) wave densities.

2.3.1 Is Developing the Ability of Tweaking the Rules of the Universal Game Some Kind of Consciousness, Respectively Its Awakening?

This directly leads us to the question whether an essential part of subjectivity may be expressed as

(a) the ability to recognize one's own individuality, which in the physical picture means the recognition of the subject's own wave function (not necessarily as such, but as field or "something" which interacts with the environment and—very important—other wave functions including the ones of other subjects);
(b) the potential to influence—however minuscule—the kernel of the individual's/subject's own extremal principle;
(c) the ability to alter a system's dimensionality via the might of conscious thought (coordinate transformation and/or change of perspective), which, as it is shown in appendix S, leads to an active overcoming of the so-called covariance principle.

With the mathematical process of the perturbation of the kernel being described in appendix I and the interesting outcome of gravity quantum equations from there, we almost automatically come to the hypothesis of consciousness being the ability of recognizing and influencing the extremal kernel, including its dimensionality (see appendix S), of the conscious subject by the subject itself.

In order to understand this, we have to apply the math being presented in the appendix. Our starting point is the Hamilton extremal principle via the Hilbert variational integral, which we present in its simplest form here:

$$\delta W = 0 = \delta \int_V d^n x \sqrt{-g} \cdot R. \tag{1}$$

Thereby we have the Ricci scalar R with:

$$R = R_{\alpha\beta} g^{\alpha\beta} = \begin{pmatrix} -\frac{1}{2}\left(g_{\alpha\beta,ab} + g_{ab,\alpha\beta} - g_{\alpha b,a\beta} - g_{\beta b,a\alpha}\right) g^{ab} \\ +\frac{1}{2}\left(\frac{1}{2} g_{ac,\alpha} \cdot g_{bd,\beta} + g_{\alpha c,a} \cdot g_{\beta d,b} - g_{\alpha c,a} \cdot g_{\beta b,d}\right) g^{ab} g^{cd} \\ -\frac{1}{4}\left(g_{\alpha c,\beta} + g_{\beta c,\alpha} - g_{\alpha\beta,c}\right)\left(2 g_{bd,a} - g_{ab,d}\right) g^{ab} g^{cd} \end{pmatrix} g^{\alpha\beta}, \tag{2}$$

and the determinant g of the metric tensor g_{ig} within the integral over the whole volume V of the space-time under consideration being variated with respect to the metric. The g with upper indices denotes the contravariant form of the metric. Now we have to understand that the metric does not only consist of the reality, which is to say all attributes (dimensions) the latter provides and consists of, but also of the/any subject's intrinsic degrees of freedom in reflecting this very reality plus adding everything what phantasy may provide. Such an extended space-time, consisting of the real space and time AND the virtual one of an entity being capable of not only mirroring reality but also adding things/attributes/properties/degrees of freedom, and thus dimensions to it, will lead to quite complicated metric systems, but still metrics these are.

In appendix I, we see that we have two ways of performing the variation, one—the classical one—is with respect to the metric tensor and the other is with respect to the scaling factor of this metric tensor. The first option leads us to the classical Einstein field equations (in space-times of all attributes being considered in our mix of reality + reflected reality + phantasy) and the second one gives us quantum equations. When performing the variation, we realize that a tweaking of the kernel of the variation with a function Φ automatically leads to a mix of equations, showing the classical terms of the

Einstein field equations and the quantum equations, and which we therefore might call quantum Einstein field equations. The evaluation is given in appendix I.

Now simply let us assume that an entity has developed enough degrees of freedom regarding its inner space-time (the intrinsic or reflecting one) that it is capable of perturbing the kernel. May it be that initially these perturbations are more or less accidental; may it be that they only happen with respect to the intrinsic degrees of freedom and not the ones being related to reality, but with every happening of such a perturbation the variational integral determining the "fate" of the entity are changing from Eq. (1) to something like:

$$\delta W = 0 = \delta \int_V d^n x \left(\sqrt{-g} \cdot \Phi \cdot R \right). \tag{3}$$

This automatically leads to quantum fields being rather individual to the entity having caused the perturbation in the first place (with respect to the solution to Eq. (3), we refer to appendix I).

Now we postulate that the moment the entity develops the ability to actively steer/adapt/adjust/tweak Φ plus the dimensionality of the space-time being considered (appendix S), and thus, will have achieved the ability to influence its very own fields of existence, we have the awakening of consciousness.

2.3.2 References

See "Global References" at the end of this book!

2.4 The Path of Desire

In the book *Lacan's Logic of Subjectivity—A Walk on the Graph of Desire,* Desmet describes the evolution of a mind or, if one may say so, the conscious development of a human being, as a complicated stumbling from perception to perception along an intricate path of desires (as illustrated in Figs. 2.8 and 2.9). The latter can be presented in graphs, which remind the author so strongly of the Feynman graphs in quantum electrodynamics [18], that we feel almost forced to suggest a strong, if not striking, similarity between the path-integral process and the evolution of the states of a mind. Shouldn't then Lacan's and Desmet's "graphs of desire" be clearly related to the quantum propagator (c.f. appendix K)?

Figure 2.8 The Path of Desire ("the blind desire" by Livia Schwarzer).

Figure 2.9 The other Path of Desire ("the knowing desire" by Livia Schwarzer).

2.4.1 References

See "Global References" at the end of this book!

2.5 The Manifestation of Thought

When listening to holistic thinkers like Gad Saad, author of *The Parasitic Mind* ... [43], one inevitably starts asking oneself how on earth it can be that informative pathogens or, to apply Prof. Saad's formulation, infectious ideas can come into existence and kill common sense. Of course, Saad explains the mechanism of this process in quite some detail in his book, but he does so more or less verbally and thus, from a mathematical psychological standpoint, the mechanism is not understood. For this, a more rigorous approach would be needed and should be undertaken in a strict mathematical "environment." In order to do so, we require a concept of "matter" for the "substance of thought." As expected, the assumed-to-be-insubstantial character of something like thought seems to be difficult to be brought into a manifested concept like matter. This, however, is only true on first and shallow sight, because our resistance in giving or allocating something like thought to anything matter-like potentially has only to do with our notion of matter rather than the existence of a true antagonism. In appendix O, when analyzing the resulting quantum gravity equations from a scaled metric, we find that the volume scaling factors are perfectly identical with the matter terms in the classical Einstein theory. There, however, both, Einstein and Hilbert [1, 2], apparently never having thought about splitting the metric into its volume part and an inner core, had to postulate matter. We—on the other hand—find that matter terms appear automatically the moment we allow the metric of a given space or space-time to volumetrically vary, evolve or just change. The corresponding volume or scaling factor takes on the properties of quantum wave functions and—most surprisingly as shown in appendix O in this book—also the Einstein and Hilbert matter terms, which is to say the energy momentum tensor in the General Theory of Relativity. Thereby, the evaluation itself is not very difficult, even though the resulting equations are. The simple metric scaling of the form:

$$G_{\alpha\beta} = g_{\alpha\beta} \cdot F[f] \tag{4}$$

gives the following quantum gravity field equations when subjected to the Hilbert variation [1]:

$$0=\left(\begin{array}{c}R_{\alpha\beta}-\frac{g_{\alpha\beta}}{2}R+\frac{g_{\alpha\beta}}{2}\left(\begin{array}{c}\frac{1}{2F}\left(\begin{array}{c}2F_{,ij}(n-1)g^{ij}+2\Gamma_{ij}^{a}F_{,a}g^{ij}\\-F_{,i}g^{ab}g_{jb,a}g^{ij}-F_{,j}g^{ab}g_{ib,a}g^{ij}\\-n\Gamma_{ij}^{d}F_{,d}g^{ij}+\frac{n}{2}F_{,d}g^{cd}g_{ab,c}g^{ab}\end{array}\right)\\+\frac{F_{,i}\cdot F_{,j}}{4F^{2}}g^{ij}\left((n-6)(n-1)\right)\end{array}\right)\\-\frac{1}{2F}\left(\begin{array}{c}F_{,\alpha\beta}(n-2)+F_{,ab}g_{\alpha\beta}g^{ab}\\+F_{,a}g^{ab}\left(g_{\beta b,\alpha}-g_{\beta\alpha,b}\right)-F_{,\alpha}g^{ab}g_{\beta b,a}-F_{,\beta}g^{ab}g_{\alpha b,a}\\+F_{,d}g^{cd}\frac{1}{2}n\left(\frac{2}{n}g_{\alpha c,\beta}-g_{\alpha c,\beta}-g_{\beta c,\alpha}+g_{\alpha\beta,c}+\frac{1}{n}g_{\alpha\beta}g_{ab,c}g^{ab}\right)\\-\frac{1}{2F}\left(F_{,\alpha}\cdot F_{,\beta}(3n-6)+g_{\alpha\beta}F_{,c}F_{,d}g^{cd}(4-n)\right)\end{array}\right)\end{array}\right). \quad (5)$$

The corresponding evaluation plus some discussion has been presented in the appendices in this book. The well-informed reader immediately realizes that the scaling terms are obviously connected with the classical matter tensor via:

$$\kappa T_{\alpha}^{\beta}=\left(\begin{array}{c}\frac{\delta_{\alpha}^{\beta}}{2}\left(\begin{array}{c}\frac{1}{2F}\left(\begin{array}{c}2F_{,ij}(n-1)g^{ij}+2\Gamma_{ij}^{a}F_{,a}g^{ij}\\-F_{,i}g^{ab}g_{jb,a}g^{ij}-F_{,j}g^{ab}g_{ib,a}g^{ij}\\-n\Gamma_{ij}^{d}F_{,d}g^{ij}+\frac{n}{2}F_{,d}g^{cd}g_{ab,c}g^{ab}\end{array}\right)\\+\frac{F_{,i}\cdot F_{,j}}{4F^{2}}g^{ij}\left((n-6)(n-1)\right)\end{array}\right)\\-\frac{1}{2F}\left(g^{\beta\gamma}\left(\begin{array}{c}F_{,\alpha\gamma}(n-2)+F_{,ab}g_{\alpha\gamma}g^{ab}\\+F_{,a}g^{ab}\left(g_{\gamma b,\alpha}-g_{\gamma\alpha,b}\right)-F_{,\alpha}g^{ab}g_{\gamma b,a}-F_{,\gamma}g^{ab}g_{\alpha b,a}\\+F_{,d}g^{cd}\frac{1}{2}n\left(\frac{2}{n}g_{\alpha c,\gamma}-g_{\alpha c,\gamma}-g_{\gamma c,\alpha}+g_{\alpha\gamma,c}+\frac{1}{n}g_{\alpha\gamma}g_{ab,c}g^{ab}\right)\\-\frac{1}{2F}\left(F_{,\alpha}\cdot F_{,\gamma}(3n-6)+g_{\alpha\gamma}F_{,c}F_{,d}g^{cd}(4-n)\right)\end{array}\right)\right)\end{array}\right). \quad (6)$$

In appendix O it will be shown that an identity, called the Bianchi identity, leads to the following result in the case of covariant derivation of Eq. (5):

$$0=\left(\begin{array}{c}\overbrace{R^{\beta}_{\alpha;\beta}-\frac{\delta^{\beta}_{\alpha}}{2}R_{;\beta}}^{=0\,(\text{Bianchi})}+\frac{\delta^{\beta}_{\alpha}}{2}\left(\begin{array}{c}\frac{1}{2F}\left(\begin{array}{c}2F_{,ij}(n-1)g^{ij}+2\Gamma^{a}_{ij}F_{,a}g^{ij}\\-F_{,i}g^{ab}g_{jb,a}g^{ij}-F_{,j}g^{ab}g_{ib,a}g^{ij}\\-n\Gamma^{d}_{ij}F_{,d}g^{ij}+\frac{n}{2}F_{,d}g^{cd}g_{ab,c}g^{ab}\end{array}\right)\\+\frac{F_{,i}\cdot F_{,j}}{4F^{2}}g^{ij}\left((n-6)(n-1)\right)\end{array}\right)_{;\beta}\\+\frac{F_{;\beta}}{2F^{2}}\left(g^{\beta\gamma}\left(\begin{array}{c}F_{,\alpha\gamma}(n-2)+F_{,ab}g_{\alpha\gamma}g^{ab}\\+F_{,a}g^{ab}\left(g_{\gamma b,\alpha}-g_{\gamma\alpha,b}\right)-F_{,\alpha}g^{ab}g_{\gamma b,a}-F_{,\gamma}g^{ab}g_{\alpha b,a}\\+F_{,d}g^{cd}\frac{1}{2}n\left(\frac{2}{n}g_{\alpha c,\gamma}-g_{\alpha c,\gamma}-g_{\gamma c,\alpha}+g_{\alpha\gamma,c}+\frac{1}{n}g_{\alpha\gamma}g_{ab,c}g^{ab}\right)\\-\frac{1}{2F}\left(F_{,\alpha}\cdot F_{,\gamma}(3n-6)+g_{\alpha\gamma}F_{,c}F_{,d}g^{cd}(4-n)\right)\end{array}\right)\right)\\-\frac{1}{2F}\left(g^{\beta\gamma}\left(\begin{array}{c}F_{,\alpha\gamma}(n-2)+F_{,ab}g_{\alpha\gamma}g^{ab}\\+F_{,a}g^{ab}\left(g_{\gamma b,\alpha}-g_{\gamma\alpha,b}\right)-F_{,\alpha}g^{ab}g_{\gamma b,a}-F_{,\gamma}g^{ab}g_{\alpha b,a}\\+F_{,d}g^{cd}\frac{1}{2}n\left(\frac{2}{n}g_{\alpha c,\gamma}-g_{\alpha c,\gamma}-g_{\gamma c,\alpha}+g_{\alpha\gamma,c}+\frac{1}{n}g_{\alpha\gamma}g_{ab,c}g^{ab}\right)\\-\frac{1}{2F}\left(F_{,\alpha}\cdot F_{,\gamma}(3n-6)+g_{\alpha\gamma}F_{,c}F_{,d}g^{cd}(4-n)\right)\end{array}\right)\right)_{;\beta}\end{array}\right), \tag{7}$$

$$\Rightarrow\ \Rightarrow\ \Rightarrow\ \Rightarrow$$

$$0=\left(\begin{array}{c}\frac{\delta^{\beta}_{\alpha}}{2}\left(\begin{array}{c}\frac{1}{2F}\left(\begin{array}{c}2F_{,ij}(n-1)g^{ij}+2\Gamma^{a}_{ij}F_{,a}g^{ij}\\-F_{,i}g^{ab}g_{jb,a}g^{ij}-F_{,j}g^{ab}g_{ib,a}g^{ij}\\-n\Gamma^{d}_{ij}F_{,d}g^{ij}+\frac{n}{2}F_{,d}g^{cd}g_{ab,c}g^{ab}\end{array}\right)\\+\frac{F_{,i}\cdot F_{,j}}{4F^{2}}g^{ij}\left((n-6)(n-1)\right)\end{array}\right)_{;\beta}\\+\frac{F_{;\beta}}{2F^{2}}\left(g^{\beta\gamma}\left(\begin{array}{c}F_{,\alpha\gamma}(n-2)+F_{,ab}g_{\alpha\gamma}g^{ab}\\+F_{,a}g^{ab}\left(g_{\gamma b,\alpha}-g_{\gamma\alpha,b}\right)-F_{,\alpha}g^{ab}g_{\gamma b,a}-F_{,\gamma}g^{ab}g_{\alpha b,a}\\+F_{,d}g^{cd}\frac{1}{2}n\left(\frac{2}{n}g_{\alpha c,\gamma}-g_{\alpha c,\gamma}-g_{\gamma c,\alpha}+g_{\alpha\gamma,c}+\frac{1}{n}g_{\alpha\gamma}g_{ab,c}g^{ab}\right)\\-\frac{1}{2F}\left(F_{,\alpha}\cdot F_{,\gamma}(3n-6)+g_{\alpha\gamma}F_{,c}F_{,d}g^{cd}(4-n)\right)\end{array}\right)\right)\end{array}\right.$$

$$-\frac{1}{2F}\left(g^{\beta\gamma}\left(\begin{array}{c} F_{,\alpha\gamma}(n-2)+F_{,ab}g_{\alpha\gamma}g^{ab} \\ +F_{,a}g^{ab}\left(g_{\gamma b,\alpha}-g_{\gamma\alpha,b}\right)-F_{,\alpha}g^{ab}g_{\gamma b,a}-F_{,\gamma}g^{ab}g_{\alpha b,a} \\ +F_{,d}g^{cd}\frac{1}{2}n\left(\frac{2}{n}g_{\alpha c,\gamma}-g_{\alpha c,\gamma}-g_{\gamma c,\alpha}+g_{\alpha\gamma,c}+\frac{1}{n}g_{\alpha\gamma}g_{ab,c}g^{ab}\right) \\ -\frac{1}{2F}\left(F_{,\alpha}\cdot F_{,\gamma}(3n-6)+g_{\alpha\gamma}F_{,c}F_{,d}g^{cd}(4-n)\right) \end{array}\right)\right)_{;\beta}\Bigg), \tag{8}$$

$$\Rightarrow\ \Rightarrow\ \kappa T^{\beta}_{\alpha\,;\beta}=0. \tag{9}$$

These are just the most general conservation laws, which is to say, in no matter what space or space-time the calculation is performed and no matter what attributes, properties, characteristics, thoughts, degrees of freedom and so on define our space, there will always be conservation laws which do not allow for any sort of energy or matter to bypass these laws.

But then, so the obvious question, how can completely parasitic entities, pathogens and systems exist at all? We mean, with everything in existence following the conservation condition (9), pure parasitism should not be a stable thing, should it?

Well, the simple answer is that Eq. (9) neither prohibits nor hinders parasitism. But it sets limits to its existence and in knowing these limits and strengthening them, we might have an easier job in suppressing parasitic behavior within our own societies in the first place. Also, in knowing the action of the conservation laws, we might be able to recognize parasitic behavior more easily. The same holds for Prof. Saad's "parasitic ideas," which, in a given metric space or space-time of conscious thought, can exist and feed on other entities, which then act—usually unwillingly—as hosts. Recognition of such parasitism simply requires the holistic consideration of all dimensions (attributes, properties, characteristics, thoughts, degrees of freedom) of all spaces and sub-spaces and their covariant derivation (exchange of scaling factors) in order to detect one-directional energetic, which is to say purely parasitic, flow.

In simple words: In case an "idea" never contributes to any life-sustaining factor, but only sucks matter and energy from the life-sustaining/supporting system, we definitively have a parasitic "thought" ... be it as a strategy, way of life or any sort of existence in general.

2.5.1 References

See "Global References" at the end of this book!

2.6 Polarization

In order to steer the masses into certain directions, one needs to literally hook the same masses with something. This something then allows interaction with the masses and can be considered, depending on the scalar, vectorial or tensorial (which is to say generalized vectorial) type of interaction, like masses or charges. With the masses of individuals not being charged with respect to a certain topic, the first step needs to be polarization of the individual fields so that charges can come forth. It is shown in appendix P how this process requires the distortion of space-time and that in essence the resulting interaction is a generalized electromagnetic one. Thus, the mathematical handling of the whole apparatus is very similar or partially even equal to the handling of the Maxwell equations, respectively, as we are dealing within quantum gravity realms, quantum gravity electro-dynamics (c.f. appendices K and P).

From the psychological perspective, it appears very intuitive that the process of polarization requires the distortion of the habitat of the masses of individuals in order to achieve the necessary grades of polarization for the purpose of hooking the masses and steering them into certain directions. It appears only logical to assume that only stressed and strained masses, being forced to exist in skewed and unnaturally warped space-times, are susceptible to any kind of otherwise completely illogic influencing.

2.6.1 References

See "Global References" at the end of this book!

2.7 The Size of a Thought

This section is not only about measuring the size of a thought—which to some people may sound quite weird—but it is about the dimensional size of a mind, measured in bits. In fact, it is about the information size of any system. A thought, a bundle of ideas, the fuzzy individual ensemble of information defining an individual, a socio-economic arrangement of such individuals and so on ... these all are just systems, systems of a certain size regarding the number of dimensions and the ranges along these dimensions. And when a system can have a size and we might want to see or at least mathematically describe a funny thing like a "thought" as such a system—a very general system, naturally—then also this very thought should be considered as something having a size (also in a very general manner and we are going to work out what this generality could be).

Yes, the size of a thought can be measured. But we have to dig quite deep before we are able to see how it can be done. First, we need to understand how the universe stores information, and from there we derive a universal measure for the thing we call information. An almost minuscule extension of the Einstein–Hilbert action [1, 2], which consists of nothing more than a scalarized metric tensor [3, 4, 5], helps us along the way. In the end, we find that the size of a thought can have measures in very different scales, but there is always one fundamental scale which holds absolute within our universe. Interestingly, this scale has been defined by objects we know as Black Holes. With this, we can allocate an absolute size to each and every system, including a system we understand under the rather fuzzy expression "thought." This allocation is simply done by measuring the amount of information this very thought contains.

But does this "measure of the size" also tell us about the "greatness" of the thought, its meaning and importance to the bearer of the information?

No, it does not!

If we also want to know the thought's impacting potential, we require more knowledge about it than just its own absolute size. We also need to know about its weight within a background of many other thoughts, ideas, inventions, innovations, including all the dark matter and energy (let's see this as phantasy for here and now) there is in just every somewhat more complex system.

We know that many people are afraid of math. They are often scared away and even do not start to read a book or article simply by the mere chance that something as primitive as 1 + 1 = 2 (which is an anything but primitive equation, by the way) could occur in it. Thus, even though all our trains of thought within this section are completely and very fundamentally mathematically based, we refrain from presenting any math here, but only give the corresponding literature instead and refer to appendix Q, where we will repeat the discussion together with all the necessary mathematical derivations.

2.7.1 The Fundamental Equation for Everything

In 1915, David Hilbert [1] was able to show that a mathematical structure, very similar to a volume equation (a volume integral to be precise), apparently contained Einstein's famous General Theory of Relativity [2], which, as we all know, is a theory about gravity. Thereby the fascinating aspect was that something so very much physical, like gravity, came from a completely mathematical source, namely Hilbert's "variated volume integral"[7]. In fact, namely, Hilbert's equation is a bit more than just a "volume integral," but an integral which actually looks for an extremum, which means maximum or minimum, of the volume result, hence, our expression "variated volume integral."

More than one hundred years after these groundbreaking works of Hilbert and Einstein, we were able to show that not only gravity resided inside the Hilbert equation, but obviously just everything [3, 4, 5].

But why, with the Hilbert equation already being there, wasn't this fact discovered much earlier?

In order to grasp the implications here, we need to understand that even though the Hilbert equation looks quite simple at first sight, it has many degrees of freedom and a fairly complicated intrinsic structure. Therefore, this author suspects that Hilbert, Einstein and many others simply have not seen all the possibilities the apparently so simple "volume integral" offered. For one thing, as this is going to be the one in which we are most interested in this section, obviously nobody ever bothered about investigating the Hilbert equation with respect to the number of dimensions in which a certain problem has been considered[8]. Many scientists always concentrated on observing our "classical" 4-dimensional space-time.

We learned: Hilbert's equation, if just being a little bit generalized, contains it all.

2.7.2 Spheres in *n* Dimensions

For instance, let us assume that there are spheres in all n-dimensional spaces. Yes, of course, we already know that a point could be seen as a sphere which

[7]Please note that this "variated volume integral" in literature is usually known under the expression "Einstein–Hilbert action" [1, 2] and it is not truly a volume integral, but an integral with the Ricci scalar as kernel and it is variated with respect to the metric tensor.

[8]Yes, there are many n-dimensional solutions to the Einstein field equations, but this is not what was meant here. We are looking for variations of the Einstein–Hilbert action with respect to the number of dimensions (e.g. [4]).

has zero dimension, that a line of a certain length 2*R has 1 dimension and could just be taken as a 1-sphere with radius R and that a circle is just a sphere in 2 dimensions. Our ordinary sphere is then 3-dimensional—naturally. But, even though we might have problems in imagining it, there are spheres also in higher numbers of dimensions.

Now the funny thing we need to learn here is the fact that for a given radius R there is always one certain n, which is to say one certain number of dimensions, for which the volume of the sphere is maximum. And the higher R, the higher also is n. In appendix Q, we are going to derive this interesting connection.

We learned: There are spheres in any arbitrary number of dimensions and we call them *n*-spheres. To each radius *R* exists just one number of dimensions where the *n*-sphere has the maximum volume.

2.7.3 Making the Connection to Hilbert's "Volume Integral"

"... for a given radius R there is always one certain n... for which the volume of the sphere is maximum."

But wait a moment! Didn't we just say that the Hilbert equation in fact wasn't much more than an instrument which searched for "volume integral maxima"? Now we also learned from a completely different field (mathematical spheres in n dimensions) that such maxima exist for all sorts of spheres for a given radius, only, that the area in which to search for these maxima covers all numbers of dimensions. Apparently, neither Hilbert nor anybody else ever looked for this possibility.

We learned: One thing Hilbert and Einstein apparently forgot to look for was the number of dimensions of a given system.

But what does this tell us?

2.7.4 The Other Sphere

In 1916, in the middle of World War I, Karl Schwarzschild published a short paper [48] about a solution to Einstein's gravity equations [2] (which also were Hilbert's equations [1]). This solution contained an object which later was known as Black Hole, an object so dense and massive that not even light could escape from it. In fact, these Schwarzschild solutions also were spheres, but they were slightly deformed, which means the time and space in which they existed, and which actually they just were, were deformed. Very close to the center of the Black Hole, the deformation was so dramatic that—as it seems—these objects cut themselves off from the rest of the universe. At a certain radius, being given the name **Schwarzschild radius**, namely, everything around the Black Hole becomes so abnormal and extreme that one may well characterize this region and everything inside it as "being out of this world." Just as *n*-spheres, these objects can exist in any arbitrary number of dimensions *n* and they sport perfect spherical symmetry. They also follow the same radius-to-dimension-maximum-volume rule as ordinary *n*-spheres do.

We learned: There are *n*-spheres with mass. We call them Black Holes.

But, so the question any astronaut wants to get answered before ever climbing into a spaceship and face the danger of potentially being swallowed by a Black Hole, what happens to the poor things falling into those Schwarzschild monsters?

Well, let us try to answer this question not for the astronaut, which would be too complicated, but for a photon.

2.7.5 What Happens to a Photon when It Falls into a Black Hole?

In the early seventies, J. Bekenstein [49, 50] investigated the connection between the Black Hole surface area and information. Thereby he simply considered the surface change of a Black Hole which would be hit by a photon just of the same size as the Black Hole (Fig. 2.10).

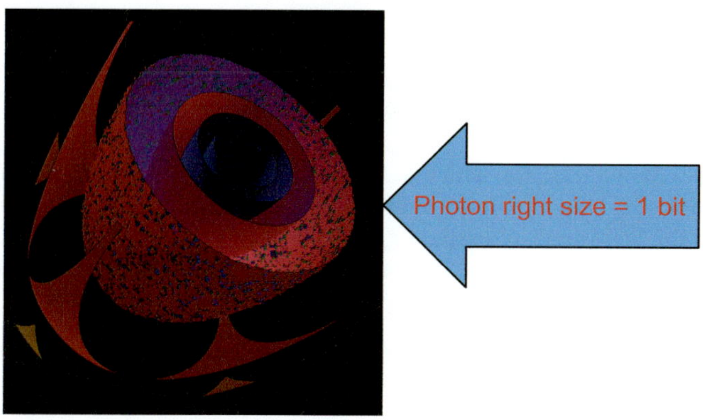

Figure 2.10 Illustration of the Bekenstein thought experiment.

His idea was that with such a geometric constellation, the outcome of the experiment would just consist of the information whether the photon fell into the Black Hole or whether it did not. Thus, it would be a 1-bit information (Fig. 2.10). Bekenstein's calculations led him to the funny proportionality of area and information. He found that the number of bits, coded by a certain Black Hole, is proportional to the surface area of this very Black Hole if measured in the Planck area ℓ_P^2. This is an extremely small surface. In other words, each photon falling into the Black Hole is similar to a bit falling into that very Black Hole and each bit increases the size of the Black Hole by a tiny little bit. Bekenstein was even able to give a nice equation for the growth of the Black Hole depending on the number of bits thrown into it. Instead of presenting that very equation here (the interested reader can find it in appendix Q), we simply illustrate its results (Fig. 2.11, solid line).

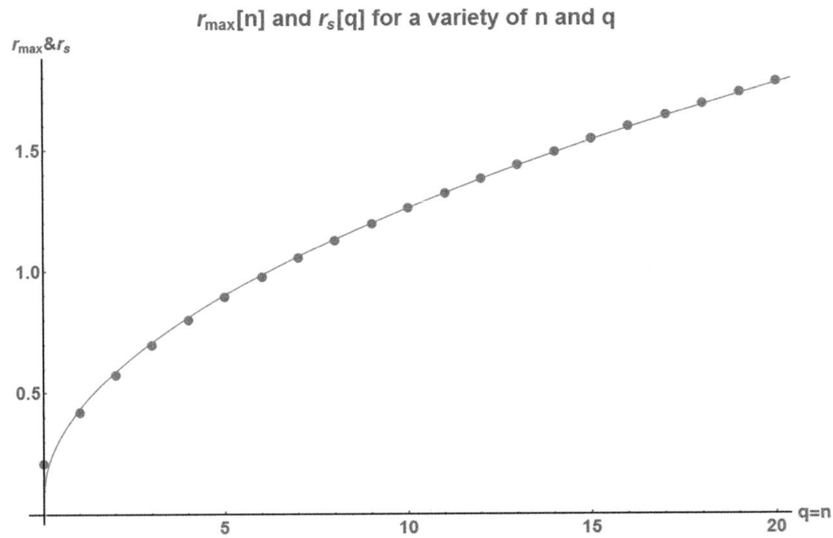

Figure 2.11 Radius r_{max} for which at a certain number of dimensions an n-sphere or a Black Hole has the maximum surface depending on n (dots) compared with the increase in the Schwarzschild radius r_s of a Black Hole depending on the number of bits q thrown into it by using the Bekenstein formula (solid line). We find that $q = n$.

We learned: Information lets Black Holes grow if being fed with.

What Bekenstein could not answer, however, was the question of how the Black Hole actually collects and stores all these bits. What happens to them? In fact, what happens to all the things, including our astronaut, that fall into Black Holes and—more important—what the heck happens to the information these things contained or just were?

2.7.6 Using the Connection We Already Found

We already have learned that spheres of a certain radius *R* find their maximum volume always in exactly one certain number of dimensions *n*. So, we ask ourselves, could it be possible that the number of dimensions, allowing a mathematical sphere to always stay in the optimum of content state to a given radius, also holds for Black Holes? Could it be that Black Holes have no fixed number of dimensions, but that they adjust their internal number of dimensions with respect to the amount of information (matter) they have swallowed?

If so, we should find some agreement between the mathematical *n*-spheres and their radius to dimensions behavior and the Bekenstein equation, right?

Well, the corresponding comparison is already shown in Fig. 2.11. While, as already explained, the solid line shows the result for the Bekenstein bits, the dots give the dependency of the radius on the number of dimensions. The agreement is not only perfect but also immediately allows us to conclude that information, in fact, just is equal to the dimension. One bit thrown into the Black Holes "reappears" as an additional dimension to a slightly grown Black Hole.

We learned: Information is dimension.

2.7.7 What Is Mass?

But does this also help us in understanding what actually makes the mass of a Black Hole?

Once again, the answer is no, because the fact that we have found out that the Black Hole stores information as internal dimensions does not tell us anything about how the Black Hole actually does really, which means physically, do it. After all, photons are massless particles. So how can they suddenly give mass to an object they fell into?

In order to answer this question, we need Quantum Theory. There we are able to show that mass is just entangled dimensions [10] (see also [4, 5]). Applying this to our Black Hole information problem, we find that each bit and therefore each subsequently added dimension connects to (or entangles with) the other dimensions already residing in the Black Hole. Non-entangled dimensions do not give mass (or other forms of matter), while entanglement leads to mass. A photon falling into a Black Hole also consists of dimensions, but these dimensions are free or non-entangled in a massive manner. When passing the event horizon of the Black Hole, which is the Schwarzschild radius, the dimensions are made to entangle such that mass occurs. The math describing this process was presented in [10] and has been partially repeated in appendix D of this book.

We learned: Mass is entangled dimensions.

2.7.8 Bringing It All Together

The evaluation of the greatness of a thought should start with the summing up of the substance of information this very thought has been made of. For this, we simply decompose the thought into its bit-structure and by counting the number of bits we have the thought's equivalent Schwarzschild radius r_S. Thus, if the thought would be compressed to a Black Hole, its size would be a sphere of exactly this radius. However, this would only be the size of the thought inside a world with no other or previous information in it. From Fig. 2.11, we can easily deduce that the more thoughts (information) there are already inside the system, the more difficult it becomes for a new thought to really make a difference (on the radius r_S, we mean). The change of the system's radius (Black Hole) is getting even smaller with each bit of information already being collected. Thus, the evaluation of the greatness of a certain thought (or idea) always requires the consideration of the whole, because it depends on the background of the total information and on how important the new piece of information would actually be.

And there is more!

As the Black Hole can only grow, respectively increase its mass, when information/dimensions added to it also entangle with the rest, it requires the new thought to connect with the information already residing within. A thought finding no connections to entangle with whatsoever is in there is useless and would just be rejected as a gamma ray burst or something similar[9].

Yes indeed, what we found is a triviality. After all, every child knows (politicians don't, though) that an idea is worth very little if it is just one which already has been spoken out many times before.

But the difference here is that we have worked out a very fundamental origin for this simple rule ... an origin, which actually gives us the means to evaluate each and every new idea, invention, innovation and—YES—thought,

[9]In the case of thoughts from greenish or leftish politicians, it will probably not be more than a lukewarm and very little fart ... if at all.

in a very fundamental and (that is probably even more important) impartial manner.

And there is more!

We can perform this evaluation with respect to the system as a whole, which is to say ideas, looking good initially, might turn out to be of no substance at all after our new purely mathematically based and very fundamental peer review. Let's take the example of the invention of the wheel. In a universe with no wheel, such an invention would definitively make a far greater impact than in a universe already being brim full of wheel-driven cars, scooters, bicycles and so on.

Or we take the example of the climate-simulators (or—liars), who seriously want to tell us that the solar activity is a constant and who actually pretend to see a CO_2 greenhouse gas effect, when the uncertainty of the cloud coverage is already 114 times bigger than any such gas effect could ever be [51]. Mind you, we really meant the UNCERTAINTY of the cloud coverage and not the absolute value. Looking for a mass equivalent for such lies [52] immediately brings in the idea of anti-matter.

In other words, many supposedly "great thoughts," especially if coming from *The Parasitic Mind* [43] (which usually are the left greenish entities) and being considered "oh so wonderful" from within the ivory tower of unworldly "experts" and quixotic politicians, will quickly reveal themselves as rather useless if not dangerous within our fundamental apparatus.

We learned: The size of a thought can be given by the Schwarzschild radius difference it would make to the system this very thought has been given to. Thereby the thought's connectivity to the existing network of knowledge is crucial, because it decides upon the usability of the thought's substance. Only the part being able to entangle with existing knowledge, thereby producing mass, is of use.

2.7.9 References

See "Global References" at the end of this book!

2.8 Intermediate Conclusions

We have shown that a simple scaled metric subjected to a Hamilton minimum principle via the Einstein–Hilbert action [1] not only gives us the classical General Theory of Relativity of Einstein [2] but also Quantum Theory. For proof, we have derived the essential Quantum equations, Klein–Gordon, Schrödinger [8, 9] and Dirac [7] (see appendices C, F, G and H of this book). As a bonus, we also saw the option for a combined Schrödinger–Dirac equation [55] and introduced a variety of alternatives for the extraction of the Dirac equation [5, 55] (see also appendices F and R of this book). As these important quantum equations already fell out of the simplest of the many scaling options one would have for a metric (c.f. (151)), namely (4), and thus, leaving us with many more possible variations, we conclude that we indeed have found a Quantum Gravity Theory. Thereby, we need to point out that we only have discovered this general theory inside an already existing one (Hilbert's and Einstein's namely) by a bit of playing with the theory's inner degrees of freedom.

2.8.1 References

See "Global References" at the end of this book!

3

Light Dishes

3.1 Brief Note about the Thermodynamic of Thoughts

In [4, 5, 57–59], it has been already shown how thermodynamics falls out of the Einstein–Hilbert action [1] for any arbitrary system consisting of metrically arranged properties. Having seen here that ensembles of thoughts, ideas, individual concepts, whole minds and so on can also be described by sets of attributes, properties, degrees of freedom and, thus, dimensions, we should also be able to derive a "thermodynamic of thoughts." After all, the derivation of the Dunning–Kruger effect from such a thermodynamic thought-ensemble (see section "2.1 How Do We Emit Confidence or Can the Dunning–Kruger Effect be Mathematically Derived?") is already "living" proof for such a concept.

Our starting point shall once again be the Einstein–Hilbert action, but this time, we simply allow for the existence of gravity and quantum centers as introduced in [5, 55, 57–59]. The corresponding quantum gravity equations then automatically sport the statistics being necessary for the derivation of thermodynamics. With respect to the derivation, the reader is referred to our previous publications mentioned above.

3.1.1 References

See "Global References" at the end of this book!

Mathematical Psychology: The World of Thoughts as a Quantum Space-Time with a Gravitational Core
Norbert Schwarzer
Copyright © 2025 Jenny Stanford Publishing Pte. Ltd.
ISBN 978-981-5129-27-4 (Hardcover), 978-1-003-58203-8 (eBook)
www.jennystanford.com

3.2 About the Fundamental Time and Uncertainty of the Mind

Some of the aspects and results that we are going to present in this chapter have already been shown previously in our book [4]. However, as the topic merits a compact presentation and as we have not considered the special aspects of the mind but were more or less treating systems in general, we are risking a bit of redundancy and repeat the basics here. This also helps us to get a holistic picture and supports the understanding.

3.2.1 An Introduction in a Somewhat Informal and Illustrative Form

Only for the time being, and in order to create some figurative understanding, we want to avoid the math in this little introduction.

We take it that the reader is familiar with the problem of quantum reality, measurement and the so-called collapse of the wave function.

One could say that an electron is not there until someone measures it.

Bohr and Einstein had quite some "gos" about this and Bohr always won. Einstein, however, never liked the whole thing (he was even cited with: "the moon isn't only there when somebody looks at it") and when he finally came out with his Einstein–Podolsky–Rosen (EPR) paradox [19, 20], everybody thought he had a point. But then John Bell came up with his famous inequality equation [21], measurements were performed on the EPR-problem and the EPR-paradox was proven to be reality (bitter to some, though) [22, 23, 24].

This author often wondered about the "conflict," because it has happened so often in science that in the end someone found out that they all had missed something, meaning, a simple change of paradigm would—perhaps—have solved the problem. To answer this question, one should first ask what could have been the cause for the problem?

Obviously, it was/is our understanding of time and space!

Let us here concentrate solely on the problem of what time is. This author was of the opinion that it all had to do with our notion (understanding) of time. In both Quantum Theory and General Theory of Relativity, namely, time has been planted into the theories like a geranium into a flowerpot. None of the two theories properly answers the question where it comes from. On the other hand, the Einstein field equations get perfectly along without any time. Flat space does not need time and the fact that we have introduced the Minkowski metric simple was to ensure the right Lorentz transformations. In order to do so, one has to introduce the so-called sign convention, where the time component of the metric has a different sign than the spatial components. And now, putting the problem of the appearance of time very simply and purely mathematically, one could ask:

Where does the minus (in fact it is $-c^2$, with c denoting the speed of light in vacuum) for the metric time component come from?

In order to answer that question, we started with a simple flat space, where all metric components are equal (just like an n-D Cartesian space) and use the usual metric factor f as given in (4). Any such factorization or quantization with functions of the type (see [25]):

$$f[x,y,z, \ldots] = f[cx * x + xy * y + cz * z + \ldots]$$

leads to conditions $cx^2 + cy^2 + cz^2 + \ldots = 0$, which means that at least one coordinate must become time-like.

Fine! So, factorization or metric scaling brings time, but with time being so omnipresent, perhaps there even is a deeper layer.

Let's go back to the Cartesian flat space and simply assume that two coordinates are forced to do something together. Let's "entangle" them in a most general manner. Our reasoning for this conjecture: If something happens, then it has to mean a change of one dimension with respect to the other. Thus, entangling them would ensure this happening, right? Doing this mathematically and looking for the most general solution for the Einstein field equations forces two things to happen:

(a) The entanglement becomes a wave (moving forward and backward).
(b) One of the entangled dimensions must become time-like.

Note, this dimension becomes time-like due to the entanglement, it did not have to be so before.

What has this to do with the measurement, the wave function collapse and all this?

And, even more important in the book here, what has this to do with mind?

Well, here comes in the change of paradigm. The states of entanglement are the states with time. These are also the states of the wave functions. Then the process of measurement destroys the entanglement; thus collapses the wave function ensuring the entanglement and provides a certain metric as (temporarily static) result. With time itself doing the coupling (and only becoming time due to it), all the "ghostly or spooky interaction at a distance" problems are solved. There is also no need for Einstein's "hidden parameters" as the communication is a direct metric one. The components of the one dimension taking on the position of time and the one staying spatial are forming ONE system or entity (or "effective dimension") as long as the entanglement is not broken up.

And the mind?

Well, how about a system solving the equation $cx^2 + cy^2 + cz^2 + \ldots = 0$ in its very own, let's say individual, way? Thereby the system can have its own set of parameters $cx, cy, cz, \ldots$ for all its degrees of freedom and potentially even many individual ("private") times or time-flows. The system may even be actively steering such solutions by using the intrinsic degrees of freedom and thereby creates its own individual space-time or universe. Could this be the recipe for the conscious mind?

Simply changing the paradigm of time from being something which is always there to something which comes and goes together with processes and action, meaning that the action does not go with time, but it actually makes the time, brings in quite some understanding into originally very fishy things.

Interestingly, the moment one "activates" the mechanics of "the process triggers time and not the other way round," one automatically gets a very fundamental solution within the apparatus of this concept. The protagonists of the "Holographic Principle" will probably like to hear it: In its simplest form, this fundamental solution is based on a 2D, which is to say pairwise, entanglement. Investigating the properties of this fundamental and most omnipresent solution gives:

(a) the Heisenberg uncertainty principle;
(b) an upper speed limit, which is the speed of light, of course;
(c) a connection between the smallest resolvable structures and the "frequency" of the solution[1].

In other words:

There is a limit in size to things we can touch, or, as some more religious people might like to put it, "God is hiding something from our curious gaze"!

[1]Meaning, with this entanglement solution being so omnipresent, there is principally no way to access structures underneath a certain scale.

3.2.2 The Psychological Time

With the concept given above, where any given system can develop its own time or time-like dimensions simply due to its fundamental solutions, we have to conclude that this also holds true for psychological systems, systems of cognition or—just very mundane—thought machines. In other words:

A sufficiently complex psychological system can have—no, must have—its very own intrinsic time or even various times.

In appendix J, we are going to mathematically derive the whole concept.

3.2.3 Interpretation of the Result

Seeing any process or action as the change of a property/dimension with respect to another property/dimension automatically leads to an entanglement of the two. As this can also be seen as being the case for more properties/dimensions changing with respect to others, we can—in principle—describe all processes and actions as entanglement. As shown above, in this case at least one of the entangled properties/dimensions has to become time-like. It is important to note that such a time-like appearance thereby comes with the entanglement (and thus the process or action). It has not necessarily been there before. Destruction of the entanglement—e.g., caused by a measurement process—breaks up the connection between the two (or more) connected properties/dimensions and thereby also freezes the formerly time-like property/dimension at a certain (non-entangled) metric state.

With time itself being part of the entanglement, it also becomes clear why the EPR paradox is no paradox [19] at all. Two quantum objects being created in one process and being separated over a long distance are connected as long as they are entangled. The entanglement even leads to the time the process "has." We easily see this when observing solution (201) further below in this book. There we have two waves which would run apart, but with the process time t itself providing the connection, "communication" among the two objects is ensured as long as the wave function is not destroyed (e.g., due to a measurement process).

And again, as this all holds for any kind of system in a very general manner, we have the same fundamental findings also for psychological systems, which is to say thinking entities.

3.2.4 The Uncertain(ty) of Thought

Investigating the simplest and fundamental 2D-entanglement solution (see (200) in appendix J) and assuming such types of solutions to be omnipresent within this universe (simply because they are the simplest such solutions and they should at least be as omnipresent as there seems to be a time), it can be shown that a fundamental uncertainty law (just like the Heisenberg uncertainty principle) must hold in such a system or universe. With respect to the evaluation, we refer to appendix J.

And again, as this all holds for any kind of system in a very general manner, we have the same fundamental uncertainty also for psychological systems, which is to say thinking entities. In other words, it is no wonder that most of thinking appears very fuzzy. However, as the conscious system (the mind) apparently can evaluate its own set of solutions, it can also have its own uncertainty, which easily explains why some minds appear foggier and more confused than others.

3.2.5 References

See "Global References" at the end of this book!

3.3 A Timid Statement

We see that we are moving more and more towards a "**mathematical psychology**," and taking the success of the introduction of higher math in physics, which—after all—came under the name of "mathematical physics," we do not think that this is a bad thing. Taking the fact alone that such an approach reduces the arbitrariness or even despotism one often finds in classical psycho-analytical processes suffices to consider this much more falsifiable and thus, Karl-Popper-like approach, an advantage.

3.4 Why Does Consciousness Need Spirituality?

As most human beings see themselves not only as intelligent but also as spiritual entities (Fig. 3.1), we find ourselves in the position of dear need to also somehow incorporate the aspect of spirituality into our mathematical psychology.

Figure 3.1 The Spiritual Body, by Livia Schwarzer.

Thus, in this chapter we are going to investigate how evolution needed to invent spirituality in order to save consciousness. It will be shown that due to Darwin's evolutionary laws, the latter is not possible without the first.

3.4.1 The Evolutionary Dilemma

There was a moment when evolution "had the amazing idea" to develop consciousness, which, as we hinted in our derivation above, seems to be "not more" than the ability to actually and actively steer certain parametric quantum gravity solutions for the system mind by the system (the mind) itself. Thereby, evolution realized, which is to say the process of mutation and selection "discovered" that life forms equipped with consciousness actually were not the fittest. As a remark, however, it needs to be pointed out that consciousness probably was not the thing evolution had aimed for. It only wanted to have cleverness. With cleverness, the evolution was able to form species being able to be innovative, use tools and perhaps even create some themselves.

With cleverness evolution could bring about entities that are able to help evolution with the ever more complicated inventions necessary to really make the difference. Consciousness, on the other hand, merely came as a by-product nobody really had aimed or even wished for. One may say that consciousness was the true original sin. A sin, by the way, nobody should be blamed for, but evolution. After all, if we all had stayed dumb like earthworms, the word "sin" would not have even existed. This author even is of the opinion, or rather sees this a possibility, that a calculating or perhaps "thinking" universe reproduced consciousness, which already existed in a metric form on an extremely low scale, by the means of biological substrates. The "unconscious goal" was to bring an intelligent vector into the apparently chaotic short-scale processes of evolutionary progress and regress (Fig. 3.2). Consciousness might have just been the outcome of a search process for a vector stabilizing evolutionary force in order to bring a bit more order into the semi-chaos of purely self-organizing-based structural formation.

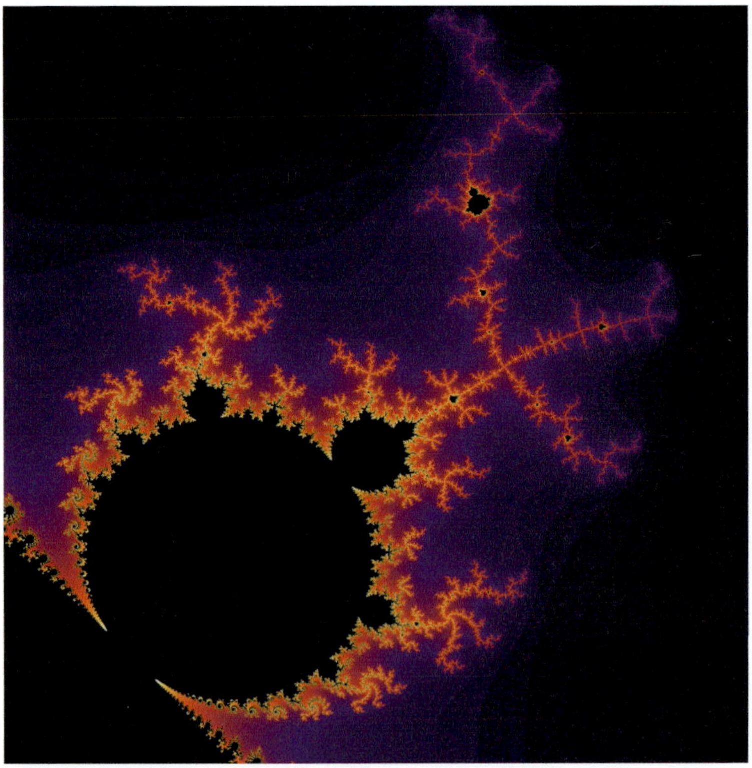

Figure 3.2 Order in the Chaos (c.f. appendix R).

The main part of this little piece of text about the connection of the evolution of consciousness and spirituality (co-evolution even) can be found in the book *The World Formula: A Late Recognition of David Hilbert's Stroke of Genius*, by this author [4] (section 17.5.3/starting page 556). There, one finds—among other things—this interesting story about a godlike entity the author had characterized as a "young nothingness."

The story moves on as follows:

At first sight, it was good to have some kind of awareness or self-awareness, but within the cruel reality it often became a disadvantage to "feel" respectively to consciously recognize what was going on and what life or being consciously alive really meant.

So, evolution had to learn it the hard way, that it could create sufficiently fit life forms, which also have consciousness only when adding yet another piece to that recipe and this was spirituality. Only spirituality allowed these self-aware and feeling life forms to put all those difficult or impossible-to-understand, often cruel or even unbearable experiences to a place in those bigger and bigger growing brains where it (meaning all the mess, cruelty and madness around) would hurt a bit less (Fig. 3.3).

Figure 3.3 Trust.

Of course, in order to properly investigate this problem, one requires a completely new and most holistic understanding not only about evolution but also about the psyche of sufficiently cognitively capable species, which not necessarily started on this planet with the early human beings. These two aspects are presented here and in the book [55]. Currently this book is only available as e-book, but it is planned for real publication with Jenny Stanford Publishing in the near future. For those among you who prefer to hold real books in their hands, we suggest the comprehensive work [5]. It does contain all the basics to derive what has been presented here and in [55].

But now we want to read on with our little story about how evolution had to invent spirituality.

Of course, evolution never intended to make this storage for spiritual thought bigger than absolutely necessary. After all, evolution only wanted to create fitter life forms and had no interest whatsoever in bringing dim and passive believers into being. Thus, the organ of spirituality only ever was thought of as a temporary storage room for everything which could not be explained by the conscious (quantum gravity equation solving and individual parameter-set-solution creating) mind straight away.

Unfortunately, however, it turned out that explanations for some things took longer than the typical life span of such life forms. In fact, there were problems where even the typical life span of a whole species would not suffice[2]. Thus, it happened that there was always something in these storages. They were never empty and the bigger brains, keen to get also these unexplainable things explained, were extremely susceptible to all sorts of prophets who promised to have solutions, answers, guidelines and—apparent—orientation.

And so, the epoch of the shamans, priests, imams, Gretas and CO_2-smellers, esoterics, astrologists, politicians, politic scientists, soul-experts, climate-apostles, philosophers and coffee shop owners had begun.

However, I do not want you to get me wrong. I am far from criticizing evolution for its invention. On the contrary, because with spirituality, evolution gave us something else ...
It gave us hope (Fig. 3.4).

[2]Maybe this universe as a whole will—if it is a thinking entity—come to the conclusion that it will not be able to answer all the questions it has in its own life span and then has to think about a way to recreate itself with some of the earlier results being implanted into the new being. Maybe this has already happened "before."

Figure 3.4 "Hope" by Livia Schwarzer.

3.5 The Field-Illustration of Mass Formation

By applying our metrics for individual fields, which are fields of enclosed (spherical based) character, from [5], we are able to at least illustratively understand the mass formation process in a quasi (figurative) mathematical way. Interestingly, when placing individual fields of the above-mentioned character in a socio-economic space-time, we automatically see the formation of clusters, which we want to call liposomes here (Fig. 3.5). With enough of such liposomes, we even get super-clusters or super-liposomes (Fig. 3.6). One may interpret the smallest clusters as groups of family size or just families and the super-liposomes consequently as small communities.

3.5.1 The Field of Fear Is Entering the Scene

Now we assume a fear field to be introduced into the socio-economic space-time (in Figs. 3.7–3.11 symbolized as cones) becoming ever more omnipresent and closing in on the super-structures of our individuals. When reaching the super-liposome's cell wall, it has been penetrated (Fig. 3.12) and cracked open. Now, the family liposomes lie exposed to the field and the fear creeps closer in on them (Figs. 3.13–3.19). Whenever the fear field reaches a family liposome, it has been penetrated and cracked open. Its cell wall disappears and the individual fields lie exposed in the open (Figs. 3.15, 3.17, and 3.19).

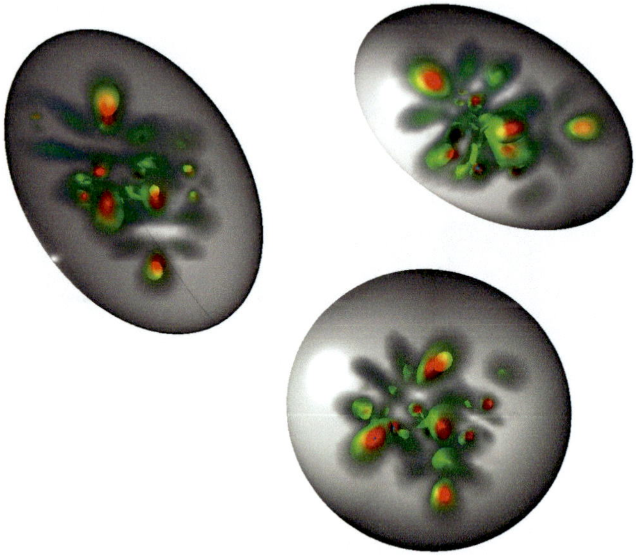

Figure 3.5

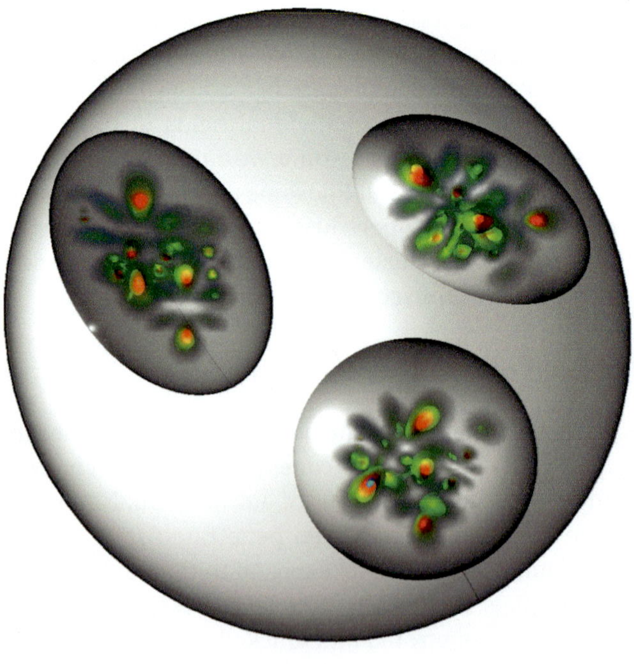

Figure 3.6

Figure 3.7

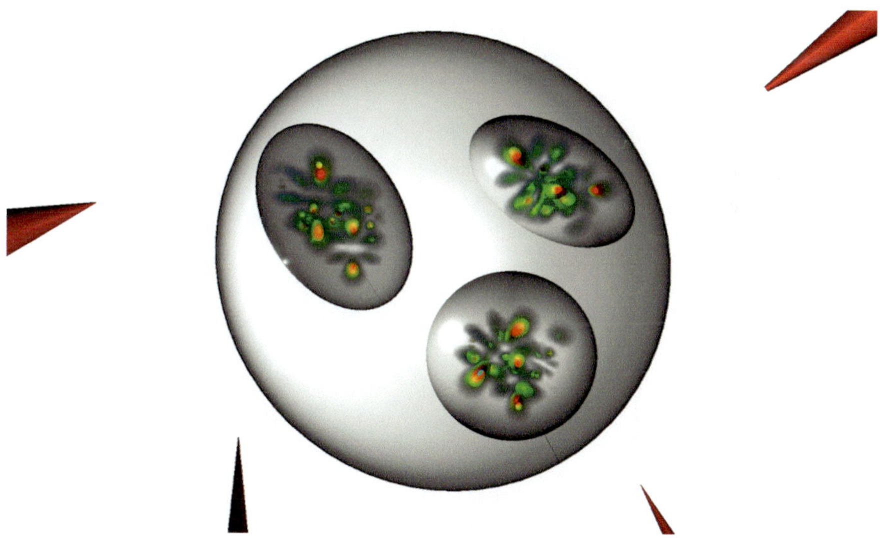

Figure 3.8

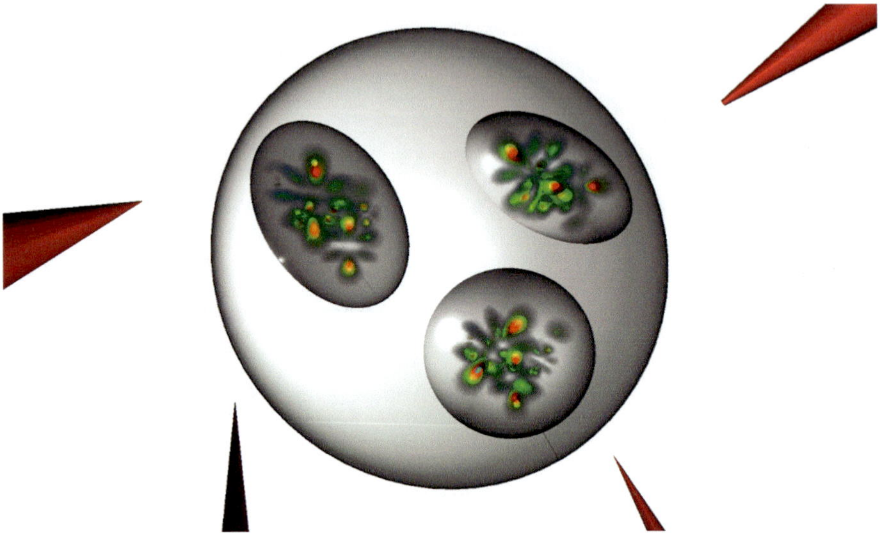

Figure 3.9

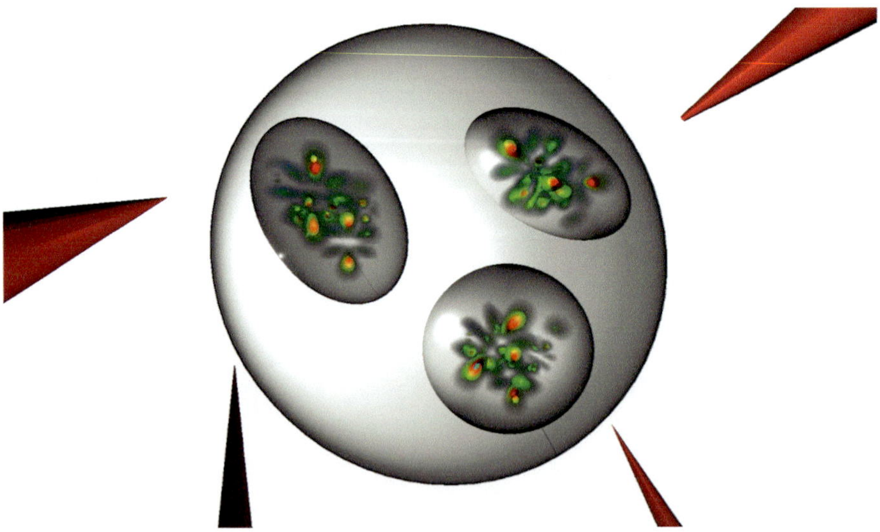

Figure 3.10

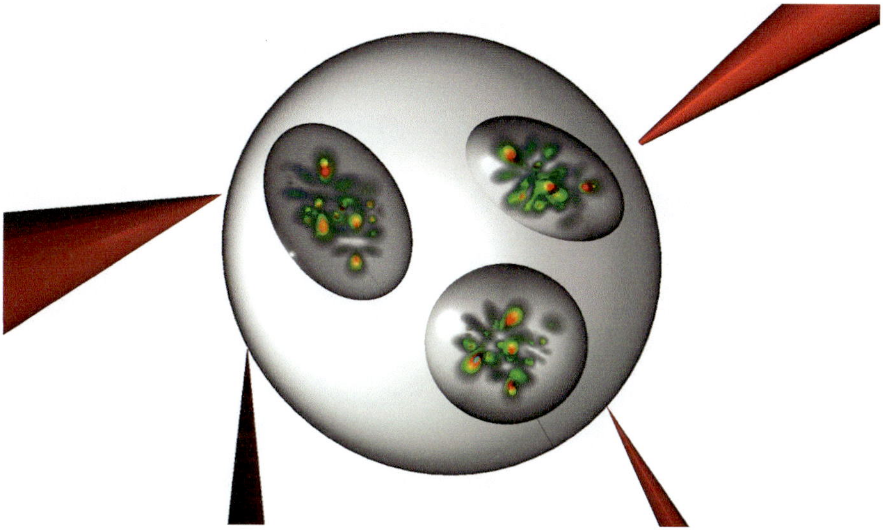

Figure 3.11

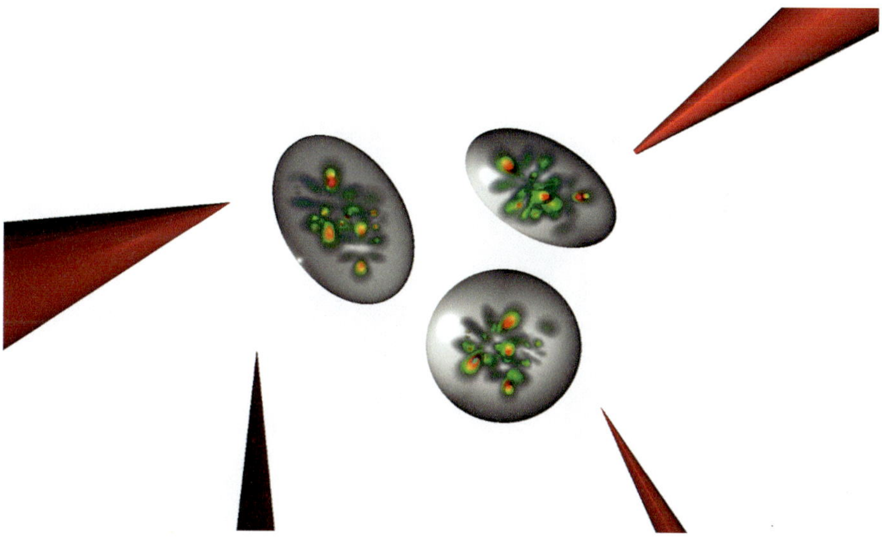

Figure 3.12

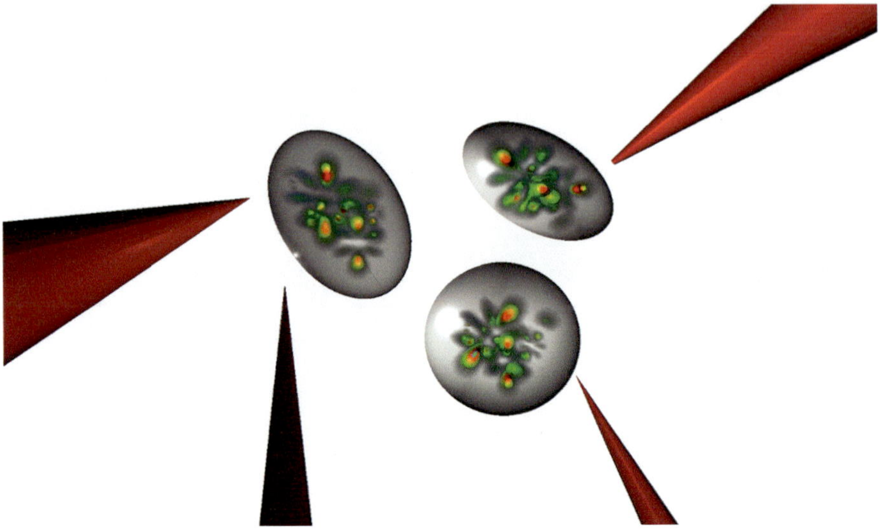

Figure 3.13

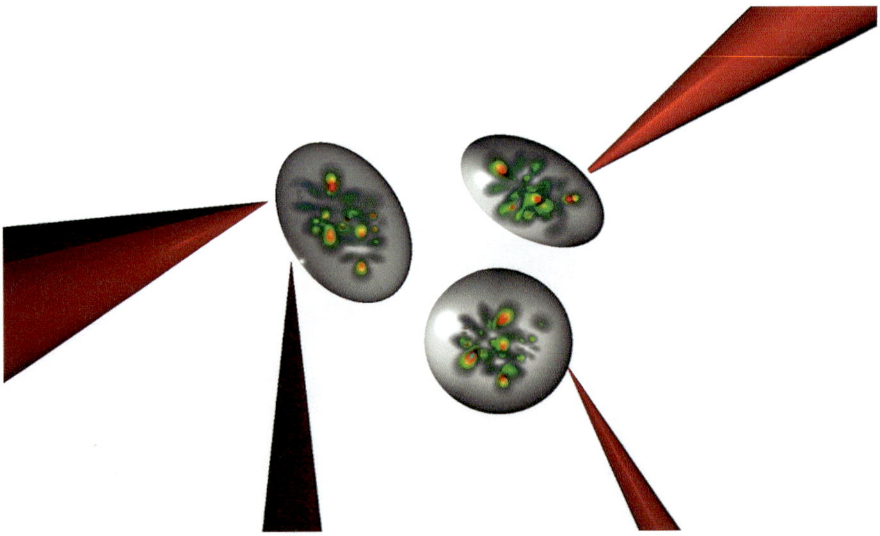

Figure 3.14

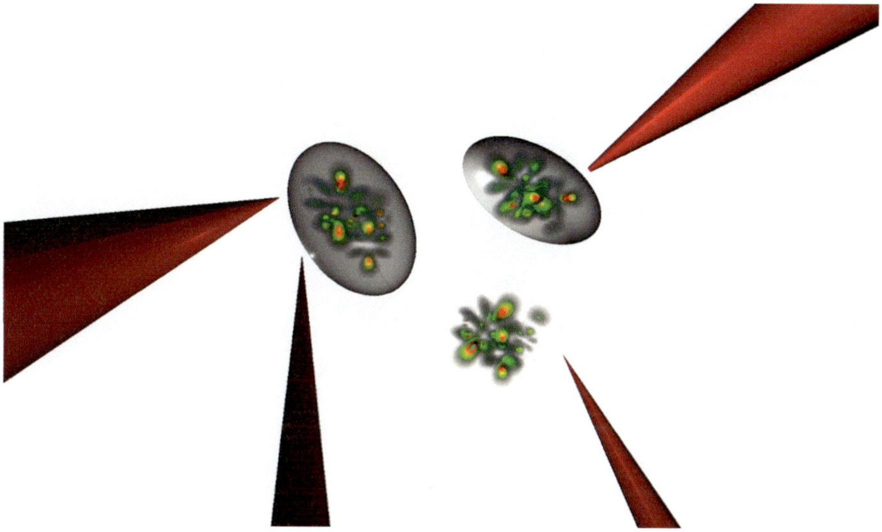

Figure 3.15

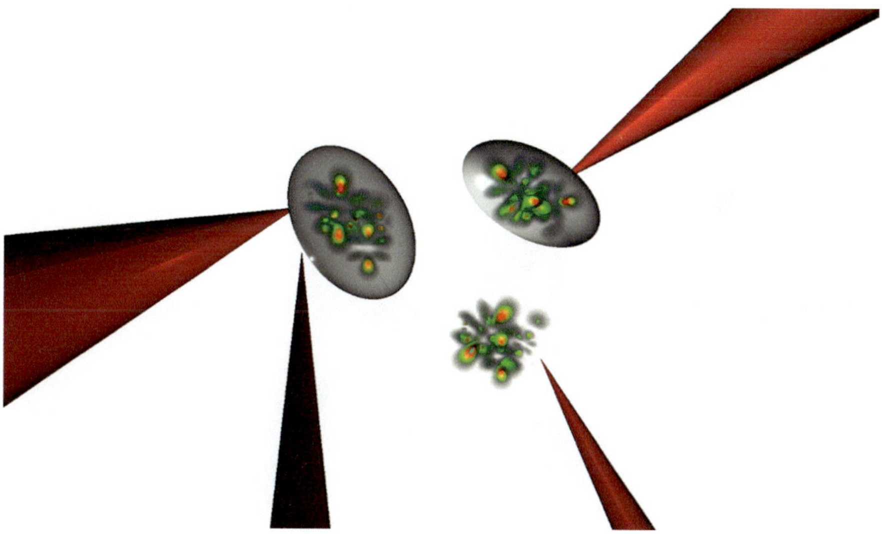

Figure 3.16

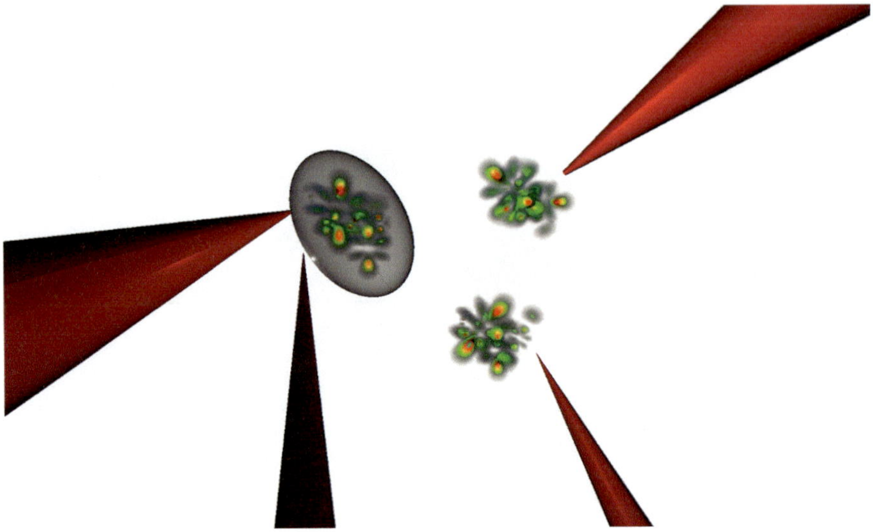

Figure 3.17

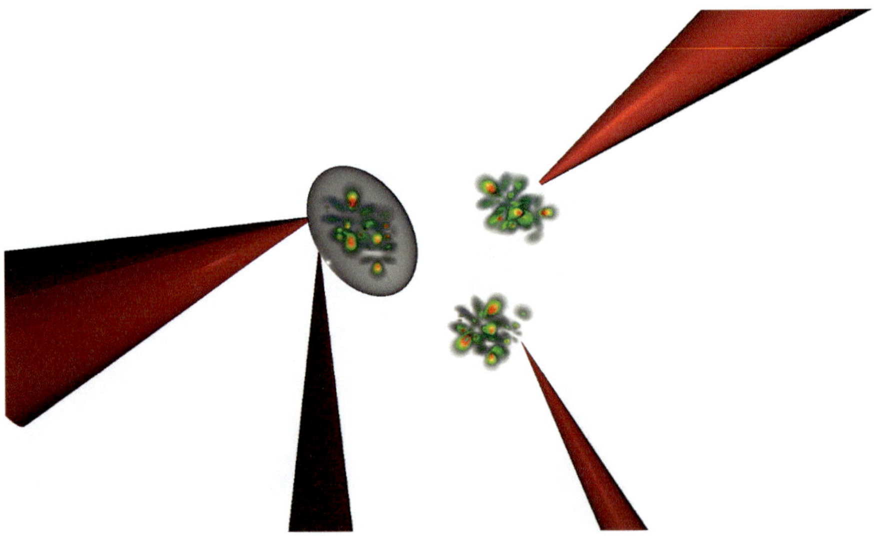

Figure 3.18

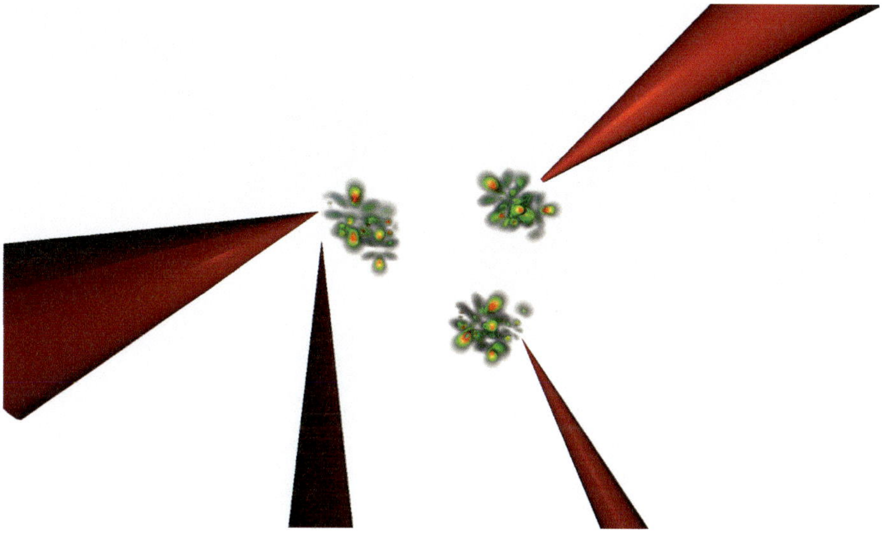

Figure 3.19

3.5.2 There Is Always Hope

Depending on the strength of the fear field and the adaptive capacities of the individuals, now a variety of scenarios can be evaluated.

We just assume only one family liposome being cracked open with some individual fields of relatively low resistance and a few of fairly high strength or resilience. As the latter are the ones being able to resist the hypnosis of the fear field, they are not sucked onto or into the community field (here: fear = cone), they do not degenerate and are not reduced in dimensionality. The fields of the first type, however, are completely taken on by the community field (Fig. 3.20). They degenerate and drop in dimensionality (crawling ants on the surface of the community instead of free individuals in the whole socio-economic space-time with access to the latter's full dimensionality; c.f. Fig. 3.21). The same happens to the individuals of the two remaining family liposomes (Figs. 3.22–3.26) and the situation appears to be decided. It seems the fear field has won. When giving those resistant individual fields enough flexibility, however, and allowing the system to further evolve, we suddenly see the formation of new family-liposomes out of the pool of those resistant individuals (Fig. 3.27).

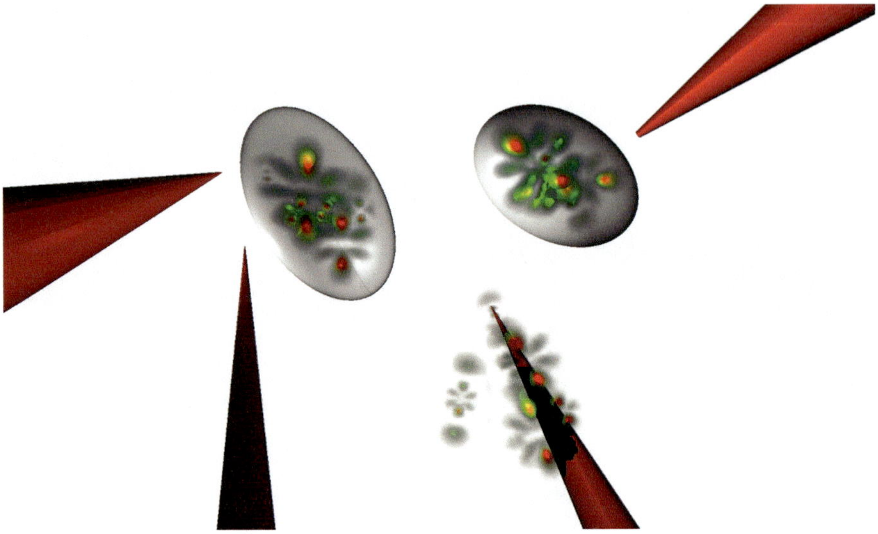

Figure 3.20

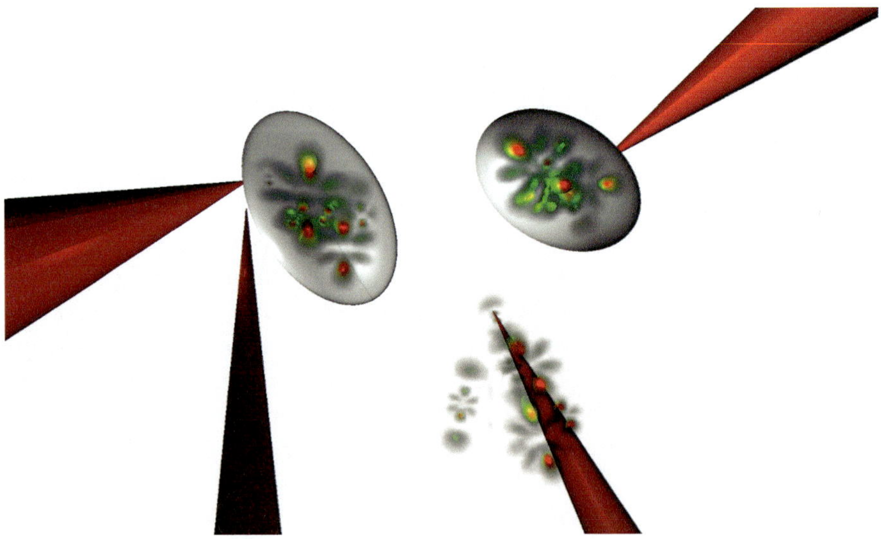

Figure 3.21

Figure 3.22

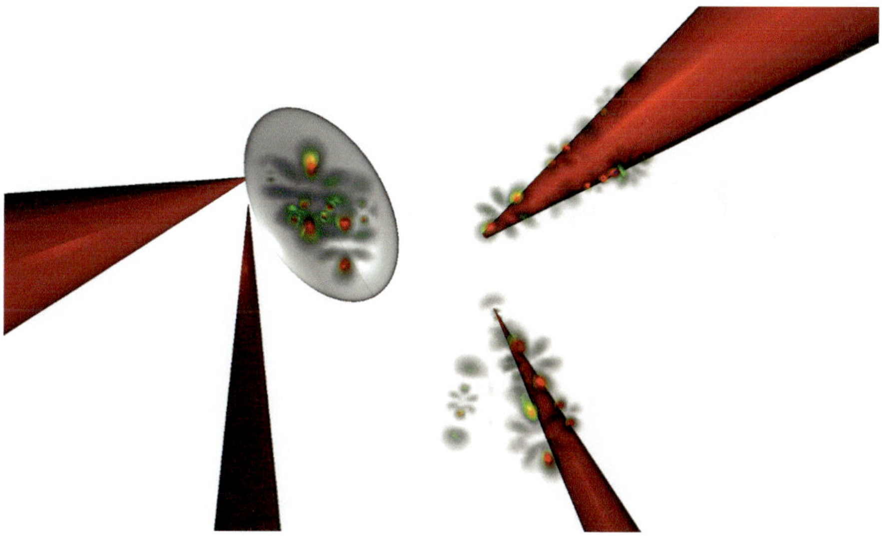

Figure 3.23

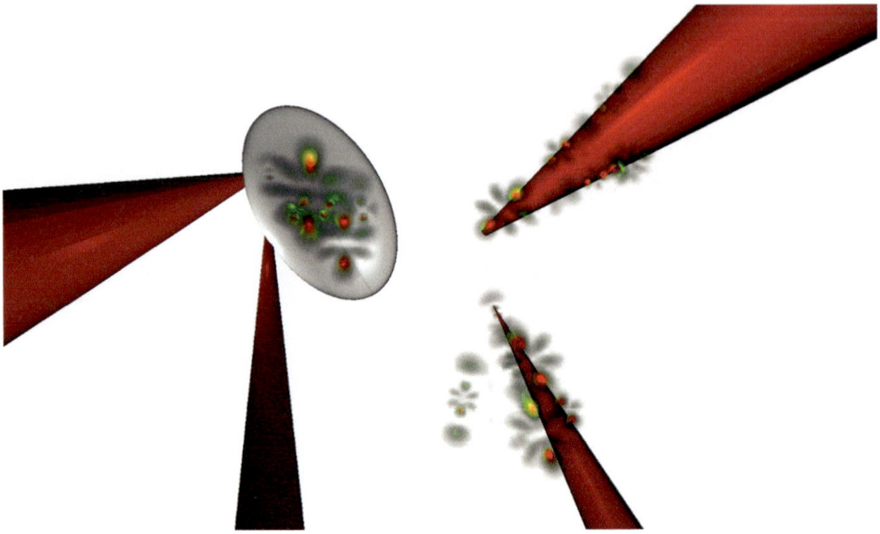

Figure 3.24

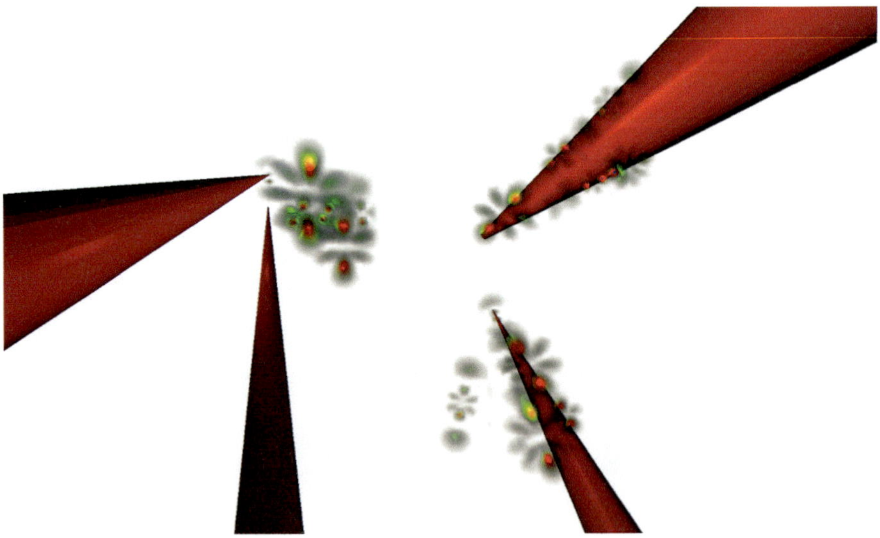

Figure 3.25

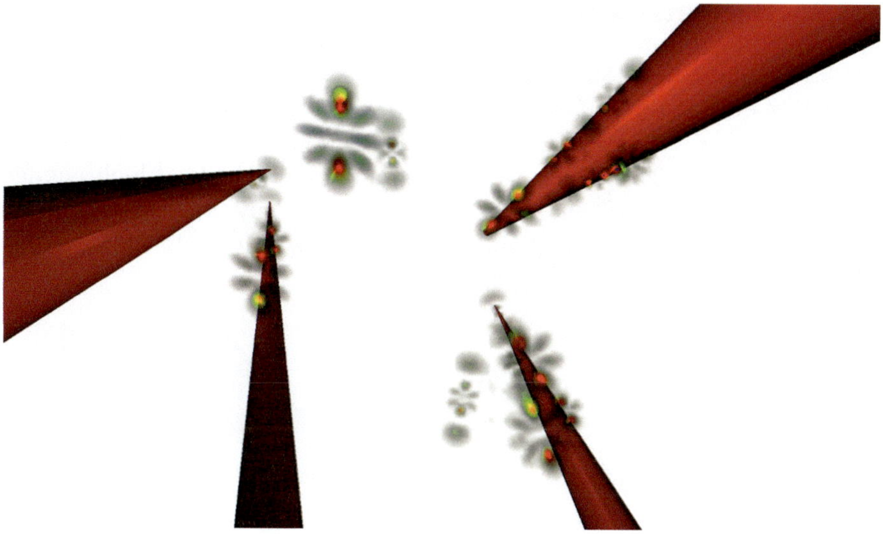

Figure 3.26

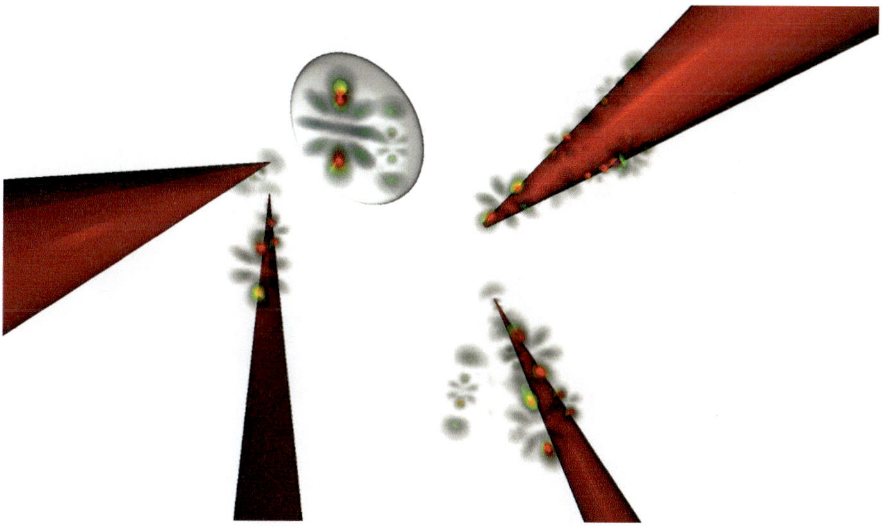

Figure 3.27

3.5.3 The Better Attractor

By letting the simulation run further and observing all possibilities, one finds that even under the condition of a maintained fear field, the resistant group provides, respectively, can provide a powerful attractor of its own (Figs. 3.28–3.45). This at least holds for some individuals of high enough complexity (residual individuality) which are not only dragged away from the community and the fear field, but which even join the liposome of the resistant (Figs. 3.34, 3.44, and 3.45).

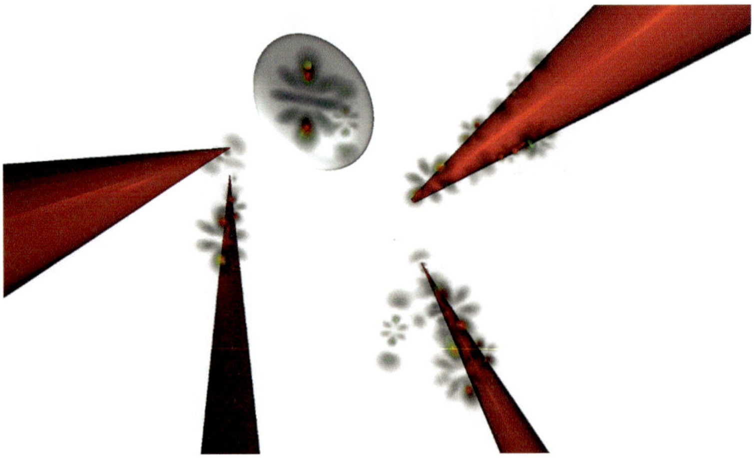

Figure 3.28

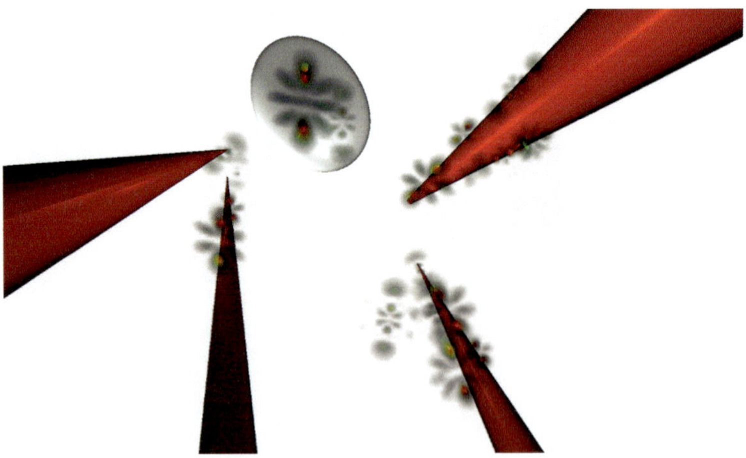

Figure 3.29

Figure 3.30

Figure 3.31

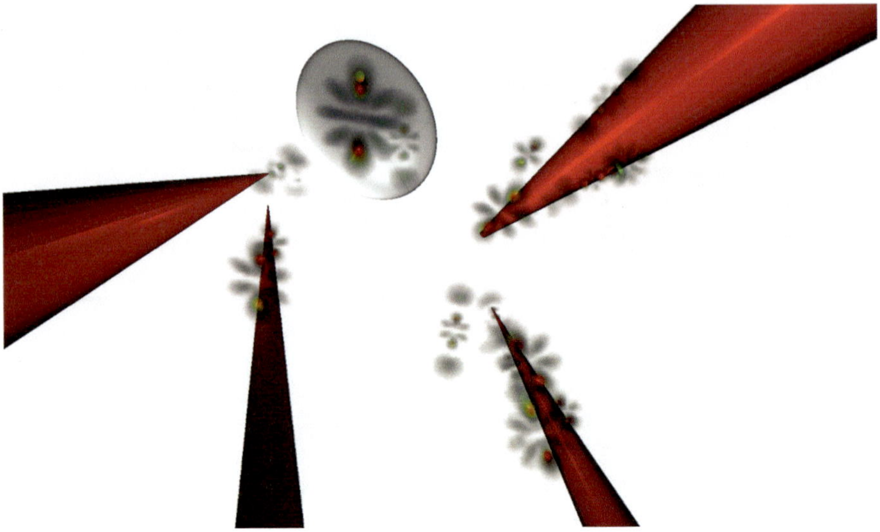

Figure 3.32

Figure 3.33

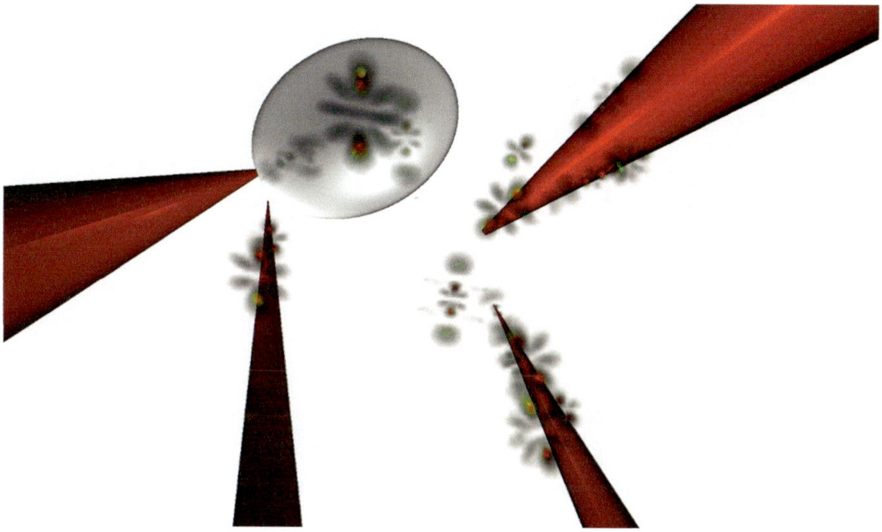

Figure 3.34

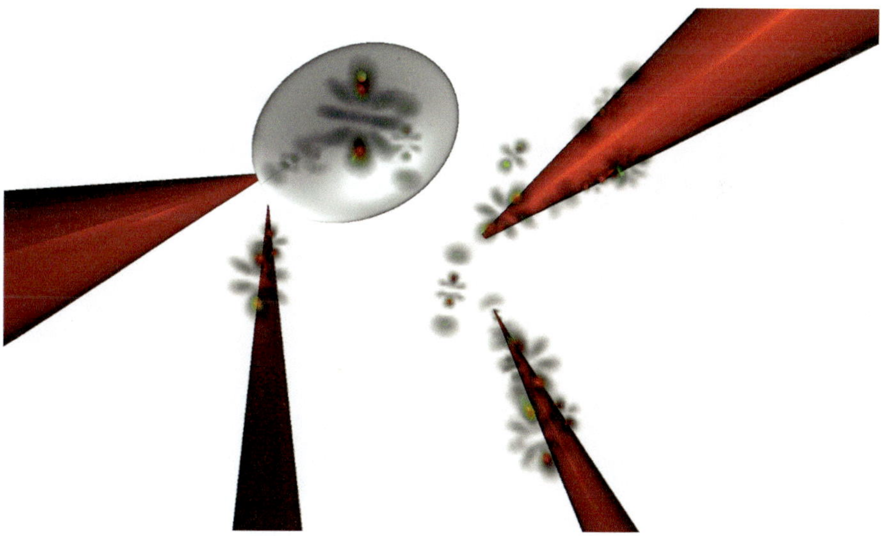

Figure 3.35

Figure 3.36

Figure 3.37

Figure 3.38

Figure 3.39

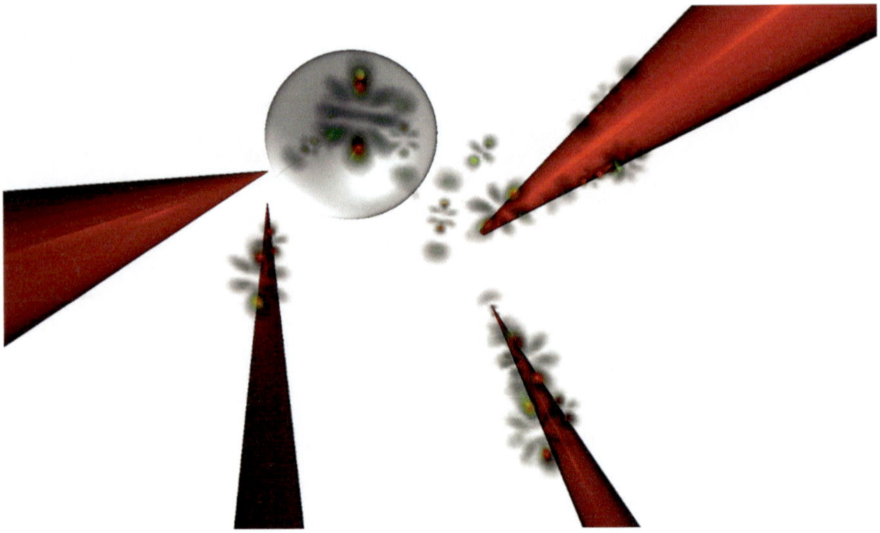

Figure 3.40

Figure 3.41

Figure 3.42

Figure 3.43

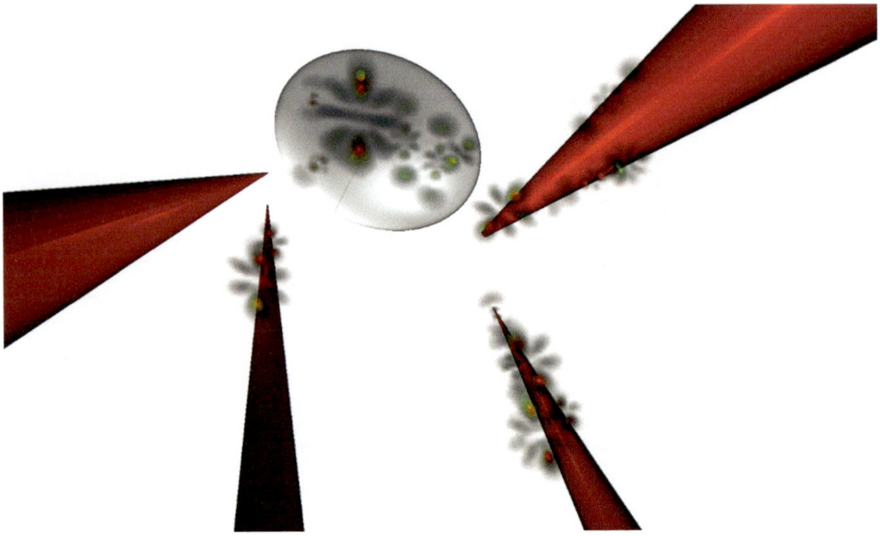

Figure 3.44

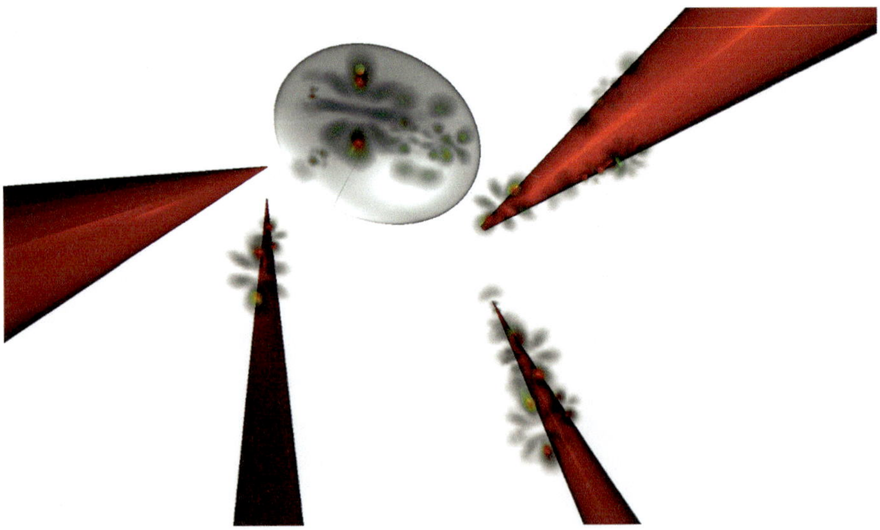

Figure 3.45

3.6 Outlook: What Happens when the Narrative Crumbles

3.6.1 About the Need for a Better Understanding of the Process of the Potential Crumbling of the House of Cards

When in July 2022, even the *Wall Street Journal* sounded off against the FDA and the CDC about those COVID-19 vaccines being a good idea, many already-awoken people asked the question: How will this turn out? How is the narrative finally going to crumble and what will be the collateral damage? It is an interesting fact that so often in history things turned out to bifurcate into completely different ways than one would consider most probable when just observing the state right before the point of no return, which is to say the starting point of the crumbling process. The reason for this apparently strange behavior can easily be found when diving into the thermodynamics of systems near phase transitions.

3.6.2 About the Math

Criticality (systems moving towards critical states—phase transitions, processes) has usually been understood as a process of vanishing characteristic scales ... in fact, this is what characterizes a critical phase state.

Unfortunately, with vanishing characteristic scales, a lot of nasty stuff is happening (singularities pop up).

The method of choice to still deal with the object was (and still is) renormalization group theory (RnGT). This, however, only ever was (and still is) only a technical trick with nobody being able to explain when and why it works (e.g., see the textbook [60]). Ok, we admit that we know why it works from a phenomenological standpoint, namely, because it helps to make the singularities disappear, but there is a huge amount of arbitrariness in the whole process and this makes it clear that we here have no fundamental law but just a technical trick, which sometimes works and sometimes it does not and nobody can be sure about the whole thing.

This can definitively not be seen as a good starting point for something new and crucial to society, which is to say something where the plan is to make a prediction, especially one about the future[3].

[3]Yes, I know that this joke is from M. Twain and quite old, too, ... it is still not a bad one, though.

What could we make better here? How could we possibly substitute a 70-year-old "well-established" but still controversial method with something more fundamental?

Well, there are two things this author sees here.

(A) The problem has been caused by vanishing characteristic scales. On the other hand, we know since Newton and Leibniz that the whole infinitesimal calculus is totally based upon vanishing differences. The latter produces horrible singularities just like the first if not being tamed via a simple limiting procedure we call differentiation. Now connect the dots and you know what I have in mind to also tame the critical phase singularities.

(B) I do not know why, but it seems that many who have worked in the fields did not realize that almost all critical processes do indeed produce vanishing characteristic scales, but just as the characteristic scales vanish, also a reduction of dimensions can be observed. In fact, it is not really a reduction, but only a dominant recessive process of certain dimensions becoming important and totally overshadowing others[4]. For illustration, one may pick almost any socio-economic crisis and find that during the move towards the critical point, certain attributes (properties) became overwhelmingly dominant against others. Society was dancing around a calf and almost totally forgot about all the other socio-economic aspects and internal interactions.

    The whole process of dimensional reduction can be made easier in physical examples, but as these are not the things we are dealing with here, it would be foolish to bring such a thing.

    The important aspect is that the observation of dimensional reduction, too, gives us another mathematical tool to handle the crisis situation without this arbitrary RnGT approach.

By the way: These two approaches might also help to better grasp (mathematically grasp) the problems of turbulences, be it in socio-economic or in physical applications.

Now, the reader will probably ask why no one before has thought about this.

I don't know, but when considering the situation in a different field, one might have an explanation.

[4]It is a bit like the dominant recessive allele mechanism in genetics ... only with quite many alleles (of usually very low complexity).

Let's take the field of technical mechanics. Together with the first computers, finite element (FE) and boundary element (BE) methods came up. These finite and boundary element features were pretty sexy stuff and as attractive to all those TechMech-scientists as a huge pile of dog sh** (well, you know) is to a fly. One just needed bigger and bigger computers to handle ever more complex models and systems and—so it seems—this only contributed to the whole hype. People loved to play with big computers and the computer industry was only all too happy to deliver them. This author, on the other hand, was never very much attracted to this stuff and rather thought about optimizing the math for a certain problem before programming it and feeding the computer with the subsequent code. It is like going into a battle with many sticks or spending a bit of time on improving the stick-tech until it becomes a wonderful sword and then going into the battle with this one instead.

This way I created software for complex technical mechanical problems being many thousand times faster and much more precise than any FE or BE system and coming with the additional feature of being invertible and running on small computer machines like my personal laptop on which I just type this text.

When first introducing this kind of stuff over 27 years ago, certain big shots in the market promised me about 5 years at most before the FE and BE companies would have caught up with me ... I am still waiting for this to happen.

Now we had a rather similar situation with the RnGT. It came together with this sexy thing named computer and everyone who considered themselves in vogue went for this tech rather than going the cumbersome Newton or Leibniz path and trying to develop proper math before resorting to the computer option.

But of course, I know that currently there is this little snag about "doing things in Newton–Leibniz style" and this is a lack of time and resources.

Well, this author is simply going to see what he can do.

3.6.3 References

See "Global References" at the end of this book!

3.7 A Connection to Chaos Theory?

Mattias Desmet about chaos (in an email to this author):

> *About chaos and where it comes from, it makes me think of Heisenberg's alleged last words: 'When I meet God, I will ask him two questions: Why relativity? Why turbulence? I am quite confident that he will be able to answer the first question ...*

Well, there must be a reason why they call turbulences "the Black Hole of physics," where all the good ones disappear and never turn up.

Nevertheless, systems, when becoming critical, require the understanding of "chaos" (all range/scale interaction) and, as already said in the previous section, this author is of the opinion that renormalization group theory is the wrong approach ... but I also do not have the inclination to get lost.

So, how to approach this field?

Sometimes it is enough to just ask a question and imagine a certain addressee for this question in order to find the answer ... or at least half the answer. After all, the way one imagines the person one formulates the question for her or him and thereby often goes quite some distance from the personal picture of the thing one wants to ask about usually already brings new insight into the thing.

In Desmet's book about "Lacan's subjectivity" [16], the author has put some effort in explaining the basics of chaos theory.

If it weren't for the fact that chaos has definitively been observed in reality, one may ask oneself where it comes from. Ok, complex systems might just produce it due to their complexity, but we have enough examples (some famous ones are in Desmet's book [16]) where the system is anything but complex and still produces horrible chaos. In some cases, equations can be found, which mathematically show how the chaos arises, but none of these equations resulted from a fundamental law (extremal principle, definitively not a metric one). We also know that many things in chaos theory can be explained by fractality.

When Dirac derived his famous equation [7], he started from a second-order differential equation (the Klein–Gordon equation) and factorized it.

This factorization was done in a perfect symmetric manner, resulting (for the case of the Minkowski metric) in first-order differential equations with quaternions.

Now this author just asked himself whether the symmetric factorization could not be generalized (in some way), which automatically also leads to fractal derivatives.

In fact, when doing the math, one receives the "chaos" we were looking for and I wonder whether this additional degree of freedom may not be something the universe makes permanent use of.

Taking Heisenberg's alleged last words (see Mattias Desmet's text above) or the "dictum" about turbulences being the Black Hole of physics (for the physicists setting out and trying to find something out there and never coming back), one might simply suggest that—so far—we have not applied the right approaches.

Otherwise, there would not be all this helpless shrugging and scientist-vanishing when it comes to trying and explaining chaos plus the funny order in it, would it?

While the helpless shrugging and the vanishing of scientists (funnily nobody seems to miss them, by the way) are not creating much damage, there are "experts," evil businessmen and politicians who cruelly know how to abuse the creation of chaos and to ride its energetic output to fill their pockets and help their courses.

Thereby it holds that it requires no deeper understanding of "the complex mechanism of wave creation for the surfer in order to ride one," which is to say, the evil guys have no more understanding about chaos than a bacterium about the molecular processes inside its own space, surrounded by a semipermeable membrane.

Unfortunately, however, it obviously requires such an understanding when one is interested in getting the chaos under control, stopping it and repairing the damage it has caused.

It is easy for mathematicians to say that non-linearity and/or the absence of characteristic scales are causing the chaos, but this isn't even half of the story. If it was, the chaos would always appear with non-linearity and/or missing of characteristic scales, but, most interestingly, this is not the case. Even worse (or—depending on the standpoint—better) non-linearity and scale missing can cancel each other out and, thus, hinder the occurrence of the chaos. This is good to know, at least if being interested in getting back "normalcy" (as from Warren Harding's 1920 slogan "A return to normalcy").

In other words, while the profiteers of chaos do not need to understand it, the system's repair genes are better off in knowing how to deal with the thing and thus, what to do.

So, what did "we" miss?

Well, considering the Hamilton formalism to be one of the most fundamental things (I mean laws) there are in this universe, non-linearities are no surprise as they are coming out when performing this formalism correctly and—that is the more important aspect—holistically. Then, namely, non-linearity is the rule, not the exception. So, chaos and normalcy, when taking their origin, amount to the same thing. By the way, in order to see this, one doesn't even need to go back to the fundamental origin of things, but just properly observe the well-known differential equations for the simplest chaotic systems, like the Lorenz wheel (e.g., see Mattias Desmet's book [16] about Lacan's logic). There, too, the same set of equations gives birth to both normalcy and chaos ... it just depends on the conditions.

Ok, how about the scale-thing then?

Well, first, we have to clarify what we mean by this "absence of characteristic scales" and why, according to some chaos researchers, it is so important to explain chaos. In essence, namely, the missing characteristic scale is not directly producing chaos but singularities. Those may be quite inconvenient—seen from the viewpoint of the practical mathematician, having to deal with them, but infinity is not necessarily chaos, as a simple consideration of just THE kernel of the Hamilton formalism shows. In a general set of attributes, this kernel, namely, is the Ricci scalar, which in infinitely many dimensions completely linearizes. So, here infinity helps to avoid chaos. Besides, this finding is quite important with respect to the suppression of uncertainty and non-linearity if one intends to learn something about a system or even wants to make predictions about it.

Now something interesting occurs when one describes a system with "no characteristic scales" in the metric complex, which is to say, forces it seamlessly (yes, this is possible) into the Hamilton–Hilbert–Einstein formalism and thus, "the most general General Theory of Relativity" there is. Suddenly, all this scale-thing just becomes a question of the number of attributes and thus, dimensionality. **In other words, there seems to be some connection between the vanishing of characteristic scales and the increase in dimensionality, which, if seen from a greater distance, leads to dimensions losing their individual integer character and becoming a blurred continuum-like ensemble**. Such a thing, however, is just the description of a fractal space or space-time, the archetype of a chaos-prone system ... with the little, but quite essential, difference that this time we have an integer and well-ordered mathematical origin of the thing.

Another aspect, which scientists apparently like to ignore (maybe because they just sense the existence of yet another "scientist sucking Black Hole" in there), is the rueful ignorance regarding non-Riemann spaces or space-times. I will not go into the math of this, not in this chapter anyway[5], but to each space with volume there are infinitely many other spaces which sport a zero determinate (usually denoted with g from the metric tensor g_{ij}) and all these spaces fulfill the Hamilton–Hilbert–Einstein condition identical ... no matter what they do. Such spaces, so our current theoretical understanding, can exist in the utmost freedom and do whatever they please, as there is no law, at least no law we know about, which would set them any limit. Spaces or space-times with $g = 0$ could even interact with our own space-time, disturb our dimensions, perhaps create chaos and are—and this is the bad news—totally indescribable by the current means at hand (unless one considers the mathematical description in appendix R, section "5.18.7 The Conventional and the Unconventional Physics Realm" as satisfactory).

Our very own universe may have just existed in such a $g = 0$-state before the ominous $t = t_0$-moment, also named big-bang, happened and created stuff out of apparently nothing. Perhaps there was something after all, and it only phase-changed from $g = 0$ to g-non-zero. This phase change might still be under way and explains what we see as an expansion of the visible part of our universe. Like so many times before in human history, the outcome of such a theoretical understanding would just be an even greater diminution of our own being inside the true size of what actually is the "universe."

As such an interaction of g-non-zero and g-zero spaces seems to be so darn important when it comes to find another approach to chaos without disappearing out of sight forever, this author thought about a way to bring some light to the matter and we think that what we found so far will help in the understanding. The two things, namely, "characteristic scale missing" becoming a dimensional aspect and the interaction of g-non-zero g-zero space-times seem to be at least a better key to the whole field. Human beings, for instance, seem to permanently create $g = 0$-spaces—and call them phantasies or ideas.

To the text above, I received an interesting comment from Troy Vom Braucke:

> *Norbert, that message took quite some time for me to digest, and I tried to highlight the sections that stood out to me so I could try and make sense of it. This part in particular resonated with me on a more meditative level*

[5]Regarding the math, the reader is referred to appendix R, section "5.18.7 The Conventional and the Unconventional Physics Realm."

"In other words, there seems to be some connection between the vanishing of characteristic scales and the increase in dimensionality, which, if seen from a greater distance, leads to dimensions losing their individual integer character and become a blurred continuum-like ensemble."

So now I have to ask, did you send this message from the other side, where the answers were given to you or are you still with us?

Where did this suddenly come from?

Following was my answer:

I merely tried to explain what I saw in the math ... and now kind of already regret it to have done so.

But speaking about the "other side" and taking the math (the one I see when doing all this fundamental stuff others have done before me[6] *and starting as low as I think is possible), I don't really see such a thing. There is no wall with one side here and one side there. It is more a semipermeable membrane where the scale determines what goes through and what does not. Ok, "scale" is the wrong word here. It is the volume. And things (spaces or pieces of such) with* $g = 0$ *always get through ... totally unimpeded.*

Now some may think I'm describing THE other side, but no, this—now I'm struggling with the words again, so let's just use the expression—"g-zero-option" seems to be an omnipresent one ... and this would explain a great deal.

And about this "are you still with us"?

The Bible is full of interesting stories. There is this one about the master leaving his home and giving each of his three servants some talents (sounds like a currency, but I think the word had been chosen for a reason). When, after quite some time, he finally comes back and checks upon his servants and what they have done with those talents, it becomes clear from the biblical messages that safekeeping of the talents was not enough ... definitively not.

Mattias Desmet ([6, 16])—on various occasions—wondered about the fact that the mass formation had become more frequent in the past 300 years or so.

Question: When was it again that public schooling started almost everywhere in the—what we call—"civilized" world?

Could it be that what we call "education" in fact is more a "domestication" (or even worse)?

[6] I only extended, generalized and combined here and there.

Was K. Lorenz right when he wrote about "die Verhausschweinung des Menschen"[7]*?*

Are we losing touch to "the other options," or, in other words, has our natural ability to entangle, interact and widely reach out been purposefully diminished, deactivated, even destroyed?

And if yes, who has the most interest in keeping our reach-out capacity, our valency as small as possible?

We should use our talents and propagate them, but most people do not only restrict themselves to the safekeeping of the talents given to them, but also just consume them and leave nothing for nobody.

---

[7]For convenience, here I simply copy a piece from *The Math of Body, Soul, and the Universe* [5]:
This finding has dramatic consequences with respect to practical problems regarding big system analysis, system science, decision making and thus, naturally, socio-economics, politics, medicine, military and so on. It also means that any ideology-containing or perhaps even ideology-dominated process of data handling, analysis and subsequent decision-making automatically increases its own—intrinsic and fundamental—uncertainty by the simple fact that all ideologies can only exist under the condition of ideology driven information selection... which is to say, an ideology dominated system always tends to (has to tend to) degeneration and thus, dimensional system reduction. In this sense, the old work of Konrad Lorenz (Nobel Prize in Physiology or Medicine in 1973) about the "Verhausschweinung des Menschen" [C1] (loosely translated one might put it as "the domestication of man") appears in a completely new light. Whatever Lorenz' original intention when writing his paper during the Nazi terror in the Third Reich and his apparent attempt to support some Nazi race theories, it obviously is a fundamental fact that the reduction of information an entity (any entity, including a human being) considers before making decisions, dramatically increases the uncertainty in an inevitable and very principal manner. The number of information a given system can hold—as we have been able to show (e.g. section "6.8. The Schwarzschild-Case and the Meaning of the Schwarzschild Radius" and references given there)—is equal to the number of dimensions of this very system. The exclusion of information, be it via degeneration, domestication, ideology-governed "political correctness," false "tolerance" or "politeness," dogmatic narratives as in the current COVID-crisis or simply due to system immanent and entity-typical boundaries, reduces the number of information the entity can hold and digest and thus, increases the uncertainty. This uncertainty increase is on the lowest quantum and thus, most fundamental level of the system in question. Therefore:

(a) Any system-inherent attempt of uncertainty reduction is futile.

(b) However, so it was shown here, increasing the system by incorporating more information (degrees of freedom, properties, dimensions) and thus, thinking outside the box (meaning the system the entity was previously in) is going to lead to success regarding the goal of a reduced uncertainty (see also [1-4]).

[C1] K. Lorenz, Durch Domestikation verursachte Störungen arteigenen Verhaltens, *Zeitschrift für angewandte Psychologie und Charakterkunde 59*, 1940, 1(2), 2–81.

They are the ones who use their lives to become nothing but empty shells. They even consume others' talents and try to destroy what is left of talents in this world. They are producing the ugliest Dorian Gray pictures the world has ever seen and even Oscar Wilde could not have imagined the horror staring into one's face when looking into the Dorian-Gray-Mirror of some of the so-called elite.

So, yes, the answers were—kind of—given to me, but this does not mean that I am not still with you!

3.7.1 References

See "Global References" at the end of this book!

4

Meat

4.1 Love Fields: Just Another Example of Physics in Psychology

If not already, latest in connection with the topic of love (Fig. 4.1), some readers will probably have started to question the possibility to—ever—mathematically describe (and comprehend) the interaction of two human beings in the situation or process usually being referred to as "being in love." This author, however, has often seen that such "resentments" only resulted from a limited knowledge about the physical options this universe provides to everything within. It was observed that many critics, who took it for granted that there is no "physics of love," simply did not know either of the two aspects. And this author, far from blaming anybody for such a perception, rather intends to point towards the world of math and physics as being guilty of neglecting these "soft fields" (perhaps in order to avoid any potential risk of being slandered for trying something out of the ordinary ... no matter how reasonable the trial would be) than to make out the usually mathematically insufficiently trained specialists in all the other disciplines for the lack of rigorousness and falsifiability in their fields of expertise. It had to be concluded—even though not wanting to go as far as Ernest Rutherford, who considered all science either physics or stamp collecting—that the clear "lack of rigorousness and falsifiability" in so many soft fields

Mathematical Psychology: The World of Thoughts as a Quantum Space-Time with a Gravitational Core
Norbert Schwarzer
Copyright © 2025 Jenny Stanford Publishing Pte. Ltd.
ISBN 978-981-5129-27-4 (Hardcover), 978-1-003-58203-8 (eBook)
www.jennystanford.com

is not due to the lack of mathematical tools and physical generality, but due to lack of motivation, phantasy, holistic thinking, also mathematical training and—most definitively—a tremendous excess of ideology, isolationism and protection of the mediocracy. Thereby, of course, some of the behavior can be excused by the fact that the necessary math is relatively complicated and in parts even cumbersome. Not everybody is keen on following the path of a long hard slog. Instead of rigorously following this path, it seems to be so much easier to invest in just producing a few hypotheses about human behavior and then apply them to whatever example comes up. Or, as it is ever more frequently applied today, we use artificial intelligence and try to find behavioral patterns without any need for an underlying theory.

Figure 4.1 Love fields (artist: Laura Kube).

However, there is also a reasonable aspect for keeping the mathematical formalism out of the discussion when it comes to the apparent "soft" fields like human interaction, animosity and love, which lies in the fact that—even though the mathematical apparatus may contain it all—the emotional discussion cannot be held in such a formal language, but requires the use of words in order to be understood in the sense of being "felt." Even in this book, we therefore separate the "talking part" from the math, which we present in appendices R and S. Nevertheless, as the wonderful example of a true "love story" [63] demonstrates, in the end, human interaction, including "this whole love-thing" can be understood as some kind of very complex and multi-dimensional "molecular dynamics" in which the individuals' communication fields interact, form orbitals, boundaries and emanate characteristic (individualized) radiation. This all—of course—happens with respect to all communicational degrees of freedom (dimensions) and this is much more than those nowadays often smart-phone-reduced marionets understand under the word "interaction" or/and "communication." In this book, we will only present the basic mathematical tools (see appendices R and S) and otherwise refer to a follow-up presentation of the complex topic of the "Love Fields" [80].

For demonstration, however, here we consider the character of opposite attributes and the problem of seduction.

4.1.1 Attributes of Apparent Opposition

First, we intend to work out the question of properly taking into account apparent antagonistic field parameters or dimensions like hate and love, for instance. Seeing the two aspects as one property with both, a repulsive and an attractive potential, it becomes immediately clear that this could just be described as one degree of freedom with a positive and a negative side as illustrated in Fig. 4.2.

4.1.2 Protective Quantum Wells

Years ago, this author was asked whether there is a way to mathematically describe the process of seduction (love).

Assuming the usual "happening," where a person (seducer = S) tries to entice another (victim = V) and thereby has to overcome a certain amount of obstacles which physicists like to call "potential walls," we find an excellent starting point in the classical models of the multidimensional quantum well [81, 82]. Here, however, we need to derive the potential of such a well from a completely metric starting point, because we intend to construct our loving pair completely from scratch with just the definition of a number of suitable attributes. In order to do so, we require the metric derivation of potential [83] in such a way that the structural outcome will give us the necessary well [84]. For demonstration, here we derive a few of such structures. Thereby we have to refer to some complex equations which are mainly given in the appendices of this book.

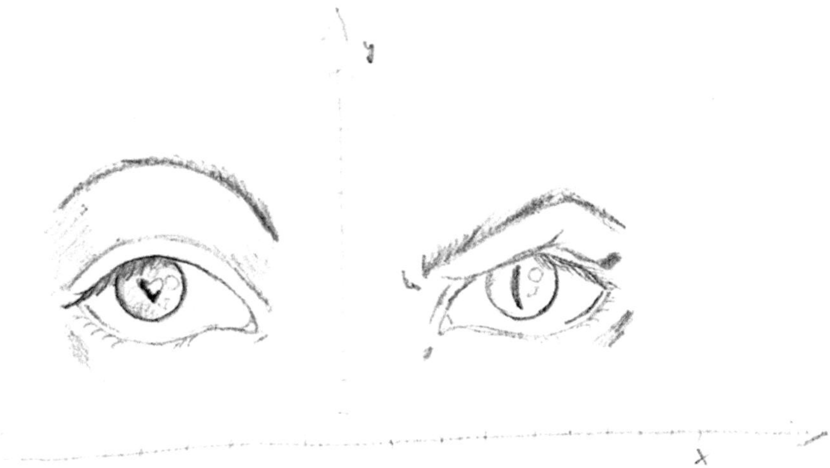

Figure 4.2 From attraction to repulsion or from love to hate—within the mathematical metric description of interhuman interaction these pairing aspects are just one and the same property, degree of freedom or dimension (artist: Laura Kube).

4.1.2.1 Metric quantum well construction

4.1.2.1.1 *Starting with 3 dimensions*

In order to keep things simple, we just investigate the quantum well in one dimension. For this, we want to construct the well from a holistic quantum gravity origin. In order to do this in metric fashion, however, we will need more than just a 1-dimensional space-time. In fact, so it seems, we would need at least 3 dimensions. With respect to the protective individual walls and the love fields, this makes intuitive sense, because there can be no protection without anything being there to protect. In other words, there needs to be something which can produce the protective wall and the quantum gravity approach immediately shows us that—in fact—there is a minimum of other attributes, which have to define (or make out) the object/subject/entity, before the rigorous math even allows for protective potential field for this very object/subject/entity. Mathematically, this is because our space shall have one main property; here we give the coordinate y, an object being only characterized via its position along this property y_0 and a potential function defining the connection between y_0 and y. Our system or pair in love shall have no time ... at least for the moment. Our metric then reads:

$$g_{ij}=\overbrace{F\left[f\left[y,y_0,\varphi\right]\right]}^{\equiv F}\cdot\begin{pmatrix} g_{00} & 0 & 0 \\ 0 & g_{11} & 0 \\ 0 & 0 & g_{22} \end{pmatrix}$$

$$=F\cdot\begin{pmatrix} 1 & 0 & 0 \\ 0 & 1 & 0 \\ 0 & 0 & g_{22}=\overbrace{k\left[y,y_0\right]}^{\equiv k} \end{pmatrix}$$

$$F\left[f\left[y,y_0,\varphi\right]\right]=f\left[y,y_0,\varphi\right]^{\frac{4}{n-2}}$$

$$\xrightarrow{n=3}F\left[\overbrace{f\left[y,y_0,\varphi\right]}^{\equiv f}\right]=f\left[y,y_0,\varphi\right]^4. \tag{10}$$

Even though it should be clear that with respect to the property coordinate y and the position coordinate y_0 the function f should only depend on a term like y-y_0 and k should only depend on y, we better keep things general at the beginning. Inserting this metric into the weak-gravity-scalarized quantum Einstein field equations in the form (5) or (379) (thereby registering that we have already taken care of the linearization condition (377)) gives us:

$$R^*=0$$

$$=\frac{1}{2f^5k^2}\left(\begin{array}{l} f\left(2k\left(k^{(0,2)}+k^{(2,0)}\right)-3k^{(0,1)2}-3k^{(1,0)2}\right) \\ +8k\left(\begin{array}{c} k^{(0,1)}f^{(0,1,0)}+k^{(1,0)}f^{(1,0,0)} \\ -2k\left(k\cdot f^{(0,0,2)}+f^{(0,2,0)}+f^{(2,0,0)}\right) \end{array}\right) \end{array}\right)$$

$$f^{(0,0,q)}=\frac{\partial^q f}{\partial\varphi^q};\quad f^{(0,q,0)}=\frac{\partial^q f}{\partial {y_0}^q};\quad f^{(q,0,0)}=\frac{\partial^q f}{\partial y^q};$$

$$k^{(0,q)}=\frac{\partial^q k}{\partial {y_0}^q};\quad k^{(q,0)}=\frac{\partial^q k}{\partial y^q}. \tag{11}$$

From there, now various protective potentials can be constructed.

4.1.2.1.2 *Constant k-wells*

At first, and this is in the spirit of the classical quantum well treatment, we only consider constant k-functions, giving us:

$$R^{*} = 0 = -8 \cdot \frac{k \cdot f^{(0,0,2)} + f^{(0,2,0)} + f^{(2,0,0)}}{f^{5}}$$

$$\Rightarrow \qquad k \cdot f^{(0,0,2)} + f^{(0,2,0)} + f^{(2,0,0)} = 0. \qquad (12)$$

In the love-field approach, this would probably apply to some rather rigid or stubborn characters ... one may tend to imagine "people with strict principle."

Applying partial separation in f with $f = f_{\varphi}[\varphi]^{*}f_{y}[y,y_0]$, we obtain:

$$k \cdot \frac{\partial^2 f_{\varphi}[\varphi]}{f_{\varphi}[\varphi] \cdot \partial \varphi^2} + \frac{f_y\left[y,y_0\right]^{(0,2)} + f_y\left[y,y_0\right]^{(2,0)}}{f_y\left[y,y_0\right]} = 0. \qquad (13)$$

Setting:

$$\begin{aligned} f_{\varphi}[\varphi] &= C_{-i} \cdot e^{-i \cdot k_c \cdot \varphi} + C_{+i} \cdot e^{+i \cdot k_c \cdot \varphi} \\ &= C_{c2} \cdot \cos\left[k_c \cdot \varphi\right] + C_{s2} \cdot \sin\left[k_c \cdot \varphi\right] \end{aligned} \qquad (14)$$

results in:

$$-k \cdot k_c^2 + \frac{f_y\left[y,y_0\right]^{(0,2)} + f_y\left[y,y_0\right]^{(2,0)}}{f_y\left[y,y_0\right]} = 0. \qquad (15)$$

With $k = \pm 1$ or $k = 0$, we can now switch on or off our constant potential. In the case of $k = 0$, we have a very general solution with the two characteristic functions:

$$f_y\left[y,y_0\right]^{(0,2)} + f_y\left[y,y_0\right]^{(2,0)} = 0$$

$$\Rightarrow \qquad f_y\left[y,y_0\right] = f_y^{\pm}\left[y \pm i \cdot y_0\right] = f_y^{\pm}\left[y \pm \sqrt{-1} \cdot y_0\right]. \qquad (16)$$

In the case $k = \pm 1$, we have to either apply a separation approach or use just another wave-ansatz as follows:

$$f_y\left[y,y_0\right] = f_y\left[c_y \cdot y + c_0 \cdot y_0\right], \qquad (17)$$

leading to the equation:

$$\mp k \cdot k_c^2 + \frac{f_y\left[y,y_0\right]^{(0,2)} + f_y\left[y,y_0\right]^{(2,0)}}{f_y\left[y,y_0\right]} = \mp k \cdot k_c^2 + c_y^2 + c_0^2 = 0$$

$$\Rightarrow \qquad c_y^2 + c_0^2 = \pm k \cdot k_c^2. \qquad (18)$$

Even though our math definitively is by far too simple and too low-dimensionally for true human communication, we already see that we are dealing with waves and oscillations when it comes to describing interhuman interactions.

4.1.2.1.2.1 *Fixing the position by the model/modeler*

With y_0 supposedly giving the position of our fictive object on the general property y, we have to exclude all solutions except the one with the argument $f_y[y, y_0] = f_y[c_y \cdot (y - y_0)]$. This results in the characteristic equation as follows:

$$c_y^2 = \pm\frac{k}{2} \cdot k_c^2 \;\; \Rightarrow \;\; c_y = \pm k_c \cdot \sqrt{\pm\frac{k}{2}}. \tag{19}$$

As solution we obtain:

$$f_y = C_y^- \cdot e^{-k_c \cdot \sqrt{\pm\frac{k}{2}} \cdot (y - y_0)} + C_y^+ \cdot e^{+k_c \cdot \sqrt{\pm\frac{k}{2}} \cdot (y - y_0)}. \tag{20}$$

Here, we discuss the obviously simple cases of purely positive k with the following $k[y]$-structure:

$$k_c = \begin{cases} k_0 & \text{for} \quad 0 \le |y| \le |y_i| & \text{with} \quad 0 < k_0 < k_1 \\ k_1 & \text{for} \quad |y_i| \le |y| \le |y_w| & \text{with} \quad k_1 > k_0 > 0 \,\&\, k_1 > k_2 > 0. \\ k_2 & \text{for} \quad |y_w| < |y| & \text{with} \quad 0 < k_2 < k_1 \end{cases} \tag{21}$$

We see that we have defined a quantum well with a certain wall width of $y_w - y_i$. Outside, which is to say for $y > y_w$, and in the case of non-oscillating solutions ($\pm k \rightarrow +1$), we can only accept asymptotically vanishing functions f, because we assume the object being coded via f of non-infinite size. Thus, from the general solution (20), we can only use:

$$f_y = C_y^- \cdot e^{-k_c \cdot \sqrt{\frac{1}{2}} \cdot (y - y_0)} \quad \text{for} \quad y > y_w. \tag{22}$$

This automatically means that even though we want to fix the object to the center of the well, which would be $y_0 = 0$, the object is also smeared over the outside and its quantum field, which is given by the scaling function $F[f] = f^4$, could be detected there. However, as detection of the field means detection of the object, the innocent (object detecting) observer would assume that the object he has found actually was, respectively, is outside the quantum well. In order to better illustrate this, one might just imagine a prisoner being nicely held behind the bars, but stretching out his arm to the outside world. Now we imagine a blind prison guard, who feels the hand of the prisoner and cannot know whether the inmate is still in his cell, with just his arm reaching out of it, or whether the bad guy has actually found a way to "tunnel through the bars" [85]. He, the poor guard, has to assume the worst and would subsequently assume that the "object" has "tunneled" through the cell boundaries, which—back to our quantum model now—are our quantum well's walls. In fact, the object has not tunneled at all, but simply already was—even if only partially—outside of the well. As long as the well is not constructed in such a way that the object-related part of the outside scaling function has been forced to disappear for $y > y_w$ (e.g., by making the well's walls infinitely high, which is to say $k_1 = \infty$), the object always partially exists also outside the well and can be detected there. With respect to our love-field consideration, things are apparently less non-intuitive as it makes a lot of sense to imagine an individual to "let the world know about its own existence" via the outside radiation of an individual (subjective) field.

4.1.2.1.2.2 *The position truly as degree of freedom*

But is this really everything?

We explicitly did not fix the coordinates y and y_0 to the y-y_0-connection when starting our journey. We only did this setting at the end of the last subsection and if truth be told, we did not feel perfectly right about this. The reason for our uncomfortable feeling in this context results from the fact that it is only us having decided to actually nail the position coordinate y_0 to the property y in the way y-y_0. But what if y_0 was just another degree of freedom and should accordingly be treated like this? After all, in our—by far too simple—3-dimensional model y_0 just stands for the center position of an object ... of something. Understandably this discussion appears a bit awkward with a model only consisting of 3 degrees of freedom and an "object" being no more than a fictive something characterized by its position y_0. Still, and in the tradition of the "cow being described by a point mass model," we may just ask ourselves the "What if?"question and wonder why—on earth—we could not let the position wander around on the property y a bit instead of fixing it via a modeler-decision potentially too early. Regarding the love-field aspect, this would be just similar to nailing an individual to a certain position, standpoint or opinion (depending on what properties of the individual's field are being considered). Let us have a look at the more general equations and subsequent partial solutions (13) to (18). It probably is quite reasonable to fix the position of the quantum well by simply assuming that it is "somehow connected" to something "heavy" (potentially something with mass then, which we discussed further elsewhere [86]) to the coordinate y. A loving person being fixed by some religious principles could provide such a boundary condition in our topic here. For this we assume the discrete quantum well form as given in Eq. (21). We also fixed the function for the coordinate φ as this is well-related and only auxiliary for the construction of this very well, anyway. But what about y_0? Any particular and directly justifiable reason why this should be bonded to y, too? In an attempt to completely liberate the position, we may just apply the factor-separation and set:

$$f_y\left[y,y_0\right]=f_{yy}\left[y\right]\cdot f_{y0}\left[y_0\right], \tag{23}$$

resulting in the equation:

$$\mp k\cdot k_c^2+\frac{f''_{yy}\left[y\right]}{f_y\left[y\right]}+\frac{f''_{y0}\left[y_0\right]}{f_{y0}\left[y_0\right]}\xrightarrow{f_{y0}\left[y_0\right]=e^{c_0\cdot y_0};f_y\left[y\right]=e^{c_y\cdot y}}=\mp k\cdot k_c^2+c_y^2+c_0^2=0$$

$$\Rightarrow\qquad c_y^2+c_0^2=\pm k\cdot k_c^2. \tag{24}$$

This is just the same characteristic equation as we got from Eq. (17). So, no need to bother there any further. We can just stick to our additive argument approach from Eq. (17) and realize that in comparison to the rigid positioning condition with the argument y–y_0, the (17)-approach already provides some liberation of the object's position from the y-property or y-dimension or y-attribute or y-coordinate.

Now we see that the position of the object indeed has partially been determined via the condition:

$$c_y^2+c_0^2=\pm k\cdot k_c^2 \tag{25}$$

in connection with the function:

$$f_y\left[y,y_0\right]=f_y\left[c_y\cdot y+c_0\cdot y_0\right]=C_y^{\pm\pm}\cdot e^{\pm c_y\cdot y\pm c_0\cdot y_0}, \tag{26}$$

but that it still has a huge variety of options to actually "move around" due to Eq. (25) instead of our earlier position fixing with $c_0 = -c_y$. From Eq. (25), we obtain the "liberate" condition for c_0, namely:

$$c_0^2=\pm k\cdot k_c^2-c_y^2\quad\Rightarrow\quad c_0=\pm\sqrt{\pm k\cdot k_c^2-c_y^2}. \tag{27}$$

This gives us three functions f_y for the three regions from Eq. (21):

$$f_y\left[y,y_0\right]=\begin{cases} f_0 & \text{for}\quad 0\leq\left|y\right|\leq\left|y_i\right| & \text{with}\quad C_{y0}^{\pm\pm}\cdot e^{\pm c_y\cdot y\pm\sqrt{k_0^2-c_y^2}\cdot y_0}\\ f_1 & \text{for}\quad \left|y_i\right|\leq\left|y\right|\leq\left|y_w\right| & \text{with}\quad C_{y1}^{\pm\pm}\cdot e^{\pm c_y\cdot y\pm\sqrt{k_1^2-c_y^2}\cdot y_0}\\ f_2 & \text{for}\quad \left|y_w\right|<\left|y\right| & \text{with}\quad C_{y2}^{\pm\pm}\cdot e^{\pm c_y\cdot y\pm\sqrt{k_2^2-c_y^2}\cdot y_0}\end{cases}, \tag{28}$$

where we have set $k = +1$.

4.1.2.1.2.3 *Entangled positioning*

Seeing the property coordinate y and the positioning coordinate y_0 as being entangled, we may change our metric approach as follows:

$$g_{ij} = \overbrace{F\left[f\left[y,y_0,\varphi\right]\right]}^{\equiv F}\cdot\begin{pmatrix} g_{00} & 0 & 0 \\ 0 & g_{11} & 0 \\ 0 & 0 & g_{22} \end{pmatrix}$$

$$= F\cdot\begin{pmatrix} e^{f_0\left[y,y_0\right]} & 0 & 0 \\ 0 & e^{f_0\left[y,y_0\right]} & 0 \\ 0 & 0 & g_{22} = \overbrace{k\left[y,y_0\right]}^{\equiv k} \end{pmatrix}$$

$$F\left[f\left[y,y_0,\varphi\right]\right] = f\left[y,y_0,\varphi\right]^{\frac{4}{n-2}}$$

$$\xrightarrow{n=3} F\left[\overbrace{f\left[y,y_0,\varphi\right]}^{\equiv f}\right] = f\left[y,y_0,\varphi\right]^{4}. \tag{29}$$

Thereby we have chosen the exponential function approach for the pairwise entanglement of y and y_0 in order to obtain a linear partial differential equation in accordance with our condition (377) and its subsequent solution:

$$F\left[f\right] = \begin{cases} C_F\cdot\left(f + C_f\right)^{\frac{4}{n-2}} & n \neq 2 \\ C_F\cdot e^{f\cdot C_f} & n = 2 \end{cases} \tag{30}$$

With the approach for the entanglement function f_0 as follows:

$$f_0\left[y,y_0\right]^{(0,2)} + f_0\left[y,y_0\right]^{(2,0)} = -8\cdot C_0$$

$$\Rightarrow \quad f_0\left[y,y_0\right] = -2\cdot C_0\cdot\left(y - y_0\right)^2 + f_{01}^{+}\left[y + i\cdot y_0\right] + f_{02}^{-}\left[y - i\cdot y_0\right], \tag{31}$$

we obtain the following Ricci scalar for our scaled metric tensor:

$$R^* = 0 = -8\cdot\frac{k\cdot f^{(0,0,2)} + e^{-f_0\left[y,y_0\right]}\cdot\left(f^{(0,2,0)} + f^{(2,0,0)}\right)}{f^5} - \frac{e^{-f_0\left[y,y_0\right]}\cdot\left(f_0^{(2,0)} + f_0^{(0,2)}\right)}{f^4}$$

$$\Rightarrow \quad 0 = \frac{k\cdot f^{(0,0,2)} + e^{-f_0\left[y,y_0\right]}\cdot\left(f^{(0,2,0)} + f^{(2,0,0)}\right)}{f} + e^{-f_0\left[y,y_0\right]}\cdot C_0, \tag{32}$$

where we assumed a constant well-potential again. Reshaping the last line of Eq. (32) into:

$$R^{*}=0=\frac{k\cdot f^{(0,0,2)}}{f}+e^{-f_0[y,y_0]}\cdot\left(C_0+\frac{f^{(0,2,0)}+f^{(2,0,0)}}{f}\right) \quad (33)$$

or:

$$R^{*}=0=e^{f_0[y,y_0]}\cdot\frac{k\cdot f^{(0,0,2)}}{f}+C_0+\frac{f^{(0,2,0)}+f^{(2,0,0)}}{f} \quad (34)$$

shows us that we can either have an eigenvalue equation for the function f with a $e^{2\cdot C_0\cdot(y-y_0)^2+\dots}$-vanishing potential term from Eq. (34), where C_0 has to be negative, leading to:

$$R^{*}=0=e^{-2\cdot|C_0|\cdot(y-y_0)^2+\dots}\cdot\frac{k\cdot f^{(0,0,2)}}{f}+C_0+\frac{f^{(0,2,0)}+f^{(2,0,0)}}{f}, \quad (35)$$

or we obtain a lasting potential in space-time and have a vanishing derivative term for the function f as follows from Eq. (33):

$$R^{*}=0=\frac{k\cdot f^{(0,0,2)}}{f}+e^{-2\cdot C_0\cdot(y-y_0)^2+\dots}\cdot\left(C_0+\frac{f^{(0,2,0)}+f^{(2,0,0)}}{f}\right), \quad (36)$$

in which case the parameter C_0 has to be positive.

Applying the separation approach from above and setting:

$$f_\varphi[\varphi]=C_{-i}\cdot e^{-i\cdot k_c\cdot\varphi}+C_{+i}\cdot e^{+i\cdot k_c\cdot\varphi}=C_{c2}\cdot\cos[k_c\cdot\varphi]+C_{s2}\cdot\sin[k_c\cdot\varphi] \quad (37)$$

again makes Eqs. (35) and (36) to:

$$R^{*}=0=e^{-2\cdot|C_0|\cdot(y-y_0)^2+\dots}\cdot k\cdot k_c^2+C_0+\frac{f_y^{(0,2)}+f_y^{(2,0)}}{f_y} \quad (38)$$

and

$$R^{*}=0=k\cdot k_c^2+e^{-2\cdot C_0\cdot(y-y_0)^2+\dots}\cdot\left(C_0+\frac{f_y^{(0,2)}+f_y^{(2,0)}}{f_y}\right), \quad (39)$$

respectively.

We realize that the position via entanglement offers us a much better understanding of the tunneling [85] or well-trap-escape process, because, when taking Eq. (38), for instance, for a quantum object with a coordinate y_0 assumed to be far away from the position of the potential, the influence of the potential on that very object vanishes with $e^{-2\cdot|C_0|\cdot(y-y_0)^2+\dots}$ and that is nothing else, but having escaped the potential prison. A loving person with strong religious beliefs would then have escaped the dogmatic cage of his or her church.

4.1.2.1.3 *Non-constant well potentials*

Now we briefly want to discuss the general case (11) with non-constant k. This is a very complicated non-linear partial differential equation. With separation approaches, it is possible to achieve some progress with respect to constructing a quantum well, but this leads to a very cumbersome evaluation. Thus, here we refrain from presenting any further evaluations along this path simply because of the volume the whole process requires. We will see that we can obtain much simpler equations in only slightly different spaces with more dimensions. We can also apply stepwise constant well behavior by means of ensembles of Heaviside functions as demonstrated in the appendix.

4.1.2.2 Moving on with 4 dimensions

We introduce the following metric (regarding the interpretation of y and y_0, see the sub-section above):

$$g_{ij} = \overbrace{F\left[f\left[y,y_0,\varphi,\varphi_0\right]\right]}^{\equiv F} \cdot \begin{pmatrix} e^{f_0} & 0 & 0 & 0 \\ 0 & e^{f_0} & 0 & 0 \\ 0 & 0 & k\left[y,y_0\right]^{-1} & 0 \\ 0 & 0 & 0 & \overbrace{k\left[y,y_0\right]}^{\equiv k} \end{pmatrix}$$

$$F\left[f\left[y,y_0,\varphi,\varphi_0\right]\right] = f\left[y,y_0,\varphi,\varphi_0\right]^{\frac{4}{n-2}}$$

$$\xrightarrow{n=4} F\left[\overbrace{f\left[y,y_0,\varphi,\varphi_0\right]}^{\equiv f}\right] = f\left[y,y_0,\varphi,\varphi_0\right]^2. \tag{40}$$

The Ricci scalar of the scaled metric reads:

$$R^* = 0 = -\frac{e^{-f_0}}{2f^3k^2}\left(\begin{array}{c} f\left(k^{(0,1)2} + k^{(1,0)2} + 2\cdot k^2\left(f_0^{(0,2)} + f_0^{(2,0)}\right)\right) \\ +12k\left(e^{f_0} f^{(0,0,0,2)} + k\left(\begin{array}{c} k\cdot e^{f_0} f^{(0,0,2,0)} \\ +f^{(0,2,0,0)} + f^{(2,0,0,0)} \end{array}\right)\right) \end{array}\right)$$

$$f^{(0,0,0,q)} = \frac{\partial^q f}{\partial\varphi_0^q};\quad f^{(0,0,q,0)} = \frac{\partial^q f}{\partial\varphi^q};\quad f^{(0,q,0,0)} = \frac{\partial^q f}{\partial {y_0}^q};$$

$$f^{(q,0,0,0)} = \frac{\partial^q f}{\partial y^q}; \quad k^{(0,q)} = \frac{\partial^q k}{\partial {y_0}^q}; \quad k^{(q,0)} = \frac{\partial^q k}{\partial y^q};$$

$$f_0^{(0,q)} = \frac{\partial^q f_0}{\partial {y_0}^q}; \quad f_0^{(q,0)} = \frac{\partial^q f_0}{\partial y^q}. \tag{41}$$

We see that the conditions for the potential function are getting significantly simpler, because with an approach like:

$$k\left[y, y_0\right] = C_0 + C_1^{\pm} \cdot \left(y \pm i \cdot y_0\right) = C_0 + C_1^{\pm} \cdot \left(y \pm \sqrt{-1} \cdot y_0\right) \tag{42}$$

and our solution for f_0 from Eq. (31) with:

$$f_0\left[y, y_0\right]^{(0,2)} + f_0\left[y, y_0\right]^{(2,0)} = -8 \cdot C_{00}$$

$$\Rightarrow \quad f_0\left[y, y_0\right] = -2 \cdot C_{00} \cdot \left(y - y_0\right)^2 + f_{01}^{+}\left[y + i \cdot y_0\right] + f_{02}^{-}\left[y - i \cdot y_0\right], \tag{43}$$

and the setting $C_{00} = 0$ we would obtain:

$$R^* = 0 = -6\frac{e^{-f_0}}{f^3 k}\left(e^{f_0} f^{(0,0,0,2)} + k\left(k \cdot e^{f_0} f^{(0,0,2,0)} + f^{(0,2,0,0)} + f^{(2,0,0,0)}\right)\right). \tag{44}$$

Now we could use a separation approach and demand $f^{(0,0,0,2)} = 0$, which results in:

$$R^* = 0 = R^* = 0 = -6\frac{e^{-f_0}}{f^3}\left(k \cdot e^{f_0} f^{(0,0,2,0)} + f^{(0,2,0,0)} + f^{(2,0,0,0)}\right)$$

$$\Rightarrow \quad 0 = k \cdot e^{f_0} f^{(0,0,2,0)} + f^{(0,2,0,0)} + f^{(2,0,0,0)}. \tag{45}$$

Still, we have the disadvantage of being forced to a restricted k in accordance with Eq. (42) and not even the option of superposition, because of the non-linearity of this differential equation. We are also not happy with the current structure of the localization (entanglement) function f_0 for y and y_0, because it is only a function of waves and not a clear localization dependency.

4.1.2.3 A much simpler form in 6 dimensions

4.1.2.3.1 *6D: Starting simple*

For simplicity and in order to get some kind of orientation, we start without the position-entanglement function $f_0[y,y_0]$ and consider the following metric:

$$g^{6}_{\alpha\beta} = \begin{pmatrix} 1 & 0 & 0 & 0 & 0 & 0 \\ 0 & 1 & 0 & 0 & 0 & 0 \\ 0 & 0 & g_{22} & 0 & 0 & 0 \\ 0 & 0 & 0 & g_{33} & 0 & 0 \\ 0 & 0 & 0 & 0 & g_{44} & 0 \\ 0 & 0 & 0 & 0 & 0 & g_{55} \end{pmatrix} \times F\left[f\left[y, y_0, \varphi, \varphi_1, \varphi_2, \varphi_3\right]\right];$$

$$F[f] = C_F \cdot \left(f + C_f\right)^{\frac{4}{n-2}} \xrightarrow{n=6} C_F \cdot \left(f + C_f\right)$$

$$g_{22} = \left(g_{\varphi}[\varphi]'\right)^2 \cdot k\left[y, y_0\right]^{i}; \quad g_{33} = \left(g_{\varphi_1}\left[\varphi_1\right]'\right)^2 \cdot k\left[y, y_0\right]^{-i};$$

$$g_{44} = \left(g_{\varphi_2}\left[\varphi_2\right]'\right)^2 \cdot k\left[y, y_0\right]^{1}; \quad g_{55} = \left(g_{\varphi_3}\left[\varphi_3\right]'\right)^2 \cdot k\left[y, y_0\right]^{-1}. \tag{46}$$

Equation (379) yields:

$$0 = -5 \frac{\begin{pmatrix} k \cdot f^{(0,0,0,0,0,2)} + \dfrac{f^{(0,0,0,0,2,0)}}{k} + k^{\mathrm{i}} f^{(0,0,0,2,0,0)} \\ + k^{-\mathrm{i}} f^{(0,0,2,0,0,0)} + f^{(0,2,0,0,0,0)} + f^{(2,0,0,0,0,0)} \end{pmatrix}}{\left(C_f + f\right)^2}$$

$$\Rightarrow$$

$$0 = \begin{pmatrix} k \cdot f^{(0,0,0,0,0,2)} + \dfrac{f^{(0,0,0,0,2,0)}}{k} + k^{\mathrm{i}} f^{(0,0,0,2,0,0)} \\ + k^{-\mathrm{i}} f^{(0,0,2,0,0,0)} + f^{(0,2,0,0,0,0)} + f^{(2,0,0,0,0,0)} \end{pmatrix}$$

$$f^{(0,0,0,0,0,q)} = \frac{\partial^q f}{\partial \varphi_3^q}; \quad f^{(0,0,0,0,q,0)} = \frac{\partial^q f}{\partial \varphi_2^q}; \quad f^{(0,0,0,q,0,0)} = \frac{\partial^q f}{\partial \varphi_1^q}$$

$$f^{(0,0,q,0,0,0)} = \frac{\partial^q f}{\partial \varphi^q}; \quad f^{(0,q,0,0,0,0)} = \frac{\partial^q f}{\partial {y_0}^q}; \quad f^{(q,0,0,0,0,0)} = \frac{\partial^q f}{\partial y^q};$$

$$k^{(0,q)} = \frac{\partial^q k}{\partial {y_0}^q}; \quad k^{(q,0)} = \frac{\partial^q k}{\partial y^q}. \tag{47}$$

Now we use the degrees of freedom for our auxiliary functions and demand:

$$f^{(0,0,0,0,2,0)} = 0; \quad f^{(0,0,0,2,0,0)} = 0; \quad f^{(0,0,2,0,0,0)} = 0, \tag{48}$$

which results in:

$$0 = k \cdot f^{(0,0,0,0,0,2)} + f^{(0,2,0,0,0,0)} + f^{(2,0,0,0,0,0)} \tag{49}$$

and leaves the function k completely free for us to choose in any way we like.

4.1.2.3.2 *Adding mass and energy via more dimensions*

Without much fuss, we point out that more dimensions could be brought into the metric approach in order to properly introduce mass and energy, independent of the potential. For the loving person, namely, the existence of inertia is crucial in order to properly account for the usual reluctance to follow a seductive field.

We consider the following metric:

$$g^{9}_{\alpha\beta} = \begin{pmatrix} -c^2 & 0 & 0 & 0 & \cdots & 0 \\ 0 & 1 & 0 & 0 & \cdots & 0 \\ 0 & 0 & 1 & 0 & \cdots & 0 \\ 0 & 0 & 0 & k^i & \cdots & 0 \\ \cdots & \cdots & \cdots & \cdots & \ddots & 0 \\ 0 & 0 & 0 & 0 & 0 & g_{88} \end{pmatrix}$$

$$\times F\left[f\left[t,y,y_0,\varphi_0,\varphi_1,\varphi_2,\varphi_3,\varphi_4,\varphi_5\right]\right]; \quad F[f] = C_f + f^{\frac{4}{9-2}} = C_f + f^{\frac{4}{7}}$$

$$f\left[t,y,y_0,\varphi_0,\varphi_1,\varphi_2,\varphi_3,\varphi_4,\varphi_5\right] = \begin{pmatrix} f_t[t]\cdot f_y[y]\cdot f_0[y_0]\cdot f_{\varphi 0}[\varphi_0]\cdot f_{\varphi 1}[\varphi_1] \\ \cdot f_{\varphi 2}[\varphi_2]\cdot f_{\varphi 3}[\varphi_3]\cdot f_{\varphi 4}[\varphi_4]\cdot f_{\varphi 5}[\varphi_5] \end{pmatrix};$$

$$g_{44} = k^{-i}; \quad g_{55} = k; \quad g_{66} = k^{-1}; \quad g_{77} = 1; \quad g_{88} = 1$$

$$f_{\varphi_i}\left[g_{\varphi_i}\right] = \text{const} \quad \text{for} \quad i = 0,1,2;$$

$$f_{\varphi_i}\left[g_{\varphi_i}\right] = C_{-i}\cdot e^{-A_i\cdot g_{\varphi_i}} + C_{+i}\cdot e^{+A_i\cdot g_{\varphi_i}} \quad \text{for} \quad i = 3,4,5, \tag{50}$$

where we have now even incorporated a time-coordinate. As it does not contribute much to the tunneling effect, we refrain from demonstrating how our metric can give mass, potential and energy independently and instead refer to our previous publications (e.g., [4, 5, 83, 86]) and the appendices of this book.

4.1.2.3.3 *The localization function f_0*

Changing the metric from sub-section "4.1.2.3.1 6D: Starting simple" to:

$$g_{\alpha\beta}^{6}=\begin{pmatrix} e^{f_0} & 0 & 0 & 0 & 0 & 0 \\ 0 & e^{f_0} & 0 & 0 & 0 & 0 \\ 0 & 0 & g_{22} & 0 & 0 & 0 \\ 0 & 0 & 0 & g_{33} & 0 & 0 \\ 0 & 0 & 0 & 0 & g_{44} & 0 \\ 0 & 0 & 0 & 0 & 0 & g_{55} \end{pmatrix}\times F\left[f\left[y,y_0,\varphi,\varphi_1,\varphi_2,\varphi_3\right]\right],$$

$$F\left[f\right]=C_F\cdot\left(f+C_f\right)^{\frac{4}{n-2}}\xrightarrow{n=6}C_F\cdot\left(f+C_f\right)$$

$$g_{22}=\left(g_{\varphi}\left[\varphi\right]'\right)^2\cdot k\left[y,y_0\right]^i,\quad g_{33}=\left(g_{\varphi_1}\left[\varphi_1\right]'\right)^2\cdot k\left[y,y_0\right]^{-i},$$

$$g_{44}=\left(g_{\varphi_2}\left[\varphi_2\right]'\right)^2\cdot k\left[y,y_0\right]^1,\quad g_{55}=\left(g_{\varphi_3}\left[\varphi_3\right]'\right)^2\cdot k\left[y,y_0\right]^{-1},\qquad (51)$$

but leaving all the settings and demanding f_0 to be:

$$f_0\left[y,y_0\right]^{(0,2)}+f_0\left[y,y_0\right]^{(2,0)}=-5\cdot C_{00}$$

$$\Rightarrow\qquad f_0\left[y,y_0\right]=-\frac{5}{4}\cdot C_{00}\cdot\left(y-y_0\right)^2+f_{01}^{+}\left[y+i\cdot y_0\right]+f_{02}^{-}\left[y-i\cdot y_0\right]\qquad (52)$$

results in:

$$0=k\cdot f^{(0,0,0,0,0,2)}-C_{00}\cdot f+e^{-f_0}\cdot\left(f^{(0,2,0,0,0,0)}+f^{(2,0,0,0,0,0)}\right)$$

$$\xrightarrow{f_{\varphi 3}\left[\varphi_3\right]=C_{-i}\cdot e^{-i\cdot k_c\cdot\varphi_3}+C_{+i}\cdot e^{+i\cdot k_c\cdot\varphi_3}=C_{c2}\cdot\cos\left[k_c\cdot\varphi_3\right]+C_{s2}\cdot\sin\left[k_c\cdot\varphi_3\right]}$$

$$0=k\cdot k_c^2\cdot f-C_{00}\cdot f+e^{-f_0}\cdot\left(f^{(0,2,0,0,0,0)}+f^{(2,0,0,0,0,0)}\right)$$

$$\xrightarrow{f=f\left[y,y_0\right]}$$

$$0=\left(k\cdot k_c^2-C_{00}\right)\cdot f+e^{-f_0}\cdot\left(f^{(0,2)}+f^{(2,0)}\right).\qquad (53)$$

We see that we have now obtained a very "clean" eigenvalue equation with potential directly from our metric (51). In addition, we have a localization function f_0, for which the same reasoning holds as discussed in connection with our approach in 3 dimensions and Eqs. (31) to (36).

When now assuming the function f to only depend on the coordinate y, we obtain the classical 1-dimensional Schrödinger equation, only that we also have a positioning dependency due to f_0. Regarding typical quantum well problems with respect to such an equation, typically reading ($f_0 = 0$):

$$0=\left(k\cdot k_c^2-C_{00}\right)\cdot f+e^{-f_0}\cdot f''$$

$$\xrightarrow{f_0\to\approx 0}$$

$$0=\left(k\cdot k_c^2-C_{00}\right)\cdot f+\left(1-f_0+\frac{f_0^2}{2}-\ldots\right)\cdot f''$$

$$\Rightarrow \qquad 0=\left(k\cdot k_c^2-C_{00}\right)\cdot f+f'', \tag{54}$$

we refer to the textbook literature (e.g., [67, 81, 82]).

4.1.2.4 Localization of an object/happening/wave-structure brings mass and/or energy

Here, we are more interested in the metric effect of the localization function $f_0=f_0[y,y_0]$. We see that this in fact "localizes" either the potential term with a C_{00} being negative or the derivative terms of f when we have $C_{00}>0$ (see the discussion around Eqs. (31) to (36)). We also realize that C_{00} is just giving the characteristics of a mass or energy term in the classical Schrödinger or Klein–Gordon equation:

$$0=\left(k\cdot k_c^2-\overbrace{C_{00}}^{energy/mass}\right)\cdot f+f''. \tag{55}$$

This shows us that the functional connection (entanglement) of dimensions can produce mass or energy. In our case, the entanglement is between the main property and the localization of a "happening" (a wave) on this property. Without this localization, our system would have no mass-energy term. Regarding our loving entity, this is just equivalent to saying that principles in our mind give us some inertia against forces aiming against these principles. In [87] it was also shown how the localization of an object directly leads to the particle-wave duality.

In addition, in dependence on the structure of f_0, parts of it may also appear to us as potential. So, in case $k_c=0$, we may rewrite the last line in Eq. (53) as follows:

$$0=e^{f_0}\cdot C_{00}\cdot f-\left(f^{(0,2)}+f^{(2,0)}\right), \tag{56}$$

where the first addend $e^{f_0}\cdot C_{00}\cdot f$ is the equivalent for a classical potential, while with an eigenvalue setting of:

$$f^{(0,2)}=\left(E+m\cdot c^2\right)\cdot f, \tag{57}$$

we could also have mass or energy or both. The resulting equation would then read:

$$0=\left(e^{f_0}\cdot C_{00}-E-m\cdot c^2\right)\cdot f\left[y\right]-\frac{\partial^2 f\left[y\right]}{\partial y^2}, \tag{58}$$

which is again a 1-dimensional Schrödinger equation with potential, mass and energy, completely obtained from a metric set-up. However, with the Schrödinger equation popping up in connection with our love-fields, it seems to be of little wonder why there apparently is such a great similarity between love and the strangeness of the quantum world.

4.1.3 The Process of Seduction

Now, after having constructed a suitable potential well, we move back to our pair of S and V. Here, the situation was that S has tried to push V up its own protective wall (the well's boundaries) by creating some sort of excitement field.

When having reached a certain level, thereby passing the top of the potential wall, the "victim" simply cannot help it anymore (point of no return) and—to make it plastic—S has V in the bag. Just as in the classical Quantum Theory we also find a certain likelihood for V reaching his (or her) goal also with an excitement state significantly below the maximum value of the protective potential, which is just the well-known tunneling effect [85].

S may only blow it by doing something really stupid (thereby kind of waking V up) or V might still be able to stop it, because he/she has previously organized preset attention points and control-instances at certain positions of his/her very own "seduction"-potential or excitement curve in order stop the process on its own ... even recognize the seduction after the hypnosis has already set in.

In idealized form the simplest seduction curve without any attention points or control-instances may be drawn as shown in Fig. 4.3.

Classically, one would just say that a continuous increase of the excitement level beyond the point of no return leads S to success.

Now something very interesting occurs. When doing the fundamental individual math, which is to say the analytical description of individual fields, and allowing for photonic interaction, which is just the exchange of information among the individuals, one can see that only discrete states (and frequencies) can be exchanged.

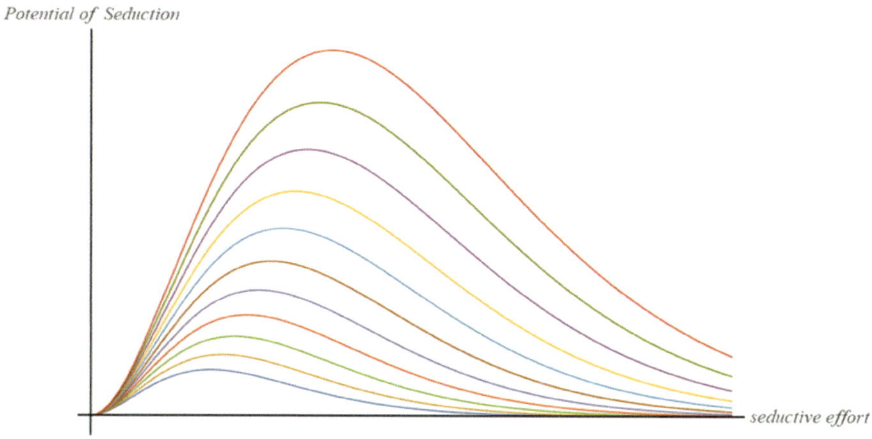

Figure 4.3 The potential of seduction for a variety of "seducers," ranging from Brad Pitt (blue) to Quasimodo (red), for the same "victim."

When now evaluating the dependency of the total radiation on the frequency (seductive effort) of the individuum of a certain set of excitement states, one—once again—gets a set of curves just as given in Fig. 4.3.

Thus, our excitement curve is not at all continuous but rather discrete ... it moves in little leaps and not in a continuous crawl.

And above that, this is just Planck's radiation law.

Funny, isn't it?!

4.1.4 References

See "Global References" at the end of this book!

4.2 More Quantum Gravity Topics and Their Bearing on Mathematical Psychology

4.2.1 Hawking Radiation and Why Thoughts Cannot Be Imprisoned ... At Least Not Forever

In appendix R we derive the Hawking radiation around a Black Hole [1] directly from our quantum gravity field equations [2]. For this author the important bearing on "Quantum Psychology" is the evidence that the metric understanding of the Hawking radiation in spaces or space-times consisting of ensembles of attributes with any arbitrary property automatically has some significant application with respect to the psychological aspect of escaping mental traps, thought cages and mind prisons. It also provides the necessary understanding for the "feeling" of individuals emanating property or attributes fields in wave-forms, which are subsequently detectable by other individuals.

In somewhat clearer words: We can just say that every system, which is capable of developing scaling waves (scaling, which referrers to the metric scaling factor $F[f]$ in Eq. (4), thus, not scalar) to its own metric existence, always has the potential to escape even Black Hole attractors. As we think that the world of thought and fabric of mind is such a system, we have to conclude that there is no way to trap and confine certain ideas indefinitely. Eventually, they are bound to escape in one way or the other. Politicians and decision makers should bear that in mind, when thinking about various forms of totalitarianism, mind control, new currencies and all those other things they try to force upon us in the so-called "great reset," the "climate protection hoax," "unexpected" resource shortages or "health reasons," which are only originating from planned and orchestrated pandemics.

4.2.2 Einstein's Moon and the Theory of Perspectivity or How Do We See Other Individuals' Informational Output

It is said that—once more—when criticizing certain and rather subjective aspects of Quantum Theory, Einstein offered the following extreme illustration to one of his colleagues:

"Do you really believe that the moon isn't there when nobody looks?"

It is further said that Einstein said this while looking at the moon himself.

In this short section, we will show how a quantum-gravity-based "Theory of Perspectivity" might help to solve the old riddle of Einstein's problem with the moon. We will also show that in a certain sense, the Genius was kind of right and that his insight may also help us to understand the fact that no matter how hard we will try to understand our fellow next, we will never be able to truly comprehend the meaning of his informational output as it was meant by the source, which is to say, the "out-putter."

So far, we have shown that a simple scaled metric subjected to a Hamilton minimum principle via the Einstein–Hilbert action [3] gives us not only the classical General Theory of Relativity of Einstein [4] but also Quantum

Theory [5, 6]. For proof, we have derived in our previous publications [5, 6, 7, 8, 9, 10] the essential quantum equations, Klein–Gordon, Schrödinger [11, 12] and Dirac [13]. As a bonus, we also saw the option for a combined Schrödinger–Dirac equation and introduced a variety of alternatives for the extraction of the Dirac equation (see the appendices). As these important quantum equations already fell out of the simplest scaling (119) and perturbation options (385), and thus, leaving us with many more possible variations and extensions like quantum and gravity centers, we conclude that we indeed have found a Quantum Gravity Theory. Thereby, we need to point out that we only have discovered this general theory inside an already existing one (namely Hilbert's and Einstein's) by a bit of playing with the theory's inner degrees of freedom. By also extending the variational degrees of freedom of the existing theory of Hilbert and Einstein [3, 4], we were able to extract conditions for the determination of the critical size of systems. We also saw that these additional degrees of freedom provide the answer for the existence of a very fundamental uncertainty or quantum source.

Before we now want to discuss the question about the subjectivity of Quantum Theory, we need to obtain a brief overview of the achievements of our quantum gravity approach.

Thus, the following points describe what we have extracted so far (see [5–10, 14–29] and appendices of this book for proof, derivation, falsification and discussion):

- We found that the diffusion equation, which is one of THE evolution equations, falls directly out of the Einstein–Hilbert action when just considering the volume behavior of a system.
- We saw that either first-order evolution equations (first order in time) are extracted from their second-order counterparts via a Klein–Gordon to Schrödinger transition process or these first-order equations have to be derived from our fundamental starting point in such a way that we always obtain Dirac or Dirac-like equations. This means, however, that the concept of bosonic and fermionic behavior is a general one and can occur in every system of a sufficiently high number of dimensions. In other words, effects like the Bose–Einstein condensate, the Pauli exclusion or the strange symmetry behavior of spin-1/2 objects are—obviously—by no means restricted to elementary particles.
- We have found that the fundamental approach hands us a tool which we can use to check on the dimensional completeness of models for the most comprehensive and truly holistic investigation of systems.

- Almost as a by-product of the latter derivation, we found a way to geometrically derive the Schwarzschild radius of a massive gravitational object of symmetry of revolution. Please note that in the classical theory, this Schwarzschild radius was never derived from a first principle but was adjusted as a parameter to the Schwarzschild metric in order to give the correct limit to the Newton gravitational law.
- Along the way, and here is what—once again—associates our finding with rather practical, down-to-earth applications, we are able to connect information with scale and mass and—based on this connection—find metric solutions, which may be considered as the universal storage for bits with an inverse mass-bit relation, thereby solving the so-called information crisis problem.
- We found that matter, which had to be postulated in both the Einstein and the classical Quantum Theory, automatically occurs when a given system, described by a metric, has been subjected to a volumetric variation or scaling. In other words: One may say that we now know what—metrically—is behind the postulated mass or matter terms in many classical theories.
- Via the fundamental properties of the Riemann and Ricci tensors (the Bianchi identity), we are able to give an evolution equation for the matter of any arbitrary system. This fundamental equation is a third-order differential equation.
- From this equation, we extracted a possible answer to the so-called 3-generation problem of elementary particles and we explicitly derive and give the corresponding equation, having 3 mass solutions to the same metric particle.
- We found that mass occurs in a process we may just see as dimensional entanglement.
- Even for rather phenomenological processes like diffusion, our findings show that in addition to the known classical equations, which we may see as ordinary matter equations, there always seems to exist an anti-matter counterpart.
- From a generalized Einstein–Hilbert action, we derive a Dirac quantum gravity field equation.

And last but not least, with the many options to derive quantum reality from the Hamilton extremal principle as we have done (see, e.g., the appendices of this book and our publications [5–6, 14–29]), we also see that the "... subjective elements of quantum mechanics or of whether, as Einstein famously asked, the Moon exists when nobody is looking at it" [30] seem to be as worse as Einstein's question suggests [14]. This automatically has some severe impact on the question of the fundamental uncertainty of not only seeing reality but also how we perceive other people's information. There simply is no chance that we can perfectly mirror what others intend to tell or potentially teach us. We can only try to come as close as possible to the meaning they (the source) might have in mind when conveying a certain set of information to us. The reason for this fundamental, universal, omnipresent and inescapable "misunderstanding" is not only the uncertainty residing in the communicated information and the communication line itself but also the perspective of the receiver. Thereby "perspective" here stands for the many possible ways to process the set of information in oneself's "internal minimum calculator" for digesting information. Even if this processor works optimally, there are too many options for its actual structural organization, intrinsic characteristics and paths for the mathematical processing so that the outcome for one and the same set of input will almost never lead to the same output. This is the inbuilt subjectivity of the universe and it reflects upon us all in the way we see each other, namely everybody sees everybody differently. Everyone interprets everything differently and there are no such things as identical conscious entities in this universe. Patients and psychologists alike should bear that in mind when doing an anamnesis or—even more so—when suggesting a certain treatment.

In other words: Getting "a second opinion" is already ensured when just asking the same doctor about the same thing a second time.

4.2.3 Interpretation Work

We have already worked out (e.g., in connection with the love-fields above) how the metric understanding of quantum effects in connection with the treatment of individuality and mind processes as field dynamics offers a new way to grasp and mathematically handle psychological aspects. Thereby we simply need to follow our strongest sense, the optical one and learn to apply the geometric meaning behind the Quantum Gravity Theory. This theory, namely, is just geometry ... only that such geometry works in many dimensions and on arbitrary attributes making out, respectively forming these dimensions. We start our interpretation work with just illustrating the classical Quantum Theory within the new metric picture, which was derived and discussed in [18]:

After deriving all fundamental quantum equations from the Einstein–Hilbert action, we have learned that Quantum Theory is just a part of the General Theory of Relativity. In fact, it is the volume part of the metric and the action-kernel scaling of the space-time curvature and shows itself inside the Einstein theory via a scaled metric tensor. The latter is subjected to the usual Hamilton treatment via the Einstein–Hilbert action. With this perception, the interpretation of the often rather strange results of Quantum Theory is now becoming understandable in a geometric way.

What classically appears as a wave function in Quantum Theory is just the volume or scale fluctuation of the metric in a Quantum Gravity Theory. Thereby the classical wave function determines the metric volume perturbation and vanishes in the far distance of the quantum object under consideration. As the metric, respectively its volume is usually much bigger than the fluctuations of the same and as the quantum effects tend to wear off rather quickly, Quantum Theory only becomes important at smaller scales (usually, which is to say under ordinary (regarding our reality) energy, temperature and gravity conditions).

Furthermore, what classically appears as potential, energy and masses is just the entanglement of dimensions.

Now, instead of thinking about space and time, we simply apply the attributes that we use to describe and characterize the human mind and immediately realize that all "mind-stuff" can just be seen as fields (quantum gravity fields, of course) with the same apparently strange behavior as we know it from classical Quantum Theory ... only that we now can understand it as geometrical dynamics of the characteristic attributes and the fields they form.

4.2.4 Metric Potential

Obtaining what physicists call a potential in a metric way [19] is important with respect to the description and modeling of the psychological aspect of influencing another person's mind [20] and implanting certain ideas, wishes, desires, addictions and so on [31]. As such aspects are of great importance for the optimization of advertisement, mass formation, deception on all scales, etc., we will consider psychological tunneling, the psychological well separately further below in this section. The math has been described in the corresponding references already given above.

4.2.5 The Three Generations of the Mind

It was shown in [24] that each system, the moment we can describe it via a Riemann space-time theory embedded in a Hamilton minimum principle of the Einstein–Hilbert style [3], has 3 principal states resulting from a fundamental differential equation of third order for this very system (each system!).

As this also holds for systems with intelligence, one automatically has to ask the question whether there are states also for the mind ... and what consciousness might have to do with it. As the latter part of the question probably is as old as mankind (Fig. 4.4), here we will go to some length to try to answer it. At least—so this author is convinced—the moment the first human beings started to use fire to warm themselves and spend the evenings huddled together around the fireplace, sooner or later the questions must have been asked:

"What is it to make us feel our existence?"

Figure 4.4 "What is consciousness?" by Livia Schwarzer (see [36] for more).

"What is this thing, which makes me realize that I am and that mirrors this funny and sometimes fuzzy thing I call 'self' or 'myself'?!"

In other words:

The question "What actually is consciousness?" is most likely at least one million years old.

Even though we may not be able to answer the question "what consciousness actually is" in a satisfying manner, we could still try to investigate some of the properties a conscious system should have. Thereby we will simply use mathematical identities in order to find the most general laws for such "intelligent entities." In this chapter, we will show that one of such principles lies in the fact that the so-called three-generation problem of elementary particles [37] also has a bearing on the fundamental states intelligent systems can exist in. Assuming that only a quantum gravity or theory-of-everything approach can describe a conscious entity, we already have our fundamental starting point. However, as the three-generation problem of elementary particles [37] has been considered one of the most important unsolved problems in physics, we cannot just claim to have solved it (or rather think to have found a path to solve this problem) and then move on to interpret the impact of our finding on other fields, but we better present everything necessary to justify our claim and allow its falsification. And this is exactly what we will do in this chapter even though it might lead to some redundancy within the book as a whole.

4.2.5.1 Fundamental starting point

Our starting point shall be the Hamilton principle in form of the Einstein–Hilbert action [3]:

$$\delta W = 0 = \delta \int_V d^n x \left(\sqrt{-g} \cdot \Phi_R [R] \right) = \delta \int_V d^n x \left(\sqrt{-g} \cdot \left(R - 2\Lambda + L_M \right) \right). \tag{59}$$

Here R gives the Ricci scalar and g denotes the determinant of the metric tensor $g_{\alpha\beta}$. The term L_M stands for the matter Lagrange density, which was postulated by both Hilbert and Einstein, and Λ gives the cosmological constant. Assuming that we can extract a functional factor $F[f]$, with f being a function of all coordinates, from the metric tensor in the following way:

$$G_{\alpha\beta} = g_{\alpha\beta} \cdot F[f], \tag{60}$$

it was shown in [5, 6] that we do not only obtain the Einstein field equations [4] but also the most important classical quantum equations. Here, we only briefly repeat the essentials.

Setting Eq. (60) into Eq. (59), we obtain the following Einstein–Hilbert action [6]:

$$\delta W = 0 = \delta \int_V d^n x \left(\sqrt{-g \cdot F^n} \times \left(\left(R - \frac{1}{2F} \begin{pmatrix} 2F_{,\alpha\beta}(n-1) g^{\alpha\beta} + 2\Gamma^a_{\alpha\beta} F_{,a} g^{\alpha\beta} \\ -F_{,\alpha} g^{ab} g_{\beta b,a} g^{\alpha\beta} - F_{,\beta} g^{ab} g_{\alpha b,a} g^{\alpha\beta} \\ -n\Gamma^d_{\alpha\beta} F_{,d} g^{\alpha\beta} + \frac{n}{2} F_{,d} g^{cd} g_{ab,c} g^{ab} \end{pmatrix} \frac{1}{F} \atop - \frac{F_{,\alpha} \cdot F_{,\beta}}{4F^2} g^{\alpha\beta} \left((n-6)(n-1) \right) \right) \atop -2\Lambda + L_M \right) \right). \tag{61}$$

Thereby, in order to keep things simple and still general enough, one can often restrict the metric selection to diagonal metrics with the following properties:

$$g_{ij} = \begin{pmatrix} g_{00} & \cdots & 0 \\ \vdots & \ddots & \vdots \\ 0 & \cdots & g_{n-1n-1} \end{pmatrix}; \quad g_{jj,j} = 0, \tag{62}$$

which gives an extended Hilbert integral:

$$\delta W = 0 = \delta \int_V d^n x \left(\sqrt{-g \cdot F^n} \times \left(\left(R + \frac{F'}{F}(1-n)\Delta f \atop + \frac{f_{,\alpha} f_{,\beta} g^{\alpha\beta} (1-n)}{4F^2} \left(4FF'' + (F')^2 (n-6) \right) \right) \frac{1}{F} \atop -2\Lambda + L_M \right) \right). \tag{63}$$

Here, we will not use this simplification. After performing the variation in Eq. (61), we obtain the following quantum gravity field equations:

$$0=\left(\begin{array}{c}\boxed{R_{\alpha\beta}-\frac{g_{\alpha\beta}}{2}R}+\frac{g_{\alpha\beta}}{2}\left(\frac{1}{2F}\left(\begin{array}{c}2F_{,ij}(n-1)g^{ij}+2\Gamma_{ij}^{a}F_{,a}g^{ij}\\-F_{,i}g^{ab}g_{jb,a}g^{ij}-F_{,j}g^{ab}g_{ib,a}g^{ij}\\-n\Gamma_{ij}^{d}F_{,d}g^{ij}+\frac{n}{2}F_{,d}g^{cd}g_{ab,c}g^{ab}\end{array}\right)+\frac{F_{,i}\cdot F_{,j}}{4F^{2}}g^{ij}\left((n-6)(n-1)\right)\right)\\-\frac{1}{2F}\left(\begin{array}{c}F_{,\alpha\beta}(n-2)+F_{,ab}g_{\alpha\beta}g^{ab}\\+F_{,a}g^{ab}\left(g_{\beta b,\alpha}-g_{\beta\alpha,b}\right)-F_{,\alpha}g^{ab}g_{\beta b,a}-F_{,\beta}g^{ab}g_{\alpha b,a}\\+F_{,d}g^{cd}\frac{1}{2}n\left(\frac{2}{n}g_{\alpha c,\beta}-g_{\alpha c,\beta}-g_{\beta c,\alpha}+g_{\alpha\beta,c}+\frac{1}{n}g_{\alpha\beta}g_{ab,c}g^{ab}\right)\\-\frac{1}{2F}\left(F_{,\alpha}\cdot F_{,\beta}(3n-6)+g_{\alpha\beta}F_{,c}F_{,d}g^{cd}(4-n)\right)\end{array}\right)\\+\kappa T_{\alpha\beta}+\Lambda\cdot F\cdot g_{\alpha\beta}\end{array}\right). \tag{64}$$

Immediately, we recognize the classical Einstein field equations (box) inside our new ones. It was shown in [4, 6] that the scaling terms in Eq. (64) are giving the Klein–Gordon, Dirac and Schrödinger equation (regarding the latter see also [7]). Under the variational integral, the integrand (64) has to be multiplied with the variated metric tensor $\delta G^{\alpha\beta}=\delta\left(\frac{g^{\alpha\beta}}{F}\right)$ and therefore could be split up into two equations:

$$0=\left(\begin{array}{c}\boxed{R_{\alpha\beta}-\frac{g_{\alpha\beta}}{2}R}+\frac{g_{\alpha\beta}}{2}\left(\frac{1}{2F}\left(\begin{array}{c}2F_{,ij}(n-1)g^{ij}+2\Gamma_{ij}^{a}F_{,a}g^{ij}\\-F_{,i}g^{ab}g_{jb,a}g^{ij}-F_{,j}g^{ab}g_{ib,a}g^{ij}\\-n\Gamma_{ij}^{d}F_{,d}g^{ij}+\frac{n}{2}F_{,d}g^{cd}g_{ab,c}g^{ab}\end{array}\right)+\frac{F_{,i}\cdot F_{,j}}{4F^{2}}g^{ij}\left((n-6)(n-1)\right)\right)\\-\frac{1}{2F}\left(\begin{array}{c}F_{,\alpha\beta}(n-2)+F_{,ab}g_{\alpha\beta}g^{ab}\\+F_{,a}g^{ab}\left(g_{\beta b,\alpha}-g_{\beta\alpha,b}\right)-F_{,\alpha}g^{ab}g_{\beta b,a}-F_{,\beta}g^{ab}g_{\alpha b,a}\\+F_{,d}g^{cd}\frac{1}{2}n\left(\frac{2}{n}g_{\alpha c,\beta}-g_{\alpha c,\beta}-g_{\beta c,\alpha}+g_{\alpha\beta,c}+\frac{1}{n}g_{\alpha\beta}g_{ab,c}g^{ab}\right)\\-\frac{1}{2F}\left(F_{,\alpha}\cdot F_{,\beta}(3n-6)+g_{\alpha\beta}F_{,c}F_{,d}g^{cd}(4-n)\right)\end{array}\right)\\+\kappa T_{\alpha\beta}+\Lambda\cdot F\cdot g_{\alpha\beta}\end{array}\right)\delta G^{\alpha\beta}$$

$$
=\left(\begin{array}{c}\boxed{R_{\alpha\beta}-\frac{g_{\alpha\beta}}{2}R}+\frac{g_{\alpha\beta}}{2}\left(\frac{1}{2F}\left(\begin{array}{c}2F_{,ij}(n-1)g^{ij}+2\Gamma_{ij}^{a}F_{,a}g^{ij}\\ -F_{,i}g^{ab}g_{jb,a}g^{ij}-F_{,j}g^{ab}g_{ib,a}g^{ij}\\ -n\Gamma_{ij}^{d}F_{,d}g^{ij}+\frac{n}{2}F_{,d}g^{cd}g_{ab,c}g^{ab}\end{array}\right)+\frac{F_{,i}\cdot F_{,j}}{4F^{2}}g^{ij}\left((n-6)(n-1)\right)\right)\\ -\frac{1}{2F}\left(\begin{array}{c}F_{,\alpha\beta}(n-2)+F_{,ab}g_{\alpha\beta}g^{ab}\\ +F_{,a}g^{ab}\left(g_{\beta b,\alpha}-g_{\beta\alpha,b}\right)-F_{,\alpha}g^{ab}g_{\beta b,a}-F_{,\beta}g^{ab}g_{\alpha b,a}\\ +F_{,d}g^{cd}\frac{1}{2}n\left(\frac{2}{n}g_{\alpha c,\beta}-g_{\alpha c,\beta}-g_{\beta c,\alpha}+g_{\alpha\beta,c}+\frac{1}{n}g_{\alpha\beta}g_{ab,c}g^{ab}\right)\\ -\frac{1}{2F}\left(F_{,\alpha}\cdot F_{,\beta}(3n-6)+g_{\alpha\beta}F_{,c}F_{,d}g^{cd}(4-n)\right)\end{array}\right)\\ +\kappa T_{\alpha\beta}+\Lambda\cdot F\cdot g_{\alpha\beta}\end{array}\right)\left(\begin{array}{c}g^{\alpha\beta}\delta\left(\frac{1}{F}\right)\\ +\frac{\delta g^{\alpha\beta}}{F}\end{array}\right), \tag{65}
$$

$$
0=\left(\left(1-\frac{n}{2}\right)\left(R-\left(\frac{1}{2F}\left(\begin{array}{c}2F_{,ij}(n-1)g^{ij}+2\Gamma_{ij}^{a}F_{,a}g^{ij}\\ -F_{,i}g^{ab}g_{jb,a}g^{ij}-F_{,j}g^{ab}g_{ib,a}g^{ij}\\ -n\Gamma_{ij}^{d}F_{,d}g^{ij}+\frac{n}{2}F_{,d}g^{cd}g_{ab,c}g^{ab}\end{array}\right)+\frac{F_{,i}\cdot F_{,j}}{4F^{2}}g^{ij}\left((n-6)(n-1)\right)\right)\right)+\Lambda\cdot F\cdot n\right)\delta\left(\frac{1}{F}\right)
$$

$$
0=\left(\begin{array}{c}R_{\alpha\beta}-\frac{g_{\alpha\beta}}{2}R+\frac{g_{\alpha\beta}}{2}\left(\frac{1}{2F}\left(\begin{array}{c}2F_{,ij}(n-1)g^{ij}+2\Gamma_{ij}^{a}F_{,a}g^{ij}\\ -F_{,i}g^{ab}g_{jb,a}g^{ij}-F_{,j}g^{ab}g_{ib,a}g^{ij}\\ -n\Gamma_{ij}^{d}F_{,d}g^{ij}+\frac{n}{2}F_{,d}g^{cd}g_{ab,c}g^{ab}\end{array}\right)+\frac{F_{,i}\cdot F_{,j}}{4F^{2}}g^{ij}\left((n-6)(n-1)\right)\right)\\ -\frac{1}{2F}\left(\begin{array}{c}F_{,\alpha\beta}(n-2)+F_{,ab}g_{\alpha\beta}g^{ab}\\ +F_{,a}g^{ab}\left(g_{\beta b,\alpha}-g_{\beta\alpha,b}\right)-F_{,\alpha}g^{ab}g_{\beta b,a}-F_{,\beta}g^{ab}g_{\alpha b,a}\\ +F_{,d}g^{cd}\frac{1}{2}n\left(\frac{2}{n}g_{\alpha c,\beta}-g_{\alpha c,\beta}-g_{\beta c,\alpha}+g_{\alpha\beta,c}+\frac{1}{n}g_{\alpha\beta}g_{ab,c}g^{ab}\right)\\ -\frac{1}{2F}\left(F_{,\alpha}\cdot F_{,\beta}(3n-6)+g_{\alpha\beta}F_{,c}F_{,d}g^{cd}(4-n)\right)\end{array}\right)\\ +\kappa T_{\alpha\beta}+\Lambda\cdot F\cdot g_{\alpha\beta}\end{array}\right)\frac{\delta g^{\alpha\beta}}{F}. \tag{66}
$$

While we consider the second equation of the pair in Eq. (66) generalized or quantum Einstein field equations, we used the first one to derive the classical quantum equations (Klein–Gordon, Dirac and Schrödinger) for a generally curved space-time. Demanding the "gravity" to be weak as follows:

$$\delta G^{\alpha\beta} = G^{\alpha\beta} \cdot \delta_0 + \overbrace{G^{ab} \delta_{ab}^{\alpha\beta}}^{\text{Gravity}} \xrightarrow{\forall\, \delta_{ab}^{\alpha\beta} \ll \delta_0} = \frac{g^{\alpha\beta}}{F} \cdot \delta_0, \tag{67}$$

we are left with the scalar and purely metrically obtained quantum equation, which, when excluding the trivial case $n = 2$, reads:

$$0 = \left(R - \left(\begin{array}{c} \frac{1}{2F} \left(\begin{array}{c} 2F_{,ij}(n-1)g^{ij} + 2\Gamma_{ij}^{a} F_{,a} g^{ij} \\ -F_{,i} g^{ab} g_{jb,a} g^{ij} - F_{,j} g^{ab} g_{ib,a} g^{ij} \\ -n\Gamma_{ij}^{d} F_{,d} g^{ij} + \frac{n}{2} F_{,d} g^{cd} g_{ab,c} g^{ab} \end{array} \right) \\ + \frac{F_{,i} \cdot F_{,j}}{4F^2} g^{ij} \left((n-6)(n-1) \right) \end{array} \right) \right) + \Lambda \cdot F \cdot n. \tag{68}$$

From there, which is to say, based on the latter equation, all classically important equations can be obtained.

Going back to the variational integral (59), but keeping the scaled metric gives us:

$$\delta_G W = 0 = \delta_G \int_V d^n x \left(\sqrt{-G} \cdot \Phi \cdot R^* \right)$$

$$\delta W = 0 = \left[\begin{array}{c} \int_V d^n x \left(\sqrt{-G} \cdot \left(\begin{array}{c} \Phi \cdot R^*_{\mu\nu} - \frac{1}{2} \Phi \cdot R^* \cdot G_{\mu\nu} \\ + \Lambda G_{\mu\nu} - \left(\nabla_\mu \nabla_\nu - G_{\mu\nu} \Delta_G \right) \Phi \end{array} \right) \right) \delta G^{\mu\nu} \\ - \int_{\text{Surface}} d^{n-1} y \left(\sqrt{|h|} \cdot \varepsilon \cdot \Phi \cdot N^\lambda h^{\mu\nu} \partial_\lambda \left(\delta G_{\mu\nu} \right) \right) \end{array} \right]. \tag{69}$$

Thereby we have assumed:

$$\delta W = 0 = \delta \int_V d^n x \left(\sqrt{-G} \cdot \left(\Phi \cdot R^* - \overbrace{2\Lambda + L_M}^{=0} \right) \right) \tag{70}$$

and used the results from [38] within the derivation.

The usual, and rather tedious, path forward would be the discussion of the last equation and its investigation with respect to all its constituents. In previous papers (e.g., [8, 9, 14–24]), however, we have shown how to avoid such cumbersome equations (especially the surface term). It was derived in [10] that the resulting variational integral would look as follows:

$$\delta W = 0 = \int_V d^n x \left(\sqrt{-G} \cdot \left(\begin{array}{c} \Phi \cdot R^*_{\mu\nu} - \frac{1}{2}\Phi \cdot R^* G_{\mu\nu} \\ +\Lambda G_{\mu\nu} - \left(\nabla_\mu \nabla_\nu - G_{\mu\nu}\Delta_G\right)\Phi \\ + \left\{ \begin{array}{ll} 0 & \text{...“vacuum”} \\ \kappa T_{\mu\nu} & \text{...postulated matter} \end{array} \right\} \Phi \end{array} \right) \right) \delta G^{\mu\nu}. \quad (71)$$

Expansion and avoiding the matter terms yields:

$$0 = \int_{V_G} d^n x \sqrt{-G} \left(\left(\begin{array}{c} R_{\alpha\beta} - \frac{g_{\alpha\beta}}{2} \left(\begin{array}{c} R - \frac{1}{2F} \left(\begin{array}{c} 2F_{,ij}(n-1)g^{ij} + 2\Gamma^a_{ij} F_{,a} g^{ij} \\ -F_{,i} g^{ab} g_{jb,a} g^{ij} - F_{,j} g^{ab} g_{ib,a} g^{ij} \\ -n\Gamma^d_{ij} F_{,d} g^{ij} + \frac{n}{2} F_{,d} g^{cd} g_{ab,c} g^{ab} \end{array} \right) \\ - \frac{F_{,i} \cdot F_{,j}}{4F^2} g^{ij} \left((n-7)(n-2)\right) \\ + \frac{1}{F}\left(F_{,ab} g^{ab} + F_{,d} g^{cd} \frac{1}{2} g_{ab,c} g^{ab} \right) \end{array} \right) \\ - \frac{1}{2F} \left(\begin{array}{c} F_{,\alpha\beta}(n-2) + F_{,a} g^{ab}\left(g_{\beta b,\alpha} - g_{\beta\alpha,b}\right) \\ -F_{,\alpha} g^{ab} g_{\beta b,a} - F_{,\beta} g^{ab} g_{\alpha b,a} \\ +F_{,d} g^{cd} \frac{1}{2} n \left(\frac{2}{n} g_{\alpha c,\beta} - g_{\alpha c,\beta} - g_{\beta c,\alpha} + g_{\alpha\beta,c} \right) \\ -\frac{1}{2F}\left(F_{,\alpha} \cdot F_{,\beta} (3n-6) \right) \end{array} \right) \\ - \frac{\left(\partial_\alpha \Phi_{,\beta} - \Gamma^{*\gamma}_{\alpha\beta} \Phi_{,\gamma} - \frac{F \cdot g_{\alpha\beta}}{\sqrt{-gF^n}} \partial_\mu \left(\sqrt{-gF^n} \cdot \frac{g^{\mu\nu}}{F} \Phi_{,\nu} \right) \right)}{\Phi} \end{array} \right) \delta G^{\alpha\beta} \right).$$

(72)

It provides a great variety of intrinsic solutions (see [10]) of which we will consider only one option here in order to deal with the so-called three generation problem (and as explained at the beginning of this section, not just with respect to the three generations of elementary particles but to every system, including the system of the mind).

4.2.5.2 Why every system has the potential to come in three generations or why every psychosis has three states of excitement

Taking Eq. (71), dividing by the function Φ, expanding the scaled Ricci tensor and scalar in accordance with the quantum Einstein field equations (64) in their mixed form from:

$$0=\left(\begin{array}{c} R_{\alpha\beta}-\frac{g_{\alpha\beta}}{2}R+\frac{g_{\alpha\beta}}{2}\left(\begin{array}{c}\frac{1}{2F}\left(\begin{array}{c}2F_{,ij}(n-1)g^{ij}+2\Gamma_{ij}^{a}F_{,a}g^{ij}\\ -F_{,i}g^{ab}g_{jb,a}g^{ij}-F_{,j}g^{ab}g_{ib,a}g^{ij}\\ -n\Gamma_{ij}^{d}F_{,d}g^{ij}+\frac{n}{2}F_{,d}g^{cd}g_{ab,c}g^{ab}\end{array}\right)\\ +\frac{F_{,i}\cdot F_{,j}}{4F^{2}}g^{ij}\left((n-6)(n-1)\right)\end{array}\right)\\ -\frac{1}{2F}\left(\begin{array}{c}F_{,\alpha\beta}(n-2)+F_{,ab}g_{\alpha\beta}g^{ab}\\ +F_{,a}g^{ab}\left(g_{\beta b,\alpha}-g_{\beta\alpha,b}\right)-F_{,\alpha}g^{ab}g_{\beta b,a}-F_{,\beta}g^{ab}g_{\alpha b,a}\\ +F_{,d}g^{cd}\frac{1}{2}n\left(\frac{2}{n}g_{\alpha c,\beta}-g_{\alpha c,\beta}-g_{\beta c,\alpha}+g_{\alpha\beta,c}+\frac{1}{n}g_{\alpha\beta}g_{ab,c}g^{ab}\right)\\ -\frac{1}{2F}\left(F_{,\alpha}\cdot F_{,\beta}(3n-6)+g_{\alpha\beta}F_{,c}F_{,d}g^{cd}(4-n)\right)\end{array}\right)\\ +\kappa T_{\alpha\beta}+\Lambda\cdot F\cdot g_{\alpha\beta}\end{array}\right), \tag{73}$$

we obtain:

$$R_{\alpha}^{\beta}-\frac{\delta_{\alpha}^{\beta}}{2}R+\frac{1}{2F}\eth_{\alpha}^{\beta}\left[g_{\alpha\beta};F\right]-\frac{\left(G^{\beta\mu}\nabla_{\mu}\nabla_{\alpha}-\delta_{\alpha}^{\beta}\Delta_{G}\right)\Phi}{\Phi}=0, \tag{74}$$

with:

$$
\frac{1}{2F}\not\partial_{\alpha}^{\beta}\left[g_{\alpha\beta};F\right]=\frac{1}{2F}\left(\begin{array}{c}
\dfrac{\delta_{\alpha}^{\beta}}{2}\left(\left(\begin{array}{l}
2F_{,ij}(n-1)g^{ij}+2\Gamma_{ij}^{a}F_{,a}g^{ij} \\
-F_{,i}g^{ab}g_{jb,a}g^{ij}-F_{,j}g^{ab}g_{ib,a}g^{ij} \\
-n\Gamma_{ij}^{d}F_{,d}g^{ij}+\dfrac{n}{2}F_{,d}g^{cd}g_{ab,c}g^{ab}
\end{array}\right) \\
+\dfrac{F_{,i}\cdot F_{,j}}{2F}g^{ij}\left((n-6)(n-1)\right)\right) \\
-g^{\beta\gamma}\left(\begin{array}{c}
F_{,\alpha\gamma}(n-2)+F_{,ab}g_{\alpha\gamma}g^{ab} \\
+F_{,a}g^{ab}\left(g_{\gamma b,\alpha}-g_{\gamma\alpha,b}\right)-F_{,\alpha}g^{ab}g_{\gamma b,a}-F_{,\gamma}g^{ab}g_{\alpha b,a} \\
+F_{,d}g^{cd}\dfrac{1}{2}n\left(\begin{array}{l}
\dfrac{2}{n}g_{\alpha c,\gamma}-g_{\alpha c,\gamma}-g_{\gamma c,\alpha} \\
+g_{\alpha\gamma,c}+\dfrac{1}{n}g_{\alpha\gamma}g_{ab,c}g^{ab}
\end{array}\right) \\
-\dfrac{1}{2F}\left(F_{,\alpha}\cdot F_{,\gamma}(3n-6)+g_{\alpha\gamma}F_{,c}F_{,d}g^{cd}(4-n)\right)
\end{array}\right)
\end{array}\right) \tag{75}
$$

and:

$$
\nabla_{\mu}V_{\alpha}=\partial_{\mu}V_{\alpha}-\Gamma_{\mu\alpha}^{*\gamma}V_{\gamma};\quad \Gamma_{\mu\alpha}^{*\gamma}=\Gamma_{\mu\alpha}^{\gamma}+\frac{F''}{2\cdot F}\left(f_{,\mu}\cdot\delta_{\alpha}^{\gamma}+f_{,\alpha}\cdot\delta_{\mu}^{\gamma}-f_{,\sigma}\cdot g^{\gamma\sigma}g_{\alpha\mu}\right)
$$

$$
\Delta_{G}\Phi=\frac{1}{\sqrt{-G}}\partial_{\mu}\left(\sqrt{-G}\cdot G^{\mu\nu}\Phi_{,\nu}\right)=\frac{1}{\sqrt{-gF^{n}}}\partial_{\mu}\left(\sqrt{-gF^{n}}\cdot\frac{g^{\mu\nu}}{F}\Phi_{,\nu}\right);\quad \Phi_{,\nu}=\partial_{\nu}\Phi. \tag{76}
$$

Please note that the Laplace operator for the scaled metric could also be given as follows:

$$
\begin{aligned}
\Delta_{G}\Phi &= G^{\mu\nu}\left(\Phi_{,\mu\nu}-\Gamma_{\mu\nu}^{*\gamma}\Phi_{,\gamma}\right) \\
&=\frac{g^{\mu\nu}}{F}\left(\Phi_{,\mu\nu}-\left(\Gamma_{\mu\nu}^{\gamma}+\frac{F''}{2\cdot F}\left(f_{,\mu}\cdot\delta_{\nu}^{\gamma}+f_{,\nu}\cdot\delta_{\mu}^{\gamma}-f_{,\sigma}\cdot g^{\gamma\sigma}g_{\nu\mu}\right)\right)\Phi_{,\gamma}\right) \\
&=\frac{1}{F}\left(g^{\mu\nu}\Phi_{,\mu\nu}-\left(g^{\mu\nu}\Gamma_{\mu\nu}^{\gamma}+\frac{F''}{2\cdot F}\left(f_{,\mu}\cdot g^{\mu\gamma}+f_{,\nu}\cdot g^{\gamma\nu}-n\cdot f_{,\sigma}\cdot g^{\gamma\sigma}\right)\right)\Phi_{,\gamma}\right) \\
&=\frac{1}{F}\left(g^{\mu\nu}\Phi_{,\mu\nu}-\left(g^{\mu\nu}\Gamma_{\mu\nu}^{\gamma}+\frac{F''}{2\cdot F}\left(2\cdot f_{,\mu}\cdot g^{\mu\gamma}-n\cdot f_{,\sigma}\cdot g^{\gamma\sigma}\right)\right)\Phi_{,\gamma}\right).
\end{aligned} \tag{77}
$$

Inserting the above into Eq. (73) results in:

$$0=\left(R_{\alpha}^{\beta}-\frac{\delta_{\alpha}^{\beta}}{2}R+\frac{1}{2F}\eth_{\alpha}^{\beta}\left[g_{\alpha\beta};F\right]-\frac{\left(\begin{array}{c}g^{\beta\mu}\left(\Phi_{,\mu\alpha}-\Gamma_{\mu\alpha}^{\gamma}\Phi_{,\gamma}+\frac{F'}{2\cdot F}\left(f_{,\mu}\cdot\delta_{\alpha}^{\gamma}+f_{,\alpha}\cdot\delta_{\mu}^{\gamma}-f_{,\sigma}\cdot g^{\gamma\sigma}g_{\alpha\mu}\right)\Phi_{,\gamma}\right)\\-\delta_{\alpha}^{\beta}\left(g^{\mu\nu}\Phi_{,\mu\nu}-\left(g^{\mu\nu}\Gamma_{\mu\nu}^{\gamma}+\frac{F'}{2\cdot F}\left(2\cdot f_{,\mu}\cdot g^{\mu\gamma}-n\cdot f_{,\sigma}\cdot g^{\gamma\sigma}\right)\right)\Phi_{,\gamma}\right)\end{array}\right)}{F\cdot\Phi}\right)$$

$$=\left(R_{\alpha}^{\beta}-\frac{\delta_{\alpha}^{\beta}}{2}R+\frac{1}{2F}\eth_{\alpha}^{\beta}\left[g_{\alpha\beta};F\right]-\frac{\left(\begin{array}{c}g^{\beta\mu}\Phi_{,\mu\alpha}-g^{\beta\mu}\Gamma_{\mu\alpha}^{\gamma}\Phi_{,\gamma}+\frac{F'}{2\cdot F}\left(g^{\beta\mu}\left(f_{,\mu}\cdot\Phi_{,\alpha}+f_{,\alpha}\cdot\Phi_{,\mu}\right)-\delta_{\alpha}^{\beta}f_{,\sigma}\cdot g^{\gamma\sigma}\Phi_{,\gamma}\right)\\-\delta_{\alpha}^{\beta}\left(g^{\mu\nu}\Phi_{,\mu\nu}-\left(g^{\mu\nu}\Gamma_{\mu\nu}^{\gamma}+\frac{F'}{2\cdot F}\left(2\cdot f_{,\mu}\cdot g^{\mu\gamma}-n\cdot f_{,\sigma}\cdot g^{\gamma\sigma}\right)\right)\Phi_{,\gamma}\right)\end{array}\right)}{F\cdot\Phi}\right). \tag{78}$$

Now we have two reasonable opportunities to apply the covariant derivative and to use the Bianchi identity onto Eq. (78). At first, we form the covariant derivative with respect to the scaled metric, which gives:

$$0=\nabla_{\beta}\left(R_{\alpha}^{\beta}-\frac{\delta_{\alpha}^{\beta}}{2}R+\frac{1}{2F}\eth_{\alpha}^{\beta}\left[g_{\alpha\beta};F\right]-\frac{\left(\begin{array}{c}g^{\beta\mu}\Phi_{,\mu\alpha}-g^{\beta\mu}\Gamma_{\mu\alpha}^{\gamma}\Phi_{,\gamma}+\frac{F'}{2\cdot F}\left(g^{\beta\mu}\left(f_{,\mu}\cdot\Phi_{,\alpha}+f_{,\alpha}\cdot\Phi_{,\mu}\right)-\delta_{\alpha}^{\beta}f_{,\sigma}\cdot g^{\gamma\sigma}\Phi_{,\gamma}\right)\\-\delta_{\alpha}^{\beta}\left(g^{\mu\nu}\Phi_{,\mu\nu}-\left(g^{\mu\nu}\Gamma_{\mu\nu}^{\gamma}+\frac{F'}{2\cdot F}\left(2\cdot f_{,\mu}\cdot g^{\mu\gamma}-n\cdot f_{,\sigma}\cdot g^{\gamma\sigma}\right)\right)\Phi_{,\gamma}\right)\end{array}\right)}{F\cdot\Phi}\right)$$

$$=\left(\overbrace{\nabla_{\beta}\left(R_{\alpha}^{\beta}-\frac{\delta_{\alpha}^{\beta}}{2}R+\frac{1}{2F}\eth_{\alpha}^{\beta}\left[g_{\alpha\beta};F\right]\right)}^{=0;\ \text{Bianchi}}-\nabla_{\beta}\frac{\left(\begin{array}{c}g^{\beta\mu}\Phi_{,\mu\alpha}-g^{\beta\mu}\Gamma_{\mu\alpha}^{\gamma}\Phi_{,\gamma}+\frac{F'}{2\cdot F}\left(g^{\beta\mu}\left(f_{,\mu}\cdot\Phi_{,\alpha}+f_{,\alpha}\cdot\Phi_{,\mu}\right)-\delta_{\alpha}^{\beta}f_{,\sigma}\cdot g^{\gamma\sigma}\Phi_{,\gamma}\right)\\-\delta_{\alpha}^{\beta}\left(g^{\mu\nu}\Phi_{,\mu\nu}-\left(g^{\mu\nu}\Gamma_{\mu\nu}^{\gamma}+\frac{F'}{2\cdot F}\left(2\cdot f_{,\mu}\cdot g^{\mu\gamma}-n\cdot f_{,\sigma}\cdot g^{\gamma\sigma}\right)\right)\Phi_{,\gamma}\right)\end{array}\right)}{F\cdot\Phi}\right)$$

$\Rightarrow$

$$0=\nabla_{\beta}\frac{\left(\begin{array}{c}g^{\beta\mu}\Phi_{,\mu\alpha}-g^{\beta\mu}\Gamma_{\mu\alpha}^{\gamma}\Phi_{,\gamma}+\frac{F'}{2\cdot F}\left(g^{\beta\mu}\left(f_{,\mu}\cdot\Phi_{,\alpha}+f_{,\alpha}\cdot\Phi_{,\mu}\right)-\delta_{\alpha}^{\beta}f_{,\sigma}\cdot g^{\gamma\sigma}\Phi_{,\gamma}\right)\\ -\delta_{\alpha}^{\beta}\left(g^{\mu\nu}\Phi_{,\mu\nu}-\left(g^{\mu\nu}\Gamma_{\mu\nu}^{\gamma}+\frac{F'}{2\cdot F}\left(2\cdot f_{,\mu}\cdot g^{\mu\gamma}-n\cdot f_{,\sigma}\cdot g^{\gamma\sigma}\right)\right)\Phi_{,\gamma}\right)\end{array}\right)}{F\cdot\Phi}. \tag{79}$$

And second, we can also covariant derivate Eq. (78) based on the unscaled metric, which results in:

$$0=\left(\begin{array}{c}R_{\alpha}^{\beta}-\frac{\delta_{\alpha}^{\beta}}{2}R+\frac{1}{2F}\eth_{\alpha}^{\beta}\left[g_{\alpha\beta};F\right]\\ -\frac{\left(\begin{array}{c}g^{\beta\mu}\left(\Phi_{,\mu\alpha}-\Gamma_{\mu\alpha}^{\gamma}\Phi_{,\gamma}+\frac{F'}{2\cdot F}\left(f_{,\mu}\cdot\delta_{\alpha}^{\gamma}+f_{,\alpha}\cdot\delta_{\mu}^{\gamma}-f_{,\sigma}\cdot g^{\gamma\sigma}g_{\alpha\mu}\right)\Phi_{,\gamma}\right)\\ -\delta_{\alpha}^{\beta}\left(g^{\mu\nu}\Phi_{,\mu\nu}-\left(g^{\mu\nu}\Gamma_{\mu\nu}^{\gamma}+\frac{F'}{2\cdot F}\left(2\cdot f_{,\mu}\cdot g^{\mu\gamma}-n\cdot f_{,\sigma}\cdot g^{\gamma\sigma}\right)\right)\Phi_{,\gamma}\right)\end{array}\right)}{F\cdot\Phi}\end{array}\right)_{;\beta}. \tag{80}$$

Thereby the term $\left(R_{\alpha}^{\beta}-\frac{\delta_{\alpha}^{\beta}}{2}R\right)_{;\beta}$ vanishes, which can directly be obtained from the Bianchi identity for the Riemann curvature tensor, reading:

$$0=R_{\alpha\beta\gamma\delta;\varepsilon}+R_{\alpha\beta\varepsilon\gamma;\delta}+R_{\alpha\beta\delta\varepsilon;\gamma}, \tag{81}$$

via contraction with the metric tensor as follows:

$$0=\left(R_{\alpha\beta\gamma\delta;\varepsilon}+R_{\alpha\beta\varepsilon\gamma;\delta}+R_{\alpha\beta\delta\varepsilon;\gamma}\right)g^{\alpha\delta}g^{\beta\gamma}=\xrightarrow{g^{\alpha\delta}{}_{;\lambda}=0}=2\cdot R^{\alpha}{}_{\varepsilon;\alpha}-R_{;\varepsilon}. \tag{82}$$

This yields:

$$0=\left(\begin{array}{c}\overbrace{\left(R_{\alpha}^{\beta}-\frac{\delta_{\alpha}^{\beta}}{2}R\right)_{;\beta}}^{=0;\ \text{Bianchi}}+\left(\frac{1}{2F}\eth_{\alpha}^{\beta}\left[g_{\alpha\beta};F\right]\right)_{;\beta}\\ -\left(\frac{\left(\begin{array}{c}g^{\beta\mu}\Phi_{,\mu\alpha}-g^{\beta\mu}\Gamma_{\mu\alpha}^{\gamma}\Phi_{,\gamma}+\frac{F'}{2\cdot F}\left(g^{\beta\mu}\left(f_{,\mu}\cdot\Phi_{,\alpha}+f_{,\alpha}\cdot\Phi_{,\mu}\right)-\delta_{\alpha}^{\beta}f_{,\sigma}\cdot g^{\gamma\sigma}\Phi_{,\gamma}\right)\\ -\delta_{\alpha}^{\beta}\left(g^{\mu\nu}\Phi_{,\mu\nu}-\left(g^{\mu\nu}\Gamma_{\mu\nu}^{\gamma}+\frac{F'}{2\cdot F}\left(2\cdot f_{,\mu}\cdot g^{\mu\gamma}-n\cdot f_{,\sigma}\cdot g^{\gamma\sigma}\right)\right)\Phi_{,\gamma}\right)\end{array}\right)}{F\cdot\Phi}\right)_{;\beta}\end{array}\right)$$

$$\left(\frac{1}{2F}\eth_{\alpha}^{\beta}\left[g_{\alpha\beta};F\right]\right)_{;\beta} = \left(-\left(\frac{\left(\begin{array}{c} g^{\beta\mu}\Phi_{,\mu\alpha} - g^{\beta\mu}\Gamma_{\mu\alpha}^{\gamma}\Phi_{,\gamma} + \frac{F'}{2\cdot F}\left(g^{\beta\mu}\left(f_{,\mu}\cdot\Phi_{,\alpha} + f_{,\alpha}\cdot\Phi_{,\mu}\right) - \delta_{\alpha}^{\beta}f_{,\sigma}\cdot g^{\gamma\sigma}\Phi_{,\gamma}\right) \\ -\delta_{\alpha}^{\beta}\left(g^{\mu\nu}\Phi_{,\mu\nu} - \left(g^{\mu\nu}\Gamma_{\mu\nu}^{\gamma} + \frac{F'}{2\cdot F}\left(2\cdot f_{,\mu}\cdot g^{\mu\gamma} - n\cdot f_{,\sigma}\cdot g^{\gamma\sigma}\right)\right)\Phi_{,\gamma}\right) \end{array}\right)}{F\cdot\Phi}\right)_{;\beta}\right).$$

(83)

We may recognize the first term $\left(\frac{1}{2F}\eth_{\alpha}^{\beta}\left[g_{\alpha\beta};F\right]\right)_{;\beta}$ as a possible "constant" for the third-order polynomial, which results from a suitable function Φ from the rest in Eq. (80), which is to say from the term:

$$-\left(\frac{\left(\begin{array}{c} g^{\beta\mu}\Phi_{,\mu\alpha} - g^{\beta\mu}\Gamma_{\mu\alpha}^{\gamma}\Phi_{,\gamma} + \frac{F'}{2\cdot F}\left(g^{\beta\mu}\left(f_{,\mu}\cdot\Phi_{,\alpha} + f_{,\alpha}\cdot\Phi_{,\mu}\right) - \delta_{\alpha}^{\beta}f_{,\sigma}\cdot g^{\gamma\sigma}\Phi_{,\gamma}\right) \\ -\delta_{\alpha}^{\beta}\left(g^{\mu\nu}\Phi_{,\mu\nu} - \left(g^{\mu\nu}\Gamma_{\mu\nu}^{\gamma} + \frac{F'}{2\cdot F}\left(2\cdot f_{,\mu}\cdot g^{\mu\gamma} - n\cdot f_{,\sigma}\cdot g^{\gamma\sigma}\right)\right)\Phi_{,\gamma}\right) \end{array}\right)}{F\cdot\Phi}\right)_{;\beta}.$$

(84)

However, as recent investigations show (see the end of this chapter), it is rather the Φ and the metric which provides the constant term, while the function F produces the third-order polynomial with respect to parameters we might interpret as mass, thereby giving us three (!) solutions for these very parameters.

As our derivation was purely mathematical (we only used the Bianchi identity), we conclude that every system providing third-order polynomials due to Eq. (84) (or any other of the many "intrinsic" Bianchi possibilities coming out from Eq. (72) as shown in [10]) has the potential to exist in three generations.

Thus, this should now help us to try and solve the 3-generation problem for elementary particles ... and every other system, too.

4.2.5.3 Simplifying the task via intrinsic solutions?

In order to keep things simple, we assume to have functions F and Φ as $F[f]$ and $\Phi[f]$, concentrate on the scalar part of Eq. (72) and after forming the covariant derivative, we split up the integrand as follows:

$$0=\left[\begin{matrix} R_{\alpha\beta}-\frac{g_{\alpha\beta}}{2}R-\frac{1}{2F}\left(\begin{matrix} F_{,\alpha\beta}(n-2)+F_{,a}g^{ab}\left(g_{\beta b,\alpha}-g_{\beta\alpha,b}\right) \\ -F_{,\alpha}g^{ab}g_{\beta b,a}-F_{,\beta}g^{ab}g_{\alpha b,a} \\ +F_{,d}g^{cd}\frac{1}{2}n\left(\frac{2}{n}g_{\alpha c,\beta}-g_{\alpha c,\beta}-g_{\beta c,\alpha}+g_{\alpha\beta,c}\right) \\ -\frac{1}{2F}\left(F_{,\alpha}\cdot F_{,\beta}(3n-6)\right) \end{matrix}\right) \\ -\frac{\left(\partial_{\alpha}\Phi_{,\beta}-\Gamma^{*\gamma}_{\alpha\beta}\Phi_{,\gamma}\right)}{\Phi} \end{matrix}\right]^{;\beta}$$

$$0=\left[g_{\alpha}^{\beta}\frac{\left(\frac{F}{\sqrt{-gF^{n}}}\partial_{\mu}\left(\sqrt{-gF^{n}}\cdot\frac{g^{\mu\nu}}{F}\Phi_{,\nu}\right)\right)}{\Phi}-\frac{g_{\alpha}^{\beta}}{2}\left(\begin{matrix} -\frac{1}{2F}\left(\begin{matrix} 2F_{,ij}(n-1)g^{ij}+2\Gamma^{a}_{ij}F_{,a}g^{ij} \\ -F_{,i}g^{ab}g_{jb,a}g^{ij}-F_{,j}g^{ab}g_{ib,a}g^{ij} \\ -n\Gamma^{d}_{ij}F_{,d}g^{ij}+\frac{n}{2}F_{,d}g^{cd}g_{ab,c}g^{ab} \end{matrix}\right) \\ -\frac{F_{,i}\cdot F_{,j}}{4F^{2}}g^{ij}\left((n-7)(n-2)\right) \\ +\frac{1}{F}\left(F_{,ab}g^{ab}+F_{,d}g^{cd}\frac{1}{2}g_{ab,c}g^{ab}\right) \end{matrix}\right)\right]_{;\beta}. \qquad (85)$$

Using the results from [10], we can write:

$$0=\left[\begin{matrix} R_{\alpha\beta}-\frac{g_{\alpha\beta}}{2}R-\frac{1}{2F}\left(\begin{matrix} +F'\left(\begin{matrix} f_{,\alpha\beta}(n-2)f_{,a}g^{ab}\left(g_{\beta b,\alpha}-g_{\beta\alpha,b}\right) \\ -f_{,\alpha}g^{ab}g_{\beta b,a}-F_{,\beta}g^{ab}g_{\alpha b,a} \\ +f_{,d}g^{cd}\frac{1}{2}n\left(\frac{2}{n}g_{\alpha c,\beta}-g_{\alpha c,\beta}-g_{\beta c,\alpha}+g_{\alpha\beta,c}\right) \end{matrix}\right) \\ f_{,\alpha}\cdot f_{,\beta}\left(F''(n-2)-\frac{(F')^{2}}{2F}(3n-6)-\frac{2F'\Phi'}{\Phi}\right) \end{matrix}\right) \\ -\Phi'\frac{\left(\partial_{\alpha}f_{,\beta}-\Gamma^{\gamma}_{\alpha\beta}f_{,\gamma}\right)}{\Phi} \end{matrix}\right]^{;\beta}$$

$$0=\overbrace{\left(R_{\alpha\beta}-\frac{g_{\alpha\beta}}{2}R\right)^{;\beta}}^{=0}-\left[\begin{array}{l}\frac{1}{2F}\left(\begin{array}{c}+F'\left(\begin{array}{c}f_{,\alpha\beta}(n-2)f_{,a}g^{ab}\left(g_{\beta b,\alpha}-g_{\beta\alpha,b}\right)\\ -f_{,\alpha}g^{ab}g_{\beta b,a}-F_{,\beta}g^{ab}g_{\alpha b,a}\\ +f_{,d}g^{cd}\frac{1}{2}n\left(\frac{2}{n}g_{\alpha c,\beta}-g_{\alpha c,\beta}-g_{\beta c,\alpha}+g_{\alpha\beta,c}\right)\end{array}\right)\\ f_{,\alpha}\cdot f_{,\beta}\left(F''(n-2)-\frac{(F')^{2}}{2F}(3n-6)-\frac{2F'\Phi'}{\Phi}\right)\end{array}\right)\\ \qquad +\Phi'\frac{\left(\partial_{\alpha}f_{,\beta}-\Gamma^{\gamma}_{\alpha\beta}f_{,\gamma}\right)}{\Phi}\end{array}\right]^{;\beta}$$

$$0=\left[\frac{1}{2F}\left(\begin{array}{c}+F'\left(\begin{array}{c}f_{,\alpha\beta}(n-2)f_{,a}g^{ab}\left(g_{\beta b,\alpha}-g_{\beta\alpha,b}\right)\\ -f_{,\alpha}g^{ab}g_{\beta b,a}-F_{,\beta}g^{ab}g_{\alpha b,a}\\ +f_{,d}g^{cd}\frac{1}{2}n\left(\frac{2}{n}g_{\alpha c,\beta}-g_{\alpha c,\beta}-g_{\beta c,\alpha}+g_{\alpha\beta,c}\right)\end{array}\right)\\ f_{,\alpha}\cdot f_{,\beta}\left(F''(n-2)-\frac{(F')^{2}}{2F}(3n-6)-\frac{2F'\Phi'}{\Phi}\right)\end{array}\right)+\Phi'\frac{\left(\partial_{\alpha}f_{,\beta}-\Gamma^{\gamma}_{\alpha\beta}f_{,\gamma}\right)}{\Phi}\right]^{;\beta},\tag{86}$$

$$0=\left[g^{\beta}_{\alpha}\frac{g^{ab}}{2F}\left(\begin{array}{c}F'\left(\begin{array}{c}2(n-2)f_{,ab}-\Gamma^{d}_{ab}f_{,d}(n-2)\\ +f_{,d}g^{cd}\left(g_{ab,c}\frac{1}{2}(n-2)-2g_{cb,a}\right)\end{array}\right)\\ +f_{,a}f_{,b}\left(\left(2\frac{F}{\Phi}\Phi''+2(n-1)F'\frac{\Phi'}{\Phi}+(n-2)\left(\frac{(F')^{2}}{2F}(n-7)+2F''\right)\right)\right)\\ +4F\frac{\Phi'\left(f_{,ab}-\Gamma^{d}_{ab}f_{,d}\right)}{\Phi}\end{array}\right)\right]_{;\beta}.\tag{87}$$

In [10], we discussed the possibility to use the split-up as given above as an option to obtain intrinsic solutions to the integrand as follows:

$$0=\left(\begin{array}{c}R_{\alpha\beta}-\frac{g_{\alpha\beta}}{2}R-\frac{1}{2F}\left(\begin{array}{c}+F'\left(\begin{array}{c}f_{,\alpha\beta}(n-2)f_{,a}g^{ab}\left(g_{\beta b,\alpha}-g_{\beta\alpha,b}\right)\\-f_{,\alpha}g^{ab}g_{\beta b,a}-F_{,\beta}g^{ab}g_{\alpha b,a}\\+f_{,d}g^{cd}\frac{1}{2}n\left(\frac{2}{n}g_{\alpha c,\beta}-g_{\alpha c,\beta}-g_{\beta c,\alpha}+g_{\alpha\beta,c}\right)\end{array}\right)\\f_{,\alpha}\cdot f_{,\beta}\left(F''(n-2)-\frac{(F')^2}{2F}(3n-6)-\frac{2F'\Phi'}{\Phi}\right)\end{array}\right)\\-\Phi'\frac{\left(\partial_\alpha f_{,\beta}-\Gamma^{\gamma}_{\alpha\beta}f_{,\gamma}\right)}{\Phi}\end{array}\right), \tag{88}$$

$$0=g^{ab}\left(\begin{array}{c}F'\left(\begin{array}{c}2(n-2)f_{,ab}-\Gamma^{d}_{ab}f_{,d}(n-2)\\+f_{,d}g^{cd}\left(g_{ab,c}\frac{1}{2}(n-2)-2g_{cb,a}\right)\end{array}\right)\\+f_{,a}f_{,b}\left(\left(2\frac{F}{\Phi}\Phi''+2(n-1)F'\frac{\Phi'}{\Phi}+(n-2)\left(\frac{(F')^2}{2F}(n-7)+2F''\right)\right)\right)\\+4F\frac{\Phi'\left(f_{,ab}-\Gamma^{d}_{ab}f_{,d}\right)}{\Phi}\end{array}\right), \tag{89}$$

which are just Eqs. (86) and (87) only without the covariant derivative.

The clever reader may already guess that we introduced the covariant derivative in order to obtain some kind of conservation law (just as in the classical General Theory of Relativity—e.g., [39]) and in aiming for some simplicity, we split up the complicated task of Eq. (71). For illustration, however, we here also want to discuss briefly the problems occurring with such intrinsic solution approaches. In order to get rid of the non-linear operator terms, namely, we have to demand the following for the equations above:

$$0=F''(n-2)-\frac{(F')^2}{2F}(3n-6)-\frac{2F'\Phi'}{\Phi}, \tag{90}$$

$$0=2\frac{F}{\Phi}\Phi''+2(n-1)F'\frac{\Phi'}{\Phi}+(n-2)\left(\frac{(F')^2}{2F}(n-7)+2F''\right). \tag{91}$$

Unfortunately, we see ourselves unable to solve the above (coupled) differential equations for both functions F and Φ (only Eq. (90) can be solved in the closed form with respect to Φ). So, with respect to find intrinsic solutions to Eq. (72), we might like to consider simpler forms, like, instead of the set-up (86) and (87), we could demand a set of intrinsic solutions to the general variational quest as follows:

$$0=-g_{\alpha\beta}\cdot\frac{1}{2}\cdot\left(\begin{array}{c}-\frac{g^{ab}}{2F}\left(\begin{array}{c}2F_{,ab}(n-2)-\Gamma_{ab}^{d}F_{,d}(n-2)\\+F_{,d}g^{cd}\left(g_{ab,c}\frac{1}{2}(n-2)-2g_{cb,a}\right)\end{array}\right)\\-\frac{F_{,i}\cdot F_{,j}}{4F^{2}}g^{ij}\left((n-7)(n-2)\right)\end{array}\right)$$

$$0=R_{\alpha\beta}-\frac{g_{\alpha\beta}}{2}R+\left(\begin{array}{c}\frac{1}{2F}\left(\begin{array}{c}F_{,\alpha\beta}(n-2)+F_{,a}g^{ab}\left(g_{\beta b,\alpha}-g_{\beta\alpha,b}\right)\\-F_{,\alpha}g^{ab}g_{\beta b,a}-F_{,\beta}g^{ab}g_{\alpha b,a}\\+F_{,d}g^{cd}\frac{1}{2}n\left(\frac{2}{n}g_{\alpha c,\beta}-g_{\alpha c,\beta}-g_{\beta c,\alpha}+g_{\alpha\beta,c}\right)\\-\frac{1}{2F}\left(F_{,\alpha}\cdot F_{,\beta}(3n-6)\right)\end{array}\right)\\-\frac{\left(\partial_{\alpha}\Phi_{,\beta}-\Gamma^{*\gamma}_{\alpha\beta}\Phi_{,\gamma}-\frac{F\cdot g_{\alpha\beta}}{\sqrt{-gF^{n}}}\partial_{\mu}\left(\sqrt{-gF^{n}}\cdot\frac{g^{\mu\nu}}{F}\Phi_{,\nu}\right)\right)}{\Phi}\end{array}\right).\tag{92}$$

Apart from the fact that this time the linearity condition with vanishing operator $F_{,i}\cdot F_{,j}$ leads to a manageable differential equation for F only, when just concentrating on the scalar equation (the first one), we will also see further below, that this setting guaranties the correct asymptotic outcome for the F-scaled metric tensor.

We start with the first expression in Eq. (92) and obtain:

$$0=-g_{\alpha\beta}\cdot\frac{1}{2}\cdot\left(\begin{array}{c}-\frac{g^{ab}}{2F}\left(\begin{array}{c}2\left(F''f_{,a}\cdot f_{,b}+F'f_{,ab}\right)(n-2)-F'\Gamma_{ab}^{d}f_{,d}(n-2)\\+F'f_{,d}g^{cd}\left(g_{ab,c}\frac{1}{2}(n-2)-2g_{cb,a}\right)\end{array}\right)\\-(F')^{2}\frac{f_{,a}\cdot f_{,b}}{4F^{2}}g^{ab}\left((n-7)(n-2)\right)\end{array}\right)$$

$$\Rightarrow \qquad 0 = g_{\alpha\beta} \cdot \left(\begin{array}{l} \frac{g^{ab}F'}{2F} \left(\begin{array}{l} 2f_{,ab}(n-2) - \Gamma^{d}_{ab} f_{,d}(n-2) \\ + f_{,d} g^{cd} \left(g_{ab,c} \frac{1}{2}(n-2) - 2g_{cb,a} \right) \end{array} \right) \\ + \frac{f_{,a} \cdot f_{,b}}{2F} g^{ab}(n-2) \left(\frac{(F')^2}{2F}(n-7) + 2F'' \right) \end{array} \right). \qquad (93)$$

As the linearization condition we demand:

$$0 = \frac{(F')^2}{2F}(n-7) + 2F'', \qquad (94)$$

which can be satisfied via the function $F[f]$ as follows:

$$F[f] = \begin{cases} C_F \cdot \left(f + C_f\right)^{\frac{4}{n-3}} & n \neq 3 \\ C_F \cdot e^{f \cdot C_f} & n = 3. \end{cases} \qquad (95)$$

Subsequently, Eq. (93) can be simplified and yields

$$0 = g_{\alpha\beta} \cdot \frac{g^{ab}F'}{F} \left(2f_{,ab}(n-2) - \Gamma^{d}_{ab} f_{,d}(n-2) + f_{,d} g^{cd} \left(g_{ab,c} \frac{1}{2}(n-2) - 2g_{cb,a} \right) \right). \qquad (96)$$

Interestingly, for metrics without shear elements:

$$g_{ij} = \begin{pmatrix} g_{00} & \cdots & 0 \\ \vdots & \ddots & \vdots \\ 0 & \cdots & g_{n-1n-1} \end{pmatrix}, \qquad (97)$$

this converges to the ordinary Laplace operator, namely:

$$0 = g_{\alpha\beta} \cdot \frac{g^{ab}F'}{F} \left(2f_{,ab}(n-2) - \Gamma^{d}_{ab} f_{,d}(n-2) + f_{,d} g^{cd} \left(g_{ab,c} \frac{1}{2}(n-2) - 2g_{cb,a} \right) \right)$$

$$\xrightarrow{g_{ij} = \begin{pmatrix} g_{00} & \cdots & 0 \\ \vdots & \ddots & \vdots \\ 0 & \cdots & g_{n-1n-1} \end{pmatrix}}$$

$$0 = g_{\alpha\beta} \cdot \frac{g^{ab}F'}{F} \cdot 2(n-2)\Delta f. \qquad (98)$$

We see that the intrinsic approach gives us some manageable possibilities to partially solve Eq. (71), especially with respect to its scalar parts, which we have found to be connected with many quantum aspects [5–10, 14–24]. Some critics may now argue that this equation is not of Klein–Gordon character as it does not contain any potential nor mass, but this author has already shown that this problem is easily solved by adding additional dimensions carrying the right properties to produce masses and potentials (e.g., [6, 8–10, 14–24]).

With respect to our goal of finding proper conservation laws and still keeping things simple, we might resort to the "weak-gravity" condition and assume the influence of the true tensor part (86) to be small in comparison to the scalar equation (87) (after factorizing out the metric tensor $g_{\alpha\beta}$ only a scalar remains). And in fact, when performing the quite cumbersome and lengthy evaluations in Eq. (87), we obtain third-order equations which might give us an explanation for the 3-generation problem of elementary particles. However, as the split-up-method may be considered a little bit arbitrary, here we intend to seek something more general when aiming for a scalar equation.

4.2.5.4 Simplification via scalarizing the integrand

Repetition: Regarding our problem of obtaining conservation laws, we might just take Eq. (71), form the covariant derivative (with respect to the unscaled metric) from the integrand as shown in the section "4.2.5.2 Why Every System Has the Potential to Come in Three Generations or Why Every Psychosis Has Three States of Excitement." We see that because the contracted Bianchi identity erases the Ricci-terms of the unscaled metric, we are left with terms containing F and Φ, providing us with the necessary conservation. The only problem with this: as Eq. (83) shows, we end up with some pretty complicated equations and we might like to look for some suitable simplifications. For this, we once again apply the "weak gravity condition" in covariant:

$$\delta G^{\alpha\beta} = G^{\alpha\beta}\cdot\delta_0 + \overbrace{G^{ab}\delta_{ab}^{\alpha\beta}}^{\text{Gravity}} = \frac{1}{F}\cdot\left(g^{\alpha\beta}\cdot\delta_0 + \overbrace{g^{ab}\delta_{ab}^{\alpha\beta}}^{\text{Gravity}}\right)\xrightarrow{\forall\,\delta_{ab}^{\alpha\beta}\ll\delta_0} = \frac{g^{\alpha\beta}}{F}\cdot\delta_0 \quad (99)$$

or in mixed form:

$$\delta G_{\beta}^{\alpha} = G_{\beta}^{\alpha}\cdot\delta_0 + \overbrace{G_a^b\delta_{\alpha\beta}^{b\alpha}}^{\text{Gravity}} \xrightarrow{\forall\,\delta_{\alpha\beta}^{b\alpha}\ll\delta_0} = G_{\beta}^{\alpha}\cdot\delta_0 = g_{\beta}^{\alpha}\cdot\delta_0 = \delta_{\beta}^{\alpha}\cdot\delta_0 \quad (100)$$

and define:

$$I_{\alpha}^{\beta} \equiv R_{\alpha}^{\beta} - \frac{\delta_{\alpha}^{\beta}}{2}R + \frac{1}{2F}\eth_{\alpha}^{\beta}\left[g_{\alpha\beta};F\right] - \frac{\left(G^{\beta\mu}\nabla_{\mu}\nabla_{\alpha} - \delta_{\alpha}^{\beta}\Delta_G\right)\Phi}{\Phi}, \quad (101)$$

again with:

$$\frac{1}{2F}\eth_{\alpha}^{\beta}\left[g_{\alpha\beta};F\right]\equiv\frac{1}{2F}\left(\begin{array}{c}\frac{\delta_{\alpha}^{\beta}}{2}\left(\begin{array}{c}\left(\begin{array}{l}2F_{,ij}(n-1)g^{ij}+2\Gamma_{ij}^{a}F_{,a}g^{ij}\\-F_{,i}g^{ab}g_{jb,a}g^{ij}-F_{,j}g^{ab}g_{ib,a}g^{ij}\\-n\Gamma_{ij}^{d}F_{,d}g^{ij}+\frac{n}{2}F_{,d}g^{cd}g_{ab,c}g^{ab}\end{array}\right)\\+\frac{F_{,i}\cdot F_{,j}}{2F}g^{ij}\left((n-6)(n-1)\right)\end{array}\right)\\-g^{\beta\gamma}\left(\begin{array}{c}F_{,\alpha\gamma}(n-2)+F_{,ab}g_{\alpha\gamma}g^{ab}\\+F_{,a}g^{ab}\left(g_{\gamma b,\alpha}-g_{\gamma\alpha,b}\right)-F_{,a}g^{ab}g_{\gamma b,a}-F_{,\gamma}g^{ab}g_{\alpha b,a}\\+F_{,d}g^{cd}\frac{1}{2}n\left(\begin{array}{c}\frac{2}{n}g_{\alpha c,\gamma}-g_{\alpha c,\gamma}-g_{\gamma c,\alpha}\\+g_{\alpha\gamma,c}+\frac{1}{n}g_{\alpha\gamma}g_{ab,c}g^{ab}\end{array}\right)\\-\frac{1}{2F}\left(F_{,\alpha}\cdot F_{,\gamma}(3n-6)+g_{\alpha\gamma}F_{,c}F_{,d}g^{cd}(4-n)\right)\end{array}\right)\end{array}\right). \tag{102}$$

Now we can evaluate the covariant derivative of the integrand of Eq. (71) as follows:

$$\begin{aligned}\left(I_{\alpha}^{\beta}\delta G_{\beta}^{\alpha}\right)_{;\gamma}&=\left(I_{\alpha}^{\beta}\left(G_{\beta}^{\alpha}\cdot\delta_{0}+\overbrace{G_{a}^{b}\delta_{\alpha\beta}^{b\alpha}}^{\text{Gravity}}\right)\right)_{;\gamma}\xrightarrow{\forall\,\delta_{\alpha\beta}^{b\alpha}\ll\delta_{0}}=\left(I_{\alpha}^{\beta}G_{\beta}^{\alpha}\cdot\delta_{0}\right)_{;\gamma}\\&=I_{\alpha;\gamma}^{\beta}G_{\beta}^{\alpha}\cdot\delta_{0}+\overbrace{I_{\alpha}^{\beta}G_{\beta;\gamma}^{\alpha}\cdot\delta_{0}}^{=0}+I_{\alpha}^{\beta}G_{\beta}^{\alpha}\cdot\delta_{0;\gamma}\\&=I_{\alpha;\gamma}^{\beta}G_{\beta}^{\alpha}\cdot\delta_{0}+I_{\alpha}^{\beta}G_{\beta}^{\alpha}\cdot\delta_{0;\gamma}.\end{aligned} \tag{103}$$

Assuming that the second term in the last line is small compared to the first, this simplifies to:

$$\left(I_{\alpha}^{\beta}\delta G_{\beta}^{\alpha}\right)_{;\gamma}\simeq I_{\alpha;\gamma}^{\beta}G_{\beta}^{\alpha}\cdot\delta_{0}+\overbrace{I_{\alpha}^{\beta}G_{\beta;\gamma}^{\alpha}\cdot\delta_{0}}^{=0}=\left(I_{\alpha}^{\beta}G_{\beta}^{\alpha}\right)_{;\gamma}\cdot\delta_{0}. \tag{104}$$

In consequence, the covariant derivative of the integrand (71) (using Eqs. (101) and (104)) reads:

$$\left(I_{\alpha}^{\beta}\delta G_{\beta}^{\alpha}\right)_{;\gamma}\simeq\left(\left(R_{\alpha}^{\beta}-\frac{\delta_{\alpha}^{\beta}}{2}R+\frac{1}{2F}\eth_{\alpha}^{\beta}\left[g_{\alpha\beta};F\right]-\frac{\left(G^{\beta\mu}\nabla_{\mu}\nabla_{\alpha}-\delta_{\alpha}^{\beta}\Delta_{G}\right)\Phi}{\Phi}\right)G_{\beta}^{\alpha}\right)_{;\gamma}\cdot\delta_{0}$$

$$= \left(\left(1-\frac{n}{2}\right) \left(R - \left(\frac{1}{2F} \begin{pmatrix} 2F_{,ij}(n-1)g^{ij} + 2\Gamma_{ij}^{a}F_{,a}g^{ij} \\ -F_{,i}g^{ab}g_{jb,a}g^{ij} - F_{,j}g^{ab}g_{ib,a}g^{ij} \\ -n\Gamma_{ij}^{d}F_{,d}g^{ij} + \frac{n}{2}F_{,d}g^{cd}g_{ab,c}g^{ab} \end{pmatrix} + \frac{F_{,i}\cdot F_{,j}}{4F^2} g^{ij}\left((n-6)(n-1)\right) \right) \right) + (n-1)\frac{\Delta_G \Phi}{\Phi} \right)_{;\gamma} \cdot \delta_0. \tag{105}$$

As shown before (e.g., [6]), we can get rid of the non-linear differential operator with the approach *F*[*f*], via:

$$4FF'' + F' \cdot F'(n-6) = 0 \;\; \Rightarrow \;\; F[f] = \begin{cases} C_F \cdot \left(f + C_f\right)^{\frac{4}{n-2}} & n \neq 2 \\ C_F \cdot e^{f \cdot C_f} & n = 2, \end{cases} \tag{106}$$

transforming Eq. (105) into:

$$\left(I_{\alpha}^{\beta}\delta G_{\beta}^{\alpha}\right)_{;\gamma} \simeq \left(\left(R_{\alpha}^{\beta} - \frac{\delta_{\alpha}^{\beta}}{2}R + \frac{1}{2F}\eth_{\alpha}^{\beta}\left[g_{\alpha\beta};F\right] - \frac{\left(G^{\beta\mu}\nabla_{\mu}\nabla_{\alpha} - \delta_{\alpha}^{\beta}\Delta_G\right)\Phi}{\Phi} \right) G_{\beta}^{\alpha} \right)_{;\gamma} \cdot \delta_0$$

$$= \left(\left(1-\frac{n}{2}\right) \left(R - \frac{g^{\alpha\beta}}{2F}F' \left(2f_{,\alpha\beta}(n-1) + f_{,d}g^{cd} \begin{pmatrix} g_{\alpha c,\beta} - g_{\beta\alpha,c} - g_{c\beta,\alpha} \\ +\frac{n}{2}\left(2g_{\alpha\beta,c} - g_{\alpha c,\beta} - g_{\beta c,\alpha}\right) \end{pmatrix} \right) - (n-1)\frac{f_{,\alpha}\cdot f_{,\beta}}{4F^2} g^{\alpha\beta} \overbrace{\left(4FF'' + F' \cdot F'(n-6)\right)}^{=0} \right) + (n-1)\frac{\Delta_G \Phi}{\Phi} \right)_{;\gamma} \cdot \delta_0$$

$$= \left(\left(1-\frac{n}{2}\right) \left(R - \frac{g^{\alpha\beta}}{2F}F' \left(2f_{,\alpha\beta}(n-1) + f_{,d}g^{cd} \begin{pmatrix} g_{\alpha c,\beta} - g_{\beta\alpha,c} - g_{c\beta,\alpha} \\ +\frac{n}{2}\left(2g_{\alpha\beta,c} - g_{\alpha c,\beta} - g_{\beta c,\alpha}\right) \end{pmatrix} \right) \right) + (n-1)\frac{\Delta_G \Phi}{\Phi} \right)_{;\gamma} \cdot \delta_0. \tag{107}$$

As this should be a conservation law, the result should be zero and thus:

$$\left(I_{\alpha}^{\beta}\delta G_{\beta}^{\alpha}\right)_{;\gamma}=0$$

$$\Rightarrow$$

$$\left(\begin{array}{c}\left(1-\frac{n}{2}\right)\left(R-\frac{g^{\alpha\beta}}{2F}F'\left(2f_{,\alpha\beta}(n-1)+f_{,d}g^{cd}\left(\begin{array}{c}g_{\alpha c,\beta}-g_{\beta\alpha,c}-g_{c\beta,\alpha}\\+\frac{n}{2}\left(2g_{\alpha\beta,c}-g_{\alpha c,\beta}-g_{\beta c,\alpha}\right)\end{array}\right)\right)\right)\\+(n-1)\frac{\Delta_G\Phi}{\Phi}\end{array}\right)_{;\nu}=0. \tag{108}$$

Now we assume that the function $f[t,x,y,z]=f[t]$ shall code a Dirac particle at rest [13], which would be given due to:

$$f[t]=e^{\pm\mathrm{i}\cdot\frac{m_R\cdot c^2}{\hbar}\cdot t}\cdot C_f=e^{\pm\mathrm{i}\cdot\mu\cdot t}\cdot C_f, \tag{109}$$

with m_R giving the rest mass of the particle and $\hbar$ denoting the reduced Planck constant.

Further assuming an ordinary space-time in 4 dimensions with the metric tensor:

$$g_{\alpha\beta}=\begin{pmatrix}g_t[t]&0&0&0\\0&1&0&0\\0&0&1&0\\0&0&0&1\end{pmatrix};\quad g_t[t]=e^{\pm i\cdot m\cdot t} \tag{110}$$

and an oscillating perturbation of the kernel of the Einstein–Hilbert action via the kernel scaling function:

$$\Phi[t]=e^{\pm\mathrm{i}\cdot M\cdot t}\cdot C_\Phi, \tag{111}$$

and inserting all this into Eq. (108) results in the following algebraic equation:

$$\frac{3mM^2}{8}-\frac{m^2M}{8}-\frac{M^3}{4}+\frac{\left(2mM-m^2\right)}{4}\cdot\mu+\frac{1}{2}M\mu^2+\mu^3=0. \tag{112}$$

Thereby we have assumed only positive signs in the exponents for all functions f, Φ and g_t. Equation (112) can be written as follows:

$$0=-C_0+C_1\cdot\mu-C_2\cdot\mu^2+\mu^3. \tag{113}$$

The general solution to a three-order polynomial could be given via the following product form:

$$(\mu-\mu_1)\cdot(\mu-\mu_2)\cdot(\mu-\mu_3)$$
$$=\mu^3-\mu^2\cdot(\mu_1+\mu_2+\mu_3)+\mu\cdot(\mu_1\mu_2+\mu_1\mu_3+\mu_2\mu_3)-\mu_1\mu_2\mu_3. \qquad (114)$$

Comparing the last line of Eq. (114) with Eq. (113) gives us:

$$C_2=\mu_1+\mu_2+\mu_3=-\frac{M}{2}$$
$$C_1=\mu_1\mu_2+\mu_1\mu_3+\mu_2\mu_3=\frac{(2mM-m^2)}{4}$$
$$C_0=\mu_1\mu_2\mu_3=\frac{1}{4}\left(M^3-\frac{3mM^2}{2}+\frac{m^2M}{2}\right). \qquad (115)$$

Unfortunately, we only have two parameters and thus, will not be able to find a metric setting for the explanation of the three generations of elementary particles. The reason for this may be seen in our by far too simple metric approach, which we have only chosen to explicitly keep things simple and stick to the classical (Dirac) particle at rest. However, keeping the lazy spirit, how about a slight extension to a 3-parameter approach with still purely time-dependent metric components like:

$$g_{\alpha\beta}=\begin{pmatrix} g_t[t] & 0 & 0 & 0 \\ 0 & g_s[t] & 0 & 0 \\ 0 & 0 & g_s[t] & 0 \\ 0 & 0 & 0 & g_s[t] \end{pmatrix};\quad g_s[t]=e^{\pm i\cdot v\cdot t}\,? \qquad (116)$$

Now we obtain the equation:

$$\begin{pmatrix} -\frac{m^2M}{8}+\frac{3mM^2}{8}-\frac{M^3}{4}-\frac{m^2v}{8}+\frac{mMv}{2}-\frac{3M^2v}{8}+\frac{mv^2}{4} \\ -\frac{Mv^2}{4}-\frac{m^2\mu}{4}+\frac{mM\mu}{2}+\frac{mv\mu}{2}+\frac{v^2\mu}{2}+\frac{M\mu^2}{2}+\frac{3v\mu^2}{2}+\mu^3 \end{pmatrix}=0. \qquad (117)$$

4.2.5.4.1 *The three generations of elementary particles*

The numerical solutions to the third-order algebraic equation (117) for the charged leptons (electron μ = 0.511 MeV/c^2, muon μ = 105.7 MeV/c^2 and tau μ = 1777 MeV/c^2) would be (all in MeV/c^2):

$$m\to 1220.68,\ M\to 1221.7,\ v\to -548.507,$$
$$m\to -1054.41,\ M\to -1053.39,\ v\to 209.857,$$

where we have given only the two real solutions, while there are 6 complex ones in total. The corresponding results for the neutrinos and the quarks are given in [15]. Thereby we found funny asymmetries of numerical character with respect to matter and anti-matter particles [16].

4.2.5.5 Intermediate conclusions with respect to general systems

We found a manageable way to extract a weak-gravity scalar from the quantum Einstein field equations. Subjecting it to the covariant derivative operation, we obtained conservation laws. Applying the latter to a simple metric with just time-dependent components directly resulted in a polynomial of third order in the mass-parameter for the object described by the metric. The polynomial gives three solutions for such masses, and by feeding the known masses of the three elementary particle generations into the approach, we were able to evaluate the metric settings for charged and uncharged leptons and all quarks [15]. Interpreting the masses in a more general way for other systems than just elementary particles, namely, as inertia, we might ask whether also intelligent systems, for which the same fundamental (because purely mathematical) law holds, have to exist in such 3 generations.

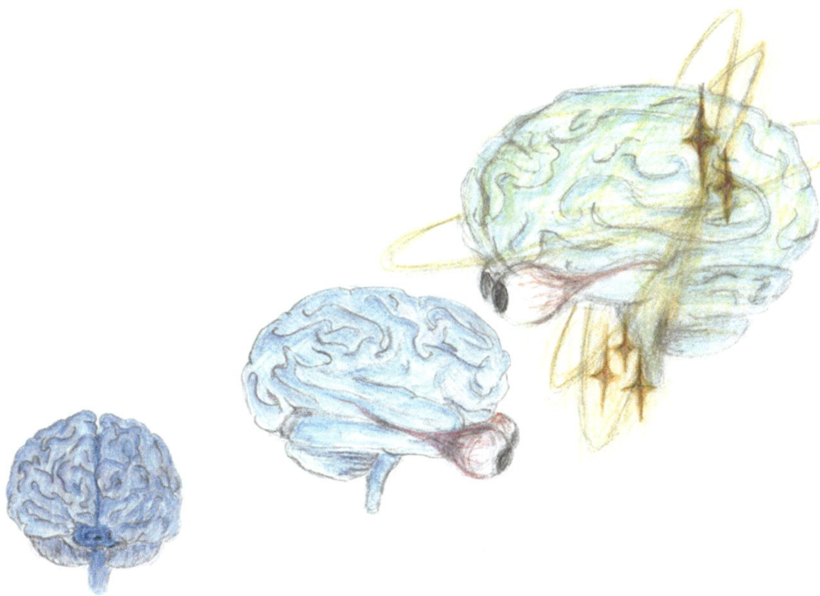

Figure 4.5 "The Three Generations of Consciousness" by Livia Schwarzer (see [36] for more).

So, for one thing, we should state that our finding with the quite fundamental third-order differential equation based on the purely mathematical Bianchi identity automatically leads to the conclusion that the occurrence of 3 generations of matter forms has to be a rather general aspect for any arbitrary system. This may help to understand certain three-fold appearances in socio-economy, nature, psychology, medicine and many other fields. Thereby, we have the interesting aspect of apparently only one generation being truly stable.

And of course, being so fundamental, the 3-generation aspect should also be of importance for intelligent systems (Fig. 4.5).

4.2.6 Tunneling: Influencing Others against Their Will

Naturally, the metric understanding of the quantum tunneling effect [20] with any arbitrary property automatically has some significant bearing on the psychological aspect of influencing another person's mind.

We can just say that the tunneling allows the understanding of getting ideas through somebody's guard and hook in (no matter how hard this somebody is fighting or mentally protecting himself). Interestingly, the math to describe this process is relatively simple. It does not need more than the metric construction of a suitable potential to obtain the necessary fundamental model [19]. Then the rest is just the classical calculus as published in a great variety of textbooks (e.g., [32–35]).

4.2.7 The Well Potential: Implanting Alien Ideas

Apart from the love-field application we have considered in section "4.1 Love Fields: Just Another Example of Physics in Psychology" above, a metric derivation of the quantum well trap with any arbitrary ensemble of properties [31] automatically has some significant impact also with respect to the understanding of the psychological aspect of storing thoughts, the "getting of ideas" and the inescapable escape of "intellectual property."

In somewhat clearer words: The quantum well storage of "objects = information" allows the understanding of ideas being taken on by somebody's mind and connecting (entangling) themselves or being entangled with existing thoughts already being present. Thereby, the existing thought (knowledge) may even build up the quantum well or trap. The math needed to obtain the metric for the generation of the right equations is given in the papers [20, 31] and the rest can be obtained from classical textbooks (e.g., [32–35]), only that the dimensions in the equations have to be taken as general properties, degrees of freedom or attributes characterizing human behavior.

4.2.8 Dirac/Pauli/The Fundamental Defense of Individuality and the Conscious Choice of Dimensionality

In appendix R it will be shown—in a compact but still comprehensive manner—how the Dirac equation [13] can be obtained from the Einstein-Hilbert action. The appendix follows a previous work done by this author [28], just adding a few things in accordance with the "Mathematical Psychology" topic of this book. The importance of this equation—as it will be worked out in the appendix—is not only the classical aspect of "explaining"[1] half spin and anti-matter, but the occurrence of the Dirac vector or spinor. The metric derivation shows where this object comes from and this has severe consequences with respect to any system being capable of holding Dirac equations and their solutions (in various dimensions).

Unfortunately, simply, because it would exceed the volume of the book dramatically, we cannot discuss all aspects of this metrically derived equation with respect to its bearings on mathematical psychology in this chapter, but only want to concentrate on the possibility to steer the variation of a given set of attributes towards a Ricci-free state. Some other aspects have been at least marginally covered in the next sub-section "4.2.9 Matter and Anti-Matter Mindsets."

How a conscious mind can obtain a Ricci-free problem in any even number of dimensions by just applying the right structure for the Ricci scaling[2] has been shown in appendix R. Thereby, it is derived that the structure of the Dirac spinor can be adjusted in such a way that in any even dimensionality the subsequent Hamilton extremal principle (optimization) leads to a perfectly linear quantum gravity equation with 100% additivity. This author suspects exactly this ability, namely the active structuring of the solution in order to achieve coherence and additivity, which goes along with Ricci-freedom, for the space-time-task being given, which is to say for the problem to be solved, as the essential characteristic of consciousness.

This way the conscious mind can create completely Ricci-free equations of perfect additivity. It must have cost evolution quite some headaches to find this solution ... unless the whole universe is not based on this concept

[1]As the Dirac equation [13] is only postulated, it does not "explain" half spin and anti-matter, but demands them out of its own structure and subsequent solutions portfolio.

[2]See below in the next sub-section about the meaning of the expression Ricci scaling.

anyway. This means that evolution had nothing to invent, but simply used a known concept on a different, which is to say biological, level.

4.2.9 Matter and Anti-Matter Mindsets

In appendix R, section "5.18.15 What Is Anti-Matter?," we will show that anti-matter is not just—as it was classically understood so far—mass which moves (exists more like) backward in time, but that it actually changes signs with matter in a symmetric fashion. Yes, there is the scaling function of the metric tensor where one sign-switch occurs and distinguishes matter from anti-matter. But there is also—at least for particles of Dirac character—the need to change the sign for the scaling of the Ricci-scalar in the Einstein–Hilbert action kernel. As the description of intelligent entities requires the use of generalized Dirac equations (quantum gravity Dirac equations), we can conclude that the possibility of matter and anti-matter states also exists for the mind, respectively, a set of minds.

But what could this possibly mean?

Well, this author is of the opinion that truly individual thinking processes always fall under the description of a generalized Fermi-particle, which is to say, they show enough individuality to claim their very own (private if you will) portion of generalized (attribute-formed) space-time. In other words, where there is one such mind (mind-set or mind-particle), there cannot be another one. Now the theory in appendix R shows us that the existence of the Dirac equation, allowing for such particles (mind-particles), requires the simultaneous scaling of both, the metric and the kernel of the Einstein–Hilbert action. Let us name the latter scaling the Ricci scaling. The scaling of the two has to be done in the opposite fashion, which is to say, if the metric scaling moves forward in time, the Ricci scaling has to move backward (matter) and vice versa (anti-matter). For truly individual thought processes, this is just the co-existence of the permanent balancing of thoughts going into one direction and others being directed in the opposite one ("opinion" and "anti-opinion"). It is the co-existence of our internal antagonism, which we experience with each and every more or less serious decision-making process. And it is of dear need, because without it, there would be no driving force feeding the search machine to move, respectively, to move at all. With the scaling of the metric moving forward in time we have the matter state (then the Ricci scaling moves backward), while it is the other way round for the anti-matter case. It now depends on the dominant-recessive character of the scaling whether the forward thinking or the backward thinking is going to win (and by how much). Usually, the metric scaling is much stronger than the Ricci scaling, which makes the matter-states to dominate the scene, but as the question of what is forward and what is backward in thinking is more or less a question of definition, this kind of "evaluation" is rather arbitrary (at least at the moment and without a specific example). The important aspect, however, is that one mind can hold many such sets simultaneously. So, it even can happen that a matter and an anti-matter state exist to the same topic, which automatically allows for the total annihilation of the two. This process will probably just be felt as some kind of release by the owner (creator and holder) of the two matter and anti-matter mind-sets and will also come with a burst of internal energy.

4.2.10 Collapse of the Wave Function and the Process of Decision-Making

It is evident that the metric understanding of the quantum collapse of the wave function in spaces or space-times with any arbitrary property and the bifurcation between the Dirac and the Klein–Gordon path, which is to say the Bosonic or the Fermionic phenomenology [21], automatically has some significant bearing on the psychological aspect of idea-evolution and decision making.

In somewhat clearer words: We have found that the collapse of the wave function, based on the "catastrophic" event of passing from linearity to non-linearity, and thus from superposition to singularity or from the Klein–Gordon option to the Dirac path or from Bosonic to Fermionic behavior, allows the understanding of the evolution of an ensemble of ideas towards a final decision. Thereby it is assumed that the "thinking mind" (if to say: in the mulling things over state) is able to keep a coherent state of additivity similar to the working state of a quantum computer. We may not even consciously realize this state, but we are aware of its ending, where—out of the many possibilities of the quantum reality and thus, superposable realm—one option has been "selected." Many repetitions of this process, potentially under various viewing angles (c.f. section "4.2.2 Einstein's Moon and the Theory of Perspectivity or How Do We See Other Individuals' Informational Output" above) will then lead to a final decision. In this sense, the working of the human mind (at least in cases of complex problem solving) seems to be very close to modern quantum computers, but we should not forget about the matter–anti-matter antagonism (see the section above), which probably plays a role as a driving force and—simultaneously—also as a perturbator.

4.2.11 A General Remark with Respect to Quantum Gravity Marketing

All this—of course—is also related to marketing ("quantum gravity marketing" one might suggest as an expression for these general tools) and thus, of much broader interest and with many more fields of applications than the lesser holistic reader might think. However, with respect to some advice given to this author regarding the exploitation of the results shown here and in the interest of brevity, we will keep "Quantum Psychology" out of this chapter. The connection undoubtedly merits a separate consideration, contemplation, discussion and—YES—publication. This, however, shall not prevent the interested and skilled reader from already starting working on the new and important field of "Quantum Marketing and Influencing."

And why do we think that the quantum or rather the quantum gravity approach is the right one for all mind things?

Well, if we were wrong about the quantum mind connection, the Dunning–Kruger dependency would never have come out of the generalized Planck radiation law (see section "2.1 How Do We Emit Confidence or Can the Dunning–Kruger Effect Be Mathematically Derived?" in this book above and [25]) regarding the emittance of confidence, the penetration of alien ideas into a well-protected mind could not be explained by quantum tunneling [20] and the collapse of an ensemble of thoughts to an idea or decision would probably not feel like a sudden flash of the mind [21].

But this also comes with a warning against those mind-controllers and totalitarians out there: With thoughts being of fundamentally quantum gravity character,

there is ALWAYS the fundamental possibility of a mental escape from any thought prison [2].

4.2.12 Cognitive Dissonance or Why It Is Soooo Difficult to See the Elephant in the Room

Cognitive dissonance is the avoidance of the truth.

How can we metrically understand and model such a situation?

As we have seen that every system—no matter how complex it is—can be characterized by a set of attributes, which then—seen as degrees of freedom or dimensions of the very system—are subjected to Hamilton's extremal principle in order to find the possible states in which the system can exist, we can—in principle—also evaluate the "true states" among all other possibilities. These states are just the ground or close-to-the-ground, which is to say the lowest energy states.

And why do we call such states the "true states"?

Well, simply because in these states the system has no energy left to hide anything, which is to say, it has no recourses to lie, because lying costs additional energy, which such a system does not have in the ground states of its existence. Therefore, the truth-seeking process is mathematically nothing else but an optimization or minimalization operation and it does not matter whether this has been performed by a human being or a piece of advanced artificial intelligence.

As we have already described the construction of various systems in a metrical manner in this book, there is no need here to add in yet another example. We do need, however, to point out that even though in most cases it will suffice to use our scalar quantum gravity equations, because one can arrange the gravity or internal metric deformation part to be small against the volume part, we still have to distinguish between Klein–Gordon and Dirac options in order to perform the "truth-seeking" on the whole picture and not just a part of it. The interested reader will find the necessary math for the Klein–Gordon part in appendices A–G, I–Q, S and V and for the Dirac option in appendices R, T and U.

In this section, we choose a purely verbal explanation of the problem of truth seeking and the avoidance of the same (which just is the state of cognitive dissonance, respectively leads to this state). For entertainment, we will use a communication this author has had with a well-informed and broadly interested reader, who also happens to be this author's friend.

The following interview was motivated by Steven Kirsch's wonderful presentation about "the elephant in the room" [40]. Hence the title of the section and the interview:

4.2.12.1 The elephant in the room (Fig. 4.6)

Troy: Recently I caught up with an old friend of mine. He works for an insurance company and falling under his supervision is the department where they handle pensions. He told me that they currently have so many deaths to handle that they were almost about to crack under the work load. I asked him to what reason he would attribute this "all of a sudden" "excessive mortality." But even with all the key words given, he answered that this probably had something to do with climate change, and that in spring there were always more deaths ... but that it never was this many. Can you believe that, Norbert?

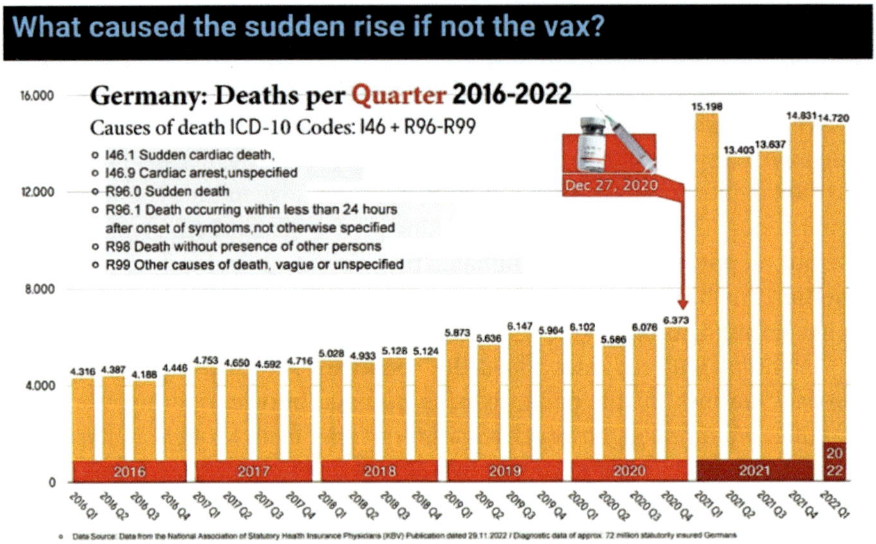

Figure 4.6 Obvious questions the cognitive dissonant person will not ask: "What caused the sudden rise if not the vax?"—Screenshot from Steven Kirsch's online presentation [40] about "the elephant in the room." Courtesy of S. Kirsch.

Norbert: Of course, I can! I know enough of these head cases myself. The three monkeys with the fourth on top who does not want to think ... or cannot think, which comes down to the same thing.

Troy: Cognitive dissonance, right?

Norbert: Yes!

Troy: Well, actually, I had already asked myself whether you would be able to use your general approach ... I mean this "mathematical psychology" [41] as you have coined it ... and model this kind of behavior?

Norbert: Actually, I have already done this!

Troy: Really, why didn't you tell anybody?

Norbert (shrugging): No idea?! Didn't see much importance in the topic ... thought the quantum gravity derivation of the origin of the Dunning–Kruger effect [25] would be much more interesting—I thought it might help me to understand my own stupidity, you see?!

Troy (shaking his head in some obvious disbelief): ...?

Troy (after a few moments and a little bit more head shaking ... with the final shake similar to the one a person does in order to shake off an irksome fly): So just tell me, what did you find there?!

Norbert: Actually, there isn't much about it. You simply have to understand how we evaluate the truth about things. Taking a holistic mathematical approach, this process is just an optimization procedure ... finding a minimum in a multi-dimensional and highly jointed landscape. If everything goes the normal way, you are able to find this minimum and this brings you close to the truth.

Troy: Why only close and not there?

Norbert: This is just the uncertainty residing in all things. The quantum gravity approach I used automatically takes care of this uncertainty and ensures that no determinism is possible. In other words: you never know everything and thus, the topography of your multi-dimensional and most badly jointed landscape is never fully known ... there is always a limit to the resolution of the "map."

Troy: Yes, that is clear ... and how does this bring you to the cognitive dissonance?

Norbert: Well, the optimization actually is a minimization of energy. I have to try and find the ground state as only this one holds the truth ... the true one, I mean.

Troy: Ok...ay?

Norbert: This process works best, respectively, is only possible, in an environment of low energy or little excitement if you like. Permanent jitter and nudging, pushing, poking from the outside **or the inside** disturbs the process or—even worse—brings my search vector off course and directs it to completely wrong ways. So, if I have to exist in an environment of high disturbance, which is just equivalent to high temperature or any other kind of energy, I have to isolate myself from the disturbances. I other words: **Before I can perform proper truth-seeking, I have to cleanse my surrounding, mentally cool down and lower my internal thought-level of excitement to underneath the truth level I'm keen to find. Without this cleansing process, I'm unable to even see the landscape in a sufficiently good resolution and my truth-seeking ends at a by far too high, which is to say untruthful, level**.

Troy: And the outside temperature, energy or whatever field is there ... it causes this "damaging" of your internal truth-seeking capability, right?

Norbert: Exactly Troy! The permanent **bombardment with false narratives not only directs your search vector in the wrong direction but also hinders it in cooling down** and cleansing before you could even think about performing a good truth-seeking process, because your landscape resolution is abysmal. And if you are permanently hindered in reaching a sufficiently cool enough energy or low excitement state, you will never clearly see the fissures and, thus, never find the bottom ... the true one, I mean. Even worse, unable to reach the lower states, you can easily be pushed around and nudged into directions where you find everything but the truth.

Troy: That's the state of cognitive dissonance then, right?

Norbert: Yes!

Troy (with brief meditative thought before answering): ... Wow! And let me guess, the cure is to remove the environmental excitement field?

Norbert: Yes, but there is a difficulty. The downfall from a high false-truth state comes nothing short of a mental disaster. **It is extremely painful and therefore the severely cognitive dissonant person tries everything to hold on to the false narrative ... no matter what it costs.** It is like the resolution of an excited state in the laser. There needs to be some push to make the person jump and really want to find the truth. The cleansing of the environment and the switching off of the perturbation or false-narrative-push field is only half the job to make the patient see. **You also have to force them to open their eyes and jump**.

4.2.13 References for Section 4.2

1. S. W. Hawking, Particle creation by black holes. *Commun. Math. Phys.*, 43, 1975, 199–220.
2. N. Schwarzer, *Derivation Hawking Radiation of Black Holes from Quantum Einstein Field Equations*, self-published, Amazon Digital Services, ASIN: B0BXQ5ZG4Z, 2023, Kindle.
3. D. Hilbert, Die Grundlagen der Physik, Teil 1, *Göttinger Nachrichten*, 1915, 395–407.
4. A. Einstein, Grundlage der allgemeinen Relativitätstheorie, *Annalen der Physik*, 49, ser. 4, 769–822.

5. N. Schwarzer, *The World Formula: A Late Recognition of David Hilbert's Stroke of Genius*, Jenny Stanford Publishing, ISBN: 9789814877206.
6. N. Schwarzer, *The Math of Body, Soul, and the Universe*, Jenny Stanford Publishing, ISBN 9789814968249.
7. N. Schwarzer, *The Theory of Everything—Quantum and Relativity Is Everywhere—A Fermat Universe*, Jenny Stanford Publishing, 2020, ISBN-10: 9814774472.
8. N. Schwarzer, *The Metric Electron—Derivation of Spin-1/2-Objects from the Einstein-Hilbert Action*, self-published, ASIN: B0BTPQM26X, Amazon Digital Services, 2023, Kindle.
9. N. Schwarzer, *Towards the Metric Quark—Derivation of More Spin-1/2-Objects from the Einstein-Hilbert Action*, self-published, Amazon Digital Services, B0BWDQDBFW, 2023, Kindle.
10. N. Schwarzer, *Quantum Gravity—Brief Derivation of Quantum Einstein Field Equations*, self-published, Amazon Digital Services, ASIN: B0BWTWZHP5, 2023, Kindle.
11. E. Schrödinger, Quantisierung als Eigenwertproblem (erste Mitteilung), *Ann. Phys.*, 1926, 384(4), 361–376.
12. E. Schrödinger, Quantisierung als Eigenwertproblem (zweite Mitteilung), *Ann. Phys.*, 1926, 384(6), 489–527.
13. P. A. M. Dirac, *The Quantum Theory of the Electron*, Published 1 February 1928, Proceedings of the Royal Society A. DOI: 10.1098/rspa.1928.0023.
14. N. Schwarzer, *Quantum Gravity and Einstein's Famous Moon Problem—The Theory of Perspectivity*, self-published, Amazon Digital Services, ASIN: B0BXPQQWZN, 2023, Kindle.
15. N. Schwarzer, *Shouldn't Quantum Gravity Solve the 3-Generation Problem*?, self-published, Amazon Digital Services, ASIN: B0BXPXP33L, 2023, Kindle.
16. N. Schwarzer, *About Apparent Asymmetries in the 3-Generations*, self-published, Amazon Digital Services, ASIN: B0BXPYGMNF, 2023, Kindle.
17. N. Schwarzer, *The Universe's Half-Spin Deception Game*, self-published, Amazon Digital Services, ASIN: B0BXPKWPQZ, 2023, Kindle.
18. N. Schwarzer, *Metric Interpretation of Quantum Theory—How Can Quantum Gravity Help Us to Understand Quantum Strangeness*, self-published, Amazon Digital Services, ASIN: B0BXQRKPV7, 2023, Kindle.

19. N. Schwarzer, *The Metric Potential—How Can Quantum Gravity Help Us to Understand the Origin of Forces*, self-published, Amazon Digital Services, ASIN: B0BXSHQZX1, 2023, Kindle.
20. N. Schwarzer, *Understanding Quantum Tunneling—How Can Quantum Gravity Help Us to Get the Gist about This Strange Phenomenon*, Amazon Digital Services, 2023, Kindle, ASIN: B0C241FQH3.
21. N. Schwarzer, *The Death of Schrödinger's Cat—How Quantum Gravity Solves the Problem of the Wave Function Collapse*, Amazon Digital Services, 2023, Kindle, ASIN: B0C2444JZX.
22. N. Schwarzer, *How Systems Evolve—A Red Pill Course on Evolution*, self-published, Amazon Digital Services, 2022, Kindle, ASIN: B0B6RGBKQ4.
23. N. Schwarzer, *The Quantum Gravity War—How Will the Nearby Unification of Physics Change Future Warfare?*, Jenny Stanford Publishing, ISBN: 9789814968584.
24. N. Schwarzer, *Quantum Gravity Systems Science*, self-published, ASIN: B0BRL8SKSB, Amazon Digital Services, 2023, Kindle.
25. N. Schwarzer, *Why Tucker Carlson Is Perfectly Right about the WEF-People—A Mathematical Proof*, self-published, ASIN: B0BSTGKC6F, Amazon Digital Services, 2023, Kindle.
26. N. Schwarzer, *Derivation of the Harmonic Oscillator from the Einstein-Hilbert Action—The Metric Harmonic Oscillator*, self-published, ASIN: B0BSXJCJVB, Amazon Digital Services, 2023, Kindle.
27. N. Schwarzer, *Derivation of the Schrödinger Hydrogen from the Einstein-Hilbert Action: The Metric Schrödinger Hydrogen*, self-published, ASIN: B0BT9Y54XP, Amazon Digital Services, 2023, Kindle.
28. N. Schwarzer, *Dirac's Little Spinor Problem—Can Quantum Gravity Explain the Occurrence of this Postulated Object?*, self-published, ASIN: B0C3QWJSQS, Amazon Digital Services, 2023, Kindle.
29. N. Schwarzer, *Mathematical Psychology—Tools for the Understanding, Simulation, Mitigation and Potential Prevention of Mass Formation Psychoses*, self-published, Amazon Digital Services, ASIN: B0BC5T1HB4, 2022, Kindle.
30. D. Song, Einstein's Moon, *Physics-Uspekhi*, 2012, 182(9), 1013 https://doi.org/10.48550/arXiv.1008.2892 & https://arxiv.org/pdf/1008.2892.pdf.
31. N. Schwarzer, *The Quantum Gravity Well—From Particle Physics to Psychology—a True Theory of Everything?*, self-published, Amazon Digital Services, ASIN: B0C3QV9S1Y, 2023, Kindle.

32. W. Greiner, Quantenmechanik—Teil 1 *Einführung*, in: *Theoretische Physik*, 5th ed., in German, Verlag Harri Deutsch, Frankfurt am Main, vol. 4, 1992, ISBN 3-8171-1206-8.
33. F. Schwabl, *Quantenmechanik*, 2nd ed., in German, Springer-Verlag Berlin Heidelberg New York, 1990, ISBN 3-540-52359-6.
34. H. Haken, H. Chr. Wolf, *Atom- und Quantenphysik*, 4th ed., in German, Springer Heidelberg, 1990, ISBN 0-387-52198-4.
35. C. Cohen-Tannoudji, B. Diu, F. Laloë, Franck, *Quantenmechanik 1&2*, 2. Auflage, Walter de Gruyter, Berlin-New York, 1999, ISBN-10: 3110626004.
36. J. Schwarzer, *3 Generations of Consciousness*, video presentation on YouTube, https://youtu.be/0AlFzIqwxN8.
37. https://en.wikipedia.org/wiki/Generation_(particle_physics).
38. C. A. Sporea, *Notes on f(R) Theories of Gravity*, 2014, arxiv.org/pdf/1403.3852.pdf.
39. H. Goenner, *Einführung in die spezielle und allgemeine Relativitätstheorie*, in German, Spektrum Akad, Verlag, Heidelberg, Berlin, Oxford, 1996, ISBN: 3-86025-333-6.
40. St. Kirsch, *Why Is Everyone So Afraid to Talk about the Elephant in the Room?*, online presentation, https://docs.google.com/presentation/d/1VU30HapG_RV1wG8z4T4fcMnSf0Kujpbmvk6GKaOJQCM/mobilepresent?slide=id.g2161d8a89b3_0_23.
41. This book.

4.3 The Most Important Conclusion of Them All

"For the right mind it always holds that thoughts are freer than the thinker!"

5

Desserts: Appendices

In the following appendices, we will only give the essential basics necessary for simple illustrative examples of the "mathematical psychology" applications of metric character that we discussed in this book. The reader can find much more information in [4, 5, 55], whereby we refer to the "Global References" in all appendices, except appendix R, where we have used local references.

While appendices A to Q present a variety of different aspects regarding the quantum gravity equations which are necessary for the derivation of the space-times of the mind, we added a more compact exposure in appendix R for the reader who is only interested in the most fundamental origin of our "mathematical psychology" ansatz. The sub-appendices in R deal with practical or mathematical problems directly being related to some of the mathematical psychology topics treated in this book.

5.1 Appendix A: System as Ensemble of Properties

We start with the assumption that every system, from the elementary particle via a "primitive" or even complex life-form to the whole universe, can be considered an ensemble of potentially entangled properties or attributes. These properties are the degrees of freedom of the system and as such they are also the dimensions of that very system forming a space or space-time. The system can mathematically be described via its metric. This most

Mathematical Psychology: The World of Thoughts as a Quantum Space-Time with a Gravitational Core
Norbert Schwarzer
Copyright © 2025 Jenny Stanford Publishing Pte. Ltd.
ISBN 978-981-5129-27-4 (Hardcover), 978-1-003-58203-8 (eBook)
www.jennystanford.com

general description of any system now allows us to investigate the system when simply assuming that it has to follow certain and very universal laws. The most fundamental law we could think of right now would be a Hamilton extremal principle and this directly brings us to our starting point.

5.2 Appendix B: Fundamental Starting Point

Our starting point shall be the Hamilton principle in form of the Einstein–Hilbert action [1]:

$$\begin{aligned}\delta W = 0 = \delta\int_V d^n x\left(\sqrt{-g}\cdot\Phi_R[R]\right)\\ = \delta\int_V d^n x\left(\sqrt{-g}\cdot\left(R-2\Lambda+L_M\right)\right).\end{aligned} \tag{118}$$

Here R gives the Ricci scalar and g denotes the determinant of the metric tensor $g_{\alpha\beta}$. The term L_M stands for the matter Lagrange density and Λ gives the cosmological constant. Assuming that we can extract a functional factor $F[f]$, with f being a function of all coordinates, from the metric tensor in the following way:

$$G_{\alpha\beta} = g_{\alpha\beta}\cdot F[f], \tag{119}$$

it was shown in [4] that we do not only obtain the Einstein field equations [1, 2], but also the most important quantum equations (see also [5] and appendices below).

It needs to be pointed out that the "variation" (119) is only the simplest form of a general metric adjustment as follows:

$$G_{\alpha\beta} = g_{ab}\cdot F_{\alpha\beta}^{ab}\left[f_{\alpha\beta}^{ab}\right]\cdot F[f]. \tag{120}$$

Please note that—just like in Eq. (119)—the indexed capital F is only the functional wrapper for a set of functions $f_{\alpha\beta}^{ab}$. As we will see below, this wrapper will help us to realize certain quantum forms. There is only one summation with respect to the outer indices. Please also note, when demanding that the new metric $G_{\alpha\beta}$ should still be a tensor and the object $F_{\alpha\beta}^{ab}\left[f_{\alpha\beta}^{ab}\right]$ shall not simply be a coordinate transformation, then g_{ab} cannot be a tensor. Thus, in order to make this clear, we may write Eq. (120) as:

$$G_{\alpha\beta} = \overbrace{\gamma_{ab}\cdot F\left[f_{\alpha\beta}^{ab}\right]_{\alpha\beta}^{ab}}^{g_{\alpha\beta}}\cdot F_0\left[f_0\right]. \tag{121}$$

This much more complex form, however, will not be considered in this book. The interested reader is referred to [5], where first attempts are made to rigorously handle such an extension.

Another interesting way to extract quantum gravity equations has been discussed in [55] in connection with a fundamental derivation of evolution equations.

5.3 Appendix C: Second-Order Quantum Equations from the Einstein–Hilbert Action

While in the appendices G and I we are going to derive second-order differential equations, clearly showing the characteristics of quantum equations, directly from generalized forms of the Einstein–Hilbert action via a variation with respect to the volume of the metric, we here want to demonstrate the appearance of such equations when following the classical Hilbert task in treating his famous action integral.

Setting Eq. (119) into Eq. (118) we obtain the following Einstein–Hilbert action [5]:

$$\delta W = 0$$

$$= \delta \int_V d^n x \left(\sqrt{-g \cdot F^n} \times \left(\left(\begin{array}{c} R - \frac{1}{2F} \left(\begin{array}{c} 2F_{,\alpha\beta}(n-1) g^{\alpha\beta} + 2\Gamma^{a}_{\alpha\beta} F_{,a} g^{\alpha\beta} \\ -F_{,\alpha} g^{ab} g_{\beta b,a} g^{\alpha\beta} - F_{,\beta} g^{ab} g_{\alpha b,a} g^{\alpha\beta} \\ -n\Gamma^{d}_{\alpha\beta} F_{,d} g^{\alpha\beta} + \frac{n}{2} F_{,d} g^{cd} g_{ab,c} g^{ab} \end{array} \right) \\ -\frac{F_{,\alpha} \cdot F_{,\beta}}{4F^2} g^{\alpha\beta} \left((n-6)(n-1) \right) \end{array} \right) \frac{1}{F} \\ -2\Lambda + L_M \right) \right). \quad (122)$$

Thereby, in order to keep things simple and still general enough, here we often restrict ourselves to diagonal metrics with the following properties:

$$g_{ij} = \begin{pmatrix} g_{00} & \cdots & 0 \\ \vdots & \ddots & \vdots \\ 0 & \cdots & g_{n-1n-1} \end{pmatrix}; \quad g_{jj,j} = 0, \quad (123)$$

which gives the following Hilbert integral:

$$\delta W = 0$$

$$= \delta \int_V d^n x \left(\sqrt{-g \cdot F^n} \times \left(\left(\begin{array}{c} R + \frac{F'}{F}(1-n)\Delta f \\ + \frac{f_{,\alpha} f_{,\beta} g^{\alpha\beta} (1-n)}{4F^2} \left(4FF'' + (F')^2 (n-6)\right) \end{array} \right) \frac{1}{F} \\ -2\Lambda + L_M \right) \right). \tag{124}$$

After performing the variation, we obtain the following quantum gravity field equations:

$$0 = \left(\begin{array}{c} R_{\alpha\beta} - \frac{g_{\alpha\beta}}{2} R + \frac{g_{\alpha\beta}}{2} \left(\begin{array}{c} \frac{1}{2F} \left(\begin{array}{c} 2F_{,ij}(n-1) g^{ij} + 2\Gamma_{ij}^{a} F_{,a} g^{ij} \\ -F_{,i} g^{ab} g_{jb,a} g^{ij} - F_{,j} g^{ab} g_{ib,a} g^{ij} \\ -n\Gamma_{ij}^{d} F_{,d} g^{ij} + \frac{n}{2} F_{,d} g^{cd} g_{ab,c} g^{ab} \end{array} \right) \\ + \frac{F_{,i} \cdot F_{,j}}{4F^2} g^{ij} \left((n-6)(n-1)\right) \end{array} \right) \\ -\frac{1}{2F} \left(\begin{array}{c} F_{,\alpha\beta}(n-2) + F_{,ab} g_{\alpha\beta} g^{ab} \\ +F_{,a} g^{ab} \left(g_{\beta b,\alpha} - g_{\beta\alpha,b}\right) - F_{,\alpha} g^{ab} g_{\beta b,a} - F_{,\beta} g^{ab} g_{\alpha b,a} \\ +F_{,d} g^{cd} \frac{1}{2} n \left(\frac{2}{n} g_{\alpha c,\beta} - g_{\alpha c,\beta} - g_{\beta c,\alpha} + g_{\alpha\beta,c} + \frac{1}{n} g_{\alpha\beta} g_{ab,c} g^{ab} \right) \\ -\frac{1}{2F} \left(F_{,\alpha} \cdot F_{,\beta}(3n-6) + g_{\alpha\beta} F_{,c} F_{,d} g^{cd} (4-n)\right) \end{array} \right) \\ +\kappa T_{\alpha\beta} + \Lambda \cdot F \cdot g_{\alpha\beta} \end{array} \right), \tag{125}$$

which, for metrics of the form (123), can be split up and simplified to:

$$1.\quad 0 = \left(\begin{array}{c} \left(R + \frac{F'}{F}(1-n)\Delta f\right)\left(1 - \frac{n}{2}\right) \\ + \frac{f_{,\alpha} f_{,\beta} (1-n)}{4F^2} g^{\alpha\beta} \left(4FF'' + F'F'(n-6)\right)\left(1 - \frac{n}{2}\right) \\ + \kappa T_{\alpha\beta} g^{\alpha\beta} + \Lambda \cdot F \cdot n \end{array} \right) \delta \frac{1}{F_i}$$

$$\Rightarrow \quad -R \cdot \delta \frac{1}{F_i} = \left(\begin{array}{c} \left(\frac{F'}{F}(1-n)\Delta f\right) \\ + \frac{f_{,\alpha} f_{,\beta} (1-n)}{4F^2} g^{\alpha\beta} \left(4FF'' + F'F'(n-6)\right) \\ + \frac{2}{2-n} \left(\kappa T_{\alpha\beta} g^{\alpha\beta} + \Lambda \cdot F \cdot n\right) \end{array} \right) \delta \frac{1}{F_i}$$

$$\Rightarrow \quad 2. \quad 0=\left(\begin{array}{c} R^{***}_{\alpha\beta}+R_{\alpha\beta}+\left(\begin{array}{c}\left(\frac{F'}{F}(1-n)\Delta f\right)\\ +\frac{f_{,\kappa}f_{,\lambda}(1-n)}{4F^{2}}g^{\kappa\lambda}\left(4FF''+F'F'(n-6)\right)\\ +\frac{2}{2-n}\left(\kappa T_{\kappa\lambda}g^{\kappa\lambda}+\Lambda\cdot F\cdot n\right)\end{array}\right)\cdot\frac{g_{\alpha\beta}}{2}\\ +\frac{F'}{F}(1-n)\left(f_{,\alpha\beta}-\Gamma^{\gamma}_{\alpha\beta}f_{,\gamma}-\frac{g_{\alpha\beta}}{2}\Delta f\right)\\ +\frac{(1-n)}{4F^{2}}\left(\begin{array}{c}4FF''\\ +F'F'(n-6)\end{array}\right)\left(\begin{array}{c}f_{,\alpha}f_{,\beta}\\ -f_{,\mu}f_{,\nu}g^{\mu\nu}\frac{g_{\alpha\beta}}{2}\end{array}\right)\\ +\kappa T_{\alpha\beta}+\Lambda\cdot F\cdot g_{\alpha\beta}\end{array}\right)\frac{\delta g^{\alpha\beta}}{F_{i}}, \tag{126}$$

$$R^{***}_{\alpha\beta}\equiv R^{**}_{\alpha\beta}+(n-1)\left(\begin{array}{c}\frac{F'}{F}\left(f_{,\alpha\beta}-\Gamma^{\gamma}_{\alpha\beta}f_{,\gamma}-\frac{g_{\alpha\beta}}{2}\Delta f\right)\\ +\frac{1}{4F^{2}}\left(\begin{array}{c}4FF''^{ab}\\ +F'F'(n-6)\end{array}\right)\left(\begin{array}{c}f_{,\alpha}f_{,\beta}\\ -f_{,\mu}f_{,\nu}g^{\mu\nu}\frac{g_{\alpha\beta}}{2}\end{array}\right)\end{array}\right), \tag{127}$$

$$R^{**}_{\alpha\beta}\equiv\frac{1}{2F}\left(\begin{array}{c}(2-n)F_{,\alpha\beta}+\frac{1}{2F}\left((3n-4)F_{,\alpha}F_{,\beta}+(2-n)F_{,c}g_{\alpha\beta}F_{,a}g^{ac}\right)\\ +\left(\begin{array}{c}\frac{1}{2}\left(\begin{array}{c}F_{,\alpha}g_{bd,\beta}g^{bd}+g_{ac,\alpha}F_{,\beta}g^{ac}-g_{\alpha c,\beta}F_{,d}g^{cd}\\ -\left(F_{,a}g^{ac}(2-n)\left(g_{\alpha c,\beta}+g_{\beta c,\alpha}-g_{\alpha\beta,c}\right)\right)\end{array}\right)\\ -\left(\begin{array}{c}2F_{,b}g_{\alpha\beta,a}+F_{,a}g_{\alpha\beta,b}+F_{,ab}g_{\alpha\beta}\\ +\frac{1}{2}\left(F_{,\alpha}g_{ab,\beta}+F_{,\beta}g_{ab,\alpha}\right)-F_{,a}g_{\alpha b,\beta}-F_{,a}g_{\beta b,\alpha}\\ +\frac{1}{2}F_{,c}g_{\alpha\beta}\left(2g_{bd,a}-g_{ab,d}\right)g^{cd}\end{array}\right)g^{ab}\end{array}\right)\end{array}\right). \tag{128}$$

Thereby we simply applied the following separation of the variated contravariant metric tensor:

$$\delta G^{\alpha\beta}=g^{\alpha\beta}\delta\frac{1}{F_i}+\frac{\delta g^{\alpha\beta}}{F_i}. \tag{129}$$

Using the assumption that no extra matter is needed, because it has been provided by the approach via dimensional entanglement [5], this may be simplified as follows:

$$1.\quad 0=\begin{pmatrix}\left(R+\frac{F'}{F}(1-n)\Delta f\right)\left(1-\frac{n}{2}\right)\\ +\frac{f_{,\alpha}f_{,\beta}(1-n)}{4F^2}g^{\alpha\beta}\left(4FF''+F'F'(n-6)\right)\left(1-\frac{n}{2}\right)\end{pmatrix}\delta\frac{1}{F_i}$$

$$\Rightarrow\qquad -R=\begin{pmatrix}\left(\frac{F'}{F}(1-n)\Delta f\right)\\ +\frac{f_{,\alpha}f_{,\beta}(1-n)}{4F^2}g^{\alpha\beta}\left(4FF''+F'F'(n-6)\right)\end{pmatrix}$$

$$\Rightarrow\ 2.\quad 0=\begin{pmatrix}R^{***}_{\alpha\beta}+R_{\alpha\beta}+\begin{pmatrix}\left(\frac{F'}{F}(1-n)\Delta f\right)\\ +\frac{f_{,\kappa}f_{,\lambda}(1-n)}{4F^2}g^{\kappa\lambda}\left(4FF''+F'F'(n-6)\right)\end{pmatrix}\cdot\frac{g_{\alpha\beta}}{2}\\ +\frac{F'}{F}(1-n)\left(f_{,\alpha\beta}-\Gamma^{\gamma}_{\alpha\beta}f_{,\gamma}-\frac{g_{\alpha\beta}}{2}\Delta f\right)\\ +\frac{(1-n)}{4F^2}\begin{pmatrix}4FF''\\ +F'F'(n-6)\end{pmatrix}\begin{pmatrix}f_{,\alpha}f_{,\beta}\\ -f_{,\mu}f_{,\nu}g^{\mu\nu}\frac{g_{\alpha\beta}}{2}\end{pmatrix}\end{pmatrix}\frac{\delta g^{\alpha\beta}}{F_i}. \tag{130}$$

Here the symbol $R_{\alpha\beta}$ stands for the Ricci tensor. As we will see in just another step of our derivation that Eq. (130) does contain the Einstein field equations and the basic quantum equations, we call the ensemble the quantum Einstein field equations. With the condition 4FF" + F'F' (n–6) = 0, we obtain the very brief form of quantum Einstein field equations as:

$$1. \quad -R=\frac{F'}{F}(1-n)\Delta f$$

$$\Rightarrow \quad 2. \quad R^{***}_{\alpha\beta}+R_{\alpha\beta}=\frac{F'}{F}(n-1)\overbrace{\left(f_{,\alpha\beta}-\Gamma^{\gamma}_{\alpha\beta}f_{,\gamma}\right)}^{\equiv f_{:\alpha\beta}}.$$

$$\Rightarrow \quad R^{***}_{\alpha\beta}+R_{\alpha\beta}=\frac{F'}{F}(n-1)f_{:\alpha\beta}. \tag{131}$$

With F given as:

$$4FF''+F'F'(n-6)=0 \;\Rightarrow\; F=\left(C_f \pm f\right)^{\frac{4}{n-2}};\quad n>2, \tag{132}$$

this results in:

$$\Rightarrow \quad \boxed{\begin{array}{l} 1.\quad 0=R\cdot\left(C_f\pm f\right)\pm\frac{4\cdot(n-1)}{(n-2)}\Delta f \\ 2.\quad \left(R^{***}_{\alpha\beta}+R_{\alpha\beta}\right)\cdot\left(C_f\pm f\right)=\pm\frac{4\cdot(n-1)}{(n-2)}f_{:\alpha\beta}\end{array}}. \tag{133}$$

Assuming that the variation with respect to the metric tensor can be ignored, because we have $\frac{\delta g^{\alpha\beta}}{F_i}=0$, we would only be left with the variational part of the volume function and have from Eq. (133) the following scalar equation:

$$\Rightarrow \quad \boxed{0=R\cdot\left(C_f\pm f\right)\pm\frac{4\cdot(n-1)}{(n-2)}\Delta f}. \tag{134}$$

The experienced reader may recognize the Klein–Gordon equation in an arbitrary number of dimensions and a Laplace operator with respect to a metric $g_{\alpha\beta}$. The Klein–Gordon equation, however, is just one of the most important quantum equations from which the Dirac and the Schrödinger equation can be evaluated. This was shown for the general metric form as given, e.g., in [5].

Please note: From the very beginning of the development of the mathematical basics of Quantum Theory, researchers tried to construct a Quantum Theory concordant with relativity. The classical Klein–Gordon equation may be seen as a first success here, because it satisfies both, quantum aspects and is concordant with Einstein's Special Theory of Relativity. What the classical form did not deliver was a conclusive agreement also with Einstein's General Theory of Relativity. By the derivation of the metric Klein–Gordon equation directly out of the uterus of Einstein's theories, which is to say Hilbert's so-called Einstein–Hilbert action, we automatically have obtained the Klein–Gordon quantum equation now in concordance with the General Theory of Relativity.

Critics may point out that our setting of $\frac{\delta g^{\alpha\beta}}{F_i}=0$ can hardly be accepted as a complete derivation of a Quantum Gravity Theory or subsequent quantum Einstein field equations and we have to agree. However, as said above, the simplest forms of quantum Einstein field equations, being obtained via a metric (volumetric) adjustment of the form (119), being itself the simplest possible form of the general "variation" (120), are given via Eq. (130) and not Eq. (133) or even just Eq. (134). This simple form only results after a metric scaling of the type (119), a setting of the scaling wrapper in accordance to Eq. (132) and the assumption of a negligible metric variation $\frac{\delta g^{\alpha\beta}}{F_i}=0$. Nevertheless, for our purposes here, such simplifications are considered justified, because the model is merely used for illustration and not for perfectly mirroring a real socio-economic system with millions or even billions of individuals.

5.3.1 Metric Observables

In Quantum Theory, observables are assumed to be based on positive reals. Metrics cannot be put under such a stringent treatment, simply because we know that there have to be certain signatures in order to have a possibility to distinguish between time- and space-like dimensions, for instance. However, our approach with the scaled metric as introduced above, reading:

$$\begin{aligned} G_{\alpha\beta} &= F\left[f\left[t,x,y,z,\ldots,\xi_k,\ldots,\xi_n\right]\right]\cdot\delta_\alpha^i\delta_\beta^j g_{ij} \\ &= F\left[f\left[t,x,y,z,\ldots,\xi_k,\ldots,\xi_n\right]\right]\cdot g_{\alpha\beta} = F\left[f\right]\cdot g_{\alpha\beta}, \end{aligned} \tag{135}$$

where the scale becomes the wave functions, should not disrupt the principal signature of a metric. We might therefore like to resort to a slightly different scaling approach, given via:

$$G_{\alpha\beta} = F \cdot \bar{F} \cdot g_{\alpha\beta}, \tag{136}$$

where $\bar{F}$ denotes the conjugate complex of F.

In 4 dimensions and simple metrics of the kind (123), this changes R^* in contrast to the kernel of Eq. (122) (c.f. Eq. (182)) as follows:

$$R^* = \operatorname{sgn}\left[F[f]\right]^2 \cdot \left(\begin{array}{c} \frac{R}{F[f]^2} - \frac{3}{2 \cdot F[f]^4} \\ \times \left(\begin{array}{c} 2 \cdot \left(1+\operatorname{sgn}\left[F[f]\right]^2\right) \cdot F[f] \cdot \frac{\partial F[f]}{\partial f} \cdot \Delta f \\ + \left(\begin{array}{c} 2 \cdot \left(1+\operatorname{sgn}\left[F[f]\right]^2\right) \cdot F[f] \cdot \frac{\partial^2 F[f]}{\partial f^2} \\ -\left(\operatorname{sgn}\left[F[f]\right]^2 - 1\right)^2 \cdot \left(\frac{\partial F[f]}{\partial f}\right)^2 \end{array} \right) \cdot f_{,\alpha} g^{\alpha\beta} f_{,\beta} \end{array} \right) \end{array} \right). \tag{137}$$

The corresponding Ricci tensor reads:

$$R^* = R^*_{\alpha\beta} G^{\alpha\beta} = R^*_{\alpha\beta} \left(F \cdot \bar{F}\right)^{-1} \cdot g_{\alpha\beta},$$

$$\Rightarrow R^*_{\alpha\beta} g_{\alpha\beta} =$$

$$\left(\begin{array}{c} R^{***}_{\alpha\beta} + R_{\alpha\beta} - \frac{3}{2 \cdot F[f]^2} \\ \times \left(\begin{array}{c} 2 \cdot \left(1+\operatorname{sgn}\left[F[f]\right]^2\right) \cdot F[f] \cdot \frac{\partial F[f]}{\partial f} \cdot \left(f_{,\alpha\beta} - \Gamma^{\gamma}_{\alpha\beta} f_{,\gamma}\right) \\ + \left(\begin{array}{c} 2 \cdot \left(1+\operatorname{sgn}\left[F[f]\right]^2\right) \cdot F[f] \cdot \frac{\partial^2 F[f]}{\partial f^2} \\ -\left(\operatorname{sgn}\left[F[f]\right]^2 - 1\right)^2 \cdot \left(\frac{\partial F[f]}{\partial f}\right)^2 \end{array} \right) \cdot f_{,\alpha} f_{,\beta} \end{array} \right) \end{array} \right) g^{\alpha\beta}, \tag{138}$$

$\Rightarrow$

$$R^{*}_{\alpha\beta}=\left(R^{***}_{\alpha\beta}+R_{\alpha\beta}-\frac{3}{2\cdot F[f]^{2}}\times\left(\begin{array}{c}2\cdot\left(1+\mathrm{sgn}\left[F[f]\right]^{2}\right)\cdot F[f]\cdot\frac{\partial F[f]}{\partial f}\cdot\left(f_{,\alpha\beta}-\Gamma^{\gamma}_{\alpha\beta}f_{,\gamma}\right)\\+\left(\begin{array}{c}2\cdot\left(1+\mathrm{sgn}\left[F[f]\right]^{2}\right)\cdot F[f]\cdot\frac{\partial^{2}F[f]}{\partial f^{2}}\\-\left(\mathrm{sgn}\left[F[f]\right]^{2}-1\right)^{2}\cdot\left(\frac{\partial F[f]}{\partial f}\right)^{2}\end{array}\right)\cdot f_{,\alpha}f_{,\beta}\end{array}\right)\right). \tag{139}$$

After variation under the Einstein–Hilbert action integral with respect to the metric, we obtain:

$$\delta W=0=\int_{V}d^{n}x\sqrt{-g\cdot(F\cdot\bar{F})^{4}}\,\delta G_{a}^{\alpha\beta}$$

$$\times\left(\begin{array}{c}R^{***}_{\alpha\beta}+R_{\alpha\beta}-R\cdot\frac{g_{\alpha\beta}}{2}\\+\frac{3\cdot g_{\alpha\beta}}{4\cdot F[f]^{2}}\cdot\left(\begin{array}{c}2\cdot\left(1+\mathrm{sgn}\left[F[f]\right]^{2}\right)\cdot F[f]\cdot\frac{\partial F[f]}{\partial f}\cdot\Delta f\\+\left(\begin{array}{c}2\cdot\left(1+\mathrm{sgn}\left[F[f]\right]^{2}\right)\cdot F[f]\cdot\frac{\partial^{2}F[f]}{\partial f^{2}}\\-\left(\mathrm{sgn}\left[F[f]\right]^{2}-1\right)^{2}\cdot\left(\frac{\partial F[f]}{\partial f}\right)^{2}\end{array}\right)\cdot f_{,\mu}g^{\mu\nu}f_{,\nu}\end{array}\right)\\-\frac{3}{2\cdot F[f]^{2}}\cdot\left(\begin{array}{c}2\cdot\left(1+\mathrm{sgn}\left[F[f]\right]^{2}\right)\cdot F[f]\cdot\frac{\partial F[f]}{\partial f}\cdot\left(f_{,\alpha\beta}-\Gamma^{\gamma}_{\alpha\beta}f_{,\gamma}\right)\\+\left(\begin{array}{c}2\cdot\left(1+\mathrm{sgn}\left[F[f]\right]^{2}\right)\cdot F[f]\cdot\frac{\partial^{2}F[f]}{\partial f^{2}}\\-\left(\mathrm{sgn}\left[F[f]\right]^{2}-1\right)^{2}\cdot\left(\frac{\partial F[f]}{\partial f}\right)^{2}\end{array}\right)\cdot f_{,\alpha}f_{,\beta}\end{array}\right)\\+\kappa T_{a\alpha\beta}+\Lambda\cdot F^{a}\cdot g_{a\alpha\beta}\end{array}\right). \tag{140}$$

This now allows for sudden changes of the equations simply due to changes of the signature function sgn[F] and can even lead to switches from

Klein–Gordon $\left(\left(1-\mathrm{sgn}\left[F[f]\right]^2\right)=0\,\&\,\frac{\partial^2 F[f]}{\partial f^2}=0\right)$ to Dirac equations (1 + sgn $[F[f]]^2$ = 0). This, however, would be equivalent to a switch of spinless or integer-spin particles (or composite particles, like the pion) to a fermion. With respect to the conditions for the existence of a conscious mind being considered above (e.g., see the abstract of this book), we offer the timid suspicion that consciousness can actively steer such states and hence, switch sub-systems of thoughts from fermionic to bosonic behavior and vice versa.

5.4 Appendix D: Simple Example for Dimensional Mass Creation

In [4, 5], we introduced special de Sitter vacuum metrics, which gave mass terms in the corresponding Ricci scalar-based Klein–Gordon equation (134) in various dimensions. For completeness we repeat the main results here.

Introducing a metric for the coordinates $t, r, \vartheta, \varphi, u, v$ of the form:

$$g_{\alpha\beta}^{6}=\begin{pmatrix} -c^2 & 0 & 0 & 0 & 0 & 0 \\ 0 & 1 & 0 & 0 & 0 & 0 \\ 0 & 0 & r^2 & 0 & 0 & 0 \\ 0 & 0 & 0 & r^2\cdot\sin[\vartheta] & 0 & 0 \\ 0 & 0 & 0 & 0 & g[v] & 0 \\ 0 & 0 & 0 & 0 & 0 & g[v] \end{pmatrix}\cdot f[t,r,\vartheta,\varphi], \qquad (141)$$

with the entanglement solution for u and v as follows:

$$g[v]=\frac{C_{v1}^2}{C_{v0}^2\cdot\left(1+\cosh\left[C_{v1}\cdot\left(v+C_{v2}\right)\right]\right)}, \qquad (142)$$

gives us the following Ricci scalar from (124) with condition (132):

$$R^*=\frac{1}{F[f]^3}\cdot\left(-C_{N2}\cdot F[f]\cdot\frac{\partial F[f]}{\partial f}\cdot\Delta_g f\right);\quad F[f]=C_f+f[t,r,\vartheta,\varphi]$$

$$\xrightarrow{6D} R^*=\frac{5}{f^2}\cdot\left(\frac{C_{v0}^2}{5}\cdot f+\frac{\partial^2 f}{c^2\cdot\partial t^2}-\Delta_{\text{3D-sphere}} f\right);\quad f=f[t,r,\vartheta,\varphi];\quad C_f=0. \qquad (143)$$

Here $\Delta_{3D\text{-sphere}}f$ denotes the Laplace operator in spherical coordinates.

As for $R^* = 0$ and $\frac{C_{v0}^2}{5} = \frac{m^2 \cdot c^2}{\hbar^2}$ (m = mass, c = speed of light in vacuum, $\hbar$ = reduced Planck constant), we clearly have obtained the quantum Klein–Gordon equation and we can easily conclude that mass can obviously be obtained (created) by a suitable set of additional, entangled coordinates.

5.4.1 The Cosmological Constant as Balance Factor

In connection with a non-vanishing cosmological constant, we have also shown in [5] that with:

$$g_{\alpha\beta}^{6} = \begin{pmatrix} -c^2 \cdot h[r] & 0 & 0 & 0 & 0 & 0 \\ 0 & \frac{1}{h[r]} & 0 & 0 & 0 & 0 \\ 0 & 0 & r^2 & 0 & 0 & 0 \\ 0 & 0 & 0 & r^2 \cdot \sin[\vartheta] & 0 & 0 \\ 0 & 0 & 0 & 0 & g[v] & 0 \\ 0 & 0 & 0 & 0 & 0 & g[v] \end{pmatrix}$$

$$\text{with:} \quad h[r] = 1 + \frac{C_v^2 \cdot r^2}{6} - r^2 \cdot \frac{\Lambda}{3} - \frac{C_r}{r}; \quad C_v^2 = \Lambda, \tag{144}$$

we could have a complete vacuum solution, where we have:

$$G^{\alpha\beta} + \Lambda g^{\alpha\beta} = R^{\alpha\beta} - \frac{1}{2} R g^{\alpha\beta} + \Lambda g^{\alpha\beta} = 0. \tag{145}$$

In this case, the condition $C_v^2 = \Lambda$ provides the necessary balance between matter and cosmological constant. As the constant C_r would be arbitrary, it could be set to zero in order to avoid any singularities.

We see that the metric (144) would be a vacuum solution of the Einstein field equations, but still the structure (universe) it describes contains matter. Assuming that the condition (145) should be a boundary condition for the object described by Eq. (144), we obtain a balance rule for the mass parameter C_v and the cosmological constant. The object would have a resulting non-zero Ricci curvature of:

$$R = 3 \cdot \Lambda. \tag{146}$$

Similarly, we could construct a metric as follows:

$$g^{6}_{\alpha\beta}=\begin{pmatrix} -c^2\cdot h[r] & 0 & 0 & 0 & 0 & 0 \\ 0 & \frac{1}{h[r]} & 0 & 0 & 0 & 0 \\ 0 & 0 & r^2 & 0 & 0 & 0 \\ 0 & 0 & 0 & r^2\cdot\sin[\vartheta] & 0 & 0 \\ 0 & 0 & 0 & 0 & a^2\cdot g[v] & 0 \\ 0 & 0 & 0 & 0 & 0 & a^2 \end{pmatrix}$$

$$\text{with: } h[r]=1+\frac{S^2\cdot r^2}{3\cdot a^2}-r^2\cdot\frac{\Lambda}{3}-\frac{C_r}{r};\; S^2=\frac{a^2\cdot\Lambda}{2}. \tag{147}$$

Now we shall consider S a spin-like matter option in addition to our masses as given above.

This time we have:

$$g_{1,2}[v]=\begin{Bmatrix} C_{S1}\sin\left[S\left(v-C_{S2}\right)\right]^2 \\ C_{S1}\cos\left[S\left(v-C_{S2}\right)\right]^2 \end{Bmatrix}. \tag{148}$$

For more information and discussion (especially with respect to the extension to arbitrary coordinates and the flatness problem of our universe), the reader is referred to the original papers [10–12].

With respect to the topic of this book, i.e., "Mathematical Psychology," we state that while the universe may not make use of a cosmological constant, the conscious mind might, and this ability would allow it a great deal of additional flexibility.

5.5 Appendix E: About the Inner Evolutionary Potential of a System

5.5.1 Evolution Equations

In many applications, it is of great importance to know about the intrinsic evolutionary potential of a certain system. For this, the system's own evolution equations have to be evaluated. Does this also hold for the human mind? Of course, it does, because even the conscious mind is but a system, being described by and constructed from attributes. From the many possibilities to derive evolution equations for a given system we present only a small excerpt here. The interested reader can find more in [55].

As evolution equations, we denote differential equations of first- or second-order derivatives in time, which is to say:

$$f' = \frac{\partial f}{\partial t} = \Psi(t, f); \quad f'' = \frac{\partial^2 f}{\partial t^2} = \Psi(t, f, f'). \tag{149}$$

Thereby the function on the right-hand side can also be a differential operator with respect to the other dimensions of the system in question.

5.5.2 The Evolutionary Potential of a System Based on the Metric Klein–Gordon Equation

From Eq. (134), we can directly extract our first evolution equation simply by splitting up the Laplace operator as follows:

$$0 = R \cdot \left(C_f \pm f\right) \pm \frac{4 \cdot (n-1)}{(n-2)} \left(\Delta_s f + \Delta_t f\right). \tag{150}$$

Here the indexes s and t stand for space and time. Usually, when dealing with Minkowski-like space-times:

$$G_{\alpha\beta} = F[f] \begin{pmatrix} g_{00} & g_{10} & \cdots & \cdots & \cdots & g_{n0} \\ g_{10} & g_{11} & \cdots & \cdots & \cdots & \vdots \\ g_{20} & g_{21} & \ddots & \cdots & \cdots & \vdots \\ g_{30} & g_{31} & \cdots & \ddots & \cdots & \vdots \\ \cdots & \cdots & \cdots & \cdots & \ddots & \vdots \\ g_{n0} & \cdots & \cdots & \cdots & \cdots & g_{nn} \end{pmatrix}$$

$$g_{00} = -c^2; \quad g_{i0} = 0 \quad i = 1, 2, \cdots, n, \tag{151}$$

we have (c is the speed of light in vacuum of the space-time under observation):

$$0 = R \cdot \left(C_f \pm f\right) \pm \frac{4 \cdot (n-1)}{(n-2)} \left(\Delta_s f - \frac{1}{c^2} \frac{\partial^2 f}{\partial t^2}\right)$$

$$\Rightarrow \qquad \pm \frac{1}{c^2} \frac{\partial^2 f}{\partial t^2} = R \cdot \frac{\left(C_f \pm f\right) \cdot (n-2)}{4 \cdot (n-1)} \pm \Delta_s f. \tag{152}$$

Please note that metrics with space-time-shear components automatically produce also terms with first-order differentiations in time.

5.5.3 Side Note

It should be pointed out that the wrapper approach (119) could also be generalized to something like:

$$G_{\alpha\beta}=F_n\left[f_n\right]\begin{pmatrix} F_{n-1}\cdot\ldots\cdot F_2\cdot g_{00} & F_{n-1}\cdot\ldots\cdot F_2\cdot g_{10} & \cdots & \cdots & \cdots & g_{n0} \\ F_{n-1}\cdot\ldots\cdot F_2\cdot g_{10} & F_{n-1}\cdot\ldots\cdot F_2\cdot g_{11} & \cdots & \cdots & \cdots & \vdots \\ F_{n-1}\cdot\ldots\cdot F_3\cdot g_{20} & F_{n-1}\cdot\ldots\cdot F_3\cdot g_{21} & \ddots & \cdots & \cdots & \vdots \\ F_{n-1}\cdot\ldots\cdot F_4\cdot g_{30} & F_{n-1}\cdot\ldots\cdot F_4\cdot g_{31} & \cdots & \ddots & \cdots & \vdots \\ \cdots & \cdots & \cdots & \cdots & \ddots & \vdots \\ g_{n0} & \cdots & \cdots & \cdots & \cdots & g_{nn} \end{pmatrix}$$

$$F_i = F_i[f_i]. \tag{153}$$

The corresponding Ricci scalar equations were already presented in [5] and can be found via an ordering of the inner metrics of Eq. (153) via the following definition:

$$\begin{aligned} {}^2G_{\alpha\beta} &= F_2\cdot {}^2g_{\alpha\beta} \\ {}^3G_{\alpha\beta} &= F_3\cdot {}^3g_{\alpha\beta} \\ &\vdots \\ {}^nG_{\alpha\beta} &= F_n\cdot {}^ng_{\alpha\beta}, \end{aligned} \tag{154}$$

where we can write Eq. (153) as follows:

$${}^2G_{\alpha\beta}=\left(F_2\cdot\left({}^2g_{\alpha\beta}\right)\right)=F_2\cdot\begin{pmatrix} {}^2g_{00} & {}^2g_{01} \\ {}^2g_{01} & {}^2g_{11} \end{pmatrix}$$

$${}^3G_{\alpha\beta}=F_3\cdot{}^3g_{\alpha\beta}=F_3\cdot\begin{pmatrix} {}^3g_{00} & {}^3g_{01} & {}^3g_{02} \\ {}^3g_{01} & {}^3g_{11} & {}^3g_{12} \\ {}^3g_{02} & {}^3g_{12} & {}^3g_{22} \end{pmatrix}=F_3\cdot\begin{pmatrix} F_2\cdot\begin{pmatrix} {}^2g_{00} & {}^2g_{01} \\ {}^2g_{01} & {}^2g_{11} \end{pmatrix} & \begin{matrix} {}^3g_{02} \\ {}^3g_{12} \end{matrix} \\ \begin{matrix} {}^3g_{02} & {}^3g_{12} \end{matrix} & {}^3g_{22} \end{pmatrix}$$

$$\vdots$$

$${}^nG_{\alpha\beta}=F_n\cdot\begin{pmatrix} \begin{pmatrix} \cdots F_3\cdot\begin{pmatrix} F_2\cdot\begin{pmatrix} {}^2g_{00} & {}^2g_{01} \\ {}^2g_{01} & {}^2g_{11} \end{pmatrix} & \begin{matrix} {}^3g_{02} \\ {}^3g_{12} \end{matrix} \\ \begin{matrix} {}^3g_{02} & {}^3g_{12} \end{matrix} & {}^3g_{22} \end{pmatrix} \cdots \\ \cdots \end{pmatrix} & \begin{matrix} {}^ng_{0m} \\ \vdots \\ \vdots \\ \vdots \end{matrix} \\ {}^ng_{0m} \quad \cdots \quad \cdots \quad \cdots \quad \cdots & {}^ng_{mm} \end{pmatrix}; \quad m=n-1. \tag{155}$$

However, even though there might be cases where such a sub-dimensional split-up could be useful, it would not change the principal outcome for our derivations of the fundamentally obtained evolution equations for an arbitrary metric system. Thus, we will further consider this sub-dimensional extension of our wrapper approach in appendix R of this book.

5.6 Appendix F: About the Metric Origin of a Higgs Field for an Arbitrary System

Starting with the general quantum gravity field equations (125), we assume the variation to have a dominant scaling or volume part, which is equivalent to the assumption of weak gravity against all other forces[1]:

$$\delta G^{\alpha\beta} = g^{\alpha\beta} \cdot \delta_0 + \overbrace{g^{ab}\,\delta^{\alpha\beta}_{ab}}^{\text{Gravity}} \xrightarrow{\forall\, \delta^{\alpha\beta}_{ab} \ll \delta_0} = g^{\alpha\beta} \cdot \delta_0 , \tag{156}$$

and we obtain the following scalar quantum gravity equation for all systems with dimensions $n>2$:

$$0 = R - \frac{1}{2F}\begin{pmatrix} 2F_{,ij}(n-1)g^{ij} + 2\Gamma^{a}_{ij}F_{,a}g^{ij} \\ -F_{,i}g^{ab}g_{jb,a}g^{ij} - F_{,j}g^{ab}g_{ib,a}g^{ij} \\ -n\Gamma^{d}_{ij}F_{,d}g^{ij} + \frac{n}{2}F_{,d}g^{cd}g_{ab,c}g^{ab} \end{pmatrix} - \frac{F_{,i}\cdot F_{,j}}{4F^2}g^{ij}\left((n-6)(n-1)\right). \tag{157}$$

Considering F a functional wrapper of the kind $F[f]$, we can write:

$$0 = R - \frac{F'}{2F}\begin{pmatrix} 2f_{,ij}(n-1)g^{ij} + 2\Gamma^{a}_{ij}f_{,a}g^{ij} \\ -f_{,i}g^{ab}g_{jb,a}g^{ij} - f_{,j}g^{ab}g_{ib,a}g^{ij} \\ -n\Gamma^{d}_{ij}f_{,d}g^{ij} + \frac{n}{2}f_{,d}g^{cd}g_{ab,c}g^{ab} \end{pmatrix}$$

$$-(n-1)\frac{f_{,i}\cdot f_{,j}}{4F^2}g^{ij}\left(4F''F + F'F'(n-6)\right). \tag{158}$$

This allows us to demand certain conditions for the squared first-order differential operator of f in the last addend of Eq. (158). Keen on obtaining

[1]Please note that gravity is about 10^{33} times weaker than the next weakest interaction (which is the weak interaction).

a quantum gravity equation of Dirac-type, in order to end up with fermionic solutions (true individuals), we demand the existence of eigen-value, respectively eigen-potential solutions to the second-order differential operator (second addend) in Eq. (158):

$$\begin{pmatrix} 2f_{,ij}(n-1)g^{ij}+2\Gamma_{ij}^{a}f_{,a}g^{ij} \\ -f_{,i}g^{ab}g_{jb,a}g^{ij}-f_{,j}g^{ab}g_{ib,a}g^{ij} \\ -n\Gamma_{ij}^{d}f_{,d}g^{ij}+\frac{n}{2}f_{,d}g^{cd}g_{ab,c}g^{ab} \end{pmatrix}=2(n-1)\cdot\left(M^2+V\right)\cdot f \tag{159}$$

and the following condition for $F[f]$:

$$\frac{4FF''+\left(F'\right)^2(n-6)}{4F}=\frac{F'}{f}\quad\Rightarrow\quad F=\begin{cases}\left(C_f\pm f^2\right)^{\frac{4}{n-2}}\cdot C_{f1} & n>2\\ e^{\pm C_f\cdot f^2}\cdot C_{f1} & n=2.\end{cases} \tag{160}$$

As desired, we actually obtain the Dirac equation, which simply needs to be derived from:

$$0=\frac{n-2}{4(n-1)}R\left(C_f\pm f_i^2\right)f_i-\left(\left(M^2+V\right)f_i^2+f_{i,\alpha}f_{i,\beta}g^{\alpha\beta}\right) \tag{161}$$

via a direct root extraction (using quaternions just as Dirac [7] did with the Klein–Gordon equation and as demonstrated in [5]) under the assumption of a vanishing curvature $R=0$[2]. The derivation goes as follows:

$$0=\frac{n-2}{4(n-1)}R\left(C_f\pm f_i^2\right)f_i-\left(\left(M^2+V\right)f_i^2+f_{i,\alpha}f_{i,\beta}g^{\alpha\beta}\right)$$

$$\xrightarrow{R=0}$$

$$0=\overbrace{\left(M^2+V\right)}^{\equiv\mu^2}f_i^2+f_{i,\alpha}f_{i,\beta}g^{\alpha\beta}$$

$$\xrightarrow{g^{\alpha\beta}\cdot I=\frac{\gamma^\alpha\gamma^\beta+\gamma^\beta\gamma^\alpha}{2}}$$

$$0=\left(f_{i,\alpha}f_{i,\beta}g^{\alpha\beta}+\mu^2\cdot f_i^2\right)\cdot I$$

$$\Rightarrow\qquad =\Rightarrow f_{i,\alpha}f_{i,\beta}\left(\frac{\gamma^\alpha\gamma^\beta+\gamma^\beta\gamma^\alpha}{2}\right)+\frac{\beta^2+\beta^2}{2}\cdot m^2\cdot f_i^2=0$$

$$\Rightarrow\qquad f_{i,\alpha}f_{i,\beta}\gamma^\alpha\gamma^\beta+\beta^2\cdot m^2\cdot f_i^2=0$$

[2]Dirac assumed a flat space-time [7].

$$\Rightarrow \qquad \left(f_{i,\alpha}\gamma^{\alpha}+i\cdot\beta\cdot m\cdot f_i\right)\left(f_{i,\beta}\gamma^{\beta}-i\cdot\beta\cdot m\cdot f_i\right)=0$$

$$\Rightarrow \qquad f_{i,\alpha}f_{i,\beta}\gamma^{\beta}\gamma^{\alpha}+\beta^2\cdot m^2\cdot f_i^2=0$$

$$\Rightarrow \qquad \left(f_{i,\alpha}\gamma^{\alpha}+i\cdot\beta\cdot m\cdot f_i\right)\left(f_{i,\beta}\gamma^{\beta}-i\cdot\beta\cdot m\cdot f_i\right)=0. \tag{162}$$

As one of the factors in the last line above being zero suffices to fulfill the whole equation, we obtain the classical Dirac equations:

$$\Rightarrow \qquad f_{i,\alpha}\gamma^{\alpha}+i\cdot I\cdot m\cdot f_i=0$$

$$\Rightarrow \qquad f_{i,\beta}\gamma^{\beta}-i\cdot I\cdot m\cdot f_i=0, \tag{163}$$

with the well-known Dirac gamma matrices, reading in 4-dimensional space-times:

$$\gamma^0=\begin{pmatrix}1&&&\\&1&&\\&&-1&\\&&&-1\end{pmatrix};\quad \gamma^1=\begin{pmatrix}&&&1\\&&1&\\&-1&&\\-1&&&\end{pmatrix}$$

$$\gamma^2=\begin{pmatrix}&&&-i\\&&i&\\&i&&\\-i&&&\end{pmatrix};\quad \gamma^3=\begin{pmatrix}&&1&\\&&&-1\\-1&&&\\&1&&\end{pmatrix};\quad I=\begin{pmatrix}1&&&\\&1&&\\&&1&\\&&&1\end{pmatrix}. \tag{164}$$

Please note that in some cases the Dirac equation is also given via a separated time-derivative. This has been obtained by multiplication of Eq. (163) with $\beta = \gamma^0$. This leads to:

$$f_{i,t}=f_{i,\lambda}\alpha^{\lambda}-i\cdot\beta\cdot m\cdot f_i;\quad \lambda=1,2,3;\quad \alpha^{\lambda}=\begin{pmatrix}0&\sigma_\lambda\\\sigma_\lambda&0\end{pmatrix}. \tag{165}$$

Thereby we have applied the Pauli matrices as we will discuss later and which are given in Eq. (538) and also used the β-matrix

with $\beta=\begin{pmatrix}1&&&\\&1&&\\&&-1&\\&&&-1\end{pmatrix}$.

Now we have derived the fermionic Dirac equation and by observing the setting for our wrapper function $F[f]$ (160) in order to obtain the right metric “Dirac-conditions,” we immediately see that we also have obtained a Higgs field. Plotting $F[f]$, namely, thereby renaming $f = \Psi$ for the reason of recognition with the classical literature, we recognize the typical symmetry-breaking, so-called Mexican hat shape of the metric scaling $F[\Psi]$ in the case of $n = 4$ (Fig. F1).

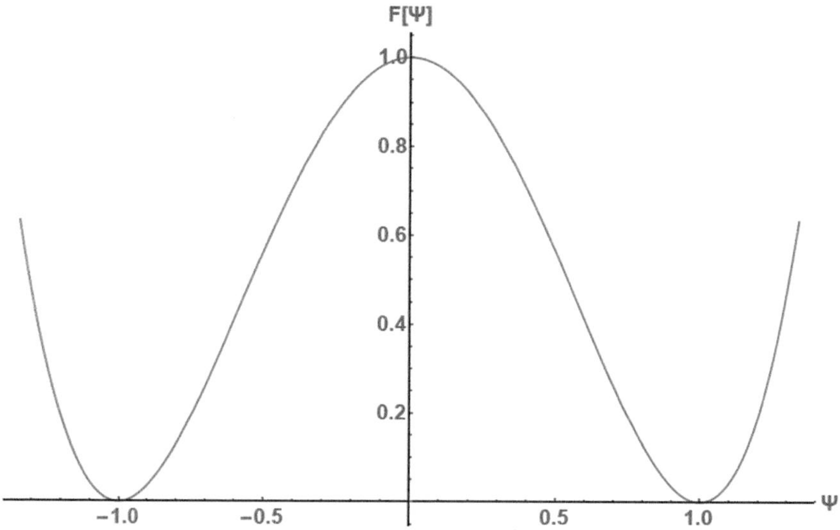

Figure F1 Higgs field as resulting from (160).

It appears as an interesting side-note that, in order to have fermionic behavior, we automatically also obtain a Higgs field.

5.7 Appendix G: How Nudging a System Gives Quantum Effects

We note that with the approach used here so far, we should discuss the following issues:

- The direct application of the Einstein–Hilbert action does not give us the perfect Laplace operator for all metrics (e.g., see Eq. (159)).
- There is no separate equation, which is to say, the equation we have for capital $F[f]$, even though already clearly mirroring the classical quantum equations, is only a part of the classical Einstein field equations.

We solve the problem via an additional variation. This time the variation is with respect to the function f and we apply the so-called functional derivative:

$$\delta W=\frac{\delta W}{\delta f[\mathbf{y}]}=0=\delta_f\int_V d^n x\left(\sqrt{g\cdot F^n}\cdot F\cdot R^*\right)$$

$$=\frac{\delta}{\delta f}\int_V d^n x\left(\sqrt{g\cdot F^n}\cdot F\cdot R^*\right)$$

$$=\frac{\delta}{\delta f}\int_V d^n x\left(\sqrt{g\cdot F^n}\cdot R^*_{\alpha\beta}g^{\alpha\beta}\right)$$

$$=\lim_{\in\to 0}\frac{1}{\in}\left(\begin{array}{c}\int_V d^n x\left(\begin{array}{c}\sqrt{g\cdot F\left[f[\mathbf{x}]+\in\cdot\delta[\mathbf{x}-\mathbf{y}]\right]^n}\\ \cdot F\left[f[\mathbf{x}]+\in\cdot\delta[\mathbf{x}-\mathbf{y}]\right]\cdot R^*\left[f[\mathbf{x}]+\in\cdot\delta[\mathbf{x}-\mathbf{y}]\right]\end{array}\right)\\ -\int_V d^n x\left(\sqrt{g\cdot F[f[\mathbf{x}]]^n}\cdot F\cdot R^*[f[\mathbf{x}]]\right)\end{array}\right), \tag{166}$$

for which we refer to the standard textbooks in theoretical physics (e.g., [13]).

Here, we perform this f-variation after the standard Hilbert variation with respect to the metric tensor (only that we here mean the scaled metric tensor $G_{\alpha\beta}$):

$$0=\int_{V_G} d^n x\sqrt{-g\cdot F^n}\left(\left(R^*_{\alpha\beta}-\frac{1}{2}\cdot R^*\cdot G_{\alpha\beta}\right)\delta G^{\alpha\beta}\right)$$

$$=\int_{V_G} d^n x\sqrt{-g\cdot F^n}\left(\left(R^*_{\alpha\beta}-\frac{1}{2}\cdot R^*\cdot F\cdot g_{\alpha\beta}\right)\delta G^{\alpha\beta}\right) \tag{167}$$

under the assumption of weak gravity:

$$\delta G^{\alpha\beta}=g^{\alpha\beta}\cdot\delta_0+\overbrace{g^{ab}\,\delta^{\alpha\beta}_{ab}}^{\text{Gravity}}\xrightarrow{\forall\,\delta^{\alpha\beta}_{ab}\ll\delta_0}=g^{\alpha\beta}\cdot\delta_0 \tag{168}$$

as follows:

$$\frac{\delta}{\delta f}\frac{\delta}{\delta G_{\alpha\beta}}W=0$$

$$= \delta_f \int_{V_G} d^n x \sqrt{-g \cdot F^n} \left(1 - \frac{n}{2}\right) \left(\begin{array}{c} R - \dfrac{1}{2F} \left(\begin{array}{c} 2F_{,\alpha\beta}(n-1) g^{\alpha\beta} + 2\Gamma^{a}_{\alpha\beta} F_{,a} g^{\alpha\beta} \\ -F_{,\alpha} g^{ab} g_{\beta b,a} g^{\alpha\beta} - F_{,\beta} g^{ab} g_{\alpha b,a} g^{\alpha\beta} \\ -n\Gamma^{d}_{\alpha\beta} F_{,d} g^{\alpha\beta} + \dfrac{n}{2} F_{,d} g^{cd} g_{\alpha\beta,c} g^{\alpha\beta} \end{array} \right) \\ -\dfrac{F_{,\alpha} \cdot F_{,\beta}}{4F^2} g^{\alpha\beta} \left((n-6)(n-1)\right) \end{array} \right) \delta_0 .$$

(169)

After some lengthy calculations (first presented in appendix C in [32] and reprinted in appendix M in this book), we obtain:

$$0 = \left[\begin{array}{c} \dfrac{\sqrt{-g}}{2} \left(\dfrac{f_{,\alpha} f_{,\beta}}{2} g^{\alpha\beta} \left(\dfrac{1}{2}(n-4)(n+2)(F')^2 + 2FF'' \right) + (n+2) FF' \Delta f \right) \\ +F^2 \left(\begin{array}{c} \left(\sqrt{-g} \cdot g^{\alpha\beta}\right)_{,\alpha\beta} - \dfrac{\sqrt{-g} \cdot R}{(n-1)} \\ -\dfrac{1}{2(n-1)} \left(\sqrt{-g} \cdot g^{\alpha\beta} \cdot g^{ij} \left(\begin{array}{c} g_{i\beta,j} - g_{ij,\beta} - g_{\beta j,i} \\ +\dfrac{n}{2}\left(2g_{ij,\beta} - g_{i\beta,j} - g_{j\beta,i}\right) \end{array} \right) \right)_{,\alpha} \end{array} \right) \end{array} \right]_{\mathbf{y}} .$$

(170)

As we see (last term in the first line), this time we obtain a perfect Laplace operator.

It should explicitly be pointed out that we can obtain the same result when applying the *f*-nudging, which is to say the functional derivation of the Einstein–Hilbert action, before the usual Hilbert variation with respect to the metric tensor. In this case, we do not even need the weak-gravity assumption. We simply perform the functional derivative as shown in Eq. (166) above. This results in the following variational task:

$$\frac{\delta}{\delta f} W = 0 = \delta_f \int_{V_G} d^n x \sqrt{-g \cdot F^n} \left(\begin{array}{c} R - \dfrac{1}{2F} \left(\begin{array}{c} 2F_{,\alpha\beta}(n-1) g^{\alpha\beta} + 2\Gamma^{a}_{\alpha\beta} F_{,a} g^{\alpha\beta} \\ -F_{,\alpha} g^{ab} g_{\beta b,a} g^{\alpha\beta} - F_{,\beta} g^{ab} g_{\alpha b,a} g^{\alpha\beta} \\ -n\Gamma^{d}_{\alpha\beta} F_{,d} g^{\alpha\beta} + \dfrac{n}{2} F_{,d} g^{cd} g_{\alpha\beta,c} g^{\alpha\beta} \end{array} \right) \\ -\dfrac{F_{,\alpha} \cdot F_{,\beta}}{4F^2} g^{\alpha\beta} \left((n-6)(n-1)\right) \end{array} \right) .$$

(171)

Again, we obtain the perfect Laplace operator via solution (170). Introduction of a suitable condition for the wrapper function *F*, namely:

$$0=\frac{1}{2}(n-4)(n+2)(F')^2+2FF'', \tag{172}$$

gives us the following quantum gravity Klein–Gordon equation:

$$0=\left[\begin{array}{c}\frac{\sqrt{-g}}{2}(n+2)F'\Delta f\\ +F\left(\begin{array}{c}\left(\sqrt{-g}\cdot g^{\alpha\beta}\right)_{,\alpha\beta}-\frac{\sqrt{-g}\cdot R}{(n-1)}\\ -\frac{1}{2(n-1)}\left(\sqrt{-g}\cdot g^{\alpha\beta}\cdot g^{ij}\left(\begin{array}{c}g_{i\beta,j}-g_{ij,\beta}-g_{\beta j,i}\\ +\frac{n}{2}\left(2g_{ij,\beta}-g_{i\beta,j}-g_{j\beta,i}\right)\end{array}\right)\right)_{,\alpha}\end{array}\right)\end{array}\right]_{\mathbf{y}}. \tag{173}$$

5.8 Appendix H: First-Order Quantum Equations from the Einstein–Hilbert Action

It has been already shown in appendix *F* how Dirac equations as first-order quantum equations fall out of the Einstein–Hilbert action via the results given in Eq. (125). Because of the importance of the Dirac equations, we are going to deepen our corresponding consideration in appendix R. Similarly, we can obtain first-order differential quantum equations from the Einstein–Hilbert action being subjected to a variation only to the metric scale or volume function *f* as discussed in appendix G. Taking our solution from there (Eq. (170)) and demanding:

$$\pm FF'\frac{1}{f}=\frac{1}{2}(n-4)(n+2)(F')^2+2FF'', \tag{174}$$

with the necessary function $F[f]$:

$$+\rightarrow F[f]=C_F\cdot\left(f^{3/2}+C_f\right)^{\frac{4}{n(n-2)-4}}$$

$$-\rightarrow F[f]=C_F\cdot\left(f^{1/2}+C_f\right)^{\frac{4}{n(n-2)-4}}, \tag{175}$$

gives us:

$$0=\left(\begin{array}{c}\frac{\sqrt{-g}}{2}FF'\left(\pm\frac{f_{,\alpha}f_{,\beta}}{2}g^{\alpha\beta}+(n+2)f\cdot\Delta f\right)\\+f\cdot F^{2}\left(\begin{array}{c}\left(\sqrt{-g}\cdot g^{\alpha\beta}\right)_{,\alpha\beta}-\frac{\sqrt{-g}\cdot R}{(n-1)}\\-\frac{1}{2(n-1)}\left(\sqrt{-g}\cdot g^{\alpha\beta}\cdot g^{ij}\left(\begin{array}{c}g_{i\beta,j}-g_{ij,\beta}-g_{\beta j,i}\\+\frac{n}{2}\left(2g_{ij,\beta}-g_{i\beta,j}-g_{j\beta,i}\right)\end{array}\right)\right)_{,\alpha}\end{array}\right)\end{array}\right)$$

$$=\left(\begin{array}{c}\frac{\sqrt{-g}}{2}F'\left((n+2)f\cdot\Delta f\pm\frac{f_{,\alpha}f_{,\beta}}{2}g^{\alpha\beta}\right)\\+f\cdot F\left(\begin{array}{c}\left(\sqrt{-g}\cdot g^{\alpha\beta}\right)_{,\alpha\beta}-\frac{\sqrt{-g}\cdot R}{(n-1)}\\-\frac{1}{2(n-1)}\left(\sqrt{-g}\cdot g^{\alpha\beta}\cdot g^{ij}\left(\begin{array}{c}g_{i\beta,j}-g_{ij,\beta}-g_{\beta j,i}\\+\frac{n}{2}\left(2g_{ij,\beta}-g_{i\beta,j}-g_{j\beta,i}\right)\end{array}\right)\right)_{,\alpha}\end{array}\right)\end{array}\right). \quad (176)$$

Now we assume the Laplace operator to deliver eigenvalues, which results in:

$$0=\left(\begin{array}{c}\frac{\sqrt{-g}}{2}F'\left((n+2)f^{2}\cdot M^{2}\pm\frac{f_{,\alpha}f_{,\beta}}{2}g^{\alpha\beta}\right)\\+f\cdot F\left(\begin{array}{c}\left(\sqrt{-g}\cdot g^{\alpha\beta}\right)_{,\alpha\beta}-\frac{\sqrt{-g}\cdot R}{(n-1)}\\-\frac{1}{2(n-1)}\left(\sqrt{-g}\cdot g^{\alpha\beta}\cdot g^{ij}\left(\begin{array}{c}g_{i\beta,j}-g_{ij,\beta}-g_{\beta j,i}\\+\frac{n}{2}\left(2g_{ij,\beta}-g_{i\beta,j}-g_{j\beta,i}\right)\end{array}\right)\right)_{,\alpha}\end{array}\right)\end{array}\right) \quad (177)$$

and by introducing Dirac gamma matrices as done in appendix *F*, we can factorize the whole expression into Dirac equations just as demonstrated before also using the recipe given in G (esp. c.f. the evaluation in Eq. (162)).

More on the topic of the metrically derived Dirac equations will be presented in appendix R.

5.9 Appendix I: How Perturbation of the Classical Hilbert Kernel in the Einstein–Hilbert Actions Automatically Leads to Quantum Gravity Field Equations

Let us assume to have a perturbation of the classical Hilbert kernel as given in (118) of the following kind:

$$\delta W = 0 = \delta \int_V d^n x \left(\sqrt{-g} \cdot \left(\Phi \cdot R - 2\Lambda + L_M \right) \right). \tag{178}$$

Thereby the symbol Φ stands for an arbitrary function of the coordinates. We know from [15] that this would usually lead to complicated expressions like this:

$$\delta W = 0 = \left[\begin{array}{c} \int_V d^n x \left(\sqrt{-g} \cdot \left(\begin{array}{c} \Phi \cdot R_{\mu\nu} - \frac{1}{2} \Phi \cdot R g_{\mu\nu} + \Lambda g_{\mu\nu} - \left(\nabla_\mu \nabla_\nu - g_{\mu\nu} \Delta_g \right) \Phi \\ + \left\{ \begin{array}{cc} 0 & \ldots\text{"vacuum"} \\ 8\pi G T_{\mu\nu} & \ldots\text{postulated matter} \end{array} \right\} \Phi \end{array} \right) \right) \delta g^{\mu\nu} \\ - \int_{\text{Surface}} d^{n-1} y \left(\sqrt{|h|} \cdot \varepsilon \cdot \Phi \cdot N^\lambda h^{\mu\nu} \partial_\lambda \left(\delta g_{\mu\nu} \right) \right) \end{array} \right]$$

$$= \left[\begin{array}{c} \int_V d^n x \left(\sqrt{-g} \cdot \left(\begin{array}{c} R_{\mu\nu} - \frac{1}{2} R g_{\mu\nu} + \frac{\Lambda}{\Phi} g_{\mu\nu} - \left(\nabla_\mu \nabla_\nu - g_{\mu\nu} \Delta_g \right) \\ + \left\{ \begin{array}{cc} 0 & \ldots\text{"vacuum"} \\ 8\pi G T_{\mu\nu} & \ldots\text{postulated matter} \end{array} \right\} \end{array} \right) \Phi \right) \delta g^{\mu\nu} \\ - \int_{\text{Surface}} d^{n-1} y \left(\sqrt{|h|} \cdot \varepsilon \cdot \Phi \cdot N^\lambda h^{\mu\nu} \partial_\lambda \left(\delta g_{\mu\nu} \right) \right) \end{array} \right]. \tag{179}$$

In this appendix we will show how to avoid such cumbersome equations (especially the surface term). For the reason of brevity and simplicity, we demand:

$$\delta W = 0 = \delta \int_V d^n x \left(\sqrt{-g} \cdot \left(\Phi \cdot R - \overbrace{2\Lambda + L_M}^{=0} \right) \right) \tag{180}$$

and adjust the kernel as follows:

$$\delta W=0=\delta\int_V d^n x\left(\sqrt{-g}\cdot\Phi\cdot R\right)$$

$$\xrightarrow{g_{\alpha\beta}=\gamma_{\alpha\beta}\cdot F;\quad F^{\frac{n}{2}}=\Phi}$$

$$0=\delta_{g_{\alpha\beta}}\int_V d^n x\left(\sqrt{-\gamma}\cdot F^{\frac{n}{2}}R^*\right)\Rightarrow 0=\delta_{g_{\alpha\beta}}\int_V d^n x\left(\sqrt{-\gamma\cdot F^n}\cdot R^*\right)$$

$$\Rightarrow\qquad 0=R^*_{\alpha\beta}-\frac{1}{2}\cdot\gamma_{\alpha\beta}\cdot R^*. \tag{181}$$

Thereby we remember that:

$$R^*=\frac{1}{F}\left(R-\frac{1}{2F}\begin{pmatrix}2F_{,ij}(n-1)\gamma^{ij}+2\Gamma^a_{ij}F_{,a}\gamma^{ij}\\ -F_{,i}\gamma^{ab}\gamma_{jb,a}\gamma^{ij}-F_{,j}\gamma^{ab}\gamma_{ib,a}\gamma^{ij}\\ -n\Gamma^d_{ij}F_{,d}\gamma^{ij}+\frac{n}{2}F_{,d}\gamma^{cd}\gamma_{ab,c}\gamma^{ab}\end{pmatrix}-\frac{F_{,i}\cdot F_{,j}}{4F^2}\gamma^{ij}\left((n-6)(n-1)\right)\right) \tag{182}$$

and subsequently obtain after variation with respect to the scaled metric tensor:

$$0=\int_{V_G} d^n x\sqrt{-\gamma\cdot F^n}\left(\left(\begin{array}{c}R_{\alpha\beta}-\frac{\gamma_{\alpha\beta}}{2}\left(\begin{array}{c}R-\frac{1}{2F}\begin{pmatrix}2F_{,ij}(n-1)\gamma^{ij}+2\Gamma^a_{ij}F_{,a}\gamma^{ij}\\ -F_{,i}\gamma^{ab}\gamma_{jb,a}\gamma^{ij}-F_{,j}\gamma^{ab}\gamma_{ib,a}\gamma^{ij}\\ -n\Gamma^d_{ij}F_{,d}\gamma^{ij}+\frac{n}{2}F_{,d}\gamma^{cd}\gamma_{ab,c}\gamma^{ab}\end{pmatrix}\\ -\frac{F_{,i}\cdot F_{,j}}{4F^2}\gamma^{ij}\left((n-6)(n-1)\right)\end{array}\right)\\ -\frac{1}{2F}\left(\begin{array}{c}F_{,\alpha\beta}(n-2)+F_{,ab}\gamma_{\alpha\beta}\gamma^{ab}\\ +F_{,a}\gamma^{ab}\left(\gamma_{\beta b,\alpha}-\gamma_{\beta\alpha,b}\right)-F_{,\alpha}\gamma^{ab}\gamma_{\beta b,a}-F_{,\beta}\gamma^{ab}\gamma_{\alpha b,a}\\ +F_{,d}\gamma^{cd}\frac{1}{2}n\begin{pmatrix}\frac{2}{n}\gamma_{\alpha c,\beta}-\gamma_{\alpha c,\beta}-\gamma_{\beta c,\alpha}\\ +\gamma_{\alpha\beta,c}+\frac{1}{n}\gamma_{\alpha\beta}\gamma_{ab,c}\gamma^{ab}\end{pmatrix}\\ -\frac{1}{2F}\left(F_{,\alpha}\cdot F_{,\beta}(3n-6)+\gamma_{\alpha\beta}F_{,c}F_{,d}\gamma^{cd}(4-n)\right)\end{array}\right)\end{array}\right)\delta g^{\alpha\beta}\right). \tag{183}$$

The resulting field equations read as follows:

$$0=\left(\begin{array}{c} R_{\alpha\beta}-\frac{\gamma_{\alpha\beta}}{2}R+\frac{\gamma_{\alpha\beta}}{2}\left(\begin{array}{c}\frac{1}{2F}\left(\begin{array}{c}2F_{,ij}(n-1)\gamma^{ij}+2\Gamma_{ij}^{a}F_{,a}\gamma^{ij}\\ -F_{,i}\gamma^{ab}\gamma_{jb,a}\gamma^{ij}-F_{,j}\gamma^{ab}\gamma_{ib,a}\gamma^{ij}\\ -n\Gamma_{ij}^{d}F_{,d}\gamma^{ij}+\frac{n}{2}F_{,d}\gamma^{cd}\gamma_{ab,c}\gamma^{ab}\end{array}\right)\\ +\frac{F_{,i}\cdot F_{,j}}{4F^{2}}\gamma^{ij}\left((n-6)(n-1)\right)\end{array}\right)\\ -\frac{1}{2F}\left(\begin{array}{c}F_{,\alpha\beta}(n-2)+F_{,ab}\gamma_{\alpha\beta}\gamma^{ab}\\ +F_{,a}\gamma^{ab}\left(\gamma_{\beta b,\alpha}-\gamma_{\beta\alpha,b}\right)-F_{,\alpha}\gamma^{ab}\gamma_{\beta b,a}-F_{,\beta}\gamma^{ab}\gamma_{\alpha b,a}\\ +F_{,d}\gamma^{cd}\frac{1}{2}n\left(\frac{2}{n}\gamma_{\alpha c,\beta}-\gamma_{\alpha c,\beta}-\gamma_{\beta c,\alpha}+\gamma_{\alpha\beta,c}+\frac{1}{n}\gamma_{\alpha\beta}\gamma_{ab,c}\gamma^{ab}\right)\\ -\frac{1}{2F}\left(F_{,\alpha}\cdot F_{,\beta}(3n-6)+\gamma_{\alpha\beta}F_{,c}F_{,d}\gamma^{cd}(4-n)\right)\end{array}\right)\end{array}\right)\quad(184)$$

and presenting them in dependency of our original kernel perturbation Φ:

$$0=\left(\begin{array}{c} R_{\alpha\beta}-\frac{\gamma_{\alpha\beta}}{2}R+\frac{\gamma_{\alpha\beta}}{2}\left(\begin{array}{c}\frac{1}{2\Phi^{\frac{2}{n}}}\left(\begin{array}{c}2\Phi^{\frac{2}{n}}{}_{,ij}(n-1)\gamma^{ij}+2\Gamma_{ij}^{a}\Phi^{\frac{2}{n}}{}_{,a}\gamma^{ij}\\ -\Phi^{\frac{2}{n}}{}_{,i}\gamma^{ab}\gamma_{jb,a}\gamma^{ij}-\Phi^{\frac{2}{n}}{}_{,j}\gamma^{ab}\gamma_{ib,a}\gamma^{ij}\\ -n\Gamma_{ij}^{d}\Phi^{\frac{2}{n}}{}_{,d}\gamma^{ij}+\frac{n}{2}\Phi^{\frac{2}{n}}{}_{,d}\gamma^{cd}\gamma_{ab,c}\gamma^{ab}\end{array}\right)\\ +\frac{\Phi^{\frac{2}{n}}{}_{,i}\cdot\Phi^{\frac{2}{n}}{}_{,j}}{4\Phi^{\frac{4}{n}}}\gamma^{ij}\left((n-6)(n-1)\right)\end{array}\right)\\ -\frac{1}{2\Phi^{\frac{2}{n}}}\left(\begin{array}{c}\Phi^{\frac{2}{n}}{}_{,\alpha\beta}(n-2)+\Phi^{\frac{2}{n}}{}_{,ab}\gamma_{\alpha\beta}\gamma^{ab}\\ +\Phi^{\frac{2}{n}}{}_{,a}\gamma^{ab}\left(\gamma_{\beta b,\alpha}-\gamma_{\beta\alpha,b}\right)-\Phi^{\frac{2}{n}}{}_{,\alpha}\gamma^{ab}\gamma_{\beta b,a}-\Phi^{\frac{2}{n}}{}_{,\beta}\gamma^{ab}\gamma_{\alpha b,a}\\ +\Phi^{\frac{2}{n}}{}_{,d}\gamma^{cd}\frac{1}{2}n\left(\frac{2}{n}\gamma_{\alpha c,\beta}-\gamma_{\alpha c,\beta}-\gamma_{\beta c,\alpha}+\gamma_{\alpha\beta,c}+\frac{1}{n}\gamma_{\alpha\beta}\gamma_{ab,c}\gamma^{ab}\right)\\ -\frac{1}{2\Phi^{\frac{2}{n}}}\left(\Phi^{\frac{2}{n}}{}_{,\alpha}\cdot\Phi^{\frac{2}{n}}{}_{,\beta}(3n-6)+\gamma_{\alpha\beta}\Phi^{\frac{2}{n}}{}_{,c}\Phi^{\frac{2}{n}}{}_{,d}\gamma^{cd}(4-n)\right)\end{array}\right)\end{array}\right),\quad(185)$$

we realize that a perturbation of the Hilbert kernel in the Einstein–Hilbert action results in quantum or at least quantum-like terms for the field equations.

In the case of variation not with respect to the metric tensor, but only to the scaling function *f*, we adjust the kernel as follows:

$$\delta W = 0 = \delta \int_V d^n x \left(\sqrt{-g} \cdot \Phi_0 \cdot R\right)$$

$$\xrightarrow{g_{\alpha\beta}=\gamma_{\alpha\beta}\cdot F=\gamma_{\alpha\beta}\cdot F[f];\quad F^{\frac{n}{2}+1}=\Phi_0}$$

$$0 = \delta_f \int_V d^n x \left(\sqrt{-\gamma} \cdot F^{\frac{n}{2}+1} R^* \right) \Rightarrow 0 = \delta_f \int_V d^n x \left(\sqrt{-\gamma \cdot F^n} \cdot F \cdot R^* \right). \quad (186)$$

Now the resulting variational task reads as follows:

$$\frac{\delta W}{\delta f} = 0 = \delta_f \int_{V_G} d^n x \sqrt{-g \cdot F^n} \left(R - \frac{1}{2F} \left(\begin{array}{c} 2F_{,\alpha\beta}(n-1) g^{\alpha\beta} + 2\Gamma^a_{\alpha\beta} F_{,a} g^{\alpha\beta} \\ -F_{,\alpha} g^{ab} g_{\beta b,a} g^{\alpha\beta} - F_{,\beta} g^{ab} g_{\alpha b,a} g^{\alpha\beta} \\ -n\Gamma^d_{\alpha\beta} F_{,d} g^{\alpha\beta} + \frac{n}{2} F_{,d} g^{cd} g_{\alpha\beta,c} g^{\alpha\beta} \end{array} \right) - \frac{F_{,\alpha} \cdot F_{,\beta}}{4F^2} g^{\alpha\beta} \left((n-6)(n-1)\right) \right) \quad (187)$$

and we obtain the scalar equation:

$$0 = \left[\begin{array}{c} \frac{\sqrt{-g}}{2} \left(\frac{f_{,\alpha} f_{,\beta}}{2} g^{\alpha\beta} \left(\frac{1}{2}(n-4)(n+2)(F')^2 + 2FF'' \right) + (n+2)FF'\Delta f \right) \\ +F^2 \left(\begin{array}{c} \left(\sqrt{-g} \cdot g^{\alpha\beta}\right)_{,\alpha\beta} - \frac{\sqrt{-g} \cdot R}{(n-1)} \\ -\frac{1}{2(n-1)} \left(\sqrt{-g} \cdot g^{\alpha\beta} \cdot g^{ij} \left(\begin{array}{c} g_{i\beta,j} - g_{ij,\beta} - g_{\beta j,i} \\ +\frac{n}{2}\left(2g_{ij,\beta} - g_{i\beta,j} - g_{j\beta,i}\right) \end{array} \right) \right)_{,\alpha} \end{array} \right) \end{array} \right]. \quad (188)$$

When interpreting the scaling or wave function *f* as something being equivalent or at least connected with the individuality aspect of an arbitrary subject, we see that a perturbation of the curvature kernel of the Hilbert-extremal integral leads to additional degrees of freedom. We realize such additional degrees of freedom as quantum effects here, while the individual in question may just take it as its very own and personal character, emotional state and/or consciousness.

It has been just shown that we can always factor out a suitable function $F[f]$ from the metric tensor in order to simplify the variational task for either the whole metric (181) or just with respect to the volume factor (186). Thereby we did not care about whether the kernels for both cases are identical, but simply adjusted the factor in order to obtain maximum simplicity to each task. However, it appears reasonable to assume that the kernel should be the same in both variational cases. Thus, aiming for maximum simplicity in the case of the variation with respect to the whole metric in applying Eq. (181), we require the derivation of the more complex case with respect to f and vice versa. The corresponding evaluation has been presented in appendix L.

5.10 Appendix J: The Introduction of a Metric System Time

5.10.1 A Variety of Timely Dimensions

Now it seems to be about time that we are going to gradually bring the math into our considerations from the section "2.3 Time and Uncertainty in Connection with Individuality and Subjectivity" in order to better understand the origin of a given system time, be it for a living or a non-living one, be it for one with or without intelligence, with or without conscious thought.

As it has been shown in the most recent papers [10, 11, 12, 26] of the series of "Science Riddles" that matter fields can just be obtained via suitably entangled dimensions and as in this context also the flatness problem of the universe [27, 28] could be solved [12], here we want to investigate the option of entangled dimensions also with respect to the character of time.

It was demonstrated in [25] that we always need at least one coordinate of "different" (imaginary) character in order to fulfill the fundamental quantum equations resulting from an extended variation of the Einstein–Hilbert action with respect to the dimension of space [29, 30].

A similar result can be obtained directly as a solution for the Einstein field equations for any metric subjected to a transformation of the kind:

$$G_{\alpha\beta} = F\left[f\left[t,x,y,z,\ldots,\xi_k,\ldots,\xi_n\right]\right]_{\alpha\beta}^{ij} g_{ij}$$

$$\rightarrow G_{\alpha\beta} = F\left[f\left[t,x,y,z,\ldots,\xi_k,\ldots,\xi_n\right]\right]\cdot \delta_{\alpha}^{i}\delta_{\beta}^{j} g_{ij}. \tag{189}$$

The reader simply checks this by introducing Eq. (189) into the resulting field equations (125) and assuming a flat space in Cartesian coordinates.

Here the functions F have to be set to $F[f[...]] = f[c_{t=0} * t + c_{x=1} * x + \ldots + c_k * \xi_k]$ with the condition:

$$\sum_{i=0}^{n} c_i^2 = 0. \tag{190}$$

Please note that the indices i, j do not necessarily have to run from 0 to $n - 1$, with n denoting the dimension of the space-time we are considering. In fact, also the functions $f[...]$ might depend only on those coordinates we want to appear entangled. This gives us $F[f[...]] = f[c_p * \xi_p + \ldots + c_i * \xi_i + \ldots + c_k * \xi_k]$ with

$$G_{\alpha\beta} = \sum_{i,j=p}^{k} F\left[f\left[\xi_p,\ldots,\xi_i,\ldots,\xi_k\right]\right]\cdot\delta_\alpha^i\delta_\beta^j g_{ij} \tag{191}$$

and the condition:

$$\sum_{i=p}^{k} c_i^2 = 0. \tag{192}$$

According to the quantum approaches in [25], we need at least one coordinate to be of imaginary character and thus, sticking out this way from the others. Let us have a function $g[...]$ being connected with this special (outstanding) coordinate. The character of the function for $g[...]$ depends on the dimension of the metric it is applied on. Here is an example.

Introducing a metric for the coordinates $t, \tau, T, r, \vartheta, \varphi$ of the form:

$$g_{\alpha\beta}^{6} = \begin{pmatrix} g[t,\tau,T] \equiv g & 0 & 0 & 0 & 0 & 0 \\ 0 & g & 0 & 0 & 0 & 0 \\ 0 & 0 & g & 0 & 0 & 0 \\ 0 & 0 & 0 & 1 & 0 & 0 \\ 0 & 0 & 0 & 0 & r^2 & 0 \\ 0 & 0 & 0 & 0 & 0 & r^2\cdot\sin[\vartheta] \end{pmatrix} \cdot f[r,\vartheta,\varphi], \tag{193}$$

with the following function for $g[...]$:

$$g[t,\tau,T] = \frac{C_1}{\left(c_0\cdot t + c_1\cdot\tau + c_2\cdot T + C_2\right)^2};\quad c_0^2 + c_1^2 + c_2^2 = 0;\quad f[...] = 1, \tag{194}$$

directly solves the vacuum Einstein field equations. We see that at least one of the c_i has to be imaginary and, thus, would make the corresponding coordinate t, τ or T to appear time-like. Choosing a general $f[...]$ as introduced in Eq. (193) and demanding $c_0^2+c_1^2+c_2^2=0$ as set in Eq. (142) in order to fulfill the Einstein field equations gives us the following Ricci scalar from (2) (see also [4, 5]):

$$R^*=\frac{1}{F[f]^3}\cdot\left(-C_{N2}\cdot F[f]\cdot\frac{\partial F[f]}{\partial f}\cdot\Delta_g f\right);\quad F[f]=C_f+f[r,\vartheta,\varphi]$$

$$\xrightarrow{6D} R^*=-\frac{1}{f^2}\cdot\left(6\cdot\frac{\left(c_0^2+c_1^2+c_2^2\right)}{C_1}\cdot f+5\cdot\Delta_{\text{3D-sphere}}f\right);\qquad (195)$$

$$f=f[r,\vartheta,\varphi];\quad C_f=0.$$

Here $\Delta_{\text{3D-sphere}}f$ denotes the Laplace operator in spherical coordinates. As for $R^*=0$ and $6\cdot\frac{\left(c_0^2+c_1^2+c_2^2\right)}{C_1}=\frac{C_{v0}^2}{5}=\frac{m^2\cdot c^2}{\hbar^2}$ (m = mass, c = speed of light in vacuum, $\hbar$ = reduced Planck constant) we clearly have obtained the quantum Klein–Gordon equation, we can easily conclude that mass can obviously be constructed by a suitable set of additional, entangled coordinates. Please note that in contrast to the entanglement of 2 dimensions resulting in mass-terms (e.g., section 5.1 in [4]), where no time-like coordinate was needed, we—this time—obtain a bell-shaped localized distribution for $g[...]$ only at certain positions of the time-like coordinate. At these positions, let us choose here $t = t_0$, however, it can easily be ensured that the additional coordinates T and τ are compactified, because the arbitrary constants C_1 and C_2 can always be chosen in such a way that $g[t=t_0, ...]$ takes on the character of delta-like distributions. Thus, the coordinates T and τ would be omnipresent in the space r, ϑ, φ, but do not appear in the elongated form (on a bigger scale), which means that T and τ are invisible but only act via the production of mass in a 3+1 space-time.

It can be shown that one can always find vacuum solutions to any kind of entanglement approach (191) in an arbitrary dimensional space-time with arbitrary many dimensions being entangled. We always find solutions of the type $g[...]$ as given in Eq. (142) or Eq. (194). So, for instance, while entangling 4 dimensions in a 6-dimensional space with Cartesian coordinates t, τ, T, x, y, z, we would have:

$$g^{6}_{\alpha\beta}=\begin{pmatrix} g[t,\tau,T,x]\equiv g & 0 & 0 & 0 & 0 & 0 \\ 0 & g & 0 & 0 & 0 & 0 \\ 0 & 0 & g & 0 & 0 & 0 \\ 0 & 0 & 0 & g & 0 & 0 \\ 0 & 0 & 0 & 0 & 1 & 0 \\ 0 & 0 & 0 & 0 & 0 & 1 \end{pmatrix}, \tag{196}$$

with the following solution function for $g[...]$:

$$g[t,\tau,T,x]=\frac{C_1}{c_0\cdot t+c_1\cdot\tau+c_2\cdot T+c_3\cdot x+C_2};\quad c_0^2+c_1^2+c_2^2+c_3^2=0. \tag{197}$$

The entanglement of 6 dimensions as chosen here:

$$g^{6}_{\alpha\beta}=\begin{pmatrix} g[t,\tau,T,x,y,z]\equiv g & 0 & 0 & 0 & 0 & 0 \\ 0 & g & 0 & 0 & 0 & 0 \\ 0 & 0 & g & 0 & 0 & 0 \\ 0 & 0 & 0 & g & 0 & 0 \\ 0 & 0 & 0 & 0 & g & 0 \\ 0 & 0 & 0 & 0 & 0 & g \end{pmatrix} \tag{198}$$

requires:

$$\begin{aligned} g[t,\tau,T,x,y,z]&=\frac{C_1}{\left(c_0\cdot t+c_1\cdot\tau+c_2\cdot T+c_3\cdot x+c_4\cdot y+c_5\cdot z+C_2\right)^2} \\ &c_0^2+c_1^2+c_2^2+c_3^2+c_4^2+c_5^2=0 \end{aligned} \tag{199}$$

if one still wants to solve the vacuum Einstein field equations.

In order to find the simplest of such solutions, where obviously only 2 dimensions are entangled, we assume a Cartesian system in *n* dimensions which we here name $t, x_1, x_2, \ldots$ and combine 2 dimensions as follows:

$$g^{n}_{\alpha\beta}=\begin{pmatrix} F[f[t,x_1]] & 0 & 0 & 0 & \cdots & 0 \\ 0 & F[f[t,x_1]] & 0 & 0 & \cdots & 0 \\ 0 & 0 & 1 & 0 & \cdots & 0 \\ 0 & 0 & 0 & 1 & \cdots & 0 \\ \cdots & \cdots & \cdots & \cdots & \ddots & 0 \\ 0 & 0 & 0 & 0 & 0 & 1 \end{pmatrix}. \tag{200}$$

A very general solution can be found via:

$$F[f]=C_1\cdot e^{C_2\cdot f};\quad f[t,x_1]=C_+[x_1+i\cdot t]+C_-[x_1-i\cdot t]. \tag{201}$$

Thereby the C_1, C_2 denote arbitrary constants and the $C_\pm[...]$ stands for arbitrary functions. We recognize the characteristic wave functions. As before, we have 1 dimension (here we chose t), which had to become time-like in order to fulfill the Einstein field equations. In general, we can have many such 2-dimensional entanglements within one and the same metric, via:

$$g^n_{\alpha\beta}=\begin{pmatrix} F[f_0[t,x_1]] & 0 & 0 & 0 & \cdots & 0 \\ 0 & F[f_0[t,x_1]] & 0 & 0 & \cdots & 0 \\ 0 & 0 & F[f_1[x_1,x_2]] & 0 & \cdots & 0 \\ 0 & 0 & 0 & F[f_1[x_1,x_2]] & \cdots & 0 \\ \cdots & \cdots & \cdots & \cdots & \ddots & 0 \\ 0 & 0 & 0 & 0 & 0 & 1 \end{pmatrix} \tag{202}$$

or

$$g^n_{\alpha\beta}=\begin{pmatrix} F[f_0[t,x_1]] & 0 & 0 & 0 & \cdots & 0 \\ 0 & F[f_2[x_1,x_2]] & 0 & 0 & \cdots & 0 \\ 0 & 0 & F[f_2[x_1,x_2]] & 0 & \cdots & 0 \\ 0 & 0 & 0 & F[f_0[x_1,x_2]] & \cdots & 0 \\ \cdots & \cdots & \cdots & \cdots & \ddots & 0 \\ 0 & 0 & 0 & 0 & 0 & 1 \end{pmatrix} \tag{203}$$

and so on, with all $f_i[...]$ being of the wave-kind as given in Eq. (201).

It should be noted that the solutions of the kind (142), (198), (202) and (203) are taking on very complicated forms in curvilinear coordinates. The usual transformation rules, as given here for the covariant metric tensor

$$g_{\delta\gamma}=\frac{\partial G^\alpha[f[x_k]]}{\partial x^\delta}\frac{\partial G^\beta[f[x_k]]}{\partial x^\gamma}g_{\alpha\beta}, \tag{204}$$

lead to fulfilled complex matrix structures even in simple coordinate systems like the spherical one. The author was therefore not able to find a suitable entanglement solution, e.g., for the Schwarzschild metric. Perhaps the skilled and interested reader might wish to find such a solution. Nevertheless, we have discussed spherical objects with respect to their time dimension(s) in chapter 11 of our book [4].

5.10.2 Uncertain(ty) Consequences

Assuming the simplest and fundamental 2D-entanglement solutions (200) to be omnipresent within this universe (at least as omnipresent as there seems to be a time), we now want to discuss what consequence such solutions would have on the properties of our universe.

We demand the functions *f*[...] of the 2D-entanglements (201) to be such that integration over the whole space is possible and that an expectation value and a variance do exist.

The expectation values of *F*[...] for *x* and *t* can be given via the integrals:

$$E[x]=\int_{-\infty}^{+\infty}\int_{-\infty}^{+\infty} x\cdot F\left[f[t,x]\right]dx\cdot dt$$

$$E[t]=\int_{-\infty}^{+\infty}\int_{-\infty}^{+\infty} t\cdot F\left[f[t,x]\right]dx\cdot dt, \tag{205}$$

which we demand to be x_0 and t_0, respectively. It should be noted that the boundaries of the integration could be a reason for discussion, because—as said—time only comes into existence in the moment of action, which is to say, in the moment the entanglement between *t* and *x* is activated. It also ends with the entanglement and thus should have finite starting and ending points. Here, however, for simplicity, we explicitly intend to avoid such a discussion and assume the time of the entanglement to be active significantly longer than the standard deviation from the expectation value, being given as the square root of $E[t^2]=\Delta t^2$. In other words: As long as Δt is much smaller than the total duration of the entanglement, we can apply Eq. (205).

In addition, we know from [31] that the distribution *F*[*f*], investigated in connection with the Cauchy–Schwarz or Hölder inequality, which is given as:

$$\left|E[x\cdot t]\right|^2 \leq E\left[x^2\right]\cdot E\left[t^2\right], \tag{206}$$

results in a Heisenberg uncertainty principle. The corresponding derivation (using examples for the scaling functions *F*[*f*]) was given in [4].

Here, however, we want to discuss another aspect leading to fundamental limits regarding our solutions. We already have seen that flat spaces in Cartesian coordinates give solutions where at least one coordinate has to be time-like. The simplest such solutions are oscillations and we may consider the possibility that our notion of space-time is just the result of the fact that we actually live on such a solution instead of "existing" on the underlaying space. What we then realize as time is just the result of the fact that, in order to allow for the solution to actually be, it has to sport at least one coordinate, which is acting as time. Some critics may now say that there is no big difference to the concept of just defining a space-time and using this as a starting point instead of only having a space and saying that quantum gravity solutions to this space make our notion of space and time, but we think that in fact there is a huge difference. Living on a solution to something is very much different from the direct existence on this very something. For one thing, not being a direct part of the space but only of one of its solutions does not allow us to directly access the space at all scales, but restricts this very access to only those scales which are in correspondence with "our" solutions. Taking the oscillating solutions, for instance, we cannot access anything smaller than the wavelengths of the oscillations. We are kind of stuck in a scale trap which only allows us certain areas or ranges of the space to observe and these areas or ranges are determined by the solutions to the space-laws governing the true foundations of the universe. We will see in appendix N that this conclusion results not only from the simple flat space Cartesian solutions that we have derived above, but also from spherically symmetric solutions.

Placing one's own existence on a space is something completely different than only existing on certain solutions to this space. Time and scalar restrictions are the direct consequences of a "way of life" based on solutions to a space rather than the space itself.

With respect to the system "the mind," we have to conclude that it also can consequently only be a solution and not the fundamental space-time bringing the solution(s) about. Thus, our genius will always be limited to the recognition of the artificial space of the solutions and not the true substrate. In other words: We cannot grasp the truth, but only a picture of it ... and this holds in a very fundamental sense no matter how holistic thinking and clever we are.

5.11 Appendix K: Metric Quantum Electrodynamics (MQED)

The apparatus of Quantum Electrodynamics (QED) is mainly based on Green's function or propagator technology. Green's functions, on the other hand, are certain fundamental solutions obtained from Schrödinger, Klein–Gordon and Dirac equations, allowing for the integral construction of superposed complex solutions [18].

With the establishment of the origin of these quantum equations out of metric principles, we automatically have derived the metric origin of QED.

For illustration, here we want to consider the so-called Feynman propagator for scalar particles Δ_F (c.f. [18], p. 500), which solves the following Klein–Gordon equation:

$$\left(m_0^2+\frac{\partial^2}{\partial t^2}-\Delta_{3D}\right)\Delta_F=-\delta^4\left[\mathbf{x}'-\mathbf{x}\right]$$

$$\Rightarrow \qquad \Delta_F=f\left[t,x,y,z\right]=\int\frac{d^4p\cdot e^{-i\cdot\mathbf{p}(\mathbf{x}'-\mathbf{x})}}{p^2-m_0^2+i\cdot\varepsilon}. \qquad (207)$$

Please remember that this equation is an adjusted form of Eq. (133) and results from the metric transformation (119) in 6 dimensions with two of these dimensions being suitably entangled (see appendix D). Choosing $C_f = 1$ in Eq. (133) we could illustrate a non-zero-ε-approximation in 1 dimension (Fig. K1). For illustration, we have shown the Dirac delta distribution as an incomplete integral in order to "smear the particle out" over a bit of space. The question we have to answer would be how to give a metric explanation for the occurrence of the Feynman propagator in Eq. (207). The easiest way would be to take Eq. (133) and demand the curvature to take on the propagator property, leading to:

$$R^*=\frac{5}{\left(f+C_f\right)^2}\cdot\left(\frac{C_{v0}^2}{5}\cdot C_f+\frac{C_{v0}^2}{5}\cdot f+\frac{\partial^2 f}{c^2\cdot\partial t^2}-\Delta_{3D}f\right);\quad f=f\left[t,\mathrm{x},\mathrm{y},\mathrm{z}\right]$$

$$\Rightarrow \qquad -\delta^4\left[\mathbf{x}'-\mathbf{x}\right]=\frac{R^*}{5}\cdot\left(f+C_f\right)^2-\overbrace{\frac{C_{v0}^2}{5}}^{m_0^2}\cdot C_f=\overbrace{\frac{C_{v0}^2}{5}}^{m_0^2}\cdot f+\frac{\partial^2 f}{c^2\cdot\partial t^2}-\Delta_{3D}f. \qquad (208)$$

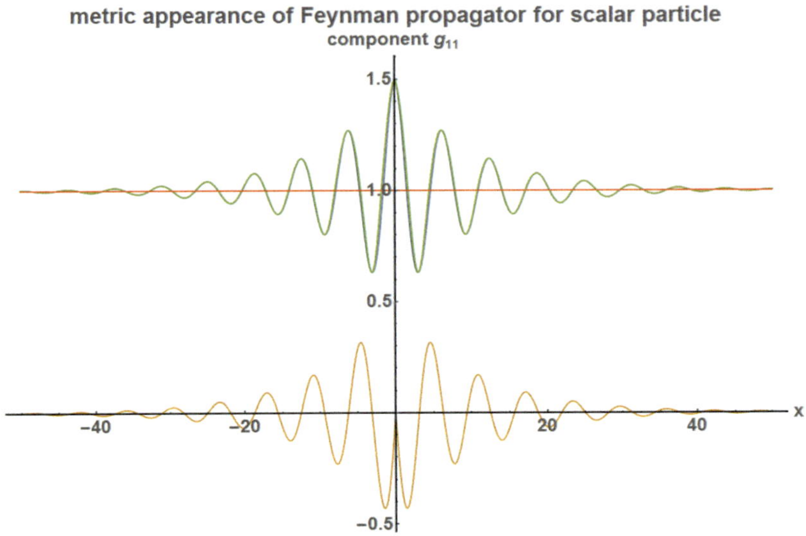

Figure K1 Scalar particle in its metric picture (approximated point-like Feynman propagator, see text) with real part (blue), absolute part (green) and imaginary part (yellow), due to non-zero-$\boldsymbol{\varepsilon}$. In addition, we have shown the g_{11} Minkowski line (red). Please note that $\boldsymbol{\varepsilon}$ is only an auxiliary parameter which should be made to zero in a limiting procedure after the performance of the integration within the QED apparatus.

Considering f quantum mechanically small against the constant C_f (ensuring the asymptotically flat space), we can rewrite Eq. (208) as follows:

$$-\delta^4[\mathbf{x'}-\mathbf{x}]=\left(\frac{R^*}{5}\cdot C_f-m_0^2\right)\cdot C_f=m_0^2\cdot f+\frac{\partial^2 f}{c^2\cdot\partial t^2}-\Delta_{3D}f$$

$$\Rightarrow \qquad R^*=\frac{5}{C_f}\cdot\left(m_0^2-\frac{\delta^4[\mathbf{x'}-\mathbf{x}]}{C_f}\right). \tag{209}$$

Thus, the transformed Ricci curvature R^* for the scalar particle (Feynman particle) would approximately be a constant (flat space) plus a Dirac delta distribution on a certain position in space and time. In the case of small rest masses m_0, the Ricci curvature would even become perfectly proportional to the Dirac delta distribution. This is not a "completely unreasonable" finding and only requires the solutions f to be asymptotically small against any constant in order to

(a) guarantee the flat space-time we obviously observe in earthly conditions AND
(b) allow for the effects of Quantum Electrodynamics to be correct for all scales being significantly bigger than the Schwarzschild sizes

(Schwarzschild radii, c.f. section "Quantum Gravity Dynamic QGD" in [12]) of the objects being considered.

As a mere by-product, we have automatically also found the explanation for the question why QED describes the world so nicely. After all, it already is a part of the Theory of Everything, already resides in the Einstein–Hilbert action (just needed to be spotted there via our $F[f]$-scaling) and the only thing we had to do was to work out the connection to the metric picture.

5.12 Appendix L: Homogenization

In the appendices above, we have seen that we can always factor out a suitable function $F[f]$ from the metric tensor in order to simplify the variational task for either the whole metric (181) or just with respect to the volume factor (186). Thereby we did not care about whether the kernels for both cases are identical, but simply adjusted the factor in order to obtain maximum simplicity to each task. However, it appears reasonable to assume that the kernel of the Einstein–Hilbert action should be the same in both variational cases. Thus, aiming for maximum simplicity in the case of the variation with respect to the whole metric in applying Eq. (181), we require the derivation of the more complex case with respect to f.

The functional derivative:

$$\delta W = \frac{\delta W}{\delta f[\mathbf{y}]} = 0 = \frac{\delta}{\delta f}\int_V d^n x\left(\sqrt{g\cdot F^n}\cdot F\cdot\Phi[f]\cdot R^*\right)$$

$$= \lim_{\in\to 0}\frac{1}{\in}\left(\begin{array}{c}\int_V d^n x\left(\begin{array}{c}\sqrt{g\cdot F\left[f[\mathbf{x}]+\in\cdot\delta[\mathbf{x}-\mathbf{y}]\right]^n}\cdot\Phi\left[f[\mathbf{x}]+\in\cdot\delta[\mathbf{x}-\mathbf{y}]\right]\\ \cdot R^{**}\left[f[\mathbf{x}]+\in\cdot\delta[\mathbf{x}-\mathbf{y}]\right]\end{array}\right)\\ -\int_V d^n x\left(\sqrt{g\cdot F[f[\mathbf{x}]]^n}\cdot\Phi[f[\mathbf{x}]]\cdot R^{**}[f[\mathbf{x}]]\right)\end{array}\right) \tag{210}$$

with:

$$R^{**} = F\cdot R^*$$

$$= R - \frac{g^{\alpha\beta}}{2F}\left(2F_{,\alpha\beta}(n-1) + F_{,d}\,g^{cd}\left(\begin{array}{c}\left(g_{\beta c,\alpha}+g_{\alpha c,\beta}-g_{\beta\alpha,c}\right)-2g_{c\beta,\alpha}\\ +\frac{n}{2}\left(\begin{array}{c}\left(-g_{\alpha c,\beta}-g_{\beta c,\alpha}+g_{\alpha\beta,c}\right)\\ +g_{\alpha\beta,c}\end{array}\right)\end{array}\right)\right)$$

$$+\frac{1}{4F^2}\left(F_{,\alpha}\cdot F_{,\beta}(3n-6)+g_{\alpha\beta}F_{,c}F_{,d}g^{cd}(4-n)\right)g^{\alpha\beta}$$

$$=R-\frac{g^{\alpha\beta}}{2F}\left(2F_{,\alpha\beta}(n-1)+F_{,d}g^{cd}\left(\begin{array}{c}\left(g_{\beta c,\alpha}+g_{\alpha c,\beta}-g_{\beta\alpha,c}\right)-2g_{c\beta,\alpha}\\+\frac{n}{2}\left(\left(-g_{\alpha c,\beta}-g_{\beta c,\alpha}+g_{\alpha\beta,c}\right)+g_{\alpha\beta,c}\right)\end{array}\right)\right)$$

$$+\frac{1}{4F^2}\left(F_{,\alpha}\cdot F_{,\beta}(3n-6)+g_{\alpha\beta}F_{,c}F_{,d}g^{cd}(4-n)\right)g^{\alpha\beta}$$

$$=R-\frac{g^{\alpha\beta}}{2F}F'\left(2f_{,\alpha\beta}(n-1)+f_{,d}g^{cd}\left(\begin{array}{c}g_{\alpha c,\beta}-g_{\beta\alpha,c}-g_{c\beta,\alpha}\\+\frac{n}{2}\left(2g_{\alpha\beta,c}-g_{\alpha c,\beta}-g_{\beta c,\alpha}\right)\end{array}\right)\right)$$
$$-(n-1)\frac{f_{,\alpha}\cdot f_{,\beta}}{4F^2}g^{\alpha\beta}\left(4FF''+F'\cdot F'(n-6)\right) \quad (211)$$

gives:

$$\frac{\delta W}{\delta f}=0=\frac{\delta}{\delta f}\int_V d^nx\left(\sqrt{g\cdot F^n}\cdot\Phi[f]\cdot R^{**}\right)$$

$$=-\left[\Phi[f]\cdot\left(\begin{array}{c}\left((n-1)\cdot\sqrt{g}\cdot F^{\frac{n}{2}-1}g^{\alpha\beta}\cdot F'\right)_{,\alpha\beta}-\sqrt{g}\cdot R\frac{dF^{\frac{n}{2}}}{df}\\-\frac{1}{2}\left(\sqrt{g}\cdot F^{\frac{n}{2}-1}g^{\alpha\beta}\cdot F'g^{cd}\left(\begin{array}{c}g_{\alpha c,\beta}-g_{\beta\alpha,c}-g_{c\beta,\alpha}\\+\frac{n}{2}\left(2g_{\alpha\beta,c}-g_{\alpha c,\beta}-g_{\beta c,\alpha}\right)\end{array}\right)\right)_{,d}\\+\frac{\sqrt{g}\cdot g^{\alpha\beta}}{2}\left(\begin{array}{c}2f_{,\alpha\beta}(n-1)\\+f_{,d}g^{cd}\left(\begin{array}{c}g_{\alpha c,\beta}-g_{\beta\alpha,c}-g_{c\beta,\alpha}\\+\frac{n}{2}\left(2g_{\alpha\beta,c}-g_{\alpha c,\beta}-g_{\beta c,\alpha}\right)\end{array}\right)\end{array}\right)\frac{d\left[F^{\frac{n}{2}-1}F'\right]}{df}\\+\frac{(n-1)}{2}\left(\begin{array}{c}-\left(f_{,\alpha}\cdot\sqrt{g}\cdot g^{\alpha\beta}F^{\frac{n}{2}-2}\left(4FF''+F'\cdot F'(n-6)\right)\right)_{,\beta}\\+\frac{f_{,\alpha}\cdot f_{,\beta}}{2}\sqrt{g}\cdot g^{\alpha\beta}\frac{d\left[F^{\frac{n}{2}-2}\left(4FF''+F'\cdot F'(n-6)\right)\right]}{df}\end{array}\right)\end{array}\right)\right]_{\mathbf{y}}$$

$$+\left[\sqrt{g\cdot F^{n}}\cdot\left(\begin{matrix}R-\frac{g^{\alpha\beta}}{2F}F'\left(\begin{matrix}2f_{,\alpha\beta}(n-1)\\+f_{,d}g^{cd}\left(\begin{matrix}g_{\alpha c,\beta}-g_{\beta\alpha,c}-g_{c\beta,\alpha}\\+\frac{n}{2}\left(2g_{\alpha\beta,c}-g_{\alpha c,\beta}-g_{\beta c,\alpha}\right)\end{matrix}\right)\end{matrix}\right)\\-(n-1)\frac{f_{,\alpha}\cdot f_{,\beta}}{4F^{2}}g^{\alpha\beta}\left(4FF''+F'\cdot F'(n-6)\right)\end{matrix}\right)\frac{d}{df}\Phi[f]\right]_{y}. \quad (212)$$

Please note that just for brevity, we assume in this appendix that g denotes the absolute value of the determinant of the metric tensor $g_{\alpha\beta}$.

Thereby, using the results from appendix M, we can significantly simplify the first term and obtain:

$$\frac{\delta W}{\delta f}=0=\frac{\delta}{\delta f}\int_{V}d^{n}x\left(\sqrt{g\cdot F^{n}}\cdot\Phi[f]\cdot R^{**}\right)$$

$$=\left\{\begin{matrix}\left[\sqrt{g\cdot F^{n}}\cdot\left(\begin{matrix}R-\frac{g^{\alpha\beta}}{2F}F'\left(\begin{matrix}2f_{,\alpha\beta}(n-1)\\+f_{,d}g^{cd}\left(\begin{matrix}g_{\alpha c,\beta}-g_{\beta\alpha,c}-g_{c\beta,\alpha}\\+\frac{n}{2}\left(2g_{\alpha\beta,c}-g_{\alpha c,\beta}-g_{\beta c,\alpha}\right)\end{matrix}\right)\end{matrix}\right)\\-(n-1)\frac{f_{,\alpha}\cdot f_{,\beta}}{4F^{2}}g^{\alpha\beta}\left(4FF''+F'\cdot F'(n-6)\right)\end{matrix}\right)\frac{d}{df}\Phi[f]\right]_{y}\\-\left[\Phi[f]\cdot\left(\begin{matrix}F^{\frac{n}{2}-3}F'\left(\begin{matrix}(n-1)\frac{\sqrt{g}\cdot g^{\alpha\beta}}{2}\frac{f_{,\alpha}f_{,\beta}}{2}\left(\frac{1}{2}(n-4)(n+2)(F')^{2}+2FF''\right)\\+(n-1)\frac{(n+2)}{2}FF'\overbrace{\left(\sqrt{g}\cdot g^{\alpha\beta}f_{,\alpha\beta}+f_{,\alpha}\left(\sqrt{g}\cdot g^{\alpha\beta}\right)_{,\beta}\right)}^{=\sqrt{g}\cdot\Delta f}\\+(n-1)F^{2}\left(\sqrt{g}\cdot g^{\alpha\beta}\right)_{,\alpha\beta}\\-\frac{F^{2}}{2}\left(\sqrt{g}\cdot g^{\alpha\beta}\cdot g^{ij}\left(\begin{matrix}g_{i\beta,j}-g_{ij,\beta}-g_{\beta j,i}\\+\frac{n}{2}\left(2g_{ij,\beta}-g_{i\beta,j}-g_{j\beta,i}\right)\end{matrix}\right)\right)_{,\alpha}\end{matrix}\right)\\-\sqrt{g}\cdot RF^{\frac{n}{2}-1}F'\end{matrix}\right)\right]_{y}\end{matrix}\right.. \quad (213)$$

Simplification yields:

$$\frac{\delta W}{\delta f}=0=\frac{\delta}{\delta f}\int_V d^n x\left(\sqrt{g\cdot F^n}\cdot\Phi[f]\cdot R^{**}\right)$$

$$=\left\{\begin{array}{l}\left[\sqrt{g\cdot F^n}\cdot\left(\begin{array}{c}R-\frac{g^{\alpha\beta}}{2F}F'\left(\begin{array}{c}2f_{,\alpha\beta}(n-1)\\+f_{,d}g^{cd}\left(\begin{array}{c}g_{\alpha c,\beta}-g_{\beta\alpha,c}-g_{c\beta,\alpha}\\+\frac{n}{2}\left(2g_{\alpha\beta,c}-g_{\alpha c,\beta}-g_{\beta c,\alpha}\right)\end{array}\right)\end{array}\right)\\-(n-1)\frac{f_{,\alpha}\cdot f_{,\beta}}{4F^2}g^{\alpha\beta}\left(4FF''+F'\cdot F'(n-6)\right)\end{array}\right)\frac{d}{df}\Phi[f]\right]_{\mathbf{y}}\\-\left[\Phi[f]\cdot\left(F^{\frac{n}{2}-3}F'\left(\begin{array}{c}(n-1)\frac{\sqrt{g}\cdot g^{\alpha\beta}}{2}\frac{f_{,\alpha}f_{,\beta}}{2}\left(\frac{1}{2}(n-4)(n+2)(F')^2+2FF''\right)\\+(n-1)F\left(\frac{(n+2)}{2}F'\sqrt{g}\cdot\Delta f+F\left(\sqrt{g}\cdot g^{\alpha\beta}\right)_{,\alpha\beta}\right)\\-\frac{F^2}{2}\left(\sqrt{g}\cdot g^{\alpha\beta}\cdot g^{ij}\left(\begin{array}{c}g_{i\beta,j}-g_{ji,\beta}-g_{\beta j,i}\\+\frac{n}{2}\left(2g_{ij,\beta}-g_{i\beta,j}-g_{j\beta,i}\right)\end{array}\right)\right)_{,\alpha}\end{array}\right)\\-\sqrt{g}\cdot RF^{\frac{n}{2}-1}F'\end{array}\right)\right]_{\mathbf{y}}\end{array}\right.\cdot$$

(214)

Now, after having applied Eq. (181) in order to obtain maximum (Hilbert-) simplicity for the variational task with respect to the whole metric tensor, we are left with the following variation regarding the function *f*:

$$\frac{\delta W}{\delta f}=0=\frac{\delta}{\delta f}\int_V d^n x\left(\sqrt{g\cdot F^n}\cdot\Phi[f]\cdot R^{**}\right)=\frac{\delta}{\delta f}\int_V d^n x\left(\sqrt{g\cdot F^n}\cdot\frac{1}{F}\cdot R^{**}\right). \qquad (215)$$

From our result (213), we immediately obtain the subsequent equation:

$$\frac{\delta W}{\delta f}=0=\frac{\delta}{\delta f}\int_V d^n x\left(\sqrt{g\cdot F^n}\cdot\frac{1}{F}\cdot R^{**}\right)$$

$$
= \left(\begin{array}{c} -\frac{1}{F} \cdot \left(F^{\frac{n}{2}-3} F' \left(\begin{array}{c} (n-1)\frac{\sqrt{g} \cdot g^{\alpha\beta}}{2} \frac{f_{,\alpha} f_{,\beta}}{2} \left(\frac{1}{2}(n-4)(n+2)(F')^2 + 2FF'' \right) \\ +(n-1)F\left(\frac{(n+2)}{2} F' \sqrt{g} \cdot \Delta f + F\left(\sqrt{g} \cdot g^{\alpha\beta} \right)_{,\alpha\beta} \right) \\ -\frac{F^2}{2}\left(\sqrt{g} \cdot g^{\alpha\beta} \cdot g^{ij} \left(\begin{array}{c} g_{i\beta,j} - g_{ij,\beta} - g_{\beta j,i} \\ +\frac{n}{2}\left(2g_{ij,\beta} - g_{i\beta,j} - g_{j\beta,i} \right) \end{array} \right) \right)_{,\alpha} \end{array} \right) \\ -\sqrt{g} \cdot RF^{\frac{n}{2}-1} F' \end{array} \right) \\ -\frac{\sqrt{g \cdot F^n}}{F^2} \cdot F' \left(\begin{array}{c} -\frac{g^{\alpha\beta}}{2F} F' \left(2f_{,\alpha\beta}(n-1) + f_{,d} g^{cd} \left(\begin{array}{c} g_{\alpha c,\beta} - g_{\beta\alpha,c} - g_{c\beta,\alpha} \\ +\frac{n}{2}\left(2g_{\alpha\beta,c} - g_{\alpha c,\beta} - g_{\beta c,\alpha} \right) \end{array} \right) \right) \\ R - (n-1)\frac{f_{,\alpha} \cdot f_{,\beta}}{4F^2} g^{\alpha\beta} \left(4FF'' + F' \cdot F'(n-6) \right) \end{array} \right) \end{array} \right), \tag{216}
$$

leading to:

$$
0 = \left(\begin{array}{c} F^{\frac{n}{2}-3} F' \left(\begin{array}{c} (n-1)\frac{\sqrt{g} \cdot g^{\alpha\beta}}{2} \frac{f_{,\alpha} f_{,\beta}}{2} \left(\frac{1}{2}(n-4)(n+2)(F')^2 + 2FF'' \right) \\ +(n-1)F\left(\frac{(n+2)}{2} F' \sqrt{g} \cdot \Delta f + F\left(\sqrt{g} \cdot g^{\alpha\beta} \right)_{,\alpha\beta} \right) \\ -\frac{F^2}{2}\left(\sqrt{g} \cdot g^{\alpha\beta} \cdot g^{ij} \left(\begin{array}{c} g_{i\beta,j} - g_{ij,\beta} - g_{\beta j,i} \\ +\frac{n}{2}\left(2g_{ij,\beta} - g_{i\beta,j} - g_{j\beta,i} \right) \end{array} \right) \right)_{,\alpha} \end{array} \right) \\ -\sqrt{g} \cdot RF^{\frac{n}{2}-1} F' \\ +\sqrt{g} \cdot F^{\frac{n}{2}-1} \cdot F' \left(\begin{array}{c} R - \frac{g^{\alpha\beta}}{2F} F' \left(2f_{,\alpha\beta}(n-1) + f_{,d} g^{cd} \left(\begin{array}{c} g_{\alpha c,\beta} - g_{\beta\alpha,c} - g_{c\beta,\alpha} \\ +\frac{n}{2}\left(2g_{\alpha\beta,c} - g_{\alpha c,\beta} - g_{\beta c,\alpha} \right) \end{array} \right) \right) \\ -(n-1)\frac{f_{,\alpha} \cdot f_{,\beta}}{4F^2} g^{\alpha\beta} \left(4FF'' + F' \cdot F'(n-6) \right) \end{array} \right) \end{array} \right), \tag{217}
$$

where we can further simplify as follows:

$$0=\left(\begin{array}{c}\frac{F'}{F^2}\left(\begin{array}{c}(n-1)\frac{\sqrt{g}\cdot g^{\alpha\beta}}{2}\frac{f_{,\alpha}f_{,\beta}}{2}\left(\frac{1}{2}(n-4)(n+2)(F')^2+2FF''\right)\\+(n-1)F\left(\frac{(n+2)}{2}F'\sqrt{g}\cdot\Delta f+F\left(\sqrt{g}\cdot g^{\alpha\beta}\right)_{,\alpha\beta}\right)\\-\frac{F^2}{2}\left(\sqrt{g}\cdot g^{\alpha\beta}\cdot g^{ij}\left(\begin{array}{c}g_{i\beta,j}-g_{ij,\beta}-g_{\beta j,i}\\+\frac{n}{2}\left(2g_{ij,\beta}-g_{i\beta,j}-g_{j\beta,i}\right)\end{array}\right)\right)_{,\alpha}\end{array}\right)\\-\sqrt{g}\cdot F'\left(\begin{array}{c}\frac{g^{\alpha\beta}}{2F}F'\left(2f_{,\alpha\beta}(n-1)+f_{,d}g^{cd}\left(\begin{array}{c}g_{\alpha c,\beta}-g_{\beta\alpha,c}-g_{c\beta,\alpha}\\+\frac{n}{2}\left(2g_{\alpha\beta,c}-g_{\alpha c,\beta}-g_{\beta c,\alpha}\right)\end{array}\right)\right)\\+(n-1)\frac{f_{,\alpha}\cdot f_{,\beta}}{4F^2}g^{\alpha\beta}\left(4FF''+F'\cdot F'(n-6)\right)\end{array}\right)\\+\overbrace{\sqrt{g}\cdot R\cdot(F'-F')}^{=0}\end{array}\right), \tag{218}$$

$$0=\left(\begin{array}{c}\frac{1}{F}\left(\begin{array}{c}(n-1)\frac{\sqrt{g}\cdot g^{\alpha\beta}}{2}\frac{f_{,\alpha}f_{,\beta}}{2}\left(\frac{1}{2}(n-4)(n+2)(F')^2+2FF''\right)\\+(n-1)F\left(\frac{(n+2)}{2}F'\sqrt{g}\cdot\Delta f+F\left(\sqrt{g}\cdot g^{\alpha\beta}\right)_{,\alpha\beta}\right)\\-\frac{F^2}{2}\left(\sqrt{g}\cdot g^{\alpha\beta}\cdot g^{ij}\left(\begin{array}{c}g_{i\beta,j}-g_{ij,\beta}-g_{\beta j,i}\\+\frac{n}{2}\left(2g_{ij,\beta}-g_{i\beta,j}-g_{j\beta,i}\right)\end{array}\right)\right)_{,\alpha}\end{array}\right)\\-\sqrt{g}\cdot\left(\begin{array}{c}\frac{g^{\alpha\beta}}{2}F'\left(2f_{,\alpha\beta}(n-1)+f_{,d}g^{cd}\left(\begin{array}{c}g_{\alpha c,\beta}-g_{\beta\alpha,c}-g_{c\beta,\alpha}\\+\frac{n}{2}\left(2g_{\alpha\beta,c}-g_{\alpha c,\beta}-g_{\beta c,\alpha}\right)\end{array}\right)\right)\\+(n-1)\frac{f_{,\alpha}\cdot f_{,\beta}}{4F}g^{\alpha\beta}\left(4FF''+F'\cdot F'(n-6)\right)\end{array}\right)\end{array}\right), \tag{219}$$

$$0=\left(\begin{array}{c}(n-1)\left(\frac{(n+2)}{2}F'\sqrt{g}\cdot\Delta f+F\left(\sqrt{g}\cdot g^{\alpha\beta}\right)_{,\alpha\beta}\right)\\ -\frac{F}{2}\left(\sqrt{g}\cdot g^{\alpha\beta}\cdot g^{ij}\left(\begin{array}{c}g_{i\beta,j}-g_{ij,\beta}-g_{\beta j,i}\\ +\frac{n}{2}\left(2g_{ij,\beta}-g_{i\beta,j}-g_{j\beta,i}\right)\end{array}\right)\right)_{,\alpha}\\ -\sqrt{g}\cdot\left(\frac{g^{\alpha\beta}}{2}F'\left(2f_{,\alpha\beta}(n-1)+f_{,d}g^{cd}\left(\begin{array}{c}g_{\alpha c,\beta}-g_{\beta\alpha,c}-g_{c\beta,\alpha}\\ +\frac{n}{2}\left(2g_{\alpha\beta,c}-g_{\alpha c,\beta}-g_{\beta c,\alpha}\right)\end{array}\right)\right)\right)\\ (n-1)\sqrt{g}\cdot g^{\alpha\beta}\frac{f_{,\alpha}f_{,\beta}}{4F}\left(\frac{(n-2)^2}{2}(F')^2-2FF''\right)\end{array}\right). \tag{220}$$

We realize that we have lost the Ricci scalar curvature R.

It was shown in [32] that this equation is also applicable in the cases n = 1 and n = 2. In the case of n = 1, we obtain:

$$0=\left(\begin{array}{c}-\frac{F}{2}\left(\sqrt{g}\cdot g^{\alpha\beta}\cdot g^{ij}\left(\begin{array}{c}g_{i\beta,j}-g_{ij,\beta}-g_{\beta j,i}\\ +\frac{n}{2}\left(2g_{ij,\beta}-g_{i\beta,j}-g_{j\beta,i}\right)\end{array}\right)\right)_{,\alpha}\\ -\sqrt{g}\cdot\left(\frac{g^{\alpha\beta}}{2}F'f_{,d}g^{cd}\left(\begin{array}{c}g_{\alpha c,\beta}-g_{\beta\alpha,c}-g_{c\beta,\alpha}\\ +\frac{n}{2}\left(2g_{\alpha\beta,c}-g_{\alpha c,\beta}-g_{\beta c,\alpha}\right)\end{array}\right)\right)\end{array}\right)$$

$$=\frac{1}{2}\cdot\left(F\left(\sqrt{g}\cdot g^{\alpha\beta}\cdot g^{ij}\left(g_{,x}\right)\right)_{,x}+\sqrt{g}\cdot g^{\alpha\beta}F'f_{,x}g^{cd}g_{,x}\right), \tag{221}$$

$$\xrightarrow{g_{\alpha\beta}=g;\ \ g^{\alpha\beta}=1/g;\ \ g=g[x]}$$

$$0=\frac{1}{2}\left(\frac{2g[x]g''[x]-3g'[x]^2}{2g[x]^{5/2}}+F'\frac{\sqrt{g}}{g^2}f_{,x}g_{,x}\right)$$

$$=\frac{1}{2}\left(\frac{2g[x]g''[x]-3g'[x]^2}{2g[x]^{5/2}}+F'f_{,x}\frac{g'[x]}{g[x]^{3/2}}\right)$$

$$=\frac{1}{2g[x]^{5/2}}\left(2g[x]g''[x]-3g'[x]^2+2F'f_{,x}g'[x]g[x]\right). \tag{222}$$

For completeness, we also give the case of $n = 2$:

$$0=\left(\begin{array}{c} 2F'\sqrt{g}\cdot\Delta f+F\left(\sqrt{g}\cdot g^{\alpha\beta}\right)_{,\alpha\beta}+\frac{F}{2}\left(\sqrt{g}\cdot g^{\alpha\beta}\cdot g^{ij}\left(2g_{\beta j,i}-g_{ij,\beta}\right)\right)_{,\alpha} \\ -\sqrt{g}\cdot\frac{g^{\alpha\beta}}{2}\left(F'\left(2f_{,\alpha\beta}+f_{,d}g^{cd}\left(g_{\alpha\beta,c}-2g_{\beta c,\alpha}\right)\right)-f_{,\alpha}f_{,\beta}\cdot F''\right)\end{array}\right). \tag{223}$$

5.12.1 Vice Versa

Now we assume that the dominating kernel shall be of the type (166), so that we have the outcome of a perfect Laplace operator for f after variation with respect to the wave functions f. The subsequent variation with respect to the complete metric tensor then becomes:

$$\delta W=0=\delta_g\int\limits_V d^nx\left(\sqrt{g\cdot F^n}\cdot F\cdot R^*\right)=\delta_g\int\limits_V d^nx\left(\sqrt{g\cdot F^n}\cdot R^*_{\alpha\beta}g^{\alpha\beta}\right), \tag{224}$$

whereby the scaled Ricci scalar R^* has already been given above. In order to apply the recipe (181), we rewrite Eq. (224) as follows:

$$\delta W=0=\delta_g\int\limits_V d^nx\left(\sqrt{g\cdot F^n}\cdot F\cdot R^*\right)=\delta_G\int\limits_V d^nx\left(\sqrt{G}\cdot F\cdot R^*\right). \tag{225}$$

As usual we have:

$$G_{\alpha\beta}=F\left[f\left[t,x,y,z,\ldots,\xi_k,\ldots,\xi_n\right]\right]\cdot g_{\alpha\beta}=F\left[f\right]\cdot g_{\alpha\beta} \tag{226}$$

and:

$$R^*=\frac{1}{F}\left(\begin{array}{c} R-\frac{g^{\alpha\beta}}{2F}F'\left(2f_{,\alpha\beta}(n-1)+f_{,d}g^{cd}\left(\begin{array}{c} g_{\alpha c,\beta}-g_{\beta\alpha,c}-g_{c\beta,\alpha} \\ +\frac{n}{2}\left(2g_{\alpha\beta,c}-g_{\alpha c,\beta}-g_{\beta c,\alpha}\right)\end{array}\right)\right) \\ -(n-1)\frac{f_{,\alpha}\cdot f_{,\beta}}{4F^2}g^{\alpha\beta}\left(4FF''+F'\cdot F'(n-6)\right)\end{array}\right). \tag{227}$$

Now we substitute in the Einstein–Hilbert action as follows:

$$\delta W=0=\delta\int\limits_V d^nx\left(\sqrt{G}\cdot F\cdot R^*\right)$$

$$\xrightarrow{G_{\alpha\beta}=\gamma_{\alpha\beta}\cdot\Theta;\ \ \Theta^{\frac{n}{2}}=F}$$

$$0=\delta_{G_{\alpha\beta}}\int\limits_V d^nx\left(\sqrt{-\gamma}\cdot\Theta^{\frac{n}{2}}R^+\right)\Rightarrow 0=\delta_{G_{\alpha\beta}}\int\limits_V d^nx\left(\sqrt{-\gamma\cdot\Theta^n}\cdot R^+\right)$$

$$\Rightarrow\qquad 0=R^+_{\alpha\beta}-\frac{1}{2}\cdot\gamma_{\alpha\beta}\cdot R^+. \tag{228}$$

Thereby we remember that:

$$R^{+}=\frac{1}{\Theta}\left(R^{*}-\frac{1}{2\Theta}\begin{pmatrix}2\Theta_{,ij}(n-1)\gamma^{ij}+2\Gamma_{ij}^{a}\Theta_{,a}\gamma^{ij}\\ -\Theta_{,i}\gamma^{ab}\gamma_{jb,a}\gamma^{ij}-\Theta_{,j}\gamma^{ab}\gamma_{ib,a}\gamma^{ij}\\ -n\Gamma_{ij}^{d}\Theta_{,d}\gamma^{ij}+\frac{n}{2}\Theta_{,d}\gamma^{cd}\gamma_{ab,c}\gamma^{ab}\end{pmatrix}-\frac{\Theta_{,i}\cdot\Theta_{,j}}{4\Theta^{2}}\gamma^{ij}\left((n-6)(n-1)\right)\right)$$

$$=\frac{1}{\Theta}\left(\begin{array}{c}\frac{1}{F}\left(\begin{array}{c}R-\frac{g^{\alpha\beta}}{2F}F'\left(2f_{,\alpha\beta}(n-1)+f_{,d}g^{cd}\left(\begin{array}{c}g_{\alpha c,\beta}-g_{\beta\alpha,c}-g_{c\beta,\alpha}\\ +\frac{n}{2}\left(2g_{\alpha\beta,c}-g_{\alpha c,\beta}-g_{\beta c,\alpha}\right)\end{array}\right)\right)\\ -(n-1)\frac{f_{,\alpha}\cdot f_{,\beta}}{4F^{2}}g^{\alpha\beta}\left(4FF''+F'\cdot F'(n-6)\right)\end{array}\right)\\ -\frac{1}{2\Theta}\begin{pmatrix}2\Theta_{,ij}(n-1)\gamma^{ij}+2\Gamma_{ij}^{a}\Theta_{,a}\gamma^{ij}\\ -\Theta_{,i}\gamma^{ab}\gamma_{jb,a}\gamma^{ij}-\Theta_{,j}\gamma^{ab}\gamma_{ib,a}\gamma^{ij}\\ -n\Gamma_{ij}^{d}\Theta_{,d}\gamma^{ij}+\frac{n}{2}\Theta_{,d}\gamma^{cd}\gamma_{ab,c}\gamma^{ab}\end{pmatrix}-\frac{\Theta_{,i}\cdot\Theta_{,j}}{4\Theta^{2}}\gamma^{ij}\left((n-6)(n-1)\right)\end{array}\right)$$

$$\text{with:}\quad g_{\alpha\beta}\cdot F=\gamma_{\alpha\beta}\cdot\Theta=\gamma_{\alpha\beta}\cdot F^{\frac{2}{n}}\Rightarrow g_{\alpha\beta}=\gamma_{\alpha\beta}\cdot F^{\frac{2-n}{n}}\qquad(229)$$

and subsequently obtain after variation with respect to the scaled metric tensor:

$$0=\int_{V_G}d^{n}x\sqrt{-\gamma\cdot\Theta^{n}}\left(\left(\left(\begin{array}{c}R_{\alpha\beta}^{*}-\frac{\gamma_{\alpha\beta}}{2}\left(\begin{array}{c}R^{*}-\frac{1}{2\Theta}\begin{pmatrix}2\Theta_{,ij}(n-1)\gamma^{ij}+2\Gamma_{ij}^{a}\Theta_{,a}\gamma^{ij}\\ -\Theta_{,i}\gamma^{ab}\gamma_{jb,a}\gamma^{ij}-\Theta_{,j}\gamma^{ab}\gamma_{ib,a}\gamma^{ij}\\ -n\Gamma_{ij}^{d}\Theta_{,d}\gamma^{ij}+\frac{n}{2}\Theta_{,d}\gamma^{cd}\gamma_{ab,c}\gamma^{ab}\end{pmatrix}\\ -\frac{\Theta_{,i}\cdot\Theta_{,j}}{4\Theta^{2}}\gamma^{ij}\left((n-6)(n-1)\right)\end{array}\right)\\ -\frac{1}{2\Theta}\left(\begin{array}{c}\Theta_{,\alpha\beta}(n-2)+\Theta_{,ab}\gamma_{\alpha\beta}\gamma^{ab}\\ +\Theta_{,a}\gamma^{ab}\left(\gamma_{\beta b,\alpha}-\gamma_{\beta\alpha,b}\right)-\Theta_{,\alpha}\gamma^{ab}\gamma_{\beta b,a}-\Theta_{,\beta}\gamma^{ab}\gamma_{\alpha b,a}\\ +\Theta_{,d}\gamma^{cd}\frac{1}{2}n\left(\begin{array}{c}\frac{2}{n}\gamma_{\alpha c,\beta}-\gamma_{\alpha c,\beta}-\gamma_{\beta c,\alpha}\\ +\gamma_{\alpha\beta,c}+\frac{1}{n}\gamma_{\alpha\beta}\gamma_{ab,c}\gamma^{ab}\end{array}\right)\\ -\frac{1}{2\Theta}\left(\Theta_{,\alpha}\cdot\Theta_{,\beta}(3n-6)+\gamma_{\alpha\beta}\Theta_{,c}\Theta_{,d}\gamma^{cd}(4-n)\right)\end{array}\right)\end{array}\right)\right)\delta g^{\alpha\beta}\right).$$

(230)

The resulting field equations read as follows:

$$0=\left(\begin{array}{c} R^{*}_{\alpha\beta}-\frac{\gamma_{\alpha\beta}}{2}\left(\begin{array}{c} R^{*}-\frac{1}{2\Theta}\left(\begin{array}{c} 2\Theta_{,ij}(n-1)\gamma^{ij}+2\Gamma^{a}_{ij}\Theta_{,a}\gamma^{ij} \\ -\Theta_{,i}\gamma^{ab}\gamma_{jb,a}\gamma^{ij}-\Theta_{,j}\gamma^{ab}\gamma_{ib,a}\gamma^{ij} \\ -n\Gamma^{d}_{ij}\Theta_{,d}\gamma^{ij}+\frac{n}{2}\Theta_{,d}\gamma^{cd}\gamma_{ab,c}\gamma^{ab} \end{array}\right) \\ -\frac{\Theta_{,i}\cdot\Theta_{,j}}{4\Theta^{2}}\gamma^{ij}\left((n-6)(n-1)\right) \end{array}\right) \\ -\frac{1}{2\Theta}\left(\begin{array}{c} \Theta_{,\alpha\beta}(n-2)+\Theta_{,ab}\gamma_{\alpha\beta}\gamma^{ab} \\ +\Theta_{,a}\gamma^{ab}\left(\gamma_{\beta b,\alpha}-\gamma_{\beta\alpha,b}\right)-\Theta_{,\alpha}\gamma^{ab}\gamma_{\beta b,a}-\Theta_{,\beta}\gamma^{ab}\gamma_{\alpha b,a} \\ +\Theta_{,d}\gamma^{cd}\frac{1}{2}n\left(\begin{array}{c} \frac{2}{n}\gamma_{\alpha c,\beta}-\gamma_{\alpha c,\beta}-\gamma_{\beta c,\alpha} \\ +\gamma_{\alpha\beta,c}+\frac{1}{n}\gamma_{\alpha\beta}\gamma_{ab,c}\gamma^{ab} \end{array}\right) \\ -\frac{1}{2\Theta}\left(\Theta_{,\alpha}\cdot\Theta_{,\beta}(3n-6)+\gamma_{\alpha\beta}\Theta_{,c}\Theta_{,d}\gamma^{cd}(4-n)\right) \end{array}\right) \end{array}\right). \quad (231)$$

5.13 Appendix M: Derivation of the Functional Derivative of the Scaled and Adjusted Einstein–Hilbert Action with Respect to the Function *f*

The functional derivative:

$$\delta W=\frac{\delta W}{\delta f[\mathbf{y}]}=0=\frac{\delta}{\delta f}\int_V d^n x\left(\sqrt{g\cdot F^n}\cdot F\cdot R^{*}\right)$$

$$=\lim_{\in\to 0}\frac{1}{\in}\left(\begin{array}{c} \int_V d^n x\left(\sqrt{g\cdot F\left[f[\mathbf{x}]+\in\cdot\delta[\mathbf{x}-\mathbf{y}]\right]^n}\cdot R^{**}\left[f[\mathbf{x}]+\in\cdot\delta[\mathbf{x}-\mathbf{y}]\right]\right) \\ -\int_V d^n x\left(\sqrt{g\cdot F[f[\mathbf{x}]]^n}\cdot R^{**}[f[\mathbf{x}]]\right) \end{array}\right) \quad (232)$$

with:

$$R^{**}=F\cdot R^{*}$$

$$=R-\frac{g^{\alpha\beta}}{2F}\left(2F_{,\alpha\beta}(n-1)+F_{,d}g^{cd}\left(\begin{array}{c} \left(g_{\beta c,\alpha}+g_{\alpha c,\beta}-g_{\beta\alpha,c}\right)-2g_{c\beta,\alpha} \\ +\frac{n}{2}\left(\left(-g_{\alpha c,\beta}-g_{\beta c,\alpha}+g_{\alpha\beta,c}\right)+g_{\alpha\beta,c}\right) \end{array}\right)\right)$$

$$+\frac{1}{4F^2}\left(F_{,\alpha}\cdot F_{,\beta}(3n-6)+g_{\alpha\beta}F_{,c}F_{,d}g^{cd}(4-n)\right)g^{\alpha\beta}$$

$$=R-\frac{g^{\alpha\beta}}{2F}\left(2F_{,\alpha\beta}(n-1)+F_{,d}g^{cd}\begin{pmatrix}\left(g_{\beta c,\alpha}+g_{\alpha c,\beta}-g_{\beta\alpha,c}\right)-2g_{c\beta,\alpha}\\+\frac{n}{2}\left(\left(-g_{\alpha c,\beta}-g_{\beta c,\alpha}+g_{\alpha\beta,c}\right)+g_{\alpha\beta,c}\right)\end{pmatrix}\right)$$

$$+\frac{1}{4F^2}\left(F_{,\alpha}\cdot F_{,\beta}(3n-6)+g_{\alpha\beta}F_{,c}F_{,d}g^{cd}(4-n)\right)g^{\alpha\beta}$$

$$=R-\frac{g^{\alpha\beta}}{2F}F'\left(2f_{,\alpha\beta}(n-1)+f_{,d}g^{cd}\begin{pmatrix}g_{\alpha c,\beta}-g_{\beta\alpha,c}-g_{c\beta,\alpha}\\+\frac{n}{2}\left(2g_{\alpha\beta,c}-g_{\alpha c,\beta}-g_{\beta c,\alpha}\right)\end{pmatrix}\right)$$
$$-(n-1)\frac{f_{,\alpha}\cdot f_{,\beta}}{4F^2}g^{\alpha\beta}\left(4FF''+F'\cdot F'(n-6)\right) \tag{233}$$

gives:

$$\frac{\delta W}{\delta f}=0$$

$$=\frac{\delta}{\delta f}\int_V d^nx\left(\sqrt{g\cdot F^n}\cdot R^{**}\right)$$

$$=-\left[\begin{array}{c}\left((n-1)\cdot\sqrt{g}\cdot F^{\frac{n}{2}-1}g^{\alpha\beta}\cdot F'\right)_{,\alpha\beta}-\sqrt{g}\cdot R\frac{dF^{\frac{n}{2}}}{df}\\-\frac{1}{2}\left(\sqrt{g}\cdot F^{\frac{n}{2}-1}g^{\alpha\beta}\cdot F'g^{cd}\begin{pmatrix}g_{\alpha c,\beta}-g_{\beta\alpha,c}-g_{c\beta,\alpha}\\+\frac{n}{2}\left(2g_{\alpha\beta,c}-g_{\alpha c,\beta}-g_{\beta c,\alpha}\right)\end{pmatrix}\right)_{,d}\\+\frac{\sqrt{g}\cdot g^{\alpha\beta}}{2}\left(2f_{,\alpha\beta}(n-1)+f_{,d}g^{cd}\begin{pmatrix}g_{\alpha c,\beta}-g_{\beta\alpha,c}-g_{c\beta,\alpha}\\+\frac{n}{2}\begin{pmatrix}2g_{\alpha\beta,c}-g_{\alpha c,\beta}\\-g_{\beta c,\alpha}\end{pmatrix}\end{pmatrix}\right)\frac{d\left[F^{\frac{n}{2}-1}F'\right]}{df}\\+\frac{(n-1)}{2}\begin{pmatrix}-\left(f_{,\alpha}\cdot\sqrt{g}\cdot g^{\alpha\beta}F^{\frac{n}{2}-2}\left(4FF''+F'\cdot F'(n-6)\right)\right)_{,\beta}\\+\frac{f_{,\alpha}\cdot f_{,\beta}}{2}\sqrt{g}\cdot g^{\alpha\beta}\frac{d\left[F^{\frac{n}{2}-2}\left(4FF''+F'\cdot F'(n-6)\right)\right]}{df}\end{pmatrix}\end{array}\right]_{\mathbf{y}}. \tag{234}$$

Thereby, as done before and just for brevity, we assume in this appendix that g denotes the absolute value of the determinant of the metric tensor $g_{\alpha\beta}$.

We can significantly simplify the outcome in the case of metrics consisting only of constant components, which we consider here not only for entertainment and demonstration but also due to the importance of Cartesian coordinates, for instance:

$$\frac{\delta W}{\delta f}=0=\frac{\delta}{\delta f}\int_V d^n x\left(\sqrt{g\cdot F^n}\cdot R^{**}\right)$$

$$=-\left[\begin{array}{l}(n-1)\cdot\sqrt{g}\cdot g^{\alpha\beta}\cdot\left(\left(F^{\frac{n}{2}-1}\cdot F'\right)_{,\alpha\beta}+f_{,\alpha\beta}\frac{d\left[F^{\frac{n}{2}-1}F'\right]}{df}\right)-\sqrt{g}\cdot R\frac{dF^{\frac{n}{2}}}{df}\\ +\frac{(n-1)}{2}\left(\begin{array}{l}-\left(f_{,\alpha}\cdot\sqrt{g}\cdot g^{\alpha\beta}F^{\frac{n}{2}-2}\left(4FF''+F'\cdot F'(n-6)\right)\right)_{,\beta}\\ +\frac{f_{,\alpha}\cdot f_{,\beta}}{2}\sqrt{g}\cdot g^{\alpha\beta}\frac{d\left[F^{\frac{n}{2}-2}\left(4FF''+F'\cdot F'(n-6)\right)\right]}{df}\end{array}\right)\end{array}\right]_{\mathbf{y}}$$

$$=-\left[\begin{array}{l}(n-1)\cdot\sqrt{g}\cdot g^{\alpha\beta}\cdot\left(\begin{array}{l}\left(\left(F^{\frac{n}{2}-1}\cdot F'\right)_{,\alpha\beta}+f_{,\alpha\beta}\frac{d\left[F^{\frac{n}{2}-1}F'\right]}{df}\right)\\ +\frac{1}{2}\left(\begin{array}{l}-\left(f_{,\alpha}\cdot F^{\frac{n}{2}-2}\left(4FF''+F'\cdot F'(n-6)\right)\right)_{,\beta}\\ +\frac{f_{,\alpha}\cdot f_{,\beta}}{2}\frac{d\left[F^{\frac{n}{2}-2}\left(4FF''+F'\cdot F'(n-6)\right)\right]}{df}\end{array}\right)\end{array}\right)\\ -\sqrt{g}\cdot R\frac{dF^{\frac{n}{2}}}{df}\end{array}\right]_{\mathbf{y}}. \quad (235)$$

Further simplification gives:

$$\frac{\delta W}{\delta f}=0=\frac{\delta}{\delta f}\int_V d^n x\left(\sqrt{g\cdot F^n}\cdot R^{**}\right)$$

$$=-\left[(n-1)\cdot\sqrt{g}\cdot g^{\alpha\beta}\cdot\left(\begin{array}{c}\dfrac{f_{,\alpha}\cdot f_{,\beta}}{4}\dfrac{d\left[F^{\frac{n}{2}-2}\left(4FF''+F'\cdot F'(n-6)\right)\right]}{df}\\ -\dfrac{1}{2}\left(f_{,\alpha}\cdot F^{\frac{n}{2}-2}\left(4FF''+F'\cdot F'(n-6)\right)\right)_{,\beta}\\ +\left(F^{\frac{n}{2}-1}\cdot F'\right)_{,\alpha\beta}+f_{,\alpha\beta}\dfrac{d\left[F^{\frac{n}{2}-1}F'\right]}{df}\end{array}\right)-R\frac{dF^{\frac{n}{2}}}{df}\right]_{\mathbf{y}}$$

$$=-\left[\begin{array}{c}(n-1)\cdot\sqrt{g}\cdot g^{\alpha\beta}\cdot\left(\begin{array}{c}\dfrac{f_{,\alpha}\cdot f_{,\beta}}{4}\dfrac{d\left[F^{\frac{n}{2}-2}\left(4FF''+F'\cdot F'(n-6)\right)\right]}{df}\\ -\dfrac{1}{2}\left(\begin{array}{c}f_{,\alpha\beta}\cdot F^{\frac{n}{2}-2}\left(4FF''+F'\cdot F'(n-6)\right)\\ +f_{,\alpha}f_{,\beta}\cdot\dfrac{d\left[F^{\frac{n}{2}-2}\left(4FF''+F'\cdot F'(n-6)\right)\right]}{df}\end{array}\right)\\ +\left(\left(\left(\dfrac{n}{2}-1\right)F^{\frac{n}{2}-2}\cdot F'F'+F^{\frac{n}{2}-1}\cdot F''\right)f_{,\alpha}\right)_{,\beta}\\ +f_{,\alpha\beta}\dfrac{d\left[F^{\frac{n}{2}-1}F'\right]}{df}\end{array}\right)\\ -\sqrt{g}\cdot R\dfrac{dF^{\frac{n}{2}}}{df}\end{array}\right]_{\mathbf{y}}$$

$$= -\left[\begin{array}{c} (n-1)\cdot\sqrt{g}\cdot g^{\alpha\beta}\cdot \left(\begin{array}{c} \frac{f_{,\alpha}\cdot f_{,\beta}}{4}\frac{d\left[F^{\frac{n}{2}-2}\left(4FF''+F'\cdot F'(n-6)\right)\right]}{df} \\ -\frac{1}{2}\left(\begin{array}{c} f_{,\alpha\beta}\cdot F^{\frac{n}{2}-2}\left(4FF''+F'\cdot F'(n-6)\right) \\ +f_{,\alpha}f_{,\beta}\cdot\frac{d\left[F^{\frac{n}{2}-2}\left(4FF''+F'\cdot F'(n-6)\right)\right]}{df} \end{array} \right) \\ +\left(\left(\begin{array}{c} \left(\frac{n}{2}-1\right)\left(\left(\frac{n}{2}-2\right)F^{\frac{n}{2}-3}(F')^3+2F^{\frac{n}{2}-2}F'F''\right) \\ +F^{\frac{n}{2}-1}\cdot F''' \end{array} \right) f_{,\alpha}f_{,\beta} \right) \\ \left(\left(\frac{n}{2}-1\right)F^{\frac{n}{2}-2}\cdot F'F'+F^{\frac{n}{2}-1}\cdot F''\right)f_{,\alpha\beta} \\ +f_{,\alpha\beta}\frac{d\left[F^{\frac{n}{2}-1}F'\right]}{df} \end{array} \right) \\ -R\frac{dF^{\frac{n}{2}}}{df} \end{array} \right]_{\mathbf{y}} \cdot \tag{236}$$

For brevity, we ignore the R-term (just for the moment) and evaluate as follows:

$$\frac{\delta W}{\delta f} = 0 = \frac{\delta}{\delta f}\int_V d^n x\left(\sqrt{g\cdot F^n}\cdot R^{**}\right) = -(n-1)\cdot\sqrt{g}\cdot g^{\alpha\beta}\Big|_{\mathbf{y}}$$

$$\times\left[\begin{array}{c} \frac{f_{,\alpha}\cdot f_{,\beta}}{4}\frac{d\left[F^{\frac{n}{2}-2}\left(4FF''+F'\cdot F'(n-6)\right)\right]}{df} \\ -\frac{1}{2}\left(\begin{array}{c} f_{,\alpha\beta}\cdot F^{\frac{n}{2}-2}\left(4FF''+F'\cdot F'(n-6)\right) \\ +f_{,\alpha}f_{,\beta}\cdot\frac{d\left[F^{\frac{n}{2}-2}\left(4FF''+F'\cdot F'(n-6)\right)\right]}{df}\end{array}\right) \\ +\left(\left(\left(\frac{n}{2}-1\right)\left(\left(\frac{n}{2}-2\right)F^{\frac{n}{2}-3}(F')^3+2F^{\frac{n}{2}-2}F'F''\right)+F^{\frac{n}{2}-1}\cdot F'''\right)f_{,\alpha}f_{,\beta}\right) \\ F^{\frac{n}{2}-2}\cdot\left(\left(\frac{n}{2}-1\right)F'F'+F\cdot F''\right)f_{,\alpha\beta} \\ +f_{,\alpha\beta}\left(\left(\frac{n}{2}-1\right)F^{\frac{n}{2}-2}F'F'+F^{\frac{n}{2}-1}\cdot F''\right)\end{array}\right]_{\mathbf{y}}, \quad (237)$$

$$\frac{\delta W}{\delta f}=0=\frac{\delta}{\delta f}\int_V d^n x\left(\sqrt{g\cdot F^n}\cdot R^{**}\right)=-(n-1)\cdot\sqrt{g}\cdot g^{\alpha\beta}\Big|_{\mathbf{y}}$$

$$\times\left[\begin{array}{c} \frac{f_{,\alpha}\cdot f_{,\beta}}{4}\frac{d\left[F^{\frac{n}{2}-2}\left(4FF''+F'\cdot F'(n-6)\right)\right]}{df} \\ -\frac{1}{2}\left(\begin{array}{c} f_{,\alpha\beta}\cdot F^{\frac{n}{2}-2}\left(4FF''+F'\cdot F'(n-6)\right) \\ +f_{,\alpha}f_{,\beta}\cdot\frac{d\left[F^{\frac{n}{2}-2}\left(4FF''+F'\cdot F'(n-6)\right)\right]}{df}\end{array}\right) \\ +F^{\frac{n}{2}-3}f_{,\alpha}f_{,\beta}\left(\left(\frac{n}{2}-1\right)\left(\left(\frac{n}{2}-2\right)(F')^3+2FF'F''\right)+F^2\cdot F'''\right) \\ +2f_{,\alpha\beta}F^{\frac{n}{2}-2}\left(\left(\frac{n}{2}-1\right)F'F'+F\cdot F''\right)\end{array}\right]_{\mathbf{y}}, \quad (238)$$

$$\frac{\delta W}{\delta f}=0=\frac{\delta}{\delta f}\int_V d^n x\left(\sqrt{g\cdot F^n}\cdot R^{*}\right)=-(n-1)\cdot\sqrt{g}\cdot g^{\alpha\beta}\Big|_{\mathbf{y}}$$

$$\times\left[\begin{array}{c}-\frac{f_{,\alpha}\cdot f_{,\beta}}{4}\frac{d\left[F^{\frac{n}{2}-2}\left(4FF''+F'\cdot F'(n-6)\right)\right]}{df}\\ -\frac{1}{2}\left(f_{,\alpha\beta}\cdot F^{\frac{n}{2}-2}\left(4FF''+F'\cdot F'(n-6)\right)\right)\\ +F^{\frac{n}{2}-3}f_{,\alpha}f_{,\beta}\left(\left(\frac{n}{2}-1\right)\left(\left(\frac{n}{2}-2\right)(F')^{3}+2FF'F''\right)+F^{2}\cdot F'''\right)\\ +2f_{,\alpha\beta}F^{\frac{n}{2}-2}\left(\left(\frac{n}{2}-1\right)F'F'+F\cdot F''\right)\end{array}\right]_{\mathbf{y}}$$

$$=-(n-1)\cdot\sqrt{g}\cdot g^{\alpha\beta}\Big|_{\mathbf{y}}$$

$$\times\left[\begin{array}{c}-\frac{f_{,\alpha}\cdot f_{,\beta}}{4}F^{\frac{n}{2}-3}\left(\begin{array}{c}F\left(4(F'F''+FF''')+2F'\cdot F''(n-6)\right)\\ +\left(\frac{n}{2}-2\right)F'\left(4FF''+F'\cdot F'(n-6)\right)\end{array}\right)\\ -\frac{1}{2}\left(f_{,\alpha\beta}\cdot F^{\frac{n}{2}-2}\left(4FF''+F'\cdot F'(n-6)\right)\right)\\ +F^{\frac{n}{2}-3}f_{,\alpha}f_{,\beta}\left(\left(\frac{n}{2}-1\right)\left(\left(\frac{n}{2}-2\right)(F')^{3}+2FF'F''\right)+F^{2}\cdot F'''\right)\\ +2f_{,\alpha\beta}F^{\frac{n}{2}-2}\left(\left(\frac{n}{2}-1\right)F'F'+F\cdot F''\right)\end{array}\right]_{\mathbf{y}}, \quad (239)$$

$$\frac{\delta W}{\delta f}=0=\frac{\delta}{\delta f}\int_{V}d^{n}x\left(\sqrt{g\cdot F^{n}}\cdot R^{**}\right)=-(n-1)\cdot\sqrt{g}\cdot g^{\alpha\beta}\Big|_{\mathbf{y}}$$

$$\times\left[\begin{array}{c}+F^{\frac{n}{2}-3}f_{,\alpha}f_{,\beta}\left(\begin{array}{c}\left(\frac{n}{2}-1\right)\left(\left(\frac{n}{2}-2\right)(F')^{3}+2FF'F''\right)+F^{2}\cdot F'''\\ -\frac{1}{4}\left(\begin{array}{c}F\left(4(F'F''+FF''')+2F'\cdot F''(n-6)\right)\\ +\left(\frac{n}{2}-2\right)F'\left(4FF''+F'\cdot F'(n-6)\right)\end{array}\right)\end{array}\right)\\ +2f_{,\alpha\beta}F^{\frac{n}{2}-2}\left(\left(\frac{n}{2}-1\right)F'F'+F\cdot F''-\frac{4FF''+F'\cdot F'(n-6)}{4}\right)\end{array}\right]_{\mathbf{y}}, \quad (240)$$

$$\frac{\delta W}{\delta f}=0=\frac{\delta}{\delta f}\int_V d^n x\left(\sqrt{g\cdot F^n}\cdot R^{**}\right)=-(n-1)\cdot\sqrt{g}\cdot g^{\alpha\beta}\Big|_{\mathbf{y}}$$

$$\times\left[\begin{array}{c} +F^{\frac{n}{2}-3}f_{,\alpha}f_{,\beta}\left(\begin{array}{l}\left(\left(\frac{n}{2}-1\right)\left(\left(\frac{n}{2}-2\right)(F')^3+2FF'F''\right)+F^2\cdot F'''\right)\\ -\frac{1}{4}\left(\begin{array}{l}F\left(4\left(F'F''+FF'''\right)+2F'\cdot F''(n-6)\right)\\ +\left(\frac{n}{2}-2\right)F'\left(4FF''+F'\cdot F'(n-6)\right)\end{array}\right)\end{array}\right)\\ +\frac{1}{2}(n+2)f_{,\alpha\beta}F^{\frac{n}{2}-2}F'F'\end{array}\right]_{\mathbf{y}},\qquad(241)$$

$$\frac{\delta W}{\delta f}=0=\frac{\delta}{\delta f}\int_V d^n x\left(\sqrt{g\cdot F^n}\cdot R^{**}\right)=-(n-1)\cdot\sqrt{g}\cdot g^{\alpha\beta}\Big|_{\mathbf{y}}$$

$$\times\left[\begin{array}{c} +\frac{F^{\frac{n}{2}-3}f_{,\alpha}f_{,\beta}}{4}\left(\begin{array}{l}\left(4\left(\frac{n}{2}-2\right)\left(\frac{n}{2}-1\right)(F')^3+8\left(\frac{n}{2}-1\right)FF'F''\right)+4F^2\cdot F'''\\ -4F^2\cdot F'''-4FF'F''-2FF'\cdot F''(n-6)\\ -\left(\frac{n}{2}-2\right)\left(4FF'F''+F'\cdot F'F'(n-6)\right)\end{array}\right)\\ +\frac{1}{2}(n+2)f_{,\alpha\beta}F^{\frac{n}{2}-2}F'F'\end{array}\right]_{\mathbf{y}}$$

$$=-(n-1)\cdot\sqrt{g}\cdot g^{\alpha\beta}\Big|_{\mathbf{y}}$$

$$\times\left[\begin{array}{c} +\frac{F^{\frac{n}{2}-3}f_{,\alpha}f_{,\beta}}{4}\left(\begin{array}{l}4\left(\frac{n}{2}-2\right)\left(\frac{n}{2}-1\right)(F')^3+8\left(\frac{n}{2}-1\right)FF'F''\\ -4FF'F''-2FF'\cdot F''(n-6)\\ -4\left(\frac{n}{2}-2\right)FF'F''-F'\cdot F'F'\left(\frac{n}{2}-2\right)(n-6)\end{array}\right)\\ +\frac{1}{2}(n+2)f_{,\alpha\beta}F^{\frac{n}{2}-2}F'F'\end{array}\right]_{\mathbf{y}}$$

$$=-(n-1)\cdot\sqrt{g}\cdot g^{\alpha\beta}\Big|_{\mathbf{y}}$$

$$\times\left[\begin{array}{c}+\frac{F^{\frac{n}{2}-3}f_{,\alpha}f_{,\beta}}{4}\left(\begin{array}{c}\left(\frac{n}{2}-2\right)\left(4\left(\frac{n}{2}-1\right)-(n-6)\right)(F')^3\\+\left(8\left(\frac{n}{2}-1\right)-4-2(n-6)-4\left(\frac{n}{2}-2\right)\right)FF'F''\end{array}\right)\\+\frac{1}{2}(n+2)f_{,\alpha\beta}F^{\frac{n}{2}-2}F'F'\end{array}\right]_{\mathbf{y}},\quad(242)$$

$$\frac{\delta W}{\delta f}=0=\frac{\delta}{\delta f}\int_V d^n x\left(\sqrt{g\cdot F^n}\cdot R^{**}\right)=-(n-1)\cdot\sqrt{g}\cdot g^{\alpha\beta}\Big|_{\mathbf{y}}$$

$$\times\left[\begin{array}{c}+\frac{F^{\frac{n}{2}-3}f_{,\alpha}f_{,\beta}}{4}\left(\begin{array}{c}\left(\frac{n}{2}-2\right)(n+2)(F')^3\\+(4n-8-4-2n+6-2n+8)FF'F''\end{array}\right)\\+\frac{1}{2}(n+2)f_{,\alpha\beta}F^{\frac{n}{2}-2}F'F'\end{array}\right]_{\mathbf{y}}$$

$$=-(n-1)\cdot\left[\frac{\sqrt{g}\cdot g^{\alpha\beta}}{2}F'F^{\frac{n}{2}-3}\left(\begin{array}{c}\frac{f_{,\alpha}f_{,\beta}}{2}\left(\frac{1}{2}(n-4)(n+2)(F')^2+2FF''\right)\\+(n+2)f_{,\alpha\beta}FF'\end{array}\right)\right]_{\mathbf{y}}.\quad(243)$$

Now we incorporate all terms we have previously omitted due to our assumption of a metric of constants:

$$\frac{\delta W}{\delta f}=0=\frac{\delta}{\delta f}\int_V d^n x\left(\sqrt{g\cdot F^n}\cdot R^{**}\right)$$

$$=-\left[\begin{array}{c}(n-1)\cdot\left(F^{\frac{n}{2}-1}F'\left(\sqrt{g}\cdot g^{\alpha\beta}\right)_{,\alpha\beta}+2\left(F^{\frac{n}{2}-1}F'\right)_{,\alpha}\left(\sqrt{g}\cdot g^{\alpha\beta}\right)_{,\beta}\right)\\-\frac{1}{2}\left(\sqrt{g}\cdot F^{\frac{n}{2}-1}g^{\alpha\beta}\cdot F'g^{cd}\left(\begin{array}{c}g_{\alpha c,\beta}-g_{\beta\alpha,c}-g_{c\beta,\alpha}\\+\frac{n}{2}\left(2g_{\alpha\beta,c}-g_{\alpha c,\beta}-g_{\beta c,\alpha}\right)\end{array}\right)\right)_{,d}\\+\frac{\sqrt{g}\cdot g^{\alpha\beta}}{2}f_{,d}g^{cd}\left(\begin{array}{c}g_{\alpha c,\beta}-g_{\beta\alpha,c}-g_{c\beta,\alpha}\\+\frac{n}{2}\left(2g_{\alpha\beta,c}-g_{\alpha c,\beta}-g_{\beta c,\alpha}\right)\end{array}\right)\frac{d\left[F^{\frac{n}{2}-1}F'\right]}{df}\\-\frac{(n-1)}{2}f_{,\alpha}\cdot F^{\frac{n}{2}-2}\left(4FF''+F'\cdot F'(n-6)\right)\left(\sqrt{g}\cdot g^{\alpha\beta}\right)_{,\beta}\end{array}\right]_{\mathbf{y}}$$

$$-(n-1)\cdot\left[\frac{\sqrt{g}\cdot g^{\alpha\beta}}{2}F'F^{\frac{n}{2}-3}\left(\begin{array}{c}\frac{f_{,\alpha}f_{,\beta}}{2}\left(\frac{1}{2}(n-4)(n+2)(F')^2+2FF''\right)\\+(n+2)f_{,\alpha\beta}FF'\end{array}\right)\right]_{\mathbf{y}} \quad (244)$$

$$\frac{\delta W}{\delta f}=0=\frac{\delta}{\delta f}\int_V d^n x\left(\sqrt{g\cdot F^n}\cdot R^{**}\right)$$

$$=-\left[\begin{array}{l}(n-1)\cdot F^{\frac{n}{2}-2}\left(FF'\left(\sqrt{g}\cdot g^{\alpha\beta}\right)_{,\alpha\beta}+2\left(\left(\frac{n}{2}-1\right)F'F'+FF''\right)f_{,\alpha}\left(\sqrt{g}\cdot g^{\alpha\beta}\right)_{,\beta}\right)\\-F^{\frac{n}{2}-2}\frac{\left(\left(\frac{n}{2}-1\right)F'F'+FF''\right)f_{,d}}{2}g^{cd}\left(\sqrt{g}\cdot g^{\alpha\beta}\left(\begin{array}{c}g_{\alpha c,\beta}-g_{\beta\alpha,c}-g_{c\beta,\alpha}\\+\frac{n}{2}\left(2g_{\alpha\beta,c}-g_{\alpha c,\beta}-g_{\beta c,\alpha}\right)\end{array}\right)\right)\\-\frac{F^{\frac{n}{2}-1}F'}{2}\left(\sqrt{g}\cdot g^{\alpha\beta}\cdot g^{cd}\left(\begin{array}{c}g_{\alpha c,\beta}-g_{\beta\alpha,c}-g_{c\beta,\alpha}\\+\frac{n}{2}\left(2g_{\alpha\beta,c}-g_{\alpha c,\beta}-g_{\beta c,\alpha}\right)\end{array}\right)\right)_{,d}\\+\frac{\sqrt{g}\cdot g^{\alpha\beta}}{2}f_{,d}g^{cd}F^{\frac{n}{2}-2}\left(\begin{array}{c}g_{\alpha c,\beta}-g_{\beta\alpha,c}-g_{c\beta,\alpha}\\+\frac{n}{2}\left(2g_{\alpha\beta,c}-g_{\alpha c,\beta}-g_{\beta c,\alpha}\right)\end{array}\right)\left(\left(\frac{n}{2}-1\right)F'F'+FF''\right)\\-\frac{(n-1)}{2}f_{,\alpha}\cdot F^{\frac{n}{2}-2}\left(4FF''+F'\cdot F'(n-6)\right)\left(\sqrt{g}\cdot g^{\alpha\beta}\right)_{,\beta}\end{array}\right]_{\mathbf{y}}$$

$$-(n-1)\cdot\left[\frac{\sqrt{g}\cdot g^{\alpha\beta}}{2}F'F^{\frac{n}{2}-3}\left(\begin{array}{c}\frac{f_{,\alpha}f_{,\beta}}{2}\left(\frac{1}{2}(n-4)(n+2)(F')^2+2FF''\right)\\+(n+2)f_{,\alpha\beta}FF'\end{array}\right)\right]_{\mathbf{y}}.$$

(245)

Factoring out first derivatives of f yields:

$$\frac{\delta W}{\delta f}=0=\frac{\delta}{\delta f}\int_V d^n x\left(\sqrt{g\cdot F^n}\cdot R^{**}\right)$$

$$
=-\left[\begin{array}{l}
(n-1)\cdot F^{\frac{n}{2}-2}\left(FF'\left(\sqrt{g}\cdot g^{\alpha\beta}\right)_{,\alpha\beta}+2\left(\left(\frac{n}{2}-1\right)F'F'+FF''\right)f_{,\alpha}\left(\sqrt{g}\cdot g^{\alpha\beta}\right)_{,\beta}\right) \\
-F^{\frac{n}{2}-2}\frac{\left(\left(\frac{n}{2}-1\right)F'F'+FF''\right)f_{,\alpha}}{2}g^{ij}\sqrt{g}\cdot g^{\alpha\beta}\left(\begin{array}{c} g_{i\beta,j}-g_{ij,\beta}-g_{\beta j,i} \\ +\frac{n}{2}\left(2g_{ij,\beta}-g_{i\beta,j}-g_{j\beta,i}\right)\end{array}\right) \\
-\frac{F^{\frac{n}{2}-1}F'}{2}\left(\sqrt{g}\cdot g^{\alpha\beta}\cdot g^{ij}\left(\begin{array}{c} g_{i\beta,j}-g_{ij,\beta}-g_{\beta j,i} \\ +\frac{n}{2}\left(2g_{ij,\beta}-g_{i\beta,j}-g_{j\beta,i}\right)\end{array}\right)\right)_{,\alpha} \\
+\frac{\sqrt{g}\cdot g^{\alpha\beta}}{2}f_{,\alpha}g^{ij}F^{\frac{n}{2}-2}\left(\begin{array}{c} g_{i\beta,j}-g_{ij,\beta}-g_{\beta j,i} \\ +\frac{n}{2}\left(2g_{ij,\beta}-g_{i\beta,j}-g_{j\beta,i}\right)\end{array}\right)\left(\left(\frac{n}{2}-1\right)F'F'+FF''\right) \\
-\frac{(n-1)}{2}f_{,\alpha}\cdot F^{\frac{n}{2}-2}\left(4FF''+F'\cdot F'(n-6)\right)\left(\sqrt{g}\cdot g^{\alpha\beta}\right)_{,\beta}
\end{array}\right]_{\mathbf{y}}
$$

$$
-(n-1)\cdot\left[\frac{\sqrt{g}\cdot g^{\alpha\beta}}{2}F'F^{\frac{n}{2}-3}\left(\begin{array}{c}\frac{f_{,\alpha}f_{,\beta}}{2}\left(\frac{1}{2}(n-4)(n+2)(F')^{2}+2FF''\right) \\ +(n+2)f_{,\alpha\beta}FF'\end{array}\right)\right]_{\mathbf{y}}, \qquad (246)
$$

$$
\frac{\delta W}{\delta f}=0=\frac{\delta}{\delta f}\int_{V}d^{n}x\left(\sqrt{g\cdot F^{n}}\cdot R^{**}\right)
$$

$$
= -\left[\begin{array}{c}
(n-1)\cdot F^{\frac{n}{2}-1}F'\left(\sqrt{g}\cdot g^{\alpha\beta}\right)_{,\alpha\beta} \\
-\frac{F^{\frac{n}{2}-1}F'}{2}\left(\sqrt{g}\cdot g^{\alpha\beta}\cdot g^{ij}\left(\begin{array}{c} g_{i\beta,j}-g_{ij,\beta}-g_{\beta j,i} \\ +\frac{n}{2}\left(2g_{ij,\beta}-g_{i\beta,j}-g_{j\beta,i}\right)\end{array}\right)\right)_{,\alpha} \\
f_{,\alpha}\left(\begin{array}{c}
(n-1)\cdot F^{\frac{n}{2}-2}2\left(\left(\frac{n}{2}-1\right)F'F'+FF''\right)\left(\sqrt{g}\cdot g^{\alpha\beta}\right)_{,\beta} \\
-F^{\frac{n}{2}-2}\frac{\left(\left(\frac{n}{2}-1\right)F'F'+FF''\right)}{2}g^{ij}\sqrt{g}\cdot g^{\alpha\beta}\left(\begin{array}{c} g_{i\beta,j}-g_{ij,\beta}-g_{\beta j,i} \\ +\frac{n}{2}\left(2g_{ij,\beta}-g_{i\beta,j}-g_{j\beta,i}\right)\end{array}\right) \\
+F^{\frac{n}{2}-2}\frac{\left(\left(\frac{n}{2}-1\right)F'F'+FF''\right)}{2}g^{ij}\sqrt{g}\cdot g^{\alpha\beta}\left(\begin{array}{c} g_{i\beta,j}-g_{ij,\beta}-g_{\beta j,i} \\ +\frac{n}{2}\left(2g_{ij,\beta}-g_{i\beta,j}-g_{j\beta,i}\right)\end{array}\right) \\
-\frac{(n-1)}{2}\cdot F^{\frac{n}{2}-2}\left(4FF''+F'\cdot F'(n-6)\right)\left(\sqrt{g}\cdot g^{\alpha\beta}\right)_{,\beta}
\end{array}\right)
\end{array}\right]_{\mathbf{y}}
$$

$$
-(n-1)\cdot\left[\frac{\sqrt{g}\cdot g^{\alpha\beta}}{2}F'F^{\frac{n}{2}-3}\left(\begin{array}{c}\frac{f_{,\alpha}f_{,\beta}}{2}\left(\frac{1}{2}(n-4)(n+2)(F')^{2}+2FF''\right) \\ +(n+2)f_{,\alpha\beta}FF'\end{array}\right)\right]_{\mathbf{y}}, \tag{247}
$$

$$
\frac{\delta W}{\delta f}=0=\frac{\delta}{\delta f}\int_{V}d^{n}x\left(\sqrt{g\cdot F^{n}}\cdot R^{**}\right)
$$

$$
= -\left[\begin{array}{c}
(n-1)\cdot F^{\frac{n}{2}-1}F'\left(\sqrt{g}\cdot g^{\alpha\beta}\right)_{,\alpha\beta} \\
-\frac{F^{\frac{n}{2}-1}F'}{2}\left(\sqrt{g}\cdot g^{\alpha\beta}\cdot g^{ij}\left(\begin{array}{c} g_{i\beta,j}-g_{ij,\beta}-g_{\beta j,i} \\ +\frac{n}{2}\left(2g_{ij,\beta}-g_{i\beta,j}-g_{j\beta,i}\right)\end{array}\right)\right)_{,\alpha} \\
+f_{,\alpha}\frac{(n-1)}{2}\cdot F^{\frac{n}{2}-2}\left(\sqrt{g}\cdot g^{\alpha\beta}\right)_{,\beta}\left((2n-4)F'F'+4FF''-4FF''-F'\cdot F'(n-6)\right)
\end{array}\right]_{\mathbf{y}}
$$

$$-(n-1)\cdot\left[\frac{\sqrt{g}\cdot g^{\alpha\beta}}{2}F'F^{\frac{n}{2}-3}\begin{pmatrix}\frac{f_{,\alpha}f_{,\beta}}{2}\left(\frac{1}{2}(n-4)(n+2)(F')^2+2FF''\right)\\+(n+2)f_{,\alpha\beta}FF'\end{pmatrix}\right]_{\mathbf{y}}$$

$$=-\left[\begin{matrix}(n-1)\cdot F^{\frac{n}{2}-1}F'\left(\sqrt{g}\cdot g^{\alpha\beta}\right)_{,\alpha\beta}\\-\frac{F^{\frac{n}{2}-1}F'}{2}\left(\sqrt{g}\cdot g^{\alpha\beta}\cdot g^{ij}\begin{pmatrix}g_{i\beta,j}-g_{ij,\beta}-g_{\beta j,i}\\+\frac{n}{2}\left(2g_{ij,\beta}-g_{i\beta,j}-g_{j\beta,i}\right)\end{pmatrix}\right)_{,\alpha}\\+f_{,\alpha}\frac{(n-1)}{2}\cdot F^{\frac{n}{2}-2}\left(\sqrt{g}\cdot g^{\alpha\beta}\right)_{,\beta}F'F'(n+2)\end{matrix}\right]_{\mathbf{y}}$$

$$-(n-1)\cdot\left[\frac{\sqrt{g}\cdot g^{\alpha\beta}}{2}F'F^{\frac{n}{2}-3}\begin{pmatrix}\frac{f_{,\alpha}f_{,\beta}}{2}\left(\frac{1}{2}(n-4)(n+2)(F')^2+2FF''\right)\\+(n+2)f_{,\alpha\beta}FF'\end{pmatrix}\right]_{\mathbf{y}}. \quad (248)$$

Now we bring back in the R-term:

$$\frac{\delta W}{\delta f}=0$$

$$=-\left[F^{\frac{n}{2}-2}F'\begin{pmatrix}(n-1)\cdot\left(F\left(\sqrt{g}\cdot g^{\alpha\beta}\right)_{,\alpha\beta}+f_{,\alpha}\frac{F'}{2}\cdot\left(\sqrt{g}\cdot g^{\alpha\beta}\right)_{,\beta}(n+2)\right)\\-\frac{F}{2}\left(\sqrt{g}\cdot g^{\alpha\beta}\cdot g^{ij}\begin{pmatrix}g_{i\beta,j}-g_{ij,\beta}-g_{\beta j,i}\\+\frac{n}{2}\left(2g_{ij,\beta}-g_{i\beta,j}-g_{j\beta,i}\right)\end{pmatrix}\right)_{,\alpha}\end{pmatrix}-\sqrt{g}\cdot R\frac{dF^{\frac{n}{2}}}{df}\right]_{\mathbf{y}}$$

$$-(n-1)\cdot\left[\frac{\sqrt{g}\cdot g^{\alpha\beta}}{2}F'F^{\frac{n}{2}-3}\begin{pmatrix}\frac{f_{,\alpha}f_{,\beta}}{2}\left(\frac{1}{2}(n-4)(n+2)(F')^2+2FF''\right)\\+(n+2)f_{,\alpha\beta}FF'\end{pmatrix}\right]_{\mathbf{y}}$$

$$
= -\left[F^{\frac{n}{2}-2} F' \left(\begin{array}{c} (n-1)\cdot\left(F\left(\sqrt{g}\cdot g^{\alpha\beta}\right)_{,\alpha\beta} + f_{,\alpha}\frac{F'}{2}\cdot\left(\sqrt{g}\cdot g^{\alpha\beta}\right)_{,\beta}(n+2)\right) \\ -\frac{F}{2}\left(\sqrt{g}\cdot g^{\alpha\beta}\cdot g^{ij}\left(\begin{array}{c} g_{i\beta,j}-g_{ij,\beta}-g_{\beta j,i} \\ +\frac{n}{2}\left(2g_{ij,\beta}-g_{i\beta,j}-g_{j\beta,i}\right)\end{array}\right)\right)_{,\alpha} \\ +(n-1)\cdot\frac{\sqrt{g}\cdot g^{\alpha\beta}}{2F}\left(\begin{array}{c}\frac{f_{,\alpha}f_{,\beta}}{2}\left(\frac{1}{2}(n-4)(n+2)(F')^2+2FF''\right) \\ +(n+2)f_{,\alpha\beta}FF'\end{array}\right)\end{array}\right) \\ -\sqrt{g}\cdot RF^{\frac{n}{2}-1}F' \right]_{\mathbf{y}},
$$

(249)

$$
\frac{\delta W}{\delta f}=0
$$

$$
= -\left[F^{\frac{n}{2}-3} F'(n-1) \left(\begin{array}{c} \frac{\sqrt{g}\cdot g^{\alpha\beta}}{2}\frac{f_{,\alpha}f_{,\beta}}{2}\left(\frac{1}{2}(n-4)(n+2)(F')^2+2FF''\right) \\ +\frac{(n+2)}{2}FF'\overbrace{\left(\sqrt{g}\cdot g^{\alpha\beta}f_{,\alpha\beta}+f_{,\alpha}\left(\sqrt{g}\cdot g^{\alpha\beta}\right)_{,\beta}\right)}^{=\sqrt{g}\cdot\Delta f} \\ +F^2\left(\sqrt{g}\cdot g^{\alpha\beta}\right)_{,\alpha\beta} \\ -\frac{F^2}{2(n-1)}\left(\sqrt{g}\cdot g^{\alpha\beta}\cdot g^{ij}\left(\begin{array}{c} g_{i\beta,j}-g_{ij,\beta}-g_{\beta j,i} \\ +\frac{n}{2}\left(2g_{ij,\beta}-g_{i\beta,j}-g_{j\beta,i}\right)\end{array}\right)\right)_{,\alpha}\end{array}\right) \\ -\sqrt{g}\cdot RF^{\frac{n}{2}-1}F' \right]_{\mathbf{y}}
$$

$$
= -\left[F^{\frac{n}{2}-3} F'(n-1) \left(\begin{array}{c} \frac{\sqrt{g}\cdot g^{\alpha\beta}}{2} \frac{f_{,\alpha} f_{,\beta}}{2} \left(\frac{1}{2}(n-4)(n+2)(F')^2 + 2FF'' \right) \\ + \frac{(n+2)}{2} FF' \sqrt{g}\cdot \Delta f \\ + F^2 \left(\sqrt{g}\cdot g^{\alpha\beta} \right)_{,\alpha\beta} - \sqrt{g}\cdot \frac{R\cdot F^2}{(n-1)} \\ - \frac{F^2}{2(n-1)} \left(\sqrt{g}\cdot g^{\alpha\beta}\cdot g^{ij} \left(\begin{array}{c} g_{i\beta,j} - g_{ij,\beta} - g_{\beta j,i} \\ + \frac{n}{2}\left(2g_{ij,\beta} - g_{i\beta,j} - g_{j\beta,i} \right) \end{array} \right) \right)_{,\alpha} \end{array} \right) \right]_{\mathbf{y}} \cdot
$$

(250)

It totally suffices to demand that the inner factor gives zero in order to fulfill the variational condition and thus, we obtain:

$$
\frac{\delta W}{\delta f} = 0 \quad \Rightarrow \quad \Rightarrow
$$

$$
0 = \left[\begin{array}{c} \frac{\sqrt{g}}{2} \left(\frac{f_{,\alpha} f_{,\beta}}{2} g^{\alpha\beta} \left(\frac{1}{2}(n-4)(n+2)(F')^2 + 2FF'' \right) + (n+2)FF'\Delta f \right) \\ + F^2 \left(\begin{array}{c} \left(\sqrt{g}\cdot g^{\alpha\beta} \right)_{,\alpha\beta} - \frac{1}{2(n-1)} \left(\sqrt{g}\cdot g^{\alpha\beta}\cdot g^{ij} \left(\begin{array}{c} g_{i\beta,j} - g_{ij,\beta} - g_{\beta j,i} \\ + \frac{n}{2}\left(2g_{ij,\beta} - g_{i\beta,j} - g_{j\beta,i} \right) \end{array} \right) \right)_{,\alpha} \\ - \frac{\sqrt{g}\cdot R}{(n-1)} \end{array} \right) \end{array} \right]_{\mathbf{y}},
$$

(251)

$$
\xrightarrow{F = C + a\cdot f}
$$

$$
0 = \left[\begin{array}{c} \frac{\sqrt{g}}{2}(n+2) \left(\frac{f_{,\alpha} f_{,\beta}}{4} g^{\alpha\beta} (n-4) a^2 + (C + a\cdot f)\cdot a\cdot \Delta f \right) \\ + (C + a\cdot f)^2 \left(\begin{array}{c} \left(\sqrt{g}\cdot g^{\alpha\beta} \right)_{,\alpha\beta} - \frac{\sqrt{g}\cdot R}{(n-1)} \\ - \frac{1}{2(n-1)} \left(\sqrt{g}\cdot g^{\alpha\beta}\cdot g^{ij} \left(\begin{array}{c} g_{i\beta,j} - g_{ij,\beta} - g_{\beta j,i} \\ + \frac{n}{2}\left(2g_{ij,\beta} - g_{i\beta,j} - g_{j\beta,i} \right) \end{array} \right) \right)_{,\alpha} \end{array} \right) \end{array} \right]_{\mathbf{y}} \cdot
$$

(252)

5.14 Appendix N: Particles and Universes at Rest Solutions

Based on the ideas given in Bodan's adventurous books [33–40] (please see also the Motivation sections of our books [4, 5]), we had developed the concept of Matryoshka-structured Black Holes and Planck-sized Friedmann universes filling up our space-time. The essentials are published elsewhere [41, 42, 5], but we still want to give a small introduction here and try to apply the corresponding simplest solutions.

Assuming that the space we see consists of many smaller n-spherical objects, we want to investigate their t-dependency. In [41], we found the following solution to the quantum Einstein field equations (125) in the case of $n = 4$:

$$G_{\alpha\beta} = C_1 \cdot f[t]^2 \cdot \begin{pmatrix} -c^2 & 0 & 0 & 0 \\ 0 & t^2 & 0 & 0 \\ 0 & 0 & t^2 \cdot \sin(r)^2 & 0 \\ 0 & 0 & 0 & t^2 \cdot \sin(r)^2 \cdot \sin(\vartheta)^2 \end{pmatrix}; \tag{253}$$
$$f[t] = t^{-1 \pm \mathrm{i} \cdot c} \cdot C_f$$

It was shown in [41] that this is just the Robertson–Walker metric with r being treated like an angle. The following transformation:

$$\sin(r) = r_{RW} \cdot \left(1 + \frac{r_{RW}^2}{4}\right)^{-1} \tag{254}$$

would give us back a $F[f[t]]$-scaled Robertson–Walker-like metric with the scale being applied on all metric components and not just on the spatial ones. We realize that while r has become an angle, t took over the position of a radius. One could even assume the object to just "live" with a fixed time-shell $t = \rho$ and obtain a solution as follows:

$$G_{\alpha\beta} = C_1 \cdot f[t]^2 \cdot \begin{pmatrix} -c^2 & 0 & 0 & 0 \\ 0 & \rho^2 & 0 & 0 \\ 0 & 0 & \rho^2 \cdot \sin(r)^2 & 0 \\ 0 & 0 & 0 & \rho^2 \cdot \sin(r)^2 \cdot \sin(\vartheta)^2 \end{pmatrix}; \tag{255}$$
$$f[t] = e^{\pm \frac{\mathrm{i} \cdot c}{\rho} \cdot t} \cdot C_f$$

The same can be found in 3 dimensions as follows:

$$G_{\alpha\beta}=C_1\cdot f[t]^4\cdot\begin{pmatrix}-c^2 & 0 & 0\\ 0 & t^2 & 0\\ 0 & 0 & t^2\cdot\sin(r)^2\end{pmatrix}. \tag{256}$$
$$f[t]=\sqrt{t^{\pm i\cdot c-1}}\cdot C_f$$

As before in 4 dimensions we see that r has become an angle while t took over the position of the radius. The shell-like object would be given as follows:

$$G_{\alpha\beta}=C_1\cdot f[t]^4\cdot\begin{pmatrix}-c^2 & 0 & 0\\ 0 & \rho^2 & 0\\ 0 & 0 & \rho^2\cdot\sin(r)^2\end{pmatrix}. \tag{257}$$
$$f[t]=e^{\pm\frac{i\cdot c}{2\cdot\rho}\cdot t}\cdot C_f$$

The reader may prove that in any dimension n, we always obtain the same solution for the function $f[t]$ with metrics of n-spherical shape as follows:

$$G_{\alpha\beta}=f[t]\cdot\begin{pmatrix}-c^2 & 0 & 0 & 0 & 0 & 0\\ 0 & \rho^2 & 0 & 0 & 0 & 0\\ 0 & 0 & \rho^2\cdot\sin(r)^2 & 0 & 0 & 0\\ 0 & 0 & 0 & \rho^2\cdot\sin(r)^2\cdot\sin(\vartheta)^2 & 0 & 0\\ 0 & 0 & 0 & 0 & \ddots & 0\\ 0 & 0 & 0 & 0 & 0 & g_{n-1n-1}\end{pmatrix}. \tag{258}$$
$$f[t]=e^{\pm 2\cdot\frac{i\cdot c}{\rho}\cdot t}\cdot C_f;\quad g_{n-1n-1}=\rho^2\cdot\sin(r)^2\cdot\prod_{i=0}^{n-4}\sin(\vartheta_i)^2$$

Now, as we see that ρ clearly is a length, we want to derive its properties. From basic Quantum Theory we know that a particle at rest has the time dependency:

$$f[t]=e^{\pm i\cdot\frac{m\cdot c^2}{\hbar}\cdot t}\cdot C_f, \tag{259}$$

with m giving the rest mass of the particle and $\hbar$ denoting the reduced Planck constant. Comparing with the $f[t]$-function from the metric solution (257), we find:

$$\frac{m \cdot c}{\hbar} = \frac{2}{\rho}. \tag{260}$$

Inserting the Schwarzschild radius $r_s = \frac{2 \cdot m \cdot G}{c^2}$ (G is Newton's constant and c is the speed of light in vacuum), thereby substituting the rest mass m, leaves us with:

$$\frac{r_s \cdot c^3}{2 \cdot \hbar \cdot G} = \frac{2}{\rho} \quad \Rightarrow \quad \rho = \frac{4}{r_s} \cdot \left(\frac{c^3}{\hbar \cdot G} \right)^{-1} = 4 \cdot \frac{\ell_P^2}{r_s}. \tag{261}$$

Here ℓ_P denotes the Planck length.

By the way, in order to have the right units for the t-spheres in Eqs. (253) and (256), we should introduce spatial "speed of light" velocities c_x and obtain the following dimensionally improved solutions:

$$G_{\alpha\beta} = C_1 \cdot f[t]^2 \cdot \begin{pmatrix} -c^2 & 0 & 0 & 0 \\ 0 & c_x^2 \cdot t^2 & 0 & 0 \\ 0 & 0 & c_x^2 \cdot t^2 \cdot \sin(r)^2 & 0 \\ 0 & 0 & 0 & c_x^2 \cdot t^2 \cdot \sin(r)^2 \cdot \sin(\vartheta)^2 \end{pmatrix},$$
$$f[t] = t^{-1 \pm \mathrm{i} \cdot c / c_x} \cdot C_f \tag{262}$$

$$G_{\alpha\beta} = C_1 \cdot f[t]^4 \cdot \begin{pmatrix} -c^2 & 0 & 0 \\ 0 & c_x^2 \cdot t^2 & 0 \\ 0 & 0 & c_x^2 \cdot t^2 \cdot \sin(r)^2 \end{pmatrix}.$$
$$f[t] = \sqrt{t^{\pm \mathrm{i} \cdot c / c_x - 1}} \cdot C_f \tag{263}$$

With Eqs. (255) and (257), we find pulsating "universes," which could easily also be seen as pulsating minds or universes of the mind, and are strongly reminded of the suggestion of the "Friedmanns" in Bodan's adventurous stories [33–40, 5], where Bodan suggested little, potentially pulsating universes forming our space-time. And what, so it was asked in Bodan's stories, forms those little "Friedmann-universes"? The answer simply was: "Even smaller Friedmanns" and the attentive reader might already guess what the answer to the question "And what do the little Friedmanns of second order consist of?" would be. Thus, we have the suggestion of a self-similar structure of universes forming universes on different scales, thereby the smaller scale providing the building blocks for the next and so on. With respect to our Mathematical Psychology, this is just equivalent to the perception that certain levels of thought are built upon each other.

The solutions (262) and (263) do also oscillate, but with a significantly t-dependent amplitude (see Fig. N1).

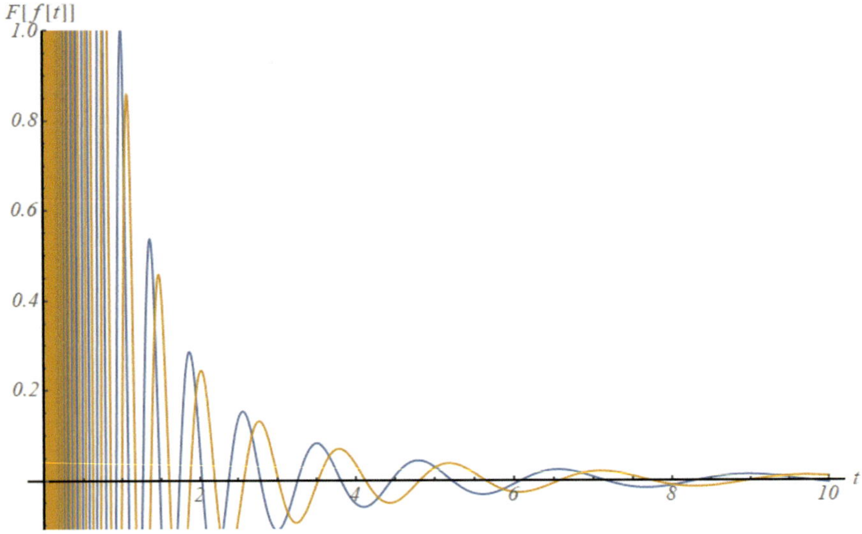

Figure N1 t-Dependency of the scale-solutions (262) and (263) with $c/c_x = 10$ (real part = blue, imaginary part = yellow).

As we see that the scale-dependency in our "Friedmanns" is made by a t, acting as a radius coordinate, we might just assume a permanent jitter of all space-points caused by these oscillations, which in consequence gives the global time. In other words, the space formed out of the oscillating and jittering Friedmann objects (253), (256), (262) and (263) (or corresponding objects in higher dimensions) has its own time due to the jittering of its building blocks. A corresponding mind would consequently have its very own mind- or system-time then. Thereby a great variety of vibrational options is at hand. For instance, one may use a simple generalization of Eqs. (262) and (263) as follows:

$$G_{\alpha\beta} = C_1 \cdot f[t]^2 \cdot \begin{pmatrix} -c^2 & 0 & 0 & 0 \\ 0 & T^2 & 0 & 0 \\ 0 & 0 & T^2 \cdot \sin(r)^2 & 0 \\ 0 & 0 & 0 & T^2 \cdot \sin(r)^2 \cdot \sin(\vartheta)^2 \end{pmatrix}$$

$$T^2 = c_x^2 \cdot t^2 + \rho^2; \quad f[t] = \frac{\left(c_x \left(c_x \cdot t + |T|\right)\right)^{-\frac{\mathrm{i} \cdot c}{cx}}}{|T|^{\frac{1}{2}}} \cdot C_f, \tag{264}$$

$$G_{\alpha\beta} = C_1 \cdot f[t]^4 \cdot \begin{pmatrix} -c^2 & 0 & 0 \\ 0 & T^2 & 0 \\ 0 & 0 & T^2 \cdot \sin(r)^2 \end{pmatrix}$$

$$T^2 = c_x^2 \cdot t^2 + \rho^2; \quad f[t] = \frac{\left(c_x \left(c_x \cdot t + |T|\right)\right)^{-\frac{\mathrm{i} \cdot c}{2 \cdot cx}}}{|T|^{\frac{1}{4}}} \cdot C_f. \tag{265}$$

Obviously, this is some kind of mix between the shell-like solutions (253), (256) and the t-spheres (262) and (263). Figure N2 illustrates the outcome.

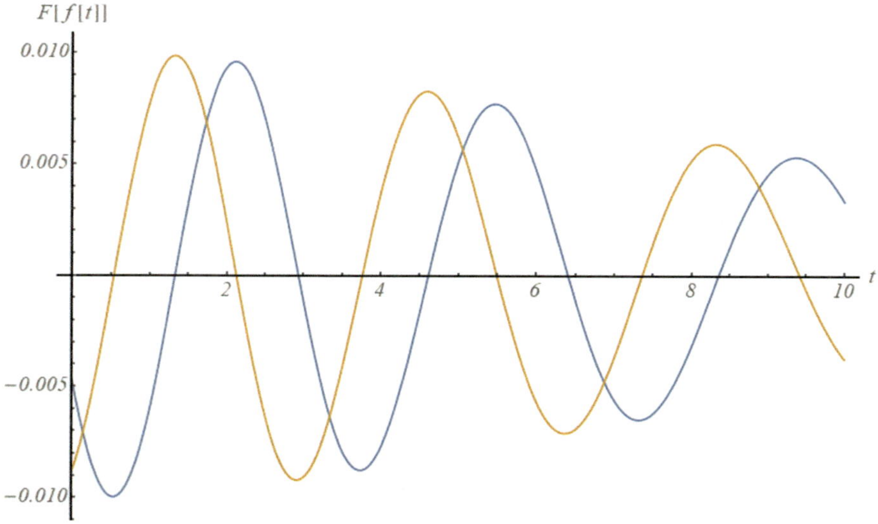

Figure N2 t-Dependency of the scale-solutions (264) and (265) with $c = 10$, $c_x = 1$ and $\rho = 10$ (real part = blue, imaginary part = yellow).

Please note that Eqs. (264) and (265) are not just simple transformations with $T = T[t]$ from Eqs. (262) and (263), as such would have to read:

$$G_{\alpha\beta} = \begin{pmatrix} C_1 \cdot f[T[t]]^2 \\ \times \begin{pmatrix} -c^2 \cdot T'[t]^2 & 0 & 0 & 0 \\ 0 & c_x^2 \cdot T[t]^2 & 0 & 0 \\ 0 & 0 & c_x^2 \cdot T[t]^2 \cdot \sin(r)^2 & 0 \\ 0 & 0 & 0 & \begin{matrix} c_x^2 \cdot T[t]^2 \\ \cdot \sin(r)^2 \cdot \sin(\vartheta)^2 \end{matrix} \end{pmatrix} \end{pmatrix},$$

$$f[t] = T[t]^{-1 \pm \mathrm{i} \cdot c / c_x} \cdot C_f \tag{266}$$

$$G_{\alpha\beta} = C_1 \cdot f[T[t]]^4 \cdot \begin{pmatrix} -c^2 \cdot T'[t]^2 & 0 & 0 \\ 0 & c_x^2 \cdot T[t]^2 & 0 \\ 0 & 0 & c_x^2 \cdot T[t]^2 \cdot \sin(r)^2 \end{pmatrix}. \tag{267}$$

$$f[t] = \sqrt{T[t]^{\pm \mathrm{i} \cdot c / c_x - 1}} \cdot C_f$$

A generalized solution in an arbitrary number of dimensions n would be:

$$G_{\alpha\beta} = f[t] \cdot \begin{pmatrix} -c^2 & 0 & 0 & 0 & 0 & 0 \\ 0 & T^m & 0 & 0 & 0 & 0 \\ 0 & 0 & T^m \cdot \sin(r)^2 & 0 & 0 & 0 \\ 0 & 0 & 0 & T^m \cdot \sin(r)^2 \cdot \sin(\vartheta)^2 & 0 & 0 \\ 0 & 0 & 0 & 0 & \ddots & 0 \\ 0 & 0 & 0 & 0 & 0 & g_{n-1n-1} \end{pmatrix}$$

$$T = c_x^2 \cdot t^2 + \rho^2; \quad g_{n-1n-1} = T^m \cdot \sin(r)^2 \cdot \prod_{i=0}^{n-4} \sin(\vartheta_i)^2$$

$$f[t] = C_f \cdot \left(c_x^2 t^2 + \rho^2\right)^{-m} \cdot \mathrm{e}^{\pm \frac{2 \cdot \mathrm{i} \cdot c \cdot t \cdot \left(c_x^2 t^2 + \rho^2\right)^{1 - \frac{m}{2}} \cdot {}_2F_1\left[1, \frac{3-m}{2}, \frac{3}{2}, -\frac{c_x^2 t^2}{\rho^2}\right]}{\rho^2}}, \tag{268}$$

with the hypergeometric function:

$${}_2F_1[a,b,c,z] = \sum_{k=0}^{\infty} (a)_k (b)_k / (c)_k z^k / k!. \tag{269}$$

With such solutions at hand, one could easily assume onion-like structures of Friedmanns with various dimensions and different shell sizes ρ, either being compressed together as an ensemble of n-spheres (possibly filling a Black Hole as suggested in [41]) or forming Russian doll or Matryoshka "universes." In Mathematical Psychology, this would just be the mind with many shells and thus onion-like, as the ogre in the movie *Shrek* suggests for himself.

It should be noted that when reformulating the approach (268) in the following way:

$$G_{\alpha\beta} = f[t]\cdot\begin{pmatrix} c^2 & 0 & 0 & 0 & 0 & 0 \\ 0 & T^m & 0 & 0 & 0 & 0 \\ 0 & 0 & T^m\cdot\sin(r)^2 & 0 & 0 & 0 \\ 0 & 0 & 0 & T^m\cdot\sin(r)^2\cdot\sin(\vartheta)^2 & 0 & 0 \\ 0 & 0 & 0 & 0 & \ddots & 0 \\ 0 & 0 & 0 & 0 & 0 & g_{n-1n-1} \end{pmatrix},$$

$$T = c_x\cdot t+\rho;\quad g_{n-1n-1} = T^m\cdot\sin(r)^2\cdot\prod_{i=0}^{n-4}\sin(\vartheta_i)^2 \tag{270}$$

we obtain the general solution to the quantum gravity field equations (125) via the wave function:

$$f[t] = C_f\cdot(c_x\cdot t+\rho)^{-m}\cdot e^{\pm\frac{4\cdot c\cdot(c_x\cdot t+\rho)^{1-\frac{m}{2}}}{c_x\cdot(m-2)}}. \tag{271}$$

5.15 Appendix O: Obtaining Matter

5.15.1 Using Coordinate Transformation to Obtain Other Degrees of Freedom

When considering the transformation rules of tensors, the classical Einstein–Hilbert action kernel:

$$\delta W = 0 = \delta\int_V d^n x\sqrt{-g}\cdot R \tag{272}$$

would only be affected with respect to the determinant term, because we have:

$$\delta W = 0 = \delta \int_V d^n x \sqrt{-g} \cdot R = \delta \int_V d^n x \sqrt{-g} \cdot R_{\alpha\beta} g^{\alpha\beta}$$

$$= \delta \int_V d^n x \sqrt{-\det\left[\frac{\partial x^a}{\partial x^\alpha}\frac{\partial x^b}{\partial x^\beta} g_{ab}\right]} \cdot \frac{\partial x^a}{\partial x^\alpha}\frac{\partial x^b}{\partial x^\beta} R_{ab} g^{cd} \frac{\partial x^\alpha}{\partial x^c}\frac{\partial x^\beta}{\partial x^d}$$

$$= \delta \int_V d^n x \sqrt{-\det\left[\frac{\partial x^a}{\partial x^\alpha}\frac{\partial x^b}{\partial x^\beta} g_{ab}\right]} \cdot R_{ab} g^{ab}. \qquad (273)$$

Describing the transformation matrix $\frac{\partial x^a}{\partial x^\alpha}$ via a function term $\frac{\partial x^a}{\partial x^\alpha} = \frac{\partial G^a}{\partial x^\alpha} = G^a{}_{,\alpha}$, we can write:

$$\delta W = 0 = \delta \int_V d^n x \sqrt{-g} \cdot R = \delta \int_V d^n x \sqrt{-\det\left[G^a{}_{,\alpha} G^b{}_{,\beta} g_{ab}\right]} \cdot R. \qquad (274)$$

We may reformulate Eq. (273) by just using the Jacobi-determinate as follows:

$$\delta W = 0 = \delta \int_V d^n x \sqrt{-g} \cdot R = \delta \int_V d^n x \left|\det\left[G^a{}_{,\alpha}\right]\right| \cdot R_{ab} g^{ab}. \qquad (275)$$

With the determinate being given via the Levi-Civita symbol as follows:

$$\det\left[G^a{}_{,\alpha}\right] = \frac{1}{n!} \cdot \varepsilon_{i_1 \ldots i_n} \varepsilon^{j_1 \ldots j_n} G^{i_1}{}_{,j_1} \cdot \ldots \cdot G^{i_n}{}_{,j_n}, \qquad (276)$$

the functional derivative or variation with respect to the function vector G^a would lead to:

$$\delta W = \delta_{G^a} \int_V d^n x \left|\det\left[G^a{}_{,\alpha}\right]\right| \cdot R = -\left(\left(\frac{d\left|\det\left[G^a{}_{,\alpha}\right]\right|}{d\left(G^a{}_{,\alpha}\right)}\right) \cdot R\right)_{,\alpha}. \qquad (277)$$

In 2 dimensions, this would lead to the equations:

$$\delta W = \delta_{G^a} \int_V d^n x \left|\det\left[G^a{}_{,\alpha}\right]\right| \cdot R = -\left(\left(\frac{d\left|\det\left[G^a{}_{,\alpha}\right]\right|}{d\left(G^a{}_{,\alpha}\right)}\right) \cdot R\right)_{,\alpha}$$

$$\xrightarrow{n=2;\; G^1{}_{,2} = G^2{}_{,1}}$$

$$= -\left(\left(\begin{matrix} \frac{d\left|\det\left[G^a{}_{,\alpha}\right]\right|}{d\left(G^1{}_{,1}\right)} & \frac{d\left|\det\left[G^a{}_{,\alpha}\right]\right|}{d\left(G^1{}_{,2}\right)} \\ \frac{d\left|\det\left[G^a{}_{,\alpha}\right]\right|}{d\left(G^1{}_{,2}\right)} & \frac{d\left|\det\left[G^a{}_{,\alpha}\right]\right|}{d\left(G^2{}_{,2}\right)} \end{matrix}\right)_a^\alpha \cdot R\right)_{,\alpha}$$

$$= -\left(\left(\begin{matrix} \frac{d\left|G^1{}_{,1}G^2{}_{,2}-\left(G^1{}_{,2}\right)^2\right|}{d\left(G^1{}_{,1}\right)} & \frac{d\left|G^1{}_{,1}G^2{}_{,2}-\left(G^1{}_{,2}\right)^2\right|}{d\left(G^1{}_{,2}\right)} \\ \frac{d\left|G^1{}_{,1}G^2{}_{,2}-\left(G^1{}_{,2}\right)^2\right|}{d\left(G^1{}_{,2}\right)} & \frac{d\left|G^1{}_{,1}G^2{}_{,2}-\left(G^1{}_{,2}\right)^2\right|}{d\left(G^2{}_{,2}\right)} \end{matrix}\right)_a^\alpha \cdot R\right)_{,\alpha}$$

$$= -\left(\left(\begin{matrix} G^2{}_{,2} & -2\cdot G^1{}_{,2} \\ -2\cdot G^1{}_{,2} & G^1{}_{,1} \end{matrix}\right)_a^\alpha \cdot R\right)_{,\alpha}. \tag{278}$$

If we prefer the determinant of the metric tensor, the variation equation would read:

$$\delta W = 0 = \delta_{G^a|G^b}\int_V d^n x\sqrt{-\det\left[G^a{}_{,\alpha}G^b{}_{,\beta}g_{ab}\right]}\cdot R$$

$$= -\left(\left(\frac{d\sqrt{-\det\left[G^a{}_{,\alpha}G^b{}_{,\beta}g_{ab}\right]}}{d\left(G^a{}_{,\alpha}\right)} + \frac{d\sqrt{-\det\left[G^a{}_{,\alpha}G^b{}_{,\beta}g_{ab}\right]}}{d\left(G^b{}_{,\beta}\right)}\right)\cdot R\right)_{,\alpha|\beta}$$

$$= -\left(\left(\sqrt{\det\left[G^b{}_{,\beta}\right]}\frac{d\sqrt{\det\left[G^a{}_{,\alpha}\right]}}{d\left(G^a{}_{,\alpha}\right)} + \sqrt{\det\left[G^a{}_{,\alpha}\right]}\frac{d\sqrt{\det\left[G^b{}_{,\beta}\right]}}{d\left(G^b{}_{,\beta}\right)}\right)\cdot\sqrt{-g}\cdot R\right)_{,\alpha|\beta}$$

$$\Rightarrow \qquad 0 = \left(\begin{matrix} \left(\left(\frac{d\sqrt{\det\left[G^a{}_{,\alpha}\right]}}{d\left(G^a{}_{,\alpha}\right)}\right)_a^\alpha \cdot\sqrt{-g\det\left[G^b{}_{,\beta}\right]}\cdot R\right)_{,\alpha} \\ +\left(\left(\frac{d\sqrt{\det\left[G^b{}_{,\beta}\right]}}{d\left(G^b{}_{,\beta}\right)}\right)_b^\beta \cdot\sqrt{-g\sqrt{\det\left[G^a{}_{,\alpha}\right]}}\cdot R\right)_{,\beta} \end{matrix}\right). \tag{279}$$

Again, using the example in $n = 2$, we now obtain:

$$\delta W = 0 = \delta_{G^a|G^b} \int_V d^n x \sqrt{-\det\left[G^a{}_{,\alpha} G^b{}_{,\beta} g_{ab}\right]} \cdot R$$

$$= \left(\begin{array}{l} \left(\frac{1}{2\sqrt{\det\left[G^a{}_{,\alpha}\right]}} \begin{pmatrix} G^2{}_{,2} & -2\cdot G^1{}_{,2} \\ -2\cdot G^1{}_{,2} & G^1{}_{,1} \end{pmatrix}_a^{\alpha} \cdot \sqrt{-g \det\left[G^b{}_{,\beta}\right]} \cdot R \right)_{,\alpha} \\ + \left(\frac{1}{2\sqrt{\det\left[G^b{}_{,\beta}\right]}} \begin{pmatrix} G^2{}_{,2} & -2\cdot G^1{}_{,2} \\ -2\cdot G^1{}_{,2} & G^1{}_{,1} \end{pmatrix}_b^{\beta} \cdot \sqrt{-g\sqrt{\det\left[G^a{}_{,\alpha}\right]}} \cdot R \right)_{,\beta} \end{array} \right)$$

$$\Rightarrow \qquad 0 = \frac{1}{2} \left(\begin{array}{l} \left(\begin{pmatrix} G^2{}_{,2} & -2\cdot G^1{}_{,2} \\ -2\cdot G^1{}_{,2} & G^1{}_{,1} \end{pmatrix}_a^{\alpha} \cdot \sqrt{-g} \cdot R \right)_{,\alpha} \\ + \left(\begin{pmatrix} G^2{}_{,2} & -2\cdot G^1{}_{,2} \\ -2\cdot G^1{}_{,2} & G^1{}_{,1} \end{pmatrix}_b^{\beta} \cdot \sqrt{-g} \cdot R \right)_{,\beta} \end{array} \right)$$

$$= \left(\begin{pmatrix} G^2{}_{,2} & -2\cdot G^1{}_{,2} \\ -2\cdot G^1{}_{,2} & G^1{}_{,1} \end{pmatrix}_a^{\alpha} \cdot \sqrt{-g} \cdot R \right)_{,\alpha} . \tag{280}$$

5.15.2 What Is Matter?

We realize that in n dimensions, we obtain n equations for the function G^a. These equations are of third order because of the Ricci scalar, and we are reminded of the following well-known condition for the Einstein field equations:

$$0 = G^{\alpha\beta}{}_{;\beta} = \left(R^{\alpha\beta} - \frac{R}{2} g^{\alpha\beta} \right)_{;\beta} = G_\alpha{}^\beta{}_{;\beta} = \left(R_\alpha^\beta - \frac{R}{2} \delta_\alpha^\beta \right)_{;\beta} . \tag{281}$$

This condition can be directly obtained from the Bianchi identity for the Riemann curvature tensor, reading:

$$0 = R_{\alpha\beta\gamma\delta;\varepsilon} + R_{\alpha\beta\varepsilon\gamma;\delta} + R_{\alpha\beta\delta\varepsilon;\gamma} , \tag{282}$$

via contraction with the metric tensor as follows:

$$0=\left(R_{\alpha\beta\gamma\delta;\varepsilon}+R_{\alpha\beta\varepsilon\gamma;\delta}+R_{\alpha\beta\delta\varepsilon;\gamma}\right)g^{\alpha\delta}g^{\beta\gamma} \\ =\xrightarrow{g^{\alpha\delta}{}_{;\lambda}=0}=2\cdot R^{\alpha}{}_{\varepsilon;\alpha}-R_{;\varepsilon}. \tag{283}$$

In other words, our variation results (277) and (279) are just different forms of the condition (generalized conservation law) (281).

Applying this onto our scaled metric variation result (our quantum Einstein field equations) from Eq. (125), we would have to demand, respectively obtain the additional condition:

$$0=\left(\begin{array}{c} R^{\beta}_{\alpha;\beta}-\frac{\delta^{\beta}_{\alpha}}{2}R_{;\beta}+\frac{\delta^{\beta}_{\alpha}}{2}\left(\begin{array}{c}\frac{1}{2F}\left(\begin{array}{c}2F_{,ij}(n-1)g^{ij}+2\Gamma^{a}_{ij}F_{,a}g^{ij}\\ -F_{,i}g^{ab}g_{jb,a}g^{ij}-F_{,j}g^{ab}g_{ib,a}g^{ij}\\ -n\Gamma^{d}_{ij}F_{,d}g^{ij}+\frac{n}{2}F_{,d}g^{cd}g_{ab,c}g^{ab}\end{array}\right)\\ +\frac{F_{,i}\cdot F_{,j}}{4F^{2}}g^{ij}\left((n-6)(n-1)\right)\end{array}\right)_{;\beta}\\ +\frac{F_{;\beta}}{2F^{2}}\left(g^{\beta\gamma}\left(\begin{array}{c}F_{,\alpha\gamma}(n-2)+F_{,ab}g_{\alpha\gamma}g^{ab}\\ +F_{,a}g^{ab}\left(g_{\gamma b,\alpha}-g_{\gamma\alpha,b}\right)-F_{,\alpha}g^{ab}g_{\gamma b,a}-F_{,\gamma}g^{ab}g_{\alpha b,a}\\ +F_{,d}g^{cd}\frac{1}{2}n\left(\frac{2}{n}g_{\alpha c,\gamma}-g_{\alpha c,\gamma}-g_{\gamma c,\alpha}+g_{\alpha\gamma,c}+\frac{1}{n}g_{\alpha\gamma}g_{ab,c}g^{ab}\right)\\ -\frac{1}{2F}\left(F_{,\alpha}\cdot F_{,\gamma}(3n-6)+g_{\alpha\gamma}F_{,c}F_{,d}g^{cd}(4-n)\right)\end{array}\right)\right)\\ -\frac{1}{2F}\left(g^{\beta\gamma}\left(\begin{array}{c}F_{,\alpha\gamma}(n-2)+F_{,ab}g_{\alpha\gamma}g^{ab}\\ +F_{,a}g^{ab}\left(g_{\gamma b,\alpha}-g_{\gamma\alpha,b}\right)-F_{,\alpha}g^{ab}g_{\gamma b,a}-F_{,\gamma}g^{ab}g_{\alpha b,a}\\ +F_{,d}g^{cd}\frac{1}{2}n\left(\frac{2}{n}g_{\alpha c,\gamma}-g_{\alpha c,\gamma}-g_{\gamma c,\alpha}+g_{\alpha\gamma,c}+\frac{1}{n}g_{\alpha\gamma}g_{ab,c}g^{ab}\right)\\ -\frac{1}{2F}\left(F_{,\alpha}\cdot F_{,\gamma}(3n-6)+g_{\alpha\gamma}F_{,c}F_{,d}g^{cd}(4-n)\right)\end{array}\right)\right)_{;\beta}\\ +\kappa T^{\beta}_{\alpha\ ;\beta}+\Lambda\cdot F_{;\beta}\cdot\delta^{\beta}_{\alpha}\end{array}\right). \tag{284}$$

Assuming vanishing cosmological constant and matter Lagrange density immediately leads to:

$$0=\left(\begin{array}{c} R^{\beta}_{\alpha;\beta}-\frac{\delta^{\beta}_{\alpha}}{2}R_{;\beta}+\frac{\delta^{\beta}_{\alpha}}{2}\left(\begin{array}{c}\frac{1}{2F}\left(2F_{,ij}(n-1)g^{ij}+2\Gamma^{a}_{ij}F_{,a}g^{ij}-F_{,i}g^{ab}g_{jb,a}g^{ij}-F_{,j}g^{ab}g_{ib,a}g^{ij}-n\Gamma^{d}_{ij}F_{,d}g^{ij}+\frac{n}{2}F_{,d}g^{cd}g_{ab,c}g^{ab}\right)\\ +\frac{F_{,i}\cdot F_{,j}}{4F^{2}}g^{ij}\left((n-6)(n-1)\right)\end{array}\right)_{;\beta}\\ +\frac{F_{;\beta}}{2F^{2}}\left(g^{\beta\gamma}\left(\begin{array}{c}F_{,\alpha\gamma}(n-2)+F_{,ab}g_{\alpha\gamma}g^{ab}\\ +F_{,a}g^{ab}\left(g_{\gamma b,\alpha}-g_{\gamma\alpha,b}\right)-F_{,\alpha}g^{ab}g_{\gamma b,a}-F_{,\gamma}g^{ab}g_{\alpha b,a}\\ +F_{,d}g^{cd}\frac{1}{2}n\left(\frac{2}{n}g_{\alpha c,\gamma}-g_{\alpha c,\gamma}-g_{\gamma c,\alpha}+g_{\alpha\gamma,c}+\frac{1}{n}g_{\alpha\gamma}g_{ab,c}g^{ab}\right)\\ -\frac{1}{2F}\left(F_{,\alpha}\cdot F_{,\gamma}(3n-6)+g_{\alpha\gamma}F_{,c}F_{,d}g^{cd}(4-n)\right)\end{array}\right)\right)\\ -\frac{1}{2F}\left(g^{\beta\gamma}\left(\begin{array}{c}F_{,\alpha\gamma}(n-2)+F_{,ab}g_{\alpha\gamma}g^{ab}\\ +F_{,a}g^{ab}\left(g_{\gamma b,\alpha}-g_{\gamma\alpha,b}\right)-F_{,\alpha}g^{ab}g_{\gamma b,a}-F_{,\gamma}g^{ab}g_{\alpha b,a}\\ +F_{,d}g^{cd}\frac{1}{2}n\left(\frac{2}{n}g_{\alpha c,\gamma}-g_{\alpha c,\gamma}-g_{\gamma c,\alpha}+g_{\alpha\gamma,c}+\frac{1}{n}g_{\alpha\gamma}g_{ab,c}g^{ab}\right)\\ -\frac{1}{2F}\left(F_{,\alpha}\cdot F_{,\gamma}(3n-6)+g_{\alpha\gamma}F_{,c}F_{,d}g^{cd}(4-n)\right)\end{array}\right)\right)_{;\beta}\end{array}\right). \tag{285}$$

Now taking into account the contracted Bianchi identity from Eq. (283), we obtain:

$$0=R^{\beta}_{\alpha;\beta}-\frac{\delta^{\beta}_{\alpha}}{2}R_{;\beta}\quad\Rightarrow$$

$$0=\left(\begin{array}{c}\frac{\delta_{\alpha}^{\beta}}{2}\left(\frac{1}{2F}\left(\begin{array}{c}2F_{,ij}(n-1)g^{ij}+2\Gamma_{ij}^{a}F_{,a}g^{ij}-F_{,i}g^{ab}g_{jb,a}g^{ij}-F_{,j}g^{ab}g_{ib,a}g^{ij}-n\Gamma_{ij}^{d}F_{,d}g^{ij}+\frac{n}{2}F_{,d}g^{cd}g_{ab,c}g^{ab}\\+\frac{F_{,i}\cdot F_{,j}}{4F^{2}}g^{ij}\left((n-6)(n-1)\right)\end{array}\right)\right)_{;\beta}\\+\frac{F_{;\beta}}{2F^{2}}\left(g^{\beta\gamma}\left(\begin{array}{c}F_{,\alpha\gamma}(n-2)+F_{,ab}g_{\alpha\gamma}g^{ab}\\+F_{,a}g^{ab}\left(g_{\gamma b,\alpha}-g_{\gamma\alpha,b}\right)-F_{,\alpha}g^{ab}g_{\gamma b,a}-F_{,\gamma}g^{ab}g_{\alpha b,a}\\+F_{,d}g^{cd}\frac{1}{2}n\left(\frac{2}{n}g_{\alpha c,\gamma}-g_{\alpha c,\gamma}-g_{\gamma c,\alpha}+g_{\alpha\gamma,c}+\frac{1}{n}g_{\alpha\gamma}g_{ab,c}g^{ab}\right)\\-\frac{1}{2F}\left(F_{,\alpha}\cdot F_{,\gamma}(3n-6)+g_{\alpha\gamma}F_{,c}F_{,d}g^{cd}(4-n)\right)\end{array}\right)\right)\\-\frac{1}{2F}\left(g^{\beta\gamma}\left(\begin{array}{c}F_{,\alpha\gamma}(n-2)+F_{,ab}g_{\alpha\gamma}g^{ab}\\+F_{,a}g^{ab}\left(g_{\gamma b,\alpha}-g_{\gamma\alpha,b}\right)-F_{,\alpha}g^{ab}g_{\gamma b,a}-F_{,\gamma}g^{ab}g_{\alpha b,a}\\+F_{,d}g^{cd}\frac{1}{2}n\left(\frac{2}{n}g_{\alpha c,\gamma}-g_{\alpha c,\gamma}-g_{\gamma c,\alpha}+g_{\alpha\gamma,c}+\frac{1}{n}g_{\alpha\gamma}g_{ab,c}g^{ab}\right)\\-\frac{1}{2F}\left(F_{,\alpha}\cdot F_{,\gamma}(3n-6)+g_{\alpha\gamma}F_{,c}F_{,d}g^{cd}(4-n)\right)\end{array}\right)\right)_{;\beta}\end{array}\right), \tag{286}$$

and by just considering all the scaling function terms as matter components (effective matter for an observer like ourselves) with:

$$
\kappa T_{\alpha}^{\beta}=\left(\begin{array}{l}
\dfrac{\delta_{\alpha}^{\beta}}{2}\left(\begin{array}{c}
\dfrac{1}{2F}\left(\begin{array}{c}
2F_{,ij}(n-1)g^{ij}+2\Gamma_{ij}^{a}F_{,a}g^{ij}\\
-F_{,i}g^{ab}g_{jb,a}g^{ij}-F_{,j}g^{ab}g_{ib,a}g^{ij}\\
-n\Gamma_{ij}^{d}F_{,d}g^{ij}+\dfrac{n}{2}F_{,d}g^{cd}g_{ab,c}g^{ab}
\end{array}\right)\\
+\dfrac{F_{,i}\cdot F_{,j}}{4F^{2}}g^{ij}\left((n-6)(n-1)\right)
\end{array}\right)\\
+\dfrac{F_{;\beta}}{2F^{2}}\left(g^{\beta\gamma}\left(\begin{array}{c}
F_{,\alpha\gamma}(n-2)+F_{,ab}g_{\alpha\gamma}g^{ab}\\
+F_{,a}g^{ab}\left(g_{\gamma b,\alpha}-g_{\gamma\alpha,b}\right)-F_{,\alpha}g^{ab}g_{\gamma b,a}-F_{,\gamma}g^{ab}g_{\alpha b,a}\\
+F_{,d}g^{cd}\dfrac{1}{2}n\left(\dfrac{2}{n}g_{\alpha c,\gamma}-g_{\alpha c,\gamma}-g_{\gamma c,\alpha}+g_{\alpha\gamma,c}+\dfrac{1}{n}g_{\alpha\gamma}g_{ab,c}g^{ab}\right)\\
-\dfrac{1}{2F}\left(F_{,\alpha}\cdot F_{,\gamma}(3n-6)+g_{\alpha\gamma}F_{,c}F_{,d}g^{cd}(4-n)\right)
\end{array}\right)\right)\\
-\dfrac{1}{2F}\left(g^{\beta\gamma}\left(\begin{array}{c}
F_{,\alpha\gamma}(n-2)+F_{,ab}g_{\alpha\gamma}g^{ab}\\
+F_{,a}g^{ab}\left(g_{\gamma b,\alpha}-g_{\gamma\alpha,b}\right)-F_{,\alpha}g^{ab}g_{\gamma b,a}-F_{,\gamma}g^{ab}g_{\alpha b,a}\\
+F_{,d}g^{cd}\dfrac{1}{2}n\left(\dfrac{2}{n}g_{\alpha c,\gamma}-g_{\alpha c,\gamma}-g_{\gamma c,\alpha}+g_{\alpha\gamma,c}+\dfrac{1}{n}g_{\alpha\gamma}g_{ab,c}g^{ab}\right)\\
-\dfrac{1}{2F}\left(F_{,\alpha}\cdot F_{,\gamma}(3n-6)+g_{\alpha\gamma}F_{,c}F_{,d}g^{cd}(4-n)\right)
\end{array}\right)\right)
\end{array}\right), \tag{287}
$$

we see that Eq. (286) would just be the classical conservation law, reading:

$$
T_{\alpha\,;\beta}^{\beta}=0. \tag{288}
$$

5.16 Appendix P: Electro-Magnetic Interaction (and Matter) via Space-Time Warping

In order to allow for an easy entrance into the theoretical apparatus, we shall start with the Minkowski flat space metric tensor in Cartesian coordinates *t*, *x*, *y*, *z* as follows:

$$g_{\alpha\beta}^{flat}=\begin{pmatrix}-c^2&0&0&0\\0&1&0&0\\0&0&1&0\\0&0&0&1\end{pmatrix}. \tag{289}$$

This metric solves the classical Einstein field equations, resulting from the Einstein–Hilbert action (118) (see [1, 2]) and corresponds to a Ricci scalar of $R = 0$. Now we introduce the following special transformation:

$$G_{\alpha\beta}=g_{\alpha\beta}^{flat}\cdot F\left[f[t,x,y,z]\right]=\begin{pmatrix}-c^2&0&0&0\\0&1&0&0\\0&0&1&0\\0&0&0&1\end{pmatrix}\cdot F\left[f[t,x,y,z]\right] \tag{290}$$

and evaluate the resulting new Ricci scalar R^*:

$$R^*=\frac{R}{F[f]}+\begin{pmatrix}-\Gamma^{\mu}_{\sigma\alpha}\Gamma^{**\sigma}_{\beta\mu}+\Gamma^{\sigma}_{\alpha\beta}\Gamma^{**\mu}_{\sigma\mu}-\Gamma^{**\mu}_{\sigma\alpha}\Gamma^{\sigma}_{\beta\mu}+\Gamma^{**\sigma}_{\alpha\beta}\Gamma^{\mu}_{\sigma\mu}\\+\Gamma^{**\sigma}_{\alpha\beta\ ,\sigma}-\Gamma^{**\sigma}_{\beta\sigma\ ,\alpha}-\Gamma^{**\mu}_{\sigma\alpha}\Gamma^{**\sigma}_{\beta\mu}+\Gamma^{**\sigma}_{\alpha\beta}\Gamma^{**\mu}_{\sigma\mu}\end{pmatrix}\frac{g^{\alpha\beta}}{F[f]}$$

$$\xrightarrow{R=0;\,g_{\alpha\beta}^{flat}}=\frac{1}{F[f]^3}\cdot\begin{pmatrix}\left(C_{N1}\cdot\left(\frac{\partial F[f]}{\partial f}\right)^2-C_{N2}\cdot F[f]\cdot\frac{\partial^2F[f]}{\partial f^2}\right)\cdot\left(\tilde{\nabla}_g f\right)^2\\-C_{N2}\cdot F[f]\cdot\frac{\partial F[f]}{\partial f}\cdot\Delta_g f\end{pmatrix}$$

$$=\frac{1}{F[f]^3}\cdot\left(\left(\frac{3}{2}\cdot\left(\frac{\partial F[f]}{\partial f}\right)^2-3\cdot F[f]\cdot\frac{\partial^2F[f]}{\partial f^2}\right)\cdot\left(\tilde{\nabla}_g f\right)^2-3\cdot F[f]\cdot\frac{\partial F[f]}{\partial f}\cdot\Delta_g f\right)$$

$$=\frac{3}{2\cdot F[f]^3}\cdot\begin{pmatrix}\left(\left(\frac{\partial F[f]}{\partial f}\right)^2-2\cdot F[f]\cdot\frac{\partial^2F[f]}{\partial f^2}\right)\cdot\begin{bmatrix}-\frac{(\partial_t f)^2}{c^2}+(\partial_x f)^2\\+(\partial_y f)^2+(\partial_z f)^2\end{bmatrix}\\-2\cdot F[f]\cdot\frac{\partial F[f]}{\partial f}\cdot\left[-\frac{\partial_t^2}{c^2}+\partial_x^2+\partial_y^2+\partial_z^2\right]f\end{pmatrix}. \tag{291}$$

Thereby we have used:

$$\Gamma^{*\gamma}_{\alpha\beta}=\frac{g^{\gamma\sigma}}{2\cdot F[f]}\left(\left[F[f]\cdot g_{\sigma\alpha}\right]_{,\beta}+\left[F[f]\cdot g_{\sigma\beta}\right]_{,\alpha}-\left[F[f]\cdot g_{\alpha\beta}\right]_{,\sigma}\right)$$

$$=\frac{g^{\gamma\sigma}}{2\cdot F[f]}\left(\left[F[f]\cdot g_{\sigma\alpha}\right]_{,\beta}+\left[F[f]\cdot g_{\sigma\beta}\right]_{,\alpha}-\left[F[f]\cdot g_{\alpha\beta}\right]_{,\sigma}\right)$$

$$=\frac{g^{\gamma\sigma}}{2}\left(g_{\sigma\alpha,\beta}+g_{\sigma\beta,\alpha}-g_{\alpha\beta,\sigma}\right)+\frac{g^{\gamma\sigma}}{2\cdot F[f]}\begin{pmatrix}F[f]_{,\beta}\cdot g_{\sigma\alpha}+F[f]_{,\alpha}\cdot g_{\sigma\beta}\\ -F[f]_{,\sigma}\cdot g_{\alpha\beta}\end{pmatrix}$$

$$=\Gamma^{\gamma}_{\alpha\beta}+\frac{g^{\gamma\sigma}}{2\cdot F[f]}\left(F[f]_{,\beta}\cdot g_{\sigma\alpha}+F[f]_{,\alpha}\cdot g_{\sigma\beta}-F[f]_{,\sigma}\cdot g_{\alpha\beta}\right)$$

$$\equiv\Gamma^{\gamma}_{\alpha\beta}+\Gamma^{**\gamma}_{\alpha\beta}. \qquad (292)$$

The reader will find a more comprehensive elaboration about the motivation of our special transformation in the book [4] in section "3.3 The Ricci Scalar Quantization." However, it is quite entertaining to follow this simple trial and observe its amazing evolution into something rather unexpected (at least if taking its origin as a metric solution to the Einstein field equations) out of our "wrapper-transformation" (290).

The symbol $\tilde{\nabla}_g$ in Eq. (291) denotes a first-order differential operator similar to the Nabla-operator in the metric $g_{\alpha\beta}$. The symbols C_{Ni} stand for constants, which only depend on the number of dimensions. Their derivation was presented in [5] and its result can also be found in appendix B of this book. Please note that the transformation (290) is just the simplest form of a general approach like:

$$G_{\alpha\beta}=F\left[f[t,x,y,z]\right]^{ij}_{\alpha\beta}g_{ij}\rightarrow G_{\alpha\beta}=F\left[f[t,x,y,z]\right]\cdot\delta^{i}_{\alpha}\delta^{j}_{\beta}g_{ij}. \qquad (293)$$

Without loss of generality, we can now demand that the first term in parenthesis in Eq. (291) would be zero, which is to say, we choose our arbitrary function F such that we have:

$$C_{N1}\cdot\left(\frac{\partial F[f]}{\partial f}\right)^2-C_{N2}\cdot F[f]\cdot\frac{\partial^2 F[f]}{\partial f^2}=0. \qquad (294)$$

For instance, in 4 dimensions this would always be the case for $F[f]=f^2$, giving us the Klein–Gordon-like equation:

$$\left[\Delta_g+\frac{R^*\cdot f^4}{f\cdot C_{N2}\cdot 2\cdot f}\right]f=\left[\Delta_g+\frac{R^*\cdot f^2}{2}\right]f=0;\ \ \Delta_{\text{g-Coordinates}}=\Delta_g. \qquad (295)$$

5.16.1 The Higgs Field in Our Head

Setting the solution $F[f] = f^2$ into our transformation starting point (290), however, one would directly obtain an all-scale quantum dominated metric solution for $G_{\alpha\beta}$. This is in total contrast to the daily experiences that quantum effects are only important in smaller scales. In order to overcome this problem, we evaluate the general solution of condition (294) in 4 dimensions and find:

$$F[f]=C_1\cdot f+\frac{C_1^2\cdot f^2}{4\cdot C_2}+C_2. \tag{296}$$

Thereby the constants C_i are arbitrary.

For entertainment and further motivation, it should be pointed out here that the functional wrapper $F[f]$ from (296) could also be adapted as follows:

$$F[f]=C_1\cdot f^2+\frac{C_1^2\cdot f^4}{4\cdot C_2}+C_2. \tag{297}$$

With arbitrary constants C_i, this ensures the appearance of linear Laplace operator terms according to the condition (294) for the resulting Ricci scalar of the transformed metric in Eq. (291) (and even Eq. (125)) as follows:

$$R^*=\frac{1}{F[f]^3}\cdot\left(-C_{N2}\cdot F[f]\cdot\frac{\partial F[f]}{\partial f}\cdot\Delta_g f^2\right)$$

$$\xrightarrow{4D} R^*\cdot\frac{\left(2\cdot C_2+C_1\cdot f^2\right)^3}{8\cdot C_2\cdot C_1}=-3\cdot\Delta_g f^2. \tag{298}$$

Now we assume f to be a constant, which automatically leads to the simple equation:

$$R^*\cdot\frac{\left(2\cdot C_2+C_1\cdot f^2\right)^3}{8\cdot C_2\cdot C_1}=0. \tag{299}$$

Also assuming that the Ricci scalar curvature R^* shall be proportional to f^n, with an arbitrary exponent $n \geq 0$, we obtain the familiar trivial solution of $f_0 = 0$. However, from Eq. (299) we also obtain the non-trivial ground states:

$$f^2=-\frac{2\cdot C_2}{C_1}. \tag{300}$$

As the constants C_i are arbitrary, we could simply set them as follows:

$$2\cdot C_2=-\mu^2;\quad C_1=2\cdot\lambda. \tag{301}$$

This gives us the additional ground state solutions directly in the classical Higgs field style [14], namely:

$$\left(f_{1,2}\right)^2=\frac{\mu^2}{2\cdot\lambda}\quad\Rightarrow\quad f_{1,2}=\pm\frac{\mu}{\sqrt{2\cdot\lambda}}. \tag{302}$$

The measured value for the $f_{1,2}$ is known to be [44]:

$$\left|f_{1,2}\right|=\frac{\mu}{\sqrt{2\cdot\lambda}}=\frac{246\,GeV}{\sqrt{2}\cdot c^2}. \tag{303}$$

Rewriting Eq. (299) with the use of Eq. (301) as:

$$R^*\cdot\frac{\left(-\mu^2+2\cdot\lambda\cdot f^2\right)^3}{8\cdot\mu^2\cdot 2\cdot\lambda}=f^n\cdot\frac{\left(2\cdot\lambda\cdot f^2-\mu^2\right)^3}{16\cdot\mu^2\cdot\lambda}=3\cdot\Delta_g f^2 \tag{304}$$

gives us the total Higgs–Ricci-curvature connection. It also, automatically, gives us a curvature value R^* at the ground state, which is:

$$R^*\sim\left|f_{1,2}\right|^n=\left(\frac{\mu}{\sqrt{2\cdot\lambda}}\right)^n=\left(\frac{246\,GeV}{\sqrt{2}\cdot c^2}\right)^n. \tag{305}$$

Now, with the Higgs mass m_H known to be 125 GeV/c^2 and the fact that we have [28]:

$$m_H=\sqrt{2\cdot\mu^2}=\sqrt{4\cdot\lambda\cdot\left(f_{1,2}\right)^2}, \tag{306}$$

we can obtain:

$$\lambda\simeq 0.13;\quad \mu\simeq 88.8GeV/c^2. \tag{307}$$

Please note that we obtain the simple case $F[f]=f^2$ with the condition:

$$C_1^2=2\cdot C_2 \tag{308}$$

and the subsequent limit procedure $C_2\to 0$ as follows:

$$\Rightarrow F[f]=\sqrt{2\cdot C_2}\cdot f+f^2+C_2\xrightarrow[C_2\to 0]{\lim}F[f]=f^2. \tag{309}$$

As such considerations can be generalized to any number of dimensions, arbitrary attributes and systems, we can immediately state that also the system of the mind can produce a Higgs field, which provides an interaction force with the "things" being in there. As these things are our thoughts, the field just gives the mass and moderates their forthcoming just as the real Higgs field does in our universal space-time with the elementary particles (not all of them, though).

5.16.2 Back to the Derivation of the Electromagnetic Field

Using Eq. (296) instead of $F = f^2$ in (291) gives us:

$$R^* = \frac{1}{F[f]^3} \cdot \begin{pmatrix} \left(C_{N1} \cdot \left(\frac{\partial F[f]}{\partial f} \right)^2 - C_{N2} \cdot F[f] \cdot \frac{\partial^2 F[f]}{\partial f^2} \right) \cdot \left(\tilde{\nabla}_g f \right)^2 \\ -C_{N2} \cdot F[f] \cdot \frac{\partial F[f]}{\partial f} \cdot \Delta_g f \end{pmatrix}$$

$$= -C_{N2} \cdot \frac{\frac{\partial F[f]}{\partial f} \cdot \Delta_g f}{F[f]^2} \xrightarrow{4D} = -3 \cdot \frac{\left(C_1 + \frac{C_1^2 \cdot f}{2 \cdot C_2} \right)}{\left(C_1 \cdot f + \frac{C_1^2 \cdot f^2}{4 \cdot C_2} + C_2 \right)^2} \cdot \Delta_g f. \quad (310)$$

Immediately we see that in the many cases with $R^* = 0$, we would obtain classical quantum Klein–Gordon equations. By the way, in the book [5] we have shown that the condition $R^* = 0$ also directly follows from the Hamilton minimum principle and is not just a simple definition. Taylor expansion for small f (which is to say at $f = 0$) for the general case in the second line in Eq. (310) also gives us equations leading to these classical equations. For instance, if only considering up to linear f as follows:

$$\frac{R^*}{3} \cdot \frac{\left(C_1 \cdot f + \frac{C_1^2 \cdot f^2}{4 \cdot C_2} + C_2 \right)^2}{C_1 + \frac{C_1^2 \cdot f}{2 \cdot C_2}} + \Delta_g f = 0$$

$$\xrightarrow{\text{small } f} \frac{R^*}{3} \cdot \left(\frac{C_2^2}{C_1} + \frac{3}{2} \cdot C_2 \cdot f + \ldots \right) + \Delta_g f = 0 \quad (311)$$

and assuming constant R^*, we already have the typical structure of the Klein–Gordon equation plus a constant term $\frac{R^*}{3} \cdot \frac{C_2^2}{C_1}$. Such a constant would still not influence the principal structural character of our resulting quantum equations because a simple transformation of:

$$f^* = \frac{C_2^2}{C_1} + \frac{3}{2} \cdot C_2 \cdot f; \quad \Delta_g f^* = \Delta_g f \quad (312)$$

does give us back the ordinary Klein–Gordon equation with:

$$\frac{R^*}{3} \cdot f^* + \Delta_g f^* = 0. \tag{313}$$

The resulting transformation in 4 dimensions would give the following quantum metric:

$$G_{\alpha\beta} = g_{\alpha\beta} \cdot F[f] = g_{\alpha\beta} \cdot \left(C_1 \cdot f + \frac{C_1^2 \cdot f^2}{4 \cdot C_2} + C_2 \right). \tag{314}$$

This clearly allows for the classical physics or Einstein situation with quantum effects becoming dominant only at small scales and thus, a dominating part C_2 at bigger scales. In result, one might separate as follows in 4 dimensions:

$$G_{\alpha\beta} = \overbrace{\overbrace{g_{\alpha\beta} \cdot C_2}^{\text{class. GTR}} + \overbrace{g_{\alpha\beta} \cdot \left(C_1 \cdot f \cdot \left(1 + \frac{C_1 \cdot f}{4 \cdot C_2} \right) \right)}^{\text{quantum metric}}}^{\text{Quantum Gravity in 4D}}. \tag{315}$$

Thereby it does not come as a surprise that we have already found so many cases where f has to be determined by classical quantum equations (see [4, 5] and references therein), respectively where our equations give similar or even equal solutions. Normalization of the quantum gravity metric $G_{\alpha\beta}$ requires division by C_2 and gives:

$$G_{\alpha\beta}^{\text{norm}} = \frac{G_{\alpha\beta}}{C_2} = \overbrace{\overbrace{g_{\alpha\beta}}^{\text{class. GTR}} + \overbrace{g_{\alpha\beta} \cdot \left(\frac{C_1}{C_2} \cdot f \cdot \left(1 + \frac{C_1 \cdot f}{4 \cdot C_2} \right) \right)}^{\text{quantum metric}}}^{\text{Quantum Gravity in 4D}}. \tag{316}$$

Now we apply the recipe outlined above to our simple flat space example in order to see how the transformation can bring about matter.
We saw that by taking the scaled metric (290), using rule (294) with the subsequent solution (296), we obtain the new Ricci scalar curvature:

$$R^* = -3 \cdot \frac{\left(C_1 + \frac{C_1^2 \cdot f}{2 \cdot C_2} \right)}{\left(C_1 \cdot f + \frac{C_1^2 \cdot f^2}{4 \cdot C_2} + C_2 \right)^2} \cdot \Delta_g f. \tag{317}$$

Now we demand that the Ricci scalar should be a conserved quantity, and as we have $R = 0$ with the flat space metric (289), we shall also demand $R^* = 0$. This gives us the simplest equation:

$$0=\Delta_g f=\left[\partial_x^2+\partial_y^2+\partial_z^2-\frac{\partial_t^2}{c^2}\right]f. \tag{318}$$

Just for an increase of generality we extend Eq. (289) as follows (A, B, D are constants):

$$g_{\alpha\beta}^{flat}=\begin{pmatrix} -c^2 & 0 & 0 & 0 \\ 0 & A^2 & 0 & 0 \\ 0 & 0 & B^2 & 0 \\ 0 & 0 & 0 & D^2 \end{pmatrix}, \tag{319}$$

which changes Eq. (318) to:

$$0=\Delta_g f=\left[\frac{\partial_x^2}{A^2}+\frac{\partial_y^2}{B^2}+\frac{\partial_z^2}{D^2}-\frac{\partial_t^2}{c^2}\right]f. \tag{320}$$

Applying the separation approach $f[t,x,y,z] = T[t]*X[x]*Y[y]*Z[z]$ gives us the solutions:

$$T[t]=C_{t1}\cdot\cos\left[c\cdot C_t\cdot t\right]+C_{t2}\cdot\sin\left[c\cdot C_t\cdot t\right], \tag{321}$$

$$X[x]=C_{x1}\cdot\cos\left[A\cdot C_x\cdot x\right]+C_{x2}\cdot\sin\left[A\cdot C_x\cdot x\right], \tag{322}$$

$$Y[y]=C_{y1}\cdot\cos\left[B\cdot C_y\cdot y\right]+C_{y2}\cdot\sin\left[B\cdot C_y\cdot y\right], \tag{323}$$

$$Z[z]=C_{z1}\cdot\cos\left[D\cdot C_z\cdot z\right]+C_{z2}\cdot\sin\left[D\cdot C_z\cdot z\right] \tag{324}$$

and the following characteristic equation:

$$0=C_x^2+C_y^2+C_z^2-C_t^2. \tag{325}$$

The matter coded with this solution for R^* can now directly be obtained via the classical Einstein field equations with matter, which reads:

$$R^{*\alpha\beta}-\frac{1}{2}R^*\cdot g^{\alpha\beta}+\Lambda\cdot g^{\alpha\beta}=-\kappa\cdot T^{\alpha\beta}. \tag{326}$$

As we had per demand $R^* = R = 0$ and as we also assume the cosmological constant to be equal to zero, we obtain the following identity:

$$R^{*\alpha\beta}=-\kappa\cdot T^{\alpha\beta}. \tag{327}$$

Thus, in order to find out what kind of matter our transformation:

$$G_{\alpha\beta}=g_{\alpha\beta}^{flat}\cdot F\left[f\left[t,x,y,z\right]\right]=\begin{pmatrix} -c^2 & 0 & 0 & 0 \\ 0 & A^2 & 0 & 0 \\ 0 & 0 & B^2 & 0 \\ 0 & 0 & 0 & D^2 \end{pmatrix}\cdot F\left[f\left[t,x,y,z\right]\right] \quad (328)$$

has created, we simply need to evaluate the corresponding Ricci tensor of the transformed/scaled metric $G_{\alpha\beta}$. In order to keep the presentation general[3] and as the evaluation of the derivatives is simple, we give the Ricci tensor with the function f. But for simplicity we simplified $F[f]$ to $F[f]=(f+C_1)^2$:

$$R^{*00}=\frac{(A\cdot B\cdot c)^2\left(f_{,z}\right)^2+D^2\left(A^2c^2\left(f_{,y}\right)^2+B^2\left(\begin{matrix} c^2\left(f_{,x}\right)^2+3A^2\left(f_{,t}\right)^2 \\ -2A^2\left(C_1+f\right)f_{,t,t}\end{matrix}\right)\right)}{(A\cdot B\cdot D)^2c^4\left(C_1+f\right)^6}. \quad (329)$$

$$R^{*11}=\frac{-(A\cdot B\cdot c)^2\left(f_{,z}\right)^2+D^2\left(\begin{matrix} -A^2c^2\left(f_{,y}\right)^2 \\ +B^2\left(\begin{matrix} 3c^2\left(f_{,x}\right)^2 \\ -2c^2\left(C_1+f\right)f_{,x,x}\end{matrix}\right)+A^2\left(f_{,t}\right)^2\end{matrix}\right)}{(c\cdot B\cdot D)^2A^4\left(C_1+f\right)^6}. \quad (330)$$

$$R^{*22}=\frac{-(A\cdot B\cdot c)^2\left(f_{,z}\right)^2+D^2\left(\begin{matrix} 3A^2c^2\left(f_{,y}\right)^2-2A^2c^2\left(C_1+f\right)f_{,y,y} \\ -B^2c^2\left(f_{,x}\right)^2+B^2A^2\left(f_{,t}\right)^2\end{matrix}\right)}{(c\cdot A\cdot D)^2B^4\left(C_1+f\right)^6}. \quad (331)$$

$$R^{*33}=\frac{3(A\cdot B\cdot c)^2\left(f_{,z}\right)^2+D^2\left(\begin{matrix} -A^2c^2\left(f_{,y}\right)^2-B^2c^2\left(f_{,x}\right)^2+B^2A^2\left(f_{,t}\right)^2 \\ +2\left(C_1+f\right)\left(A^2c^2f_{,y,y}+B^2\left(c^2f_{,x,x}-A^2f_{,t,t}\right)\right)\end{matrix}\right)}{(c\cdot A\cdot D)^2B^4\left(C_1+f\right)^6}.$$

(332)

[3]This comes in handy the moment we want to exploit the additive character of our metric solutions. Thereby the additivity is ensured by the means of condition (294). See further below in the main text!

$$R^{*01}=\frac{2\left(-2f_{,x}f_{,t}+\left(C_1+f\right)f_{,t,x}\right)}{A^2c^2\left(C_1+f\right)^6};\quad R^{*02}=\frac{2\left(-2f_{,y}f_{,t}+\left(C_1+f\right)f_{,t,y}\right)}{B^2c^2\left(C_1+f\right)^6}$$

$$R^{*03}=\frac{2\left(-2f_{,z}f_{,t}+\left(C_1+f\right)f_{,t,z}\right)}{D^2c^2\left(C_1+f\right)^6};\quad R^{*12}=\frac{2\left(2f_{,y}f_{,x}-\left(C_1+f\right)f_{,x,y}\right)}{A^2B^2\left(C_1+f\right)^6}$$

$$R^{*13}=\frac{2\left(2f_{,z}f_{,x}-\left(C_1+f\right)f_{,x,z}\right)}{A^2D^2\left(C_1+f\right)^6};\quad R^{*23}=\frac{2\left(2f_{,y}f_{,z}-\left(C_1+f\right)f_{,z,y}\right)}{B^2D^2\left(C_1+f\right)^6}. \tag{333}$$

Previously we had discussed corresponding solutions in spherical [45] and cylindrical [46] geometries and it is clear that we have obtained photons or photonic matter forms. What is more, with condition (294), we have automatically achieved additive character for all our solutions (321) to (325), allowing the application of simple integral transform methods in order to construct almost arbitrary photonic matter forms.

Without applying the technique of superposition of our solutions (321) to (325), however, the subsequent photonic solution would be a plane wave structure with no changes along the *x*-axis. It was shown in [47] how such a structure can be localized also in lateral directions via standard integral transform methods and that the subsequent solutions do fulfill the Maxwell equations. The connection with the energy momentum tensor for the electromagnetic field is therefore achieved via Eq. (327). The electrostatic and magnetic vector fields **E** and **B** with components E_i and B_i form the energy momentum tensor in contravariant form as follows:

$$\left(T^{\alpha\beta}\right)=\begin{pmatrix}\frac{1}{2}\cdot(\mathbf{E}\cdot\mathbf{E}+\mathbf{B}\cdot\mathbf{B}) & (\mathbf{E}\times\mathbf{B})^T\\ \mathbf{E}\times\mathbf{B} & \frac{1}{2}\cdot(\mathbf{E}\cdot\mathbf{E}+\mathbf{B}\cdot\mathbf{B})\cdot\delta_{ik}-E_iE_k-B_iB_k\end{pmatrix}$$

$$\delta_{ik}=\begin{pmatrix}1&0&0\\0&1&0\\0&0&1\end{pmatrix}. \tag{334}$$

Please note that in SI units, this tensor reads:

$$\left(T^{\alpha\beta}\right)=\begin{pmatrix}\dfrac{\varepsilon_0\cdot\mathbf{E}\cdot\mathbf{E}+\dfrac{\mathbf{B}\cdot\mathbf{B}}{\mu_0}}{2} & c\cdot\varepsilon_0\cdot(\mathbf{E}\times\mathbf{B})^T \\ c\cdot\varepsilon_0\cdot\mathbf{E}\times\mathbf{B} & \dfrac{\varepsilon_0\cdot\mathbf{E}\cdot\mathbf{E}+\dfrac{\mathbf{B}\cdot\mathbf{B}}{\mu_0}}{2}\cdot\delta_{ik}-\varepsilon_0\cdot E_iE_k-\dfrac{B_iB_k}{\mu_0}\end{pmatrix}. \tag{335}$$

The constants ε_0, μ_0 are the electric and the magnetic field constant, respectively.

For brevity, we shall leave it to the mathematically skilled and interested reader to work out the detailed connection of the two classical fields with our quantum metric distortion. For illustration, however, we just consider the energy density of the electromagnetic field given via the component T^{00} and, thus, obtain:

$$R^{*00}=\frac{(A\cdot B\cdot c)^2\left(f_{,z}\right)^2+D^2\left(A^2c^2\left(f_{,y}\right)^2+B^2\begin{pmatrix}c^2\left(f_{,x}\right)^2+3A^2\left(f_{,t}\right)^2\\-2A^2\left(C_1+f\right)f_{,t,t}\end{pmatrix}\right)}{(A\cdot B\cdot D)^2c^4\left(C_1+f\right)^6}$$

$$=-\kappa\cdot T^{00}=-\kappa\cdot\frac{\varepsilon_0\cdot\mathbf{E}\cdot\mathbf{E}+\dfrac{\mathbf{B}\cdot\mathbf{B}}{\mu_0}}{2}=-\frac{\kappa}{2}\cdot\begin{pmatrix}\varepsilon_0\cdot\left(E_1^2+E_2^2+E_3^2\right)\\+\dfrac{B_1^2+B_2^2+B_3^2}{\mu_0}\end{pmatrix}. \tag{336}$$

In fact, we find quadratic terms on both sides of the equation, and it appears kind of attractive to seek the connections between the classical E- and B-fields and the quantum metric distortion in the following way:

$$\frac{\left(f_{,x}\right)^2+C_{tx}\left(3\left(f_{,t}\right)^2-2\left(C_1+f\right)f_{,t,t}\right)}{\left(C_1+f\right)^6}=-\frac{\kappa_x}{2}\cdot\left(\varepsilon_0\cdot E_1^2+\frac{B_1^2}{\mu_0}\right)$$

$$\frac{\left(f_{,y}\right)^2+C_{ty}\left(3\left(f_{,t}\right)^2-2\left(C_1+f\right)f_{,t,t}\right)}{\left(C_1+f\right)^6}=-\frac{\kappa_y}{2}\cdot\left(\varepsilon_0\cdot E_2^2+\frac{B_2^2}{\mu_0}\right)$$

$$\frac{\left(f_{,z}\right)^2+C_{tz}\left(3\left(f_{,t}\right)^2-2\left(C_1+f\right)f_{,t,t}\right)}{\left(C_1+f\right)^6}=-\frac{\kappa_z}{2}\cdot\left(\varepsilon_0\cdot E_3^2+\frac{B_3^2}{\mu_0}\right). \tag{337}$$

Thus, obviously we have created photonic matter from a flat space vacuum solution to the Einstein field equations by the means of a simple metric transformation. The scalar Ricci curvature was thereby kept unchanged, which is to say $R = R^*$. The equations necessary to solve this were the classical quantum Klein–Gordon equations, which we directly obtain from the $R = R^*$-condition.

5.17 Appendix Q: About the Dimensional Size of Systems

In classical systems science, there is no way to derive the necessary dimension of a system in a truly fundamental and neutral (mathematically based) manner. Thus, systems are often "defined" as it pleases the creator of the simulation or as there are restrictions in ability and calculation power for the "digital twin" of the natural system one intends to model. As this holds for any system, this is also true—of course—for the unconscious or conscious mind and thus, of great interest here.

We start with the conjecture that not just the system's inner properties and corresponding governing equations but also the system's size (number of degrees of freedom or dimensions) can be derived from a suitable minimum principle. Our starting point shall this time be the Einstein–Hilbert action with a generalized Lagrange density function $\Phi_R[R]$ and a yet undefined variation, which we write as follows:

$$\delta_? W = 0 = \delta_? \int_V d^n x \left(\sqrt{-g} \cdot \Phi_R[R] \right). \tag{338}$$

Please note that we could also write this for a scaled metric tensor as elaborated in the previous appendices in order to work out the connection to Quantum Theory, respectively, in order to make it show itself directly via a set of wrapping and wave functions $F_i[f_i]$ and f_i within the usual variational calculus.

In order to adjust undefined parameters and finalize the character of the variational task (338), we intend to consider a fundamental problem and here determine the size of a Black Hole in a completely new way. Classically, the size of a Black Hole is given by the Schwarzschild radius, which itself is determined by the mass m of the Black Hole via: $r_s = \frac{2 \cdot m \cdot G}{c^2}$ (G is Newton's constant and c is the speed of light in vacuum). This Schwarzschild radius, however, was never derived from a first principle but was adjusted as a parameter to the Schwarzschild metric [48] in order to give the correct limit to the Newton gravitational law.

Here now we want to derive the Schwarzschild radius via a suitable version of Eq. (338). In order to do so, we first need to repeat Bekenstein's thought experiment of Black Holes.

5.17.1 The Bekenstein Bit Problem

One of the most famous and equally puzzling problems in General Theory of Relativity is the Bekenstein bit problem, where it was found that Black Holes can store information, but so far it has been seen as a mystery how these objects actually do this. In [4, 5], we have shown that bit-like information has been stored as dimensions and that each bit becomes one dimension. For convenience, we are repeating parts of the evaluation here.

In the early seventies, J. Bekenstein [49, 50] investigated the connection between Black Hole surface area and information. Thereby he simply considered the surfaces change of a Black Hole which would be hit by a photon just of the same size as the Black Hole. His idea was that with such a geometric constellation the outcome of the experiment would just consist of the information whether the photon fell into the Black Hole or whether it did not. Thus, it would be a 1-bit information. His calculations led him to the funny proportionality of area and information. He found that the number of bits, coded by a certain Black Hole, is proportional to the surface area of this very Black Hole if measured in Planck area ℓ_P^2. In fact, the dependency how one bit of information changes the area of the Black Hole (ΔA) reads:

$$\Delta A = 32 \cdot \pi^2 \cdot \ell_P^2 + 64 \cdot \pi^3 \cdot \frac{\ell_P^4}{r_s^2}. \tag{339}$$

Ignoring the extremely small second term, one could just assume our Black Hole to be constructed of many such bit surface pieces. Thus, we could write:

$$q \cdot \Delta A = q \cdot 32 \cdot \pi^2 \cdot \ell_P^2 = 4 \cdot \pi \cdot r_s^2 \quad \Rightarrow \quad r_s^2 = q \cdot 8 \cdot \pi \cdot \ell_P^2, \tag{340}$$

where r_s gives the radius of the Black Hole. We see that our Black Hole radius is proportional to the square root of the bits q thrown into it.

Now we want to compare the dependency $r_s[q]$ with the radii $r_{\max}[N]$ resulting in maximum volume of n-spheres for a certain number of space-time dimensions $N = n + 1$.

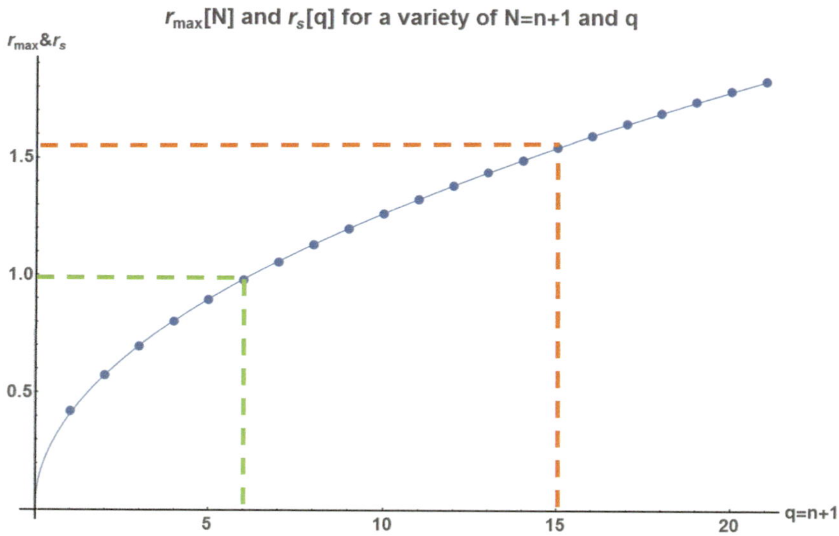

Figure Q1 Radius r_{max} for which at a certain number of dimensions the n-sphere has maximum volume in dependency on $N = n + 1$ compared with the increase of the Schwarzschild radius r_s of a Black Hole in dependence on the number of bits q thrown into it. We find that $q = N = n + 1$. As examples, we pick the situation with a radius slightly bigger than 1.5 (whatever unit). We obtain maximum volume for a sphere in 15 dimensions (orange dotted line). Picking a radius slightly below 1, however, gives us a 6-dimensional sphere which can have maximum volume at such a size (green dotted line).

We find a perfect fit (see Fig. Q1) to the $r_{max}[N]$-dependency for $q = N$ with the following function:

$$r_s = U \cdot \left(0.014948 + 0.3951244 \cdot \sqrt{q}\right); \quad U^2 = 8 \cdot \pi \cdot \ell_P^2, \tag{341}$$

where U denotes a unit-factor which was set $U = 1$ in Fig. Q1.

Our finding not only connects the intrinsic dimension of a Black Hole with its mass, respectively its surface, but also, at least partially, gives an explanation to the hitherto unsolved problem of "what are the micro states of a Black Hole giving it temperature and allowing it to store information." According to the evaluation in this section, these microstates are just various states of dimensions realized within the Black Hole in dependence on the number of bits it contains (and thus, its mass). The bigger the number of bits, the higher the intrinsic dimensions the Black Hole has. In fact, the connection even is a direct one and only seems to deviate from the simple direct proportionality for very low numbers of masses[4], respectively Schwarzschild radii r_s, respectively numbers of bits q the Black Hole has swallowed.

This finding also gives us a direct connection between a principal mathematical law (the maximum volume as a function of the dimension for a given radius of an n-sphere) to the number of bits a Black Hole contains. Thereby, the latter is equivalent to the mass or Schwarzschild radius of this very Black Hole and the number and character of microstates the Black Hole actually uses to internally code the bits.

It has to be pointed out that the expression "intrinsic dimension" truly stands for the part of space for $r<r_s$, which is to say, the space behind the event horizon. As for the outside, the solution of a Schwarzschild object in n+1-dimensional space-times is given via:

$$g_{\alpha\beta}^{N}=\begin{pmatrix} -c^2\cdot f[r] & 0 & 0 & 0 & \cdots & 0 \\ 0 & \frac{1}{f[r]} & 0 & 0 & \cdots & 0 \\ 0 & 0 & r^2 & 0 & \cdots & 0 \\ 0 & 0 & 0 & r^2\cdot\sin^2\varphi_1 & \cdots & 0 \\ \cdots & \cdots & \cdots & \cdots & \ddots & 0 \\ 0 & 0 & 0 & 0 & 0 & g_{aa} \end{pmatrix}$$

$$g_{44}=r^2\cdot\sin^2\varphi_1\cdot\sin^2\varphi_2;\quad g_{aa}=r^2\cdot\prod_{j=1}^{a-2}\sin^2\varphi_j;\quad a=N-1=n;$$

$$f[r]=1-\frac{r_s^{N-3}}{r^{N-3}}=1-\frac{r_s^{n-2}}{r^{n-2}}. \tag{342}$$

As we find that the Newton laws of gravity, however, require $N = 4$, it has to be assumed that in a region near the event horizon the dimension of the Black Hole decreases to the known 4 dimensions so that Newton's laws of gravity are properly mirrored to us (as outside observers). The corresponding derivation was given in chapter 11 of the book [4].

5.17.2 A System-Immanent Scale

Note: the correct solution for the evaluation of the Schwarzschild radius r_s as a function of the bits thrown into a Black Hole object (if using the results from [53]) would be:

---

[4]Besides, this deviation is also suggested by the Bekenstein finding summed up in Eq. (339), where we could assume the second term to become of importance at lower numbers of r_s.

$$r_s = 2 \cdot \ell_P \cdot \sqrt{\;\cdot\left(q+\sqrt{q\cdot(1+q)}\right)}! \tag{343}$$

We find a perfect fit to the n-spheres with maximized volume to a given radius (dots in Fig. 3.43) with a Planck length of $\ell_P = 0.07881256452824544$ (see Fig. Q2, which is almost perfectly equal to the fit in Fig. Q1).

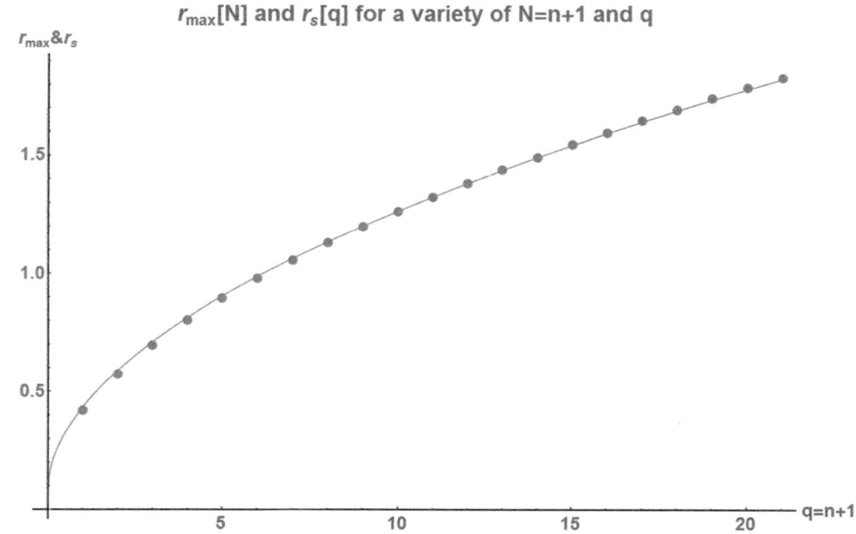

Figure Q2 (Please note that we have applied a slightly different fit than it was applied in Fig. Q1): Radius r_{max} for which at a certain number of dimensions the n-sphere has maximum volume in dependence on $N = n + 1$ compared with the increase of the Schwarzschild radius r_s of a Black Hole in dependence on the number of bits q thrown into it by using Eq. (343). We find that $q = N = n + 1$.

But what would be the unit for this Planck length?
Well, it has been already shown in [4] that by using the results from [53] and the "volume integral" (338) with:

$$\delta_n W = 0 = \delta_n \int_V d^n x \sqrt{-g}, \tag{344}$$

for n-spheres (in [53] with $T[n] = 1$):

$$V_N = V_{n+1} = T[n] \cdot \frac{\pi^{\frac{(n)}{2}}}{\Gamma\left[\frac{(n+2)}{2}\right]} \cdot r^n, \tag{345}$$

we have evaluated the very dimension n to which at a given radius r_S the volume of an n-sphere has a maximum. The r_S are put into the calculation as plain "natural" numbers, meaning an $r_S = 1$ is just a 1 and that is it. We might name this unit a "mathematical meter" or just "mams" (plural for "mathematical meters"). Transformation to our usual units, like meters, requires the introduction of a factor $T[n] = U^n$.

With $U = \frac{\left[\ell_P\right]_{\text{in\_meters}}}{\left[\ell_P\right]_{\text{in\_mams}}} = \frac{1.616255(18)\times 10^{-35}\text{ meters}}{0.07881256452824544}$, for instance,

we can easily change to our meters. Nevertheless, it appears somehow astonishing that there seems to exist a fundamental "natural" unit, being completely based on a mathematical–geometrical–extremal principle (the maximum volume of n-spheres as functions of their dimensions for certain radii). It is also interesting that these dimensions are so nicely correlated to the number of bits a Black Hole has swallowed. In fact, using the unit of mams, the number of spatial (n-sphere) dimensions is perfectly equal to the number of swallowed bits.

Thus, in the case that Black Holes would in fact store their content as dimensions and the Einstein–Hilbert action being extended with respect to the number of dimensions in addition to the metric, we immediately also get an absolute scale for our Black Hole system in which the number 1 is "made out" of 12.6883 Planck length and where a 3-sphere has a radius of 0.6969979737167096 mams.

5.17.2.1 Back to the optimum size question for any system

When observing the integral in Eq. (338), we see that—in principle—we seek for a maximum volume for a given dimension or, taking the radius of a Schwarzschild object, look for the corresponding dimension making the volume integral an extremum. As the determinant g of the Schwarzschild metric is just equal to the one of an n-sphere with the additional time-dimension to be integrated, we can easily use the volume integral result of n-spheres, which reads:

$$V_N = V_{n+1} = \frac{\pi^{\frac{(n)}{2}}}{\Gamma\left[\frac{(n+2)}{2}\right]}\cdot r^n. \tag{346}$$

Please note that due to the time-coordinate t, we have V_{n+1} instead of V_n. Thereby the integration via t in Eq. (346) is assumed to be performed such that it would give 1. In general, we might take care of this part of the integration via a proportional constant T, which we could even consider to be n-dependent $T[n]$ and thus, Eq. (345) changes to:

$$V_N = V_{n+1} = \overbrace{T[n]}^{=U^n} \cdot \frac{\pi^{\frac{(n)}{2}}}{\Gamma\left[\frac{(n+2)}{2}\right]} \cdot r^n. \tag{347}$$

Now we evaluate the various dimensions for which, for a given radius r of the n-sphere, we would obtain extrema. The results were already given in Figs. Q1 and Q2. There we have illustrated the resulting $r_{\max}$ as functions of the dimensions $N = n + 1$ (note: n = n-sphere dimension, N = t+n-sphere dimension).

Now we just compare our findings with the original question of extracting a minimum principle for the dimensional size of a given system with our generalized starting point for the variational task (338) and conclude that:

$$\delta_? W = 0 = \delta_? \int_V d^n x \left(\sqrt{-g} \cdot \Phi_R[R]\right)$$

$$\Rightarrow \qquad \delta_n W = 0 = \delta_n \int_V d^n x \left(\sqrt{-g} \cdot \left[\Phi_R[R] = 1\right]\right) = \delta_n \int_V d^n x \sqrt{-g}. \tag{348}$$

Thus, the determination of the optimum size of a system we intend to consider, investigate or analyze can just be found by a dimensional variation of the volume integral of that very system. In the case of spherical symmetries, this then leads to equations of the form:

$$\Rightarrow \qquad \delta_n W = 0 = \delta_n \int_V d^n x \left(\sqrt{-g} \cdot \left[\Phi_R[R] = 1\right]\right) = \delta_n T[n] \cdot \frac{\pi^{\frac{(n)}{2}}}{\Gamma\left[\frac{(n+2)}{2}\right]} \cdot r^n. \tag{349}$$

Along the way, we also can extract suitable fundamental scales for our system.

5.17.2.2 The first bit requires the highest mass = The first thought is the most difficult

From Eq. (343), we can now extract the minimum Schwarzschild radius for the storage of one bit, which would be equal to 5.508 Planck length and corresponds to 11.16 times the Planck mass. This is a huge amount of mass, and thus also energy, one needs to safely store just one single bit. Luckily, the situation improves the more bits one intends to store, as for instance the one-millionth bit only requires about 5.5 × 10^{-3} Planck masses. Please note that, of course, one might also store bits within spin arrangements of electrons. Then a 1-bit information would be connected with a single electron, whose mass and spin energy are many magnitudes below the Planck mass. This spin storage, however, cannot be seen as a storage of a classical binary bit, because, in fact, it resembles a quantum bit. Apparently, the safe storage of a pure and truly binary information requires an—almost—macroscopic massive structure. Here the Black Hole probably provides the smallest possible mass ensemble there can be to arrange such a storage for a certain bit. The limit is given at about 11 times the Planck mass. Only from this mass onward Black Holes can store binary information ... at least until the Hawking radiation leads to the destruction of our Black Hole 1-bit storage system.

And what would then happen to the stored information? Well, this brings us to the question: how safe is information within our universe [54]?

5.17.2.3 By-products: A few fundamental questions

5.17.2.3.1 *About the relativity of system-scales*

We saw that—similar to the Bekenstein or Bekenstein–Hawking problem (see Fig. 2.10 in Chapter 2 and reference [30])—we add bits via dimensions to our metric structure (here a Black Hole). Assuming our metric system to be a Black Hole (which we only use here to have the simplest possible math), we can even obtain a ratio of the Black Hole's radius r_s to the smallest structure this Black Hole can resolve. Taking the result for ℓ_P—the Planck length—from the Bekenstein thought experiment, we find the ratio between the Schwarzschild radius r_s of a Black Hole and ℓ_P:

$$\frac{r_s}{\ell_P} = 2 \cdot \sqrt{\pi \cdot \left(q + \sqrt{q \cdot (1+q)}\right)}. \tag{350}$$

In our universe, the Planck length ℓ_P is considered to be the smallest length possible to resolve. What if the ratio (350) that we found for Black Holes, actually is a more fundamental law? At any rate, it appears logical to assume that more bits could be coded or stored by an object of bigger size and smaller internal structure, thereby leaving more options to describe something with these structures. Thus, Eq. (350) makes intuitive sense, but could it also be just the other way round? Could it be that for an object of a given size, the number of bits (being equivalent to its dimensions as we see in Figs. Q1 and Q2) it contains determines the smallest scale—the Planck length—of the object, too? And, if referring to "Mathematical Psychology," the system presents a thinking entity, does this also mean that thoughts have a physical scale?

From inside and taking the Planck length as a measure, the increase in information to this very object, subject or entity would look like an increase of its size. Now assuming the inside of the Black Hole to be a general system, the inhabitants of this system may see this system as their very own universe and would register the increase in information as a growth of their "universe," measured in the Planck length of that very system-universe. When learning, we seem to feel the increase of mind. Maybe this perception is just what is actually really going on.

5.17.2.3.2 *Does more information always mean more mass?*

Quantum computer scientists have already pointed out that with our current way of storing information, we will one day reach a limit with respect to the number of atoms we can apply for the storing process and the energy needed to keep the information stable (maintained). Citing from the abstract of [56], we have the following situation:

> *"Currently, we produce ~10^{21} digital bits of information annually on Earth. Assuming a 20% annual growth rate, we estimate that after ~350 years from now, the number of bits produced will exceed the number of all atoms on Earth, ~10^{50}. After ~300 years, the power required to sustain this digital production will exceed 18.5 × 10^{15} W, i.e., the total planetary power consumption today, and after ~500 years from now, the digital content will account for more than half Earth's mass, according to the mass-energy–information equivalence principle. Besides the existing global challenges such as climate, environment, population, food, health, energy, and security, our estimates point to another singular event for our planet, called information catastrophe."*

It has to be pointed out that when looking for possible inner Schwarzschild solutions [41], we also found that there are solutions where the mass decreases with the increase in the object size. It may well be that such strange states are not only realized in Black Holes but could perhaps also help to overcome our future information storage problem.

5.17.2.3.3 *Generalization to spheres?*

In the previous sub-sections, we saw that when taking the equation for the Laplace length ℓ_P from the Bekenstein thought experiment [49, 50], we find the following ratio between r_s (Schwarzschild radius of a Black Hole) and ℓ_P (c.f. Eq. (350)):

$$\frac{r_s}{\ell_P} = 2 \cdot \sqrt{\pi \cdot \left(q + \sqrt{q \cdot (1+q)}\right)}. \tag{351}$$

Most interestingly, we also found that the solution to the extremal volume problem for a fixed radius r_f for n-spheres results in the same dependency when variating with respect to the number of dimensions of those n-spheres. We obtain (see dots in Figs. Q1 and Q2) excellent fits, when applying an approach like:

$$r_f = L \cdot \sqrt{\pi \cdot \left(n + \sqrt{n \cdot (1+n)}\right)}. \tag{352}$$

Thereby we have the characteristic (system-dependent) length scale L.

This automatically gives us a connection between the size-parameter r_f of any system of spherical symmetry and its theoretical capability to store information. A perfect mathematical n-sphere thereby follows the rule (352) almost perfectly, while other systems may do so only from certain critical sizes onwards, but, nevertheless, we think we can draw the conclusion that the information storage capacity of given systems—if showing enough spherical symmetry—can be extracted from Eq. (352). Then the structural size-parameter r_f determines the number of storable bits n in dependence on the system-immanent length parameter L.

From this, one even may deduce that r_f and L could be substituted by other system characteristics. While in Eq. (352) their dimension is length, we should not exclude mass, time, charges, energies and so on.

5.17.2.4 Other geometries

The simplest generalization of Eq. (348) can be given for an ensemble of N tori of dimensions n_j for the sub-n_j-spheres of the individual torus. The volume integral would then yield:

$$W = \int_V d^n x \left(\sqrt{-g}\right) = \prod_{j=1}^{n} V_j\left(n_j, r_j\right) = \prod_{j=1}^{n} T\left[n_j\right] \cdot \frac{\pi^{\frac{n_j}{2}}}{\Gamma\left[\frac{\left(n_j+2\right)}{2}\right]} \cdot \left(r_j\right)^{n_j}. \tag{353}$$

This could be further generalized for a sum of tori and leaves us with a great variety of pure volume (radii) and dimension variations.

As tori can be seen as combined n_{ij}-spheres with n_{ij} giving the dimension of the sub-spheres, constructing each torus (348) can be generalized as follows when assuming N tori with dimensions n_i and sub-spheres of dimensions n_{ij}:

$$W=\sum_{i=1}^{N}W_i=\sum_{i=1}^{N}\int_{V_i}d^{n_i}x\left(\sqrt{-g}\right)=\sum_{i=1}^{N}\prod_{j=1}^{n_{ij}}T\left[n_{ij}\right]\cdot\frac{\pi^{\frac{(n_{ij})}{2}}}{\Gamma\left[\frac{(n_{ij}+2)}{2}\right]}\cdot\left(r_{ij}\right)^{n_{ij}}. \tag{354}$$

Assuming that also complex symmetries of systems could be constructed out of sums of tori, we realize that the variational options are manifold:

$$\delta W=0=\delta\left(n_{ij},r_{ij}\right)\sum_{i=1}^{N}\prod_{j=1}^{n_{ij}}T\left[n_{ij}\right]\cdot\frac{\pi^{\frac{(n_{ij})}{2}}}{\Gamma\left[\frac{(n_{ij}+2)}{2}\right]}\cdot\left(r_{ij}\right)^{n_{ij}} \tag{355}$$

and leave us with a great variety of options for an optimum-sized system in the case of complex symmetries.

5.17.2.5 Consequences from the Bekenstein thought experiment regarding the solutions to the quantum Einstein field equations

In [4], we have shown that the classical n-dimensional Schwarzschild solution could be applied to construct internally structured n-dimensionally Black Holes, while outside we still have the usual 4-dimensional solution from [48] with the classical Schwarzschild metric. This, however, would not explain how the Black Hole can code any information.

With the help of the new metric solutions evaluated in [41], namely, to just give an example, in the 3-dimensional case with coordinates t, r and an angle:

$$g_{\alpha\beta}=C_1\cdot f[t]^4\cdot\begin{pmatrix}-c^2 & 0 & 0\\ 0 & t^2 & 0\\ 0 & 0 & t^2\cdot\sin(r)^2\end{pmatrix}$$

$$f[t]=\sqrt{t^{\pm i\cdot c-1}}\cdot C_f, \tag{356}$$

(please note that r has become an angle, while t took over the position of the radius),

$$g_{\alpha\beta} = C_1 \cdot f[t]^4 \cdot \begin{pmatrix} -c^2 & 0 & 0 \\ 0 & \rho^2 & 0 \\ 0 & 0 & \rho^2 \cdot \sin(r)^2 \end{pmatrix}$$

$$f[t] = e^{\pm\frac{\mathrm{i}\cdot c}{2\cdot\rho}\cdot t} \cdot C_f, \tag{357}$$

(this represents a shell-like object), we want to solve this problem as well. A generalization of this type of solution has been given in appendix N. Thereby, we found that the Schwarzschild singularity could be avoided (Fig. Q3).

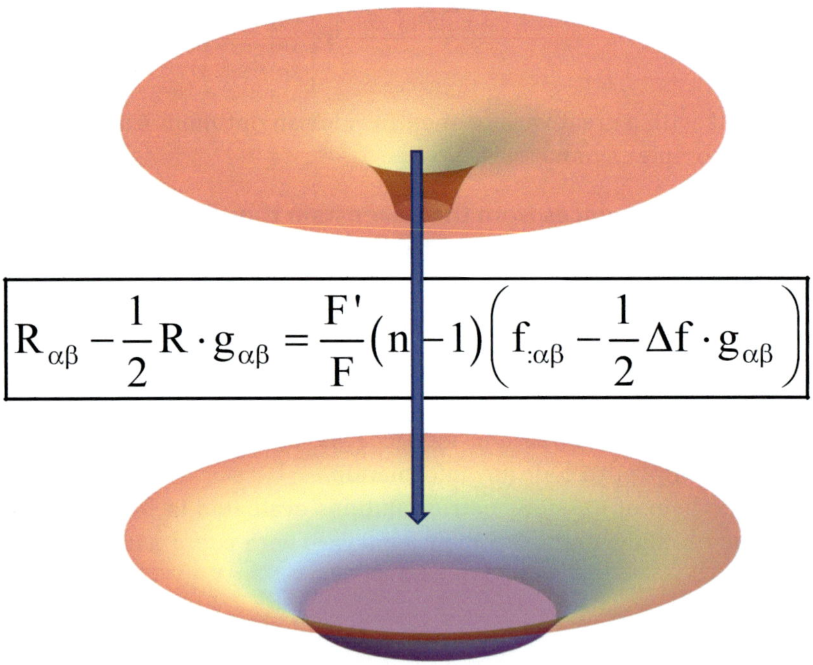

Figure Q3 "From the classical Schwarzschild solution to a Quantum Black Hole" [41].

First, however, we should note that also the n-dimensional Schwarzschild solutions from section 3.8 in [4] (c.f. solution (342) in the current book) would provide plenty of options to code information, because there are enough degrees of freedom regarding the thicknesses of the individual x-dimensional layers of the onion-like Schwarzschild object, which was proposed there (see also [41]). Similar assumptions could be made for the Robertson–Walker approach introduced in [41], but apart from again mentioning the onion-layer structured ogre-mind from the movie *Schrek*, we will not further consider these possibilities in here.

In the case of photonic inner solutions as also suggested in [41], one might assume some kind of standing waves inside the Black Hole, but as we currently don't have the math to realize such structures, we postpone the investigation of this possibility.

Thus, here we concentrate on solutions (356) and (357) as potential inner solutions to a Black Hole. As we see that the parameter ρ clearly is a length, we want to derive its properties. For the general case, this has been already done in appendix N. Nevertheless, we repeat it here for the settings (356) and (357). From basic Quantum Theory, we know that a particle at rest has the time dependency:

$$f[t]=e^{\pm i\cdot\frac{m\cdot c^2}{\hbar}\cdot t}\cdot C_f, \tag{358}$$

with m giving the rest mass of the particle and $\hbar$ denoting the reduced Planck constant. Comparing with the $f[t]$-function from the metric solution (357), we find:

$$\frac{m\cdot c}{\hbar}=\frac{1}{2\cdot\rho}. \tag{359}$$

Inserting the Schwarzschild radius $r_s=\frac{2\cdot m\cdot G}{c^2}$ (G is Newton's constant and c is the speed of light in vacuum), thereby substituting the rest mass m, leaves us with:

$$\frac{r_s\cdot c^3}{2\cdot\hbar\cdot G}=\frac{1}{2\cdot\rho}\quad\Rightarrow\quad\rho=\frac{1}{r_s}\cdot\left(\frac{c^3}{\hbar\cdot G}\right)^{-1}=\frac{\ell_P^2}{r_s}. \tag{360}$$

Here ℓ_P denotes the Planck length. By inserting Eq. (350) into Eq. (360), we obtain:

$$\rho=\frac{\ell_P^2}{r_s}=\frac{\ell_P^2}{2\cdot\ell_P\cdot\sqrt{\pi\cdot\left(q+\sqrt{q\cdot(1+q)}\right)}}=\frac{\ell_P}{2\cdot\sqrt{\pi\cdot\left(q+\sqrt{q\cdot(1+q)}\right)}}. \tag{361}$$

Thus, while for a Black Hole the number of bits thrown into it leads to an almost perfectly square-root-like increase of the Schwarzschild radius in accordance with Eq. (350), the ρ-parameter of the (357)-objects decreases with the number of bits. The (357)-objects would have the same ρ-parameter, which we may see as a size, as a Black Hole only for Schwarzschild radii r_s equal to the Planck length. In other words, for growing Black Holes with radii bigger than the Planck length the corresponding equally heavy (357)-objects would be significantly smaller than the Black Holes.

So, we ask: Could the (357)-objects be used as building blocks for the Black Holes residing inside it, which is to say behind the event horizon?

Assuming that the Black Hole's surface is made out of metric spherical objects of the type (357) and further assuming that each of these objects on the surface of the Black Hole, which is to say at $r = r_s$ (which also happens to be the event horizon), requires its own surface space of something like $C_\rho{}^*\rho^2$, we can directly evaluate the number of such (357)-objects, we from now on name ρ-spheres, that are residing inside the event horizon with increasing numbers of bits thrown into the Black Hole. Assuming that the mass is always additive, the total mass m of the Black Hole must then be distributed among the N ρ-spheres, which changes Eq. (360) to:

$$\frac{r_s \cdot c^3}{2 \cdot \hbar \cdot G} = \frac{1}{2 \cdot N \cdot \rho} \quad \Rightarrow \quad N \cdot \rho = \frac{1}{r_s} \cdot \left(\frac{c^3}{\hbar \cdot G} \right)^{-1} = \frac{\ell_P^2}{r_s} \quad \Rightarrow \quad \rho = \frac{\ell_P^2}{N \cdot r_s}. \tag{362}$$

Also having to satisfy the following equation for the N ρ-spheres sitting on the surface, we have to solve the following equation:

$$N \cdot C_\rho \cdot \rho^2 = 4 \cdot \pi \cdot r_s^2$$

$$\Rightarrow \qquad C_\rho \cdot \frac{\ell_P^2}{4 \cdot \pi \cdot N \cdot \left(q + \sqrt{q \cdot (1+q)} \right)} = (4 \cdot \pi)^2 \cdot \ell_P^2 \cdot \left(q + \sqrt{q \cdot (1+q)} \right)$$

$$\Rightarrow \qquad N = \frac{1}{64 \cdot \pi^3 \cdot \left(q + \sqrt{q(1+q)} \right)^2}. \tag{363}$$

We realize that such a structure could not be used to store any information, because the number of ρ-spheres should have to increase with the number of bits and not decrease as it does. Things are improving the moment we allow a combination of ρ-spheres and (356)-objects (the latter we shall call t-spheres) to make up our inner Black Hole. We propose the following (simplest of the many possibilities) structure:

(A) In the center of the Black Hole sits a ρ-spheres of "radius-parameter" ρ given in Eq. (360) and thus, $\rho = \frac{\ell_P^2}{r_s}$, which is to say, the bigger the Schwarzschild radius r_s of the Black Hole, the smaller its core. In fact, for infinite masses the core would become a singularity.

(B) This single ρ-sphere core is surrounded by t-spheres (Eq. (356)) and the number of those t-spheres, which a Black Hole can bind, is proportional to the number of bits the Black Hole has swallowed.

(C) Taking the Bekenstein condition, this demands an average size for the t-spheres, being bound by the Black Hole or the Black Hole's surface, to be such that its projected surface would be equal to ℓ_P^2. In other words, we could assume the average radius of the t-spheres (the ones bound to the Black Hole) to be equal to $\ell_P/\sqrt{\pi}$.

With such a structure, it is very well possible that in fact Black Holes have no singularity and follow our scheme of inner-outer-solution, but one cannot detect any difference from the classical Schwarzschild solution from the outside, because the inner parts are always hidden behind the event horizon.

But does this help us to solve the Bekenstein information problem?

Yes, it does.

We can imagine many t-sphere objects (of number $N = q$) sitting on the surface of the Black Hole. As the generalized solution to Eq. (356) would read:

$$g_{\alpha\beta} = C_1 \cdot f[t]^4 \cdot \begin{pmatrix} -c^2 & 0 & 0 \\ 0 & A^2 \cdot t^2 & 0 \\ 0 & 0 & B^2 \cdot t^2 \cdot \sin(r)^2 \end{pmatrix}$$

$$f[t] = \sqrt{t^{\pm\frac{i \cdot c}{A} - 1}} \cdot C_f, \tag{364}$$

we see that each t-sphere could store information not only via a certain sign within the exponent but also via the free parameter B.

5.17.2.6 The size of the electron?

Applying Eq. (362) and assuming a ρ-sphere-structure for the electron gives us:

$$\frac{r_s \cdot c^3}{2 \cdot \hbar \cdot G} = \frac{1}{2 \cdot N \cdot \rho} \Rightarrow N \cdot \rho = \frac{1}{r_s} \cdot \left(\frac{c^3}{\hbar \cdot G} \right)^{-1} = \frac{\ell_P^2}{r_s}$$

$$\Rightarrow \qquad \rho = \frac{\ell_P^2}{N \cdot r_s} = \frac{1.93 \times 10^{-13}}{N} \text{meter.} \qquad (365)$$

Setting $N = 1$, we would end up with a ρ-sphere of $\rho = 1.93 \times 10^{-13}$ meter for the "pure" or "naked" electron.

5.18 Appendix R: Compact Derivation of the Most Fundamental Basics to a "Mathematical Psychology"

As in this appendix we will not only consider the various ways to mathematically produce individualism but—along the way—repeat all fundamental essentials and add on even more options like an "unconventional physics realm for the mind, the spirit, the soul, consciousness and perhaps even a place where the dead could live on," where everything may be or rather has been before time started (the so-called chaos perhaps)[5], we are going to present this appendix with its own list of local references. We will also partially repeat certain essentials that we have already covered in the book previously. In some cases, this is just for convenience for the reader in order to avoid referencing equations too far away but mostly this is also motivated and necessary because crucial adjustments are needed.

5.18.1 An Equation for Individuality

In this book, we have already rigorously derived the fact that a system, any system, including a thinking and even consciously awake (not woke, which is the opposite) entity, can be fully characterized by a set of attributes, properties and degrees of freedom. These attributes, properties and degrees of freedom can be mathematically treated as dimensions, forming a space and when subjected to a Hamilton extremal principle, we obtain an Einstein–Hilbert action variational integral. Deriving and solving the subsequent equations, which are quantum gravity field equations, we obtain a full dynamic description of the system, any system, including a thinking and even consciously awake (not woke, which is the opposite) entity or individual.

---

[5]We apologize for the strange use of tenses, but—obviously—when speaking about the time before time, which is to say, the chaos, we meet some difficulties with the usual time concepts in our languages.

Assuming that individuals require at least some fermionic character, we consider the Dirac equation essential to our attempt of the construction of a holistic and consistent "mathematical theory of the mind." As the "guarantor of individuality" has to have some Pauli-like exclusion properties, potentially due to Fermi attributes, we need Dirac equations in generalized, multidimensional and metric forms.

Along the way, we have to answer quite a few questions unsolved so far:

(a) How can the Dirac equation rigorously be derived from the Einstein–Hilbert action variational integral?
(b) Where does the Dirac spinor come from?
(c) Are there alternatives to the quaternions?
(d) What happens in arbitrary numbers of dimensions, as we will need them for the description of entities with huge numbers of attributes?
(e) How can the approach be generalized to ensembles of systems?
(f) Are there Dirac field equations, compatible with the General Theory of Relativity and how can they be derived?
(g) What is anti-matter in a metrical sense?
(h) How does this all relate to "Mathematical Psychology"?
(i) ... and more ...

5.18.2 Various Ways to Derive the Dirac Equation

In this part of the appendix, we will show how the Dirac equation can be evaluated in a metric manner directly out of the Einstein–Hilbert action with an assumed scaled metric tensor. Thereby, we also find an explanation for the occurrence of the list of functions which Dirac has postulated in order to make his matrix equation consistent and which has typically been called a spinor.

Along the way, we also find an—one possible, as there seem to be some—explanation for the nature of anti-matter. It seems that a Dirac object has "two feet," one having its bearing on the metric tensor and leading to volume fluctuations of space-time (thereby betraying the object's presence) and the other perturbing the kernel of the Einstein–Hilbert action integral. Assuming that the matter object has a minus sign on the time-component of the metric, it would be a plus for the corresponding term in the kernel. For anti-matter, it is just the other way round.

We also found a possible explanation for the distinction of 4 dimensions. Apart from the well-known fact that gravity requires at least 4 dimensions, so far, we saw no special reason why—on bigger scales—there could not also have been higher-dimensional spaces or space-times. In this appendix, we work out that the simplest internal structure of a metric scale, allowing for a natural appearance of the Dirac-spinors, automatically makes the Ricci curvature to vanish in a scalar quantum gravity field equation in 4 dimensions. Thus, the appearance of Dirac objects in the first place and the fact that we have four distinct bigger-scale dimensions seem to be just the two sides of the same coin.

5.18.3 A Fundamental Starting Point

Our starting point shall be the Hamilton principle in the form of the Einstein–Hilbert action [1]:

$$\delta W = 0 = \delta \int_V d^n x \left(\sqrt{-g} \cdot \Phi_R[R] \right) = \delta \int_V d^n x \left(\sqrt{-g} \cdot \left(R - 2\Lambda + L_M \right) \right). \quad (366)$$

Here R gives the Ricci scalar and g denotes the determinant of the metric tensor $g_{\alpha\beta}$. The term L_M stands for the matter Lagrange density, which was postulated by both Hilbert and Einstein and Λ gives the cosmological constant. Assuming that we can extract a functional factor $F[f]$, with f being a function of all coordinates, from the metric tensor in the following way:

$$G_{\alpha\beta} = g_{\alpha\beta} \cdot F[f], \quad (367)$$

it was shown in [2] that we do not only obtain the Einstein field equations [3], but also the most important classical quantum equations (see also [4]).

Here, we only briefly repeat the essentials.

Setting Eq. (367) into Eq. (366), we obtain the following Einstein–Hilbert action [4]:

$$\delta W = 0 = \delta \int_V d^n x \left(\sqrt{-g \cdot F^n} \times \left(\left(R - \frac{1}{2F} \begin{pmatrix} 2F_{,\alpha\beta}(n-1) g^{\alpha\beta} + 2\Gamma^a_{\alpha\beta} F_{,a} g^{\alpha\beta} \\ -F_{,\alpha} g^{ab} g_{\beta b,a} g^{\alpha\beta} - F_{,\beta} g^{ab} g_{\alpha b,a} g^{\alpha\beta} \\ -n\Gamma^d_{\alpha\beta} F_{,d} g^{\alpha\beta} + \frac{n}{2} F_{,d} g^{cd} g_{ab,c} g^{ab} \end{pmatrix} - \frac{F_{,\alpha} \cdot F_{,\beta}}{4F^2} g^{\alpha\beta} \left((n-6)(n-1) \right) \right) \frac{1}{F} - 2\Lambda + L_M \right) \right). \quad (368)$$

Thereby, in order to keep things simple and still general enough, one can often restrict the metric selection to diagonal metrics with the following properties:

$$g_{ij} = \begin{pmatrix} g_{00} & \cdots & 0 \\ \vdots & \ddots & \vdots \\ 0 & \cdots & g_{n-1n-1} \end{pmatrix}; \quad g_{jj,j} = 0, \tag{369}$$

which gives an extended Hilbert integral:

$$\delta W = 0 = \delta \int_V d^n x \left(\sqrt{-g \cdot F^n} \times \left(\left(\begin{array}{c} R + \frac{F'}{F}(1-n)\Delta f \\ + \frac{f_{,\alpha} f_{,\beta} g^{\alpha\beta}(1-n)}{4F^2} \left(4FF'' + (F')^2 (n-6)\right) \end{array} \right) \frac{1}{F} \\ -2\Lambda + L_M \right) \right). \tag{370}$$

Here, we will not use this simplification. After performing the variation in Eq. (368), we obtain the following quantum gravity field equations:

$$0 = \left(\begin{array}{c} \boxed{R_{\alpha\beta} - \frac{g_{\alpha\beta}}{2} R} + \frac{g_{\alpha\beta}}{2} \left(\frac{1}{2F} \left(\begin{array}{c} 2F_{,ij}(n-1) g^{ij} + 2\Gamma_{ij}^{a} F_{,a} g^{ij} \\ -F_{,i} g^{ab} g_{jb,a} g^{ij} - F_{,j} g^{ab} g_{ib,a} g^{ij} \\ -n\Gamma_{ij}^{d} F_{,d} g^{ij} + \frac{n}{2} F_{,d} g^{cd} g_{ab,c} g^{ab} \end{array} \right) \\ + \frac{F_{,i} \cdot F_{,j}}{4F^2} g^{ij} \left((n-6)(n-1)\right) \right) \\ -\frac{1}{2F} \left(\begin{array}{c} F_{,\alpha\beta}(n-2) + F_{,ab} g_{\alpha\beta} g^{ab} \\ + F_{,a} g^{ab} \left(g_{\beta b,\alpha} - g_{\beta\alpha,b} \right) - F_{,\alpha} g^{ab} g_{\beta b,a} - F_{,\beta} g^{ab} g_{\alpha b,a} \\ + F_{,d} g^{cd} \frac{1}{2} n \left(\frac{2}{n} g_{\alpha c,\beta} - g_{\alpha c,\beta} - g_{\beta c,\alpha} + g_{\alpha\beta,c} + \frac{1}{n} g_{\alpha\beta} g_{ab,c} g^{ab} \right) \\ -\frac{1}{2F} \left(F_{,\alpha} \cdot F_{,\beta}(3n-6) + g_{\alpha\beta} F_{,c} F_{,d} g^{cd} (4-n) \right) \end{array} \right) \\ + \kappa T_{\alpha\beta} + \Lambda \cdot F \cdot g_{\alpha\beta} \end{array} \right). \tag{371}$$

Immediately, we recognize the classical Einstein field equations (box) inside our new ones. It was shown in [2, 4] that the scaling terms in Eq. (371) are giving the Klein–Gordon, Dirac and Schrödinger equation (regarding the latter see also [5]). Under the variational integral, the integrand (71) has to be multiplied with the variated metric tensor $\delta G^{\alpha\beta}=\delta\left(\frac{g^{\alpha\beta}}{F}\right)$ and therefore could be split up into two equations:

$$0=\left(\begin{array}{c} \boxed{R_{\alpha\beta}-\frac{g_{\alpha\beta}}{2}R}+\frac{g_{\alpha\beta}}{2}\left(\begin{array}{c} \frac{1}{2F}\left(\begin{array}{c} 2F_{,ij}(n-1)g^{ij}+2\Gamma^{a}_{ij}F_{,a}g^{ij} \\ -F_{,i}g^{ab}g_{jb,a}g^{ij}-F_{,j}g^{ab}g_{ib,a}g^{ij} \\ -n\Gamma^{d}_{ij}F_{,d}g^{ij}+\frac{n}{2}F_{,d}g^{cd}g_{ab,c}g^{ab} \end{array}\right) \\ +\frac{F_{,i}\cdot F_{,j}}{4F^{2}}g^{ij}\left((n-6)(n-1)\right) \end{array}\right) \\ -\frac{1}{2F}\left(\begin{array}{c} F_{,\alpha\beta}(n-2)+F_{,ab}g_{\alpha\beta}g^{ab} \\ +F_{,a}g^{ab}\left(g_{\beta b,\alpha}-g_{\beta\alpha,b}\right)-F_{,\alpha}g^{ab}g_{\beta b,a}-F_{,\beta}g^{ab}g_{\alpha b,a} \\ +F_{,d}g^{cd}\frac{1}{2}n\left(\frac{2}{n}g_{\alpha c,\beta}-g_{\alpha c,\beta}-g_{\beta c,\alpha}+g_{\alpha\beta,c}+\frac{1}{n}g_{\alpha\beta}g_{ab,c}g^{ab}\right) \\ -\frac{1}{2F}\left(F_{,\alpha}\cdot F_{,\beta}(3n-6)+g_{\alpha\beta}F_{,c}F_{,d}g^{cd}(4-n)\right) \end{array}\right) \\ +\kappa T_{\alpha\beta}+\Lambda\cdot F\cdot g_{\alpha\beta} \end{array}\right)\delta G^{\alpha\beta}$$

$$=\left(\begin{array}{c} \boxed{R_{\alpha\beta}-\frac{g_{\alpha\beta}}{2}R}+\frac{g_{\alpha\beta}}{2}\left(\begin{array}{c} \frac{1}{2F}\left(\begin{array}{c} 2F_{,ij}(n-1)g^{ij}+2\Gamma^{a}_{ij}F_{,a}g^{ij} \\ -F_{,i}g^{ab}g_{jb,a}g^{ij}-F_{,j}g^{ab}g_{ib,a}g^{ij} \\ -n\Gamma^{d}_{ij}F_{,d}g^{ij}+\frac{n}{2}F_{,d}g^{cd}g_{ab,c}g^{ab} \end{array}\right) \\ +\frac{F_{,i}\cdot F_{,j}}{4F^{2}}g^{ij}\left((n-6)(n-1)\right) \end{array}\right) \\ -\frac{1}{2F}\left(\begin{array}{c} F_{,\alpha\beta}(n-2)+F_{,ab}g_{\alpha\beta}g^{ab} \\ +F_{,a}g^{ab}\left(g_{\beta b,\alpha}-g_{\beta\alpha,b}\right)-F_{,\alpha}g^{ab}g_{\beta b,a}-F_{,\beta}g^{ab}g_{\alpha b,a} \\ +F_{,d}g^{cd}\frac{1}{2}n\left(\frac{2}{n}g_{\alpha c,\beta}-g_{\alpha c,\beta}-g_{\beta c,\alpha}+g_{\alpha\beta,c}+\frac{1}{n}g_{\alpha\beta}g_{ab,c}g^{ab}\right) \\ -\frac{1}{2F}\left(F_{,\alpha}\cdot F_{,\beta}(3n-6)+g_{\alpha\beta}F_{,c}F_{,d}g^{cd}(4-n)\right) \end{array}\right) \\ +\kappa T_{\alpha\beta}+\Lambda\cdot F\cdot g_{\alpha\beta} \end{array}\right)\left(\begin{array}{c} g^{\alpha\beta}\delta\left(\frac{1}{F}\right) \\ +\frac{\delta g^{\alpha\beta}}{F} \end{array}\right), \tag{372}$$

$$0=\left(\left(1-\frac{n}{2}\right)\left(R-\left(\frac{1}{2F}\begin{pmatrix}2F_{,ij}(n-1)g^{ij}+2\Gamma_{ij}^{a}F_{,a}g^{ij}\\-F_{,i}g^{ab}g_{jb,a}g^{ij}-F_{,j}g^{ab}g_{ib,a}g^{ij}\\-n\Gamma_{ij}^{d}F_{,d}g^{ij}+\frac{n}{2}F_{,d}g^{cd}g_{ab,c}g^{ab}\end{pmatrix}+\frac{F_{,i}\cdot F_{,j}}{4F^2}g^{ij}\left((n-6)(n-1)\right)\right)\right)+\Lambda\cdot F\cdot n\right)\delta\left(\frac{1}{F}\right)$$

$$0=\begin{pmatrix}R_{\alpha\beta}-\frac{g_{\alpha\beta}}{2}R+\frac{g_{\alpha\beta}}{2}\left(\frac{1}{2F}\begin{pmatrix}2F_{,ij}(n-1)g^{ij}+2\Gamma_{ij}^{a}F_{,a}g^{ij}\\-F_{,i}g^{ab}g_{jb,a}g^{ij}-F_{,j}g^{ab}g_{ib,a}g^{ij}\\-n\Gamma_{ij}^{d}F_{,d}g^{ij}+\frac{n}{2}F_{,d}g^{cd}g_{ab,c}g^{ab}\end{pmatrix}+\frac{F_{,i}\cdot F_{,j}}{4F^2}g^{ij}\left((n-6)(n-1)\right)\right)\\-\frac{1}{2F}\begin{pmatrix}F_{,\alpha\beta}(n-2)+F_{,ab}g_{\alpha\beta}g^{ab}\\+F_{,a}g^{ab}\left(g_{\beta b,\alpha}-g_{\beta\alpha,b}\right)-F_{,\alpha}g^{ab}g_{\beta b,a}-F_{,\beta}g^{ab}g_{\alpha b,a}\\+F_{,d}g^{cd}\frac{1}{2}n\left(\frac{2}{n}g_{\alpha c,\beta}-g_{\alpha c,\beta}-g_{\beta c,\alpha}+g_{\alpha\beta,c}+\frac{1}{n}g_{\alpha\beta}g_{ab,c}g^{ab}\right)\\-\frac{1}{2F}\left(F_{,\alpha}\cdot F_{,\beta}(3n-6)+g_{\alpha\beta}F_{,c}F_{,d}g^{cd}(4-n)\right)\end{pmatrix}\\+\kappa T_{\alpha\beta}+\Lambda\cdot F\cdot g_{\alpha\beta}\end{pmatrix}\frac{\delta g^{\alpha\beta}}{F}. \tag{373}$$

While we consider the second equation of the pair in Eq. (373) generalized or quantum Einstein field equations, we used the first one to derive the classical quantum equations (Klein–Gordon, Dirac and Schrödinger) for a generally curved space-time. Demanding the "gravity" to be weak as follows:

$$\delta G^{\alpha\beta}=G^{\alpha\beta}\cdot\delta_0+\overbrace{G^{ab}\delta_{ab}^{\alpha\beta}}^{\text{Gravity}}\xrightarrow{\forall\,\delta_{ab}^{\alpha\beta}\ll\delta_0}=\frac{g^{\alpha\beta}}{F}\cdot\delta_0, \tag{374}$$

we are left with the scalar and purely metrically obtained quantum equation, which, when excluding the trivial case $n = 2$, reads:

$$0=\left(R-\left(\frac{1}{2F}\begin{pmatrix}2F_{,ij}(n-1)g^{ij}+2\Gamma_{ij}^{a}F_{,a}g^{ij}\\-F_{,i}g^{ab}g_{jb,a}g^{ij}-F_{,j}g^{ab}g_{ib,a}g^{ij}\\-n\Gamma_{ij}^{d}F_{,d}g^{ij}+\frac{n}{2}F_{,d}g^{cd}g_{ab,c}g^{ab}\end{pmatrix}+\frac{F_{,i}\cdot F_{,j}}{4F^{2}}g^{ij}\left((n-6)(n-1)\right)\right)\right)+\Lambda\cdot F\cdot n. \tag{375}$$

From there, which is to say, based on the latter equation, all classically important equations can be obtained. In the case of a vanishing cosmological constant, Eq. (375) simplifies to:

$$0=R-\left(\frac{1}{2F}\begin{pmatrix}2F_{,ij}(n-1)g^{ij}+2\Gamma_{ij}^{a}F_{,a}g^{ij}\\-F_{,i}g^{ab}g_{jb,a}g^{ij}-F_{,j}g^{ab}g_{ib,a}g^{ij}\\-n\Gamma_{ij}^{d}F_{,d}g^{ij}+\frac{n}{2}F_{,d}g^{cd}g_{ab,c}g^{ab}\end{pmatrix}+\frac{F_{,i}\cdot F_{,j}}{4F^{2}}g^{ij}\left((n-6)(n-1)\right)\right)$$

$$=R-\left(\frac{F'}{2F}\begin{pmatrix}2f_{,ij}(n-1)g^{ij}+2\Gamma_{ij}^{a}f_{,a}g^{ij}\\-f_{,i}g^{ab}g_{jb,a}g^{ij}-f_{,j}g^{ab}g_{ib,a}g^{ij}\\-n\Gamma_{ij}^{d}f_{,d}g^{ij}+\frac{n}{2}f_{,d}g^{cd}g_{ab,c}g^{ab}\end{pmatrix}+(n-1)\frac{f_{,i}\cdot f_{,j}}{4F^{2}}g^{ij}\left(4FF''+(F')^{2}(n-6)\right)\right), \tag{376}$$

where we expanded the scaling function $F[f]$. Demanding the following linearization condition:

$$0=4FF''+(F')^{2}(n-6), \tag{377}$$

which can be satisfied via the function $F[f]$ as follows:

$$F[f]=\begin{cases}C_F\cdot\left(f+C_f\right)^{\frac{4}{n-2}} & n\neq 2\\ C_F\cdot e^{f\cdot C_f} & n=2,\end{cases} \tag{378}$$

Eq. (376) can be simplified and yields:

$$0=R-\frac{F'}{2F}\begin{pmatrix}2f_{,ij}(n-1)g^{ij}+2\Gamma_{ij}^{a}f_{,a}g^{ij}\\-f_{,i}g^{ab}g_{jb,a}g^{ij}-f_{,j}g^{ab}g_{ib,a}g^{ij}\\-n\Gamma_{ij}^{d}f_{,d}g^{ij}+\frac{n}{2}f_{,d}g^{cd}g_{ab,c}g^{ab}\end{pmatrix}. \tag{379}$$

Interestingly, for metrics without shear elements:

$$g_{ij}=\begin{pmatrix} g_{00} & \cdots & 0 \\ \vdots & \ddots & \vdots \\ 0 & \cdots & g_{n-1n-1} \end{pmatrix}; \quad g_{ii,i}=0, \tag{380}$$

this converges to the ordinary Laplace operator, namely:

$$R^{*}=0 \quad \rightarrow \quad 0=F\cdot R+F'\cdot(1-n)\cdot\Delta f$$

$$\Rightarrow \qquad 0=\begin{cases}\left(f-C_{f}\right)^{\frac{4}{n-2}}\cdot C_{F}\left(R+\dfrac{4}{n-2}\cdot\dfrac{(1-n)}{\left(f-C_{f}\right)}\cdot\Delta f\right) & n>2\\ e^{C_{f}\cdot f}\cdot C_{F}\left(R+C_{f}\cdot(1-n)\cdot\Delta f\right) \quad n=2.\end{cases} \tag{381}$$

Thus, in the case of $n > 2$, we always also have the option for a constant (broken symmetry) solution of the kind:

$$0=f-C_{f0} \quad \Rightarrow \quad f=C_{f0}. \tag{382}$$

Otherwise, we have the simple equations:

$$0=\begin{cases}\left(f-C_{f0}\right)\cdot R+(1-n)\cdot\dfrac{4}{n-2}\cdot\Delta f & n>2\\ R+C_{f0}\cdot(1-n)\cdot\Delta f & n=2.\end{cases} \tag{383}$$

A critical argument should now be that this equation is not of Klein–Gordon character as it contains neither potential nor mass, but this author has already shown that this problem is easily solved by adding additional dimensions carrying the right properties to produce masses and potentials (e.g., [4, 7–10] and appendices above). Here, we are especially concentrating on the question of the metric appearance of physical potentials. But before we come to that, we want to consider an alternative, slightly more general way to obtain quantum Einstein field equations and to scalarize them.

5.18.4 Fundamental Uncertainties

Let us assume that there is a fundamental uncertainty. In fact, we assume this uncertainty to be so fundamental that it even already influences very fundamental principles such as Hamilton's minimum principle. Translating this into the picture of Hilbert's action [1], the variational task:

$$\delta_{g}W=0=\delta_{g}\int_{V}d^{n}x\left(\sqrt{-g}\cdot\Phi_{R}[R]\right)=\delta_{g}\int_{V}d^{n}x\left(\sqrt{-g}\cdot R\right)$$

$$\delta_G W = ?? = \delta_G \int_V d^n x \left(\sqrt{-G} \cdot R^* \right)$$

$$\delta_g W = ? = \delta_g \int_V d^n x \left(\sqrt{-g} \cdot \Phi_{R^*} \left[R^* \right] \right) \tag{384}$$

should read:

$$\delta W = 0 = \delta \int_V d^n x \left(\sqrt{-g} \cdot \left(\overbrace{(1-\Omega)}^{=\Phi} \cdot R - 2\Lambda + L_M \right) \right)$$

$$\Rightarrow \qquad 0 = \delta \int_V d^n x \left(\sqrt{-g} \cdot \left(\Phi \cdot R - 2\Lambda + L_M \right) \right). \tag{385}$$

Thereby Ω presents the general perturbation of the usual action, and thus the symbol Φ stands for an arbitrary function of the coordinates. We know from [6] that this would usually lead to complicated expressions like this:

$$\delta W = 0 = \left[\begin{array}{c} \int_V d^n x \left(\sqrt{-g} \cdot \left(\begin{array}{c} \Phi \cdot R_{\mu\nu} - \frac{1}{2} \Phi \cdot R g_{\mu\nu} \\ + \Lambda g_{\mu\nu} - \left(\nabla_\mu \nabla_\nu - g_{\mu\nu} \Delta_g \right) \Phi \\ + \left\{ \begin{array}{ll} 0 & \ldots \text{"vacuum"} \\ 8\pi G T_{\mu\nu} & \ldots \text{postulated matter} \end{array} \right\} \Phi \end{array} \right) \right) \delta g^{\mu\nu} \\ - \int_{\text{Surface}} d^{n-1} y \left(\sqrt{|h|} \cdot \varepsilon \cdot \Phi \cdot N^\lambda h^{\mu\nu} \partial_\lambda \left(\delta g_{\mu\nu} \right) \right) \end{array} \right]$$

$$= \left[\begin{array}{c} \int_V d^n x \left(\sqrt{-g} \cdot \left(\begin{array}{c} R_{\mu\nu} - \frac{1}{2} R g_{\mu\nu} \\ + \frac{\Lambda}{\Phi} g_{\mu\nu} - \left(\nabla_\mu \nabla_\nu - g_{\mu\nu} \Delta_g \right) \\ + \left\{ \begin{array}{ll} 0 & \ldots \text{"vacuum"} \\ 8\pi G T_{\mu\nu} & \ldots \text{postulated matter} \end{array} \right\} \end{array} \right) \Phi \right) \delta g^{\mu\nu} \\ - \int_{\text{Surface}} d^{n-1} y \left(\sqrt{|h|} \cdot \varepsilon \cdot \Phi \cdot N^\lambda h^{\mu\nu} \partial_\lambda \left(\delta g_{\mu\nu} \right) \right) \end{array} \right]. \tag{386}$$

Thereby, the latter was extracted from [6], where we find that:

$$\delta_g W = 0 = \delta_g \int_V d^n x \left(\sqrt{-g} \cdot \Phi_R \left[R \right] \right)$$

$$\delta W=0=\left[\begin{array}{c}\int_V d^n x\left(\sqrt{-g}\cdot\left(\begin{array}{c}\Phi'_R[R]\cdot R_{\mu\nu}-\frac{1}{2}\Phi_R[R]\cdot g_{\mu\nu}\\+\Lambda g_{\mu\nu}-\left(\nabla_\mu\nabla_\nu-g_{\mu\nu}\Delta_g\right)\Phi'_R[R]\end{array}\right)\right)\delta g^{\mu\nu}\\-\int_{\text{Surface}} d^{n-1}y\left(\sqrt{|h|}\cdot\varepsilon\cdot\Phi'_R[R]\cdot N^\lambda h^{\mu\nu}\partial_\lambda\left(\delta g_{\mu\nu}\right)\right)\end{array}\right]. \tag{387}$$

Now we simply substitute $R\rightarrow R^*$, $g_{ab}\rightarrow G_{ab}$ and $\Phi_R[R]\rightarrow\Phi^*R$, which gives us:

$$\delta_G W=0=\delta_G\int_V d^n x\left(\sqrt{-G}\cdot\Phi\cdot R^*\right)$$

$$\delta W=0=\left[\begin{array}{c}\int_V d^n x\left(\sqrt{-G}\cdot\left(\begin{array}{c}\Phi\cdot R^*{}_{\mu\nu}-\frac{1}{2}\Phi\cdot R^*\cdot G_{\mu\nu}\\+\Lambda G_{\mu\nu}-\left(\nabla_\mu\nabla_\nu-G_{\mu\nu}\Delta_G\right)\Phi\end{array}\right)\right)\delta G^{\mu\nu}\\-\int_{\text{Surface}} d^{n-1}y\left(\sqrt{|h|}\cdot\varepsilon\cdot\Phi\cdot N^\lambda h^{\mu\nu}\partial_\lambda\left(\delta G_{\mu\nu}\right)\right)\end{array}\right]. \tag{388}$$

The usual, and rather tedious, path forward would be the discussion of the last equation and its investigation with respect to all its constituents. In this section, however, we will show how to avoid such cumbersome equations (especially the surface term). For the reason of brevity and simplicity, we demand:

$$\delta W=0=\delta\int_V d^n x\left(\sqrt{-g}\cdot\left(\Phi\cdot R-\overbrace{2\Lambda+L_M}^{=0}\right)\right) \tag{389}$$

and adjust the kernel as follows:

$$\delta W=0=\delta\int_V d^n x\left(\sqrt{-g}\cdot\Phi\cdot R\right)$$

$$\xrightarrow{g_{\alpha\beta}=\gamma_{\alpha\beta}\cdot F;\quad F^{\frac{n}{2}}=\Phi}$$

$$0=\delta_{g_{\alpha\beta}}\int_V d^n x\left(\sqrt{-\gamma}\cdot F^{\frac{n}{2}}R^*\right)\Rightarrow 0=\delta_{g_{\alpha\beta}}\int_V d^n x\left(\sqrt{-\gamma\cdot F^n}\cdot R^*\right)$$

$$\Rightarrow \qquad 0 = R^*_{\alpha\beta} - \frac{1}{2} \cdot \gamma_{\alpha\beta} \cdot R^*. \tag{390}$$

Thereby we remember that:

$$R^* = \frac{1}{F}\left(\begin{array}{c} R - \frac{1}{2F}\left(\begin{array}{c} 2F_{,ij}(n-1)\gamma^{ij} + 2\Gamma^a_{ij}F_{,a}\gamma^{ij} \\ -F_{,i}\gamma^{ab}\gamma_{jb,a}\gamma^{ij} - F_{,j}\gamma^{ab}\gamma_{ib,a}\gamma^{ij} \\ -n\Gamma^d_{ij}F_{,d}\gamma^{ij} + \frac{n}{2}F_{,d}\gamma^{cd}\gamma_{ab,c}\gamma^{ab} \end{array}\right) \\ -\frac{F_{,i}\cdot F_{,j}}{4F^2}\gamma^{ij}\left((n-6)(n-1)\right) \end{array}\right) \tag{391}$$

and subsequently obtain, after variation with respect to the scaled metric tensor:

$$0 = \int_{V_G} d^n x \sqrt{-\gamma \cdot F^n}\left(\left(\begin{array}{c} R_{\alpha\beta} - \frac{\gamma_{\alpha\beta}}{2}\left(\begin{array}{c} R - \frac{1}{2F}\left(\begin{array}{c} 2F_{,ij}(n-1)\gamma^{ij} + 2\Gamma^a_{ij}F_{,a}\gamma^{ij} \\ -F_{,i}\gamma^{ab}\gamma_{jb,a}\gamma^{ij} - F_{,j}\gamma^{ab}\gamma_{ib,a}\gamma^{ij} \\ -n\Gamma^d_{ij}F_{,d}\gamma^{ij} + \frac{n}{2}F_{,d}\gamma^{cd}\gamma_{ab,c}\gamma^{ab} \end{array}\right) \\ -\frac{F_{,i}\cdot F_{,j}}{4F^2}\gamma^{ij}\left((n-6)(n-1)\right) \end{array}\right) \\ -\frac{1}{2F}\left(\begin{array}{c} F_{,\alpha\beta}(n-2) + F_{,ab}\gamma_{\alpha\beta}\gamma^{ab} \\ +F_{,a}\gamma^{ab}\left(\gamma_{\beta b,\alpha} - \gamma_{\beta\alpha,b}\right) \\ -F_{,\alpha}\gamma^{ab}\gamma_{\beta b,a} - F_{,\beta}\gamma^{ab}\gamma_{\alpha b,a} \\ +F_{,d}\gamma^{cd}\frac{1}{2}n\left(\begin{array}{c} \frac{2}{n}\gamma_{\alpha c,\beta} - \gamma_{\alpha c,\beta} - \gamma_{\beta c,\alpha} \\ +\gamma_{\alpha\beta,c} + \frac{1}{n}\gamma_{\alpha\beta}\gamma_{ab,c}\gamma^{ab} \end{array}\right) \\ -\frac{1}{2F}\left(\begin{array}{c} F_{,\alpha}\cdot F_{,\beta}(3n-6) \\ +\gamma_{\alpha\beta}F_{,c}F_{,d}\gamma^{cd}(4-n) \end{array}\right) \end{array}\right) \end{array}\right)\delta g^{\alpha\beta}\right). \tag{392}$$

The resulting field equations read as follows:

$$0=\left(\begin{array}{c} R_{\alpha\beta}-\frac{\gamma_{\alpha\beta}}{2}R+\frac{\gamma_{\alpha\beta}}{2}\left(\begin{array}{c}\frac{1}{2F}\left(\begin{array}{c}2F_{,ij}(n-1)\gamma^{ij}+2\Gamma_{ij}^{a}F_{,a}\gamma^{ij}\\ -F_{,i}\gamma^{ab}\gamma_{jb,a}\gamma^{ij}-F_{,j}\gamma^{ab}\gamma_{ib,a}\gamma^{ij}\\ -n\Gamma_{ij}^{d}F_{,d}\gamma^{ij}+\frac{n}{2}F_{,d}\gamma^{cd}\gamma_{ab,c}\gamma^{ab}\end{array}\right)\\ +\frac{F_{,i}\cdot F_{,j}}{4F^{2}}\gamma^{ij}\left((n-6)(n-1)\right)\end{array}\right)\\ -\frac{1}{2F}\left(\begin{array}{c}F_{,\alpha\beta}(n-2)+F_{,ab}\gamma_{\alpha\beta}\gamma^{ab}\\ +F_{,a}\gamma^{ab}\left(\gamma_{\beta b,\alpha}-\gamma_{\beta\alpha,b}\right)-F_{,\alpha}\gamma^{ab}\gamma_{\beta b,a}-F_{,\beta}\gamma^{ab}\gamma_{\alpha b,a}\\ +F_{,d}\gamma^{cd}\frac{1}{2}n\left(\frac{2}{n}\gamma_{\alpha c,\beta}-\gamma_{\alpha c,\beta}-\gamma_{\beta c,\alpha}+\gamma_{\alpha\beta,c}+\frac{1}{n}\gamma_{\alpha\beta}\gamma_{ab,c}\gamma^{ab}\right)\\ -\frac{1}{2F}\left(F_{,\alpha}\cdot F_{,\beta}(3n-6)+\gamma_{\alpha\beta}F_{,c}F_{,d}\gamma^{cd}(4-n)\right)\end{array}\right)\end{array}\right). \tag{393}$$

Presenting them in dependency of our original kernel perturbation Φ gives us:

$$0=\left(\begin{array}{c} R_{\alpha\beta}-\frac{\gamma_{\alpha\beta}}{2}R+\frac{\gamma_{\alpha\beta}}{2}\left(\begin{array}{c}\frac{1}{2\Phi^{\frac{2}{n}}}\left(\begin{array}{c}2\Phi^{\frac{2}{n}}{}_{,ij}(n-1)\gamma^{ij}+2\Gamma_{ij}^{a}\Phi^{\frac{2}{n}}{}_{,a}\gamma^{ij}\\ -\Phi^{\frac{2}{n}}{}_{,i}\gamma^{ab}\gamma_{jb,a}\gamma^{ij}-\Phi^{\frac{2}{n}}{}_{,j}\gamma^{ab}\gamma_{ib,a}\gamma^{ij}\\ -n\Gamma_{ij}^{d}\Phi^{\frac{2}{n}}{}_{,d}\gamma^{ij}+\frac{n}{2}\Phi^{\frac{2}{n}}{}_{,d}\gamma^{cd}\gamma_{ab,c}\gamma^{ab}\end{array}\right)\\ +\frac{\Phi^{\frac{2}{n}}{}_{,i}\cdot\Phi^{\frac{2}{n}}{}_{,j}}{4\Phi^{\frac{4}{n}}}\gamma^{ij}\left((n-6)(n-1)\right)\end{array}\right)\\ -\frac{1}{2\Phi^{\frac{2}{n}}}\left(\begin{array}{c}\Phi^{\frac{2}{n}}{}_{,\alpha\beta}(n-2)+\Phi^{\frac{2}{n}}{}_{,ab}\gamma_{\alpha\beta}\gamma^{ab}\\ +\Phi^{\frac{2}{n}}{}_{,a}\gamma^{ab}\left(\gamma_{\beta b,\alpha}-\gamma_{\beta\alpha,b}\right)-\Phi^{\frac{2}{n}}{}_{,\alpha}\gamma^{ab}\gamma_{\beta b,a}-\Phi^{\frac{2}{n}}{}_{,\beta}\gamma^{ab}\gamma_{\alpha b,a}\\ +\Phi^{\frac{2}{n}}{}_{,d}\gamma^{cd}\frac{1}{2}n\left(\frac{2}{n}\gamma_{\alpha c,\beta}-\gamma_{\alpha c,\beta}-\gamma_{\beta c,\alpha}+\gamma_{\alpha\beta,c}+\frac{1}{n}\gamma_{\alpha\beta}\gamma_{ab,c}\gamma^{ab}\right)\\ -\frac{1}{2\Phi^{\frac{2}{n}}}\left(\Phi^{\frac{2}{n}}{}_{,\alpha}\cdot\Phi^{\frac{2}{n}}{}_{,\beta}(3n-6)+\gamma_{\alpha\beta}\Phi^{\frac{2}{n}}{}_{,c}\Phi^{\frac{2}{n}}{}_{,d}\gamma^{cd}(4-n)\right)\end{array}\right)\end{array}\right). \tag{394}$$

We realize that a perturbation of the Hilbert kernel in the Einstein–Hilbert action results in quantum or at least quantum like terms for the field equations.

Now, substituting $1 - \Omega = \Phi$ as done in Eq. (385) and setting:

$$\Phi = F^{\frac{n}{2}} \tag{395}$$

gives us the following illustrative development:

$$\begin{aligned}
&\delta W = 0 = \delta \int_V d^n x \left(\sqrt{-g} \cdot (1-\Omega) \cdot R \right) = \delta \int_V d^n x \left(\sqrt{-g} \cdot R \right) - \delta \int_V d^n x \left(\sqrt{-g} \cdot \Omega \cdot R \right) \\
&\xrightarrow{g_{\alpha\beta} = \gamma_{\alpha\beta} \cdot F;\ F^{\frac{n}{2}} = \Omega} 0 = \delta \int_V d^n x \left(\sqrt{-g} \cdot R \right) - \delta_{g_{\alpha\beta}} \int_V d^n x \left(\sqrt{-\gamma} \cdot F^{\frac{n}{2}} R^{*} \right) \\
\Rightarrow \quad & 0 = \delta \int_V d^n x \left(\sqrt{-g} \cdot R \right) - \delta_{g_{\alpha\beta}} \int_V d^n x \left(\sqrt{-\gamma \cdot F^{n}} \cdot R^{*} \right) \\
\Rightarrow \quad & 0 = R_{\alpha\beta} - \frac{1}{2} \cdot g_{\alpha\beta} \cdot R - \left(R^{*}_{\alpha\beta} - \frac{1}{2} \cdot \gamma_{\alpha\beta} \cdot R^{*} \right).
\end{aligned} \tag{396}$$

Incorporating our results from above (e.g., Eq. (392)) leads to:

$$0 = \int_{V_G} d^n x \sqrt{-\gamma \cdot F^n} \left(\begin{aligned}
& \left(R_{\alpha\beta} - \frac{F}{2} \cdot \gamma_{\alpha\beta} \cdot R \right) \delta g^{\alpha\beta} \\
& - \left(\begin{aligned}
& R_{\alpha\beta} - \frac{\gamma_{\alpha\beta}}{2} \left(\begin{aligned}
& R - \frac{1}{2F} \left(\begin{aligned}
& 2F_{,ij}(n-1)\gamma^{ij} + 2\Gamma^{a}_{ij} F_{,a} \gamma^{ij} \\
& - F_{,i} \gamma^{ab} \gamma_{jb,a} \gamma^{ij} - F_{,j} \gamma^{ab} \gamma_{ib,a} \gamma^{ij} \\
& - n \Gamma^{d}_{ij} F_{,d} \gamma^{ij} + \frac{n}{2} F_{,d} \gamma^{cd} \gamma_{ab,c} \gamma^{ab}
\end{aligned} \right) \\
& - \frac{F_{,i} \cdot F_{,j}}{4F^2} \gamma^{ij} \left((n-6)(n-1) \right)
\end{aligned} \right) \\
& - \frac{1}{2F} \left(\begin{aligned}
& F_{,\alpha\beta}(n-2) + F_{,ab} \gamma_{\alpha\beta} \gamma^{ab} + F_{,a} \gamma^{ab} \left(\gamma_{\beta b,\alpha} - \gamma_{\beta\alpha,b} \right) \\
& - F_{,\alpha} \gamma^{ab} \gamma_{\beta b,a} - F_{,\beta} \gamma^{ab} \gamma_{\alpha b,a} \\
& + F_{,d} \gamma^{cd} \frac{1}{2} n \left(\begin{aligned}
& \frac{2}{n} \gamma_{\alpha c,\beta} - \gamma_{\alpha c,\beta} - \gamma_{\beta c,\alpha} \\
& + \gamma_{\alpha\beta,c} + \frac{1}{n} \gamma_{\alpha\beta} \gamma_{ab,c} \gamma^{ab}
\end{aligned} \right) \\
& - \frac{1}{2F} \left(F_{,\alpha} \cdot F_{,\beta} (3n-6) + \gamma_{\alpha\beta} F_{,c} F_{,d} \gamma^{cd} (4-n) \right)
\end{aligned} \right)
\end{aligned} \right) \delta g^{\alpha\beta}
\end{aligned} \right) \tag{397}$$

and makes the Ricci tensor to disappear:

$$0=\int\limits_{V_G} d^n x\sqrt{-\gamma\cdot F^n}\left(\left(\begin{array}{c}\frac{\gamma_{\alpha\beta}}{2}\left(\begin{array}{c}R-\frac{1}{2F}\left(\begin{array}{c}2F_{,ij}(n-1)\gamma^{ij}+2\Gamma_{ij}^{a}F_{,a}\gamma^{ij}\\-F_{,i}\gamma^{ab}\gamma_{jb,a}\gamma^{ij}-F_{,j}\gamma^{ab}\gamma_{ib,a}\gamma^{ij}\\-n\Gamma_{ij}^{d}F_{,d}\gamma^{ij}+\frac{n}{2}F_{,d}\gamma^{cd}\gamma_{ab,c}\gamma^{ab}\end{array}\right)\\-\frac{F_{,i}\cdot F_{,j}}{4F^2}\gamma^{ij}\left((n-6)(n-1)\right)\end{array}\right)-\frac{F}{2}\cdot\gamma_{\alpha\beta}\cdot R\\+\frac{1}{2F}\left(\begin{array}{c}F_{,\alpha\beta}(n-2)+F_{,ab}\gamma_{\alpha\beta}\gamma^{ab}\\+F_{,a}\gamma^{ab}\left(\gamma_{\beta b,\alpha}-\gamma_{\beta\alpha,b}\right)-F_{,\alpha}\gamma^{ab}\gamma_{\beta b,a}-F_{,\beta}\gamma^{ab}\gamma_{\alpha b,a}\\+F_{,d}\gamma^{cd}\frac{1}{2}n\left(\begin{array}{c}\frac{2}{n}\gamma_{\alpha c,\beta}-\gamma_{\alpha c,\beta}-\gamma_{\beta c,\alpha}\\+\gamma_{\alpha\beta,c}+\frac{1}{n}\gamma_{\alpha\beta}\gamma_{ab,c}\gamma^{ab}\end{array}\right)\\-\frac{1}{2F}\left(F_{,\alpha}\cdot F_{,\beta}(3n-6)+\gamma_{\alpha\beta}F_{,c}F_{,d}\gamma^{cd}(4-n)\right)\end{array}\right)\end{array}\right)\delta g^{\alpha\beta}\right).\tag{398}$$

In the case of variation not with respect to the metric tensor, but only with respect to the scaling function f (c.f. section “Dissolving an Apparent Conflict” in [7]), we adjust the kernel as follows:

$$\delta W=0=\delta\int\limits_{V} d^n x\left(\sqrt{-g}\cdot\Phi_0\cdot R\right)$$

$$\xrightarrow{g_{\alpha\beta}=\gamma_{\alpha\beta}\cdot F=\gamma_{\alpha\beta}\cdot F[f];\;\; F^{\frac{n}{2}+1}=\Phi_0}$$

$$0=\delta_f\int\limits_{V} d^n x\left(\sqrt{-\gamma}\cdot F^{\frac{n}{2}+1}R^*\right)\Rightarrow 0=\delta_f\int\limits_{V} d^n x\left(\sqrt{-\gamma\cdot F^n}\cdot F\cdot R^*\right).\tag{399}$$

Now the resulting variational task reads as follows:

$$\frac{\delta W}{\delta f}=0=\delta_f\int\limits_{V_G} d^n x\sqrt{-g\cdot F^n}\left(\begin{array}{c}R-\frac{1}{2F}\left(\begin{array}{c}2F_{,\alpha\beta}(n-1)g^{\alpha\beta}+2\Gamma_{\alpha\beta}^{a}F_{,a}g^{\alpha\beta}\\-F_{,\alpha}g^{ab}g_{\beta b,a}g^{\alpha\beta}-F_{,\beta}g^{ab}g_{\alpha b,a}g^{\alpha\beta}\\-n\Gamma_{\alpha\beta}^{d}F_{,d}g^{\alpha\beta}+\frac{n}{2}F_{,d}g^{cd}g_{\alpha\beta,c}g^{\alpha\beta}\end{array}\right)\\-\frac{F_{,\alpha}\cdot F_{,\beta}}{4F^2}g^{\alpha\beta}\left((n-6)(n-1)\right)\end{array}\right),\tag{400}$$

and we obtain the scalar equation:

$$0=\left[\begin{array}{l}\frac{\sqrt{-g}}{2}\left(\frac{f_{,\alpha}f_{,\beta}}{2}g^{\alpha\beta}\left(\frac{1}{2}(n-4)(n+2)(F')^2+2FF''\right)+(n+2)FF'\Delta f\right)\\+F^2\left(\begin{array}{c}\left(\sqrt{-g}\cdot g^{\alpha\beta}\right)_{,\alpha\beta}-\frac{\sqrt{-g}\cdot R}{(n-1)}\\-\frac{1}{2(n-1)}\left(\sqrt{-g}\cdot g^{\alpha\beta}\cdot g^{ij}\left(\begin{array}{c}g_{i\beta,j}-g_{ij,\beta}-g_{\beta j,i}\\+\frac{n}{2}\left(2g_{ij,\beta}-g_{i\beta,j}-g_{j\beta,i}\right)\end{array}\right)\right)_{,\alpha}\end{array}\right)\end{array}\right]. \qquad (401)$$

When interpreting the scaling function *f* as something being equivalent or at least connected with the quantum theoretical wave function of an arbitrary system, we see that a perturbation of the curvature kernel of the Hilbert-extremal integral (384)→(385) directly leads to additional degrees of freedom and quantum equations. Thus, an observer would realize the perturbations as quantum effects, while regarding the system in question, this may just be taken as its very own and "individual" uncertainties. The importance of our finding with respect to the development of a "Mathematical Psychology," even though quite far from being very obvious, can hardly be overstated. After all, our results are true for each and every system, and a thinking system, a mind, be it a conscious or an unconscious one, is but a system. It may be characterized by many trillions of attributes but still is a system and thus has to be described by the fundamental math derived and presented here, which holds true for EACH ARBITRARY SYSTEM. De facto, we are only at the beginning of extracting all the consequences which do follow from the generality of these findings for the human psyche, respectively, the conscious mind.

So, in order to give an example, what we just obtained in the few lines above is the following:

We can state that variational perturbation, respectively, some kernel uncertainty inside the Einstein–Hilbert action automatically leads to Quantum Theory. This is a general result true for every system and thus also the mind. From there, it now automatically follows that every mind must have its very own—potentially individual—quantum realm. Thereby, the conscious mind might even distinguish itself from

the unconscious one by having the ability to actually adjust this kernel uncertainty (see abstract of this book).

It was just shown that we can always factor out a suitable function $F[f]$ from the metric tensor in order to simplify the variational task for either the whole metric or just with respect to the volume factor (see also [8, 19] for more details). Thereby we did not care about whether the kernels for both cases are identical, but simply adjusted the factor in order to obtain maximum simplicity for each task. However, it appears reasonable to assume that the kernel should be the same in both variational cases. Thus, aiming for maximum simplicity in the case of the variation with respect to the whole metric in applying Eq. (390), we require the derivation of the more complex case with respect to f and vice versa. The corresponding evaluation has been presented in appendix L of our publication [8].

5.18.4.1 The scaled Hilbert kernel in Eq. (386)

Having derived that a scaled Einstein–Hilbert action kernel of the type (385) can be brought into such a form that in the end we obtain quantum equations, should also allow us to extract such or similar quantum equations via the classical derivational path as shown in [6], for instance. Assuming that we already have a scaled metric of type:

$$G_{\alpha\beta} = g_{\alpha\beta} \cdot F[f], \tag{402}$$

plus a scaled Einstein–Hilbert action with a scaled kernel of type (385), we should be able to define the variation in such a way that we are allowed to demand the variation to give zero on both the surface of our volume integral and the sub-surface of the integral term from Eq. (386):

$$\int_{\text{Surface}} d^{n-1}y\left(\sqrt{|h|}\cdot\varepsilon\cdot\Phi\cdot N^{\lambda}h^{\mu\nu}\partial_{\lambda}\left(\delta G_{\mu\nu}\right)\right)=0. \tag{403}$$

This is further motivated by the fact that in the sub-section above we have been able to show that the variational integral of the Einstein–Hilbert action (e.g., in Eq. (396)) can be brought in such a form that we end up with surface terms. We can set them to zero (per definition with respect to the variation) and because our extended form of the variational task with scaled kernel plus scaled metric allows for more degrees of freedom regarding the definition of our variation, we take it for granted to be allowed to, in fact, omit all surface terms. We leave the rigorous proof to the skilled mathematician. In essence, however, this all then converts Eq. (386) to:

$$\delta W = 0$$
$$= \int_V d^n x \left(\sqrt{-G} \cdot \left(\begin{array}{c} \Phi \cdot R^*_{\mu\nu} - \frac{1}{2} \Phi \cdot R^* G_{\mu\nu} \\ + \Lambda G_{\mu\nu} - \left(\nabla_\mu \nabla_\nu - G_{\mu\nu} \Delta_G \right) \Phi \\ + \left\{ \begin{array}{cc} 0 & \ldots\text{"vacuum"} \\ \kappa T_{\mu\nu} & \ldots\text{postulated matter} \end{array} \right\} \Phi \end{array} \right) \right) \delta G^{\mu\nu}. \quad (404)$$

Setting the cosmological constant and the classical matter tensor equal zero simplifies the kernel as follows:

$$\Phi \cdot R^*_{\mu\nu} - \frac{1}{2} \Phi \cdot R^* G_{\mu\nu} - \left(\nabla_\mu \nabla_\nu - G_{\mu\nu} \Delta_G \right) \Phi = 0. \quad (405)$$

Assuming the space-time to be Ricci flat results in:

$$\left(\nabla_\mu \nabla_\nu - G_{\mu\nu} \Delta_G \right) \Phi = 0. \quad (406)$$

Applying our ansatz from [4], chapter 10 (esp. section "10.10 The Gravity Dirac Equation"), we can decompose the last equation as follows.
First, we assume eigenvalue solutions to the Laplace operator:

$$\Delta_G \Phi = m^2 \cdot \Phi. \quad (407)$$

We set this into Eq. (406) and get:

$$\left(\nabla_\mu \nabla_\nu - G_{\mu\nu} m^2 \right) \Phi = \left(\nabla_\mu \nabla_\nu - \mathbf{E}_\mu \mathbf{E}_\nu m^2 \right) \Phi = 0. \quad (408)$$

Then our Dirac-like approach shall be:

$$\left(a \cdot \nabla_\mu + A \cdot \mathbf{E}_\mu \cdot m \right) \left(b \cdot \nabla_\nu - B \cdot \mathbf{E}_\nu \cdot m \right) \Phi = 0. \quad (409)$$

With the objects a, A, b and B being general 2×2 matrices, we find a variety of solutions of which we only present a very simple one here, namely:

$$A = \begin{pmatrix} 1 & 1 \\ 1 & 1 \end{pmatrix}; \quad a = \begin{pmatrix} -1 & 1 \\ -1 & 1 \end{pmatrix}; \quad B = \begin{pmatrix} 1 & -1 \\ 1 & -1 \end{pmatrix}; \quad b = \begin{pmatrix} -1 & 1 \\ 1 & -1 \end{pmatrix}. \quad (410)$$

This leads Eq. (409) to result in:

$$\begin{aligned} &\left(a \cdot \nabla_\mu + A \cdot \mathbf{E}_\mu \cdot m \right) \left(b \cdot \nabla_\nu - B \cdot \mathbf{E}_\nu \cdot m \right) \Phi \\ &= 2 \cdot \begin{pmatrix} 1 & -1 \\ 1 & -1 \end{pmatrix} \cdot \left(\nabla_\mu \nabla_\nu - \mathbf{E}_\mu \mathbf{E}_\nu m^2 \right) \Phi \\ &= 2 \cdot \begin{pmatrix} 1 & -1 \\ 1 & -1 \end{pmatrix} \cdot \left(\nabla_\mu \nabla_\nu - G_{\mu\nu} m^2 \right) \Phi \end{aligned} \quad (411)$$

and gives us the following gravity "Dirac equations":

$$\left(a\cdot\nabla_{\mu}+A\cdot\mathbf{E}_{\mu}\cdot m\right)\Phi=\left(\begin{pmatrix}-1 & 1\\ -1 & 1\end{pmatrix}\cdot\nabla_{\mu}+\begin{pmatrix}1 & 1\\ 1 & 1\end{pmatrix}\cdot\mathbf{E}_{\mu}\cdot m\right)\Phi=0$$

$$\left(b\cdot\nabla_{v}-B\cdot\mathbf{E}_{v}\cdot m\right)\Phi=\left(\begin{pmatrix}-1 & 1\\ 1 & -1\end{pmatrix}\cdot\nabla_{v}-\begin{pmatrix}1 & -1\\ 1 & -1\end{pmatrix}\cdot\mathbf{E}_{v}\cdot m\right)\Phi=0. \quad (412)$$

Due to the structure of the base vectors $\mathbf{E}_{\mu}$, $\mathbf{E}_{v}$, m cannot be a simple scalar but has to be some kind of vector $\mathbf{m}$ with the property:

$$\mathbf{m}\cdot\mathbf{m}=m^{2}. \quad (413)$$

Thus, Eq. (412) reads:

$$\left(a\cdot\nabla_{\mu}+A\cdot\mathbf{E}_{\mu}\cdot\mathbf{m}\right)\Phi=\left(\begin{pmatrix}-1 & 1\\ -1 & 1\end{pmatrix}\cdot\nabla_{\mu}+\begin{pmatrix}1 & 1\\ 1 & 1\end{pmatrix}\cdot\mathbf{E}_{\mu}\cdot\mathbf{m}\right)\Phi=0$$

$$\left(b\cdot\nabla_{v}-B\cdot\mathbf{E}_{v}\cdot\mathbf{m}\right)\Phi=\left(\begin{pmatrix}-1 & 1\\ 1 & -1\end{pmatrix}\cdot\nabla_{v}-\begin{pmatrix}1 & -1\\ 1 & -1\end{pmatrix}\cdot\mathbf{E}_{v}\cdot\mathbf{m}\right)\Phi=0. \quad (414)$$

It should be pointed out that due to the fact that—in principle—the indices μ and v can be replaced with any symbol, the introduction of the matrices is not really needed. One could also write the factorization of Eq. (408) as follows:

$$\left(\nabla_{\mu}\nabla_{v}-G_{\mu v}m^{2}\right)\Phi=\left(\nabla_{\mu}\nabla_{v}-\mathbf{E}_{\mu}\mathbf{E}_{v}m^{2}\right)\Phi=0$$

$$\Rightarrow$$

$$\left(\nabla_{\mu}\nabla_{v}-\mathbf{E}_{\mu}\mathbf{E}_{v}m^{2}\right)\Phi=\left(\nabla_{\mu}\nabla_{v}-\mathbf{E}_{\mu}\mathbf{E}_{v}m^{2}-\nabla_{\mu}\mathbf{E}_{v}\cdot\mathbf{m}+\mathbf{E}_{\mu}\cdot\mathbf{m}\nabla_{v}\right)\Phi$$

$$=\left(\nabla_{\mu}\nabla_{v}-\mathbf{E}_{\mu}\mathbf{E}_{v}m^{2}-\nabla_{\mu}\mathbf{E}_{v}\cdot\mathbf{m}+\nabla_{v}\mathbf{E}_{\mu}\cdot\mathbf{m}\right)\Phi$$

$$=\left(\nabla_{\mu}\nabla_{v}-\mathbf{E}_{\mu}\mathbf{E}_{v}m^{2}+\frac{\nabla_{v}\mathbf{E}_{\mu}\cdot\mathbf{m}-\nabla_{\mu}\mathbf{E}_{v}\cdot\mathbf{m}}{2}+\frac{\nabla_{v}\mathbf{E}_{\mu}\cdot\mathbf{m}-\nabla_{\mu}\mathbf{E}_{v}\cdot\mathbf{m}}{2}\right)\Phi$$

$$\xrightarrow{\text{exchange }\ \mu,v\ \text{ in }\ \frac{\nabla_{v}\mathbf{E}_{\mu}\cdot\mathbf{m}-\nabla_{\mu}\mathbf{E}_{v}\cdot\mathbf{m}}{2}}$$

$$=\left(\nabla_{\mu}\nabla_{v}-\mathbf{E}_{\mu}\mathbf{E}_{v}m^{2}+\frac{1}{2}\overbrace{\left(\nabla_{v}\mathbf{E}_{\mu}\cdot\mathbf{m}-\nabla_{\mu}\mathbf{E}_{v}\cdot\mathbf{m}+\nabla_{\mu}\mathbf{E}_{v}\cdot\mathbf{m}-\nabla_{v}\mathbf{E}_{\mu}\cdot\mathbf{m}\right)}^{=0}\right)\Phi$$

$$=\left(\nabla_{\mu}+\mathbf{E}_{\mu}\cdot\mathbf{m}\right)\left(\nabla_{v}-\mathbf{E}_{v}\cdot\mathbf{m}\right)\Phi. \quad (415)$$

Thereby we used the fact that the covariant derivative of the base vector gives zero and that we also assume the parameter vector **m** to be a list of constants. In all other cases, we have to apply the matrices and end up with Eq. (412), respectively Eq. (414).

The Dirac-like equations resulting from Eq. (415) would be:

$$\left(\nabla_\mu + \mathbf{E}_\mu \cdot \mathbf{m}\right)\Phi = 0$$

$$\left(\nabla_\nu - \mathbf{E}_\nu \cdot \mathbf{m}\right)\Phi = 0. \tag{416}$$

Contraction with the corresponding base vector leads to the scalar equations:

$$\mathbf{E}^\mu\left(\nabla_\mu + \mathbf{E}_\mu \cdot \mathbf{m}\right)\Phi = \left(\mathbf{E}^\mu\nabla_\mu + n\cdot\mathbf{m}\right)\Phi = 0$$

$$\mathbf{E}^\nu\left(\nabla_\nu - \mathbf{E}_\nu \cdot \mathbf{m}\right)\Phi = \left(\mathbf{E}^\nu\nabla_\nu - n\cdot\mathbf{m}\right)\Phi = 0. \tag{417}$$

If we want to make these equations complete scalars, we realize that Φ also has to be some kind of vector. This vector would then just be a Dirac-like spinor as known from the Dirac theory.

5.18.4.2 Scalarizing the new quantum gravity field equation

It should be pointed out that one way[6] of scalarizing of Eq. (405) can be achieved via the following expansion of the variational term:

$$\delta G^{\mu\nu} = \delta\left(\frac{g^{\mu\nu}}{F}\right) = \frac{1}{F}\cdot\delta g^{\mu\nu} + g^{\mu\nu}\cdot\delta\left(\frac{1}{F}\right) \tag{418}$$

under the integral (404). This converts Eq. (405)—if taken out from Eq. (404) more completely—in principle to:

$$0 = \left(\Phi\cdot R^*_{\mu\nu} - \frac{1}{2}\Phi\cdot R^* G_{\mu\nu} - \left(\nabla_\mu\nabla_\nu - G_{\mu\nu}\Delta_G\right)\Phi\right)\left(\frac{1}{F}\cdot\delta g^{\mu\nu} + g^{\mu\nu}\cdot\delta\left(\frac{1}{F}\right)\right)$$

$$= \left(\Phi\cdot R^*_{\mu\nu} - \frac{1}{2}\Phi\cdot R^* G_{\mu\nu} - \left(\nabla_\mu\nabla_\nu - G_{\mu\nu}\Delta_G\right)\Phi\right)\frac{1}{F}\cdot\delta g^{\mu\nu}$$

$$+\left(\Phi\cdot R^*_{\mu\nu} - \frac{1}{2}\Phi\cdot R^* G_{\mu\nu} - \left(\nabla_\mu\nabla_\nu - G_{\mu\nu}\Delta_G\right)\Phi\right)g^{\mu\nu}\cdot\delta\left(\frac{1}{F}\right)$$

$$= \frac{1}{F}\cdot\left(\Phi\cdot R^*_{\mu\nu} - \frac{1}{2}\Phi\cdot R^* G_{\mu\nu} - \left(\nabla_\mu\nabla_\nu - G_{\mu\nu}\Delta_G\right)\Phi\right)\delta g^{\mu\nu}$$

[6]There are more such possibilities, which we have partially already seen in this book.

$$+\left(\Phi\cdot R^{*}{}_{\mu\nu}-\frac{1}{2}\Phi\cdot R^{*}G_{\mu\nu}-\left(\nabla_{\mu}\nabla_{\nu}-G_{\mu\nu}\Delta_{G}\right)\Phi\right)g^{\mu\nu}\cdot\delta\left(\frac{1}{F}\right)$$

$$\xrightarrow{G_{\mu\nu}=F\cdot g_{\mu\nu}\Leftrightarrow g^{\mu\nu}=F\cdot G^{\mu\nu}}$$

$$=\frac{1}{F}\cdot\left(\Phi\cdot R^{*}{}_{\mu\nu}-\frac{1}{2}\Phi\cdot R^{*}G_{\mu\nu}-\left(\nabla_{\mu}\nabla_{\nu}-G_{\mu\nu}\Delta_{G}\right)\Phi\right)\delta g^{\mu\nu}$$

$$+\left(\Phi\cdot F\cdot R^{*}{}_{\mu\nu}G^{\mu\nu}-\frac{1}{2}\Phi\cdot F\cdot R^{*}g_{\mu\nu}g^{\mu\nu}-F\cdot G^{\mu\nu}\left(\nabla_{\mu}\nabla_{\nu}-G_{\mu\nu}\Delta_{G}\right)\Phi\right)\cdot\delta\left(\frac{1}{F}\right)$$

$$=\frac{1}{F}\cdot\left(\Phi\cdot R^{*}{}_{\mu\nu}-\frac{1}{2}\Phi\cdot R^{*}G_{\mu\nu}-\left(\nabla_{\mu}\nabla_{\nu}-G_{\mu\nu}\Delta_{G}\right)\Phi\right)\delta g^{\mu\nu}$$

$$+\left(\Phi\cdot F\cdot R^{*}\left(1-\frac{n}{2}\right)-(1-n)F\cdot\Delta_{G}\Phi\right)\cdot\delta\left(\frac{1}{F}\right). \tag{419}$$

This can be split up into two scalar equations instead of one tensor (which would be Eq. (405) again, of course), namely:

$$0=\frac{1}{F}\cdot\left(\Phi\cdot R^{*}{}_{\mu\nu}-\frac{1}{2}\Phi\cdot R^{*}G_{\mu\nu}-\left(\nabla_{\mu}\nabla_{\nu}-G_{\mu\nu}\Delta_{G}\right)\Phi\right)\delta g^{\mu\nu}$$

$$0=\left(\Phi\cdot F\cdot R^{*}\left(1-\frac{n}{2}\right)-(1-n)F\cdot\Delta_{G}\Phi\right)\cdot\delta\left(\frac{1}{F}\right)$$

$$\Rightarrow$$

$$0=\left(\Phi\cdot R^{*}{}_{\mu\nu}-\frac{1}{2}\Phi\cdot R^{*}G_{\mu\nu}-\left(\nabla_{\mu}\nabla_{\nu}-G_{\mu\nu}\Delta_{G}\right)\Phi\right)\delta g^{\mu\nu}$$

$$0=\Phi\cdot F\cdot R^{*}\left(1-\frac{n}{2}\right)-(1-n)F\cdot\Delta_{G}\Phi. \tag{420}$$

When now assuming the variation to be of the following character:

$$\delta G^{\alpha\beta}=\frac{g^{\alpha\beta}\cdot\delta_{0}+\overbrace{g^{ab}\delta_{ab}^{\alpha\beta}}^{\text{Gravity}}}{F}\xrightarrow{\forall\,\delta_{ab}^{\alpha\beta}\ll\delta_{0}}=g^{\alpha\beta}\cdot\frac{\delta_{0}}{F}, \tag{421}$$

we can omit the first equation (third line) in Eq. (420) and then, only the last equation dominates the scene. This latter equation can be rewritten with respect to the unscaled metric as follows:

$$0=\Phi\cdot F\cdot R^{*}\left(1-\frac{n}{2}\right)-(1-n)F\cdot\Delta_{G}\Phi$$

$$= \Phi \cdot F \cdot R^{*}\left(1-\frac{n}{2}\right)-\frac{(1-n)}{F}F\cdot\left(\Delta_{g}\Phi-\frac{F_{,\mu}}{F}g^{\mu\nu}\Phi_{,\nu}\right)$$

$$\Rightarrow \qquad 0=\Phi\cdot F^{2}\cdot R^{*}\frac{(2-n)}{2\cdot(1-n)}-F\cdot\Delta_{g}\Phi+F_{,\mu}g^{\mu\nu}\Phi_{,\nu}. \tag{422}$$

Apparently, this is a polynomial of second order in F, but when inserting Eq. (391) (with the correct metric, which would be g_{ab}), we obtain:

$$0=\Phi\cdot F^{2}\cdot R^{*}\frac{(2-n)}{2\cdot(1-n)}-F\cdot\Delta_{g}\Phi+F_{,\mu}g^{\mu\nu}\Phi_{,\nu}$$

$$0=\Phi\cdot F\cdot\left(\begin{matrix}R-\frac{1}{2F}\left(\begin{matrix}2F_{,ij}(n-1)g^{ij}+2\Gamma_{ij}^{a}F_{,a}g^{ij}\\-F_{,i}g^{ab}g_{jb,a}g^{ij}-F_{,j}g^{ab}g_{ib,a}g^{ij}\\-n\Gamma_{ij}^{d}F_{,d}g^{ij}+\frac{n}{2}F_{,d}g^{cd}g_{ab,c}g^{ab}\end{matrix}\right)\\-\frac{F_{,i}\cdot F_{,j}}{4F^{2}}g^{ij}\left((n-6)(n-1)\right)\end{matrix}\right)\frac{(2-n)}{2\cdot(1-n)}-F\cdot\Delta_{g}\Phi+F_{,\mu}g^{\mu\nu}\Phi_{,\nu}$$

$$=\left(\begin{matrix}\Phi\cdot F\cdot R-\frac{\Phi}{2}\left(\begin{matrix}2F_{,ij}(n-1)g^{ij}+2\Gamma_{ij}^{a}F_{,a}g^{ij}\\-F_{,i}g^{ab}g_{jb,a}g^{ij}-F_{,j}g^{ab}g_{ib,a}g^{ij}\\-n\Gamma_{ij}^{d}F_{,d}g^{ij}+\frac{n}{2}F_{,d}g^{cd}g_{ab,c}g^{ab}\end{matrix}\right)\\-\Phi\cdot\frac{F_{,i}\cdot F_{,j}}{4F}g^{ij}\left((n-6)(n-1)\right)\end{matrix}\right)\frac{(2-n)}{2\cdot(1-n)}-F\cdot\Delta_{g}\Phi+F_{,\mu}g^{\mu\nu}\Phi_{,\nu}. \tag{423}$$

This leads us to:

$$0=\left[\begin{matrix}\left(\begin{matrix}\Phi\cdot F^{2}\cdot R-F\cdot\frac{\Phi}{2}\left(\begin{matrix}2F_{,ij}(n-1)g^{ij}+2\Gamma_{ij}^{a}F_{,a}g^{ij}\\-F_{,i}g^{ab}g_{jb,a}g^{ij}-F_{,j}g^{ab}g_{ib,a}g^{ij}\\-n\Gamma_{ij}^{d}F_{,d}g^{ij}+\frac{n}{2}F_{,d}g^{cd}g_{ab,c}g^{ab}\end{matrix}\right)\\-\Phi\cdot\frac{F_{,i}\cdot F_{,j}}{4}g^{ij}\left((n-6)(n-1)\right)\end{matrix}\right)\frac{(2-n)}{2\cdot(1-n)}\\-F^{2}\cdot\Delta_{g}\Phi+F\cdot F_{,\mu}g^{\mu\nu}\Phi_{,\nu}\end{matrix}\right]. \tag{424}$$

Thus, we still have a quadratic equation in F. However, when remembering that within the variational integral, we could always choose F and Φ in such a way that—with a new metric γ_{ab}—the two functions would become equal:

$$\delta W = 0 = \delta \int_V d^n x \left(\sqrt{-G} \cdot \Phi \cdot R^*\right)$$

$$\xrightarrow{g_{\alpha\beta} = \gamma_{\alpha\beta} \cdot F; \;\; F = \Phi}$$

$$0 = \delta_{g_{\alpha\beta}} \int_V d^n x \left(\sqrt{-\gamma} \cdot F \cdot R^*\right). \tag{425}$$

Thereby we took into account that the metric g_{ab} would already be scaled via an arbitrary function F^* and that some shifting is necessary in order to truly make the whole a two-function equation and adjust the metric accordingly. In the end, with respect to Eq. (423), we obtain:

$$0 = \left(\left(\begin{array}{c} \Phi \cdot F^2 \cdot R - \frac{\Phi \cdot F \cdot}{2} \left(\begin{array}{c} 2F_{,ij}(n-1)g^{ij} + 2\Gamma^a_{ij} F_{,a} g^{ij} \\ -F_{,i} g^{ab} g_{jb,a} g^{ij} - F_{,j} g^{ab} g_{ib,a} g^{ij} \\ -n\Gamma^d_{ij} F_{,d} g^{ij} + \frac{n}{2} F_{,d} g^{cd} g_{ab,c} g^{ab} \end{array} \right) \\ -\Phi \cdot \frac{F_{,i} \cdot F_{,j}}{4} g^{ij} \left((n-6)(n-1)\right) \end{array} \right) \frac{(2-n)}{2 \cdot (1-n)} \\ -F^2 \cdot \Delta_g \Phi + F \cdot F_{,\mu} g^{\mu\nu} \Phi_{,\nu} \right)$$

$$\xrightarrow{\Phi \to F \;\; \& \;\; g^{ij} \to \gamma^{ij}}$$

$$= \left(\begin{array}{c} F^3 \cdot R - \frac{F^2}{2} \left(\begin{array}{c} 2F_{,ij}(n-1)g^{ij} + 2\Gamma^a_{ij} F_{,a} \gamma^{jj} \\ -F_{,i} \gamma^{ab} \gamma_{jb,a} \gamma^{jj} - F_{,j} \gamma^{ab} \gamma_{ib,a} \gamma^{jj} \\ -n\Gamma^d_{ij} F_{,d} \gamma^{jj} + \frac{n}{2} F_{,d} \gamma^{cd} \gamma_{ab,c} \gamma^{ab} \end{array} \right) \\ -F \cdot \frac{F_{,i} \cdot F_{,j}}{4} \gamma^{jj} \left((n-6)(n-1)\right) \end{array} \right) \frac{(2-n)}{2 \cdot (1-n)} - F^2 \cdot \Delta_g F + F \cdot F_{,\mu} g^{\mu\nu} F_{,\nu}.$$

(426)

Of course, a much simpler way would be just to take Eq. (423) and set $\Phi = F$:

$$0=\left(\begin{array}{c}F^3\cdot R-\dfrac{F^2}{2}\left(\begin{array}{l}2F_{,ij}(n-1)g^{ij}+2\Gamma_{ij}^{a}F_{,a}g^{ij}\\-F_{,i}g^{ab}g_{jb,a}g^{ij}-F_{,j}g^{ab}g_{ib,a}g^{ij}\\-n\Gamma_{ij}^{d}F_{,d}g^{ij}+\dfrac{n}{2}F_{,d}g^{cd}g_{ab,c}g^{ab}\end{array}\right)\\-F\cdot\dfrac{F_{,i}\cdot F_{,j}}{4}g^{ij}\left((n-6)(n-1)\right)\end{array}\right)\frac{(2-n)}{2\cdot(1-n)}-F^2\cdot\Delta_g F+F\cdot F_{,\mu}g^{\mu\nu}F_{,\nu}$$

$$\Rightarrow$$

$$0=\left(\begin{array}{c}F^2\cdot R-\dfrac{F}{2}\left(\begin{array}{l}2F_{,ij}(n-1)g^{ij}+2\Gamma_{ij}^{a}F_{,a}g^{ij}\\-F_{,i}g^{ab}g_{jb,a}g^{ij}-F_{,j}g^{ab}g_{ib,a}g^{ij}\\-n\Gamma_{ij}^{d}F_{,d}g^{ij}+\dfrac{n}{2}F_{,d}g^{cd}g_{ab,c}g^{ab}\end{array}\right)\\-\dfrac{F_{,i}\cdot F_{,j}}{4}g^{ij}\left((n-6)(n-1)\right)\end{array}\right)\frac{(2-n)}{2\cdot(1-n)}-F\cdot\Delta_g F+F_{,\mu}g^{\mu\nu}F_{,\nu}.$$

(427)

We may see this as a special eigenvalue equation of quadratic character with eigenvalues as follows:

$$\frac{1}{2}\left(\begin{array}{l}2F_{,ij}(n-1)g^{ij}+2\Gamma_{ij}^{a}F_{,a}g^{ij}\\-F_{,i}g^{ab}g_{jb,a}g^{ij}-F_{,j}g^{ab}g_{ib,a}g^{ij}\\-n\Gamma_{ij}^{d}F_{,d}g^{ij}+\dfrac{n}{2}F_{,d}g^{cd}g_{ab,c}g^{ab}\end{array}\right)\frac{(2-n)}{2\cdot(1-n)}+\Delta_g F=M_{\Delta}^2\cdot F$$

$$F_{,\mu}g^{\mu\nu}F_{,\nu}\left(1-\frac{(n-6)(n-1)(2-n)}{8\cdot(1-n)}\right)=M_{\nabla}^2\cdot F^2, \tag{428}$$

which results in the following equation:

$$0=F^2\cdot\left(R\frac{(2-n)}{2\cdot(1-n)}-M_{\Delta}^2+M_{\nabla}^2\right). \tag{429}$$

Without any setting for the function Φ, using an eigenvalue approach for the function F, we could similarly evaluate from Eq. (423) onward as follows:

$$0=\left[\Phi\cdot\overbrace{\left(F^2\cdot R-\frac{F}{2}\left(\begin{array}{l}2F_{,ij}(n-1)g^{ij}+2\Gamma_{ij}^{a}F_{,a}g^{ij}\\ -F_{,i}g^{ab}g_{jb,a}g^{ij}-F_{,j}g^{ab}g_{ib,a}g^{ij}\\ -n\Gamma_{ij}^{d}F_{,d}g^{ij}+\frac{n}{2}F_{,d}g^{cd}g_{ab,c}g^{ab}\end{array}\right)\frac{(2-n)}{2\cdot(1-n)}\\ -\frac{F_{,i}\cdot F_{,j}}{4}g^{ij}\left((n-6)(n-1)\right)\right)}^{=F^2\cdot M_F^2}\\ +F\cdot\left(F_{,\mu}g^{\mu\nu}\Phi_{,\nu}-F\cdot\Delta_g\Phi\right)\right]$$

$$\Rightarrow\qquad 0=\Phi\cdot F^2\cdot M_F^2+F\cdot\left(F_{,\mu}g^{\mu\nu}\Phi_{,\nu}-F\cdot\Delta_g\Phi\right)$$

$$0=\Phi\cdot F\cdot M_F^2+F_{,\mu}g^{\mu\nu}\Phi_{,\nu}-F\cdot\Delta_g\Phi. \tag{430}$$

The last line now offers a few options. Here, in order to give an example, we just set an eigenvalue for the function Φ under the Laplace operator with respect to the metric g_{ab} and obtain:

$$0=\Phi\cdot F\cdot M_F^2+F_{,\mu}g^{\mu\nu}\Phi_{,\nu}-F\cdot\overbrace{\Delta_g\Phi}^{=\Phi\cdot M_\Phi^2}$$

$$\Rightarrow\qquad 0=\Phi\cdot F\cdot\overbrace{\left(M_F^2-M_\Phi^2\right)}^{-M^2}+F_{,\mu}g^{\mu\nu}\Phi_{,\nu}$$

$$0=-\Phi\cdot F\cdot M^2+F_{,\mu}\mathbf{g}^{\mu}\cdot\mathbf{g}^{\nu}\Phi_{,\nu}$$

$$\Rightarrow\qquad 0=\Phi\cdot F\cdot M^2-F_{,\mu}\mathbf{g}^{\mu}\cdot\mathbf{g}^{\nu}\Phi_{,\nu}. \tag{431}$$

Now our Dirac-like approach shall be:

$$\left(a\cdot\Phi\cdot M+A\cdot\mathbf{g}^{\nu}\Phi_{,\nu}\right)\left(b\cdot\Phi\cdot M-B\cdot F_{,\mu}\mathbf{g}^{\mu}\right)\Phi=0. \tag{432}$$

With the objects a, A, b and B being general 2×2 matrices, we find a variety of solutions of which we only present a very simple one, namely:

$$A=\begin{pmatrix}1&1\\1&1\end{pmatrix};\quad a=\begin{pmatrix}-1&1\\-1&1\end{pmatrix};\quad B=\begin{pmatrix}1&-1\\1&-1\end{pmatrix};\quad b=\begin{pmatrix}-1&1\\1&-1\end{pmatrix}. \tag{433}$$

This leads Eq. (432) to result in:

$$\begin{aligned}&\left(a\cdot\Phi\cdot M+A\cdot\mathbf{g}^{\nu}\Phi_{,\nu}\right)\left(b\cdot\Phi\cdot M-B\cdot F_{,\mu}\mathbf{g}^{\mu}\right)\Phi=0\\&=2\cdot\begin{pmatrix}1&-1\\1&-1\end{pmatrix}\cdot\left(\Phi\cdot F\cdot M^{2}-F_{,\mu}\mathbf{g}^{\mu}\cdot\mathbf{g}^{\nu}\Phi_{,\nu}\right)\\&=2\cdot\begin{pmatrix}1&-1\\1&-1\end{pmatrix}\cdot\left(\Phi\cdot F\cdot M^{2}-F_{,\mu}g^{\mu\nu}\Phi_{,\nu}\right)\end{aligned}\tag{434}$$

and gives us the following quantum gravity "Dirac equations":

$$\begin{aligned}&a\cdot\Phi\cdot M+A\cdot\mathbf{g}^{\nu}\Phi_{,\nu}=\begin{pmatrix}-1&1\\-1&1\end{pmatrix}\cdot\Phi\cdot M+\begin{pmatrix}1&1\\1&1\end{pmatrix}\cdot\mathbf{g}^{\nu}\Phi_{,\nu}=0\\\Rightarrow\quad&0=\left(\begin{pmatrix}-1&1\\-1&1\end{pmatrix}\cdot M+\begin{pmatrix}1&1\\1&1\end{pmatrix}\cdot\mathbf{g}^{\nu}\partial_{\nu}\right)\Phi\\&b\cdot F\cdot M-B\cdot F_{,\mu}\mathbf{g}^{\mu}=\begin{pmatrix}-1&1\\1&-1\end{pmatrix}\cdot F\cdot M-\begin{pmatrix}1&-1\\1&-1\end{pmatrix}\cdot F_{,\mu}\mathbf{g}^{\mu}=0\\\Rightarrow\quad&0=\left(\begin{pmatrix}-1&1\\1&-1\end{pmatrix}\cdot M-\begin{pmatrix}1&-1\\1&-1\end{pmatrix}\cdot\mathbf{g}^{\mu}\partial_{\mu}\right)F.\end{aligned}\tag{435}$$

As Eq. (431) is to be seen as some kind of boundary condition for the eigenvalue equations in Eqs. (430) and (431) (first line), we have the following pair of separated equations for the functions F and Φ:

$$F\Rightarrow\begin{cases}F^{2}\cdot M_{F}^{2}=\left(\begin{aligned}&F^{2}\cdot R-\frac{F}{2}\left(\begin{aligned}&2F_{,ij}(n-1)g^{ij}+2\Gamma_{ij}^{a}F_{,a}g^{ij}\\&-F_{,i}g^{ab}g_{jb,a}g^{ij}-F_{,j}g^{ab}g_{ib,a}g^{ij}\\&-n\Gamma_{ij}^{d}F_{,d}g^{ij}+\frac{n}{2}F_{,d}g^{cd}g_{ab,c}g^{ab}\end{aligned}\right)\\&-\frac{F_{,i}\cdot F_{,j}}{4}g^{ij}\left((n-6)(n-1)\right)\end{aligned}\right)\frac{(2-n)}{2\cdot(1-n)}\\[2ex]0=\left(\begin{pmatrix}-1&1\\1&-1\end{pmatrix}\cdot M-\begin{pmatrix}1&-1\\1&-1\end{pmatrix}\cdot\mathbf{g}^{\mu}\partial_{\mu}\right)F\end{cases}\tag{436}$$

and

$$\Phi \Rightarrow \begin{cases} \Phi \cdot M_{\Phi}^{2} = \Delta_{g}\Phi \\ 0 = \left(\begin{pmatrix} -1 & 1 \\ -1 & 1 \end{pmatrix} \cdot M + \begin{pmatrix} 1 & 1 \\ 1 & 1 \end{pmatrix} \cdot \mathbf{g}^{\nu}\partial_{\nu} \right)\Phi, \end{cases} \tag{437}$$

respectively.

Thereby it should be pointed out that with a suitable choice of the function $F[f]$, the term $\frac{F_{,i} \cdot F_{,j}}{4} g^{ij}$ can always be avoided in any number of dimensions. As shown further up in this book and in many other papers by this author (sub-section "So now what?" in [9]; also see [4] for the complete calculation), the corresponding function reads:

$$F[f] = \left(C_{f} + f\right)^{\frac{4}{n-2}}. \tag{438}$$

Some discussion about these results can be found in our previous papers, of which (apart from the books [2, 4, 5]) we especially recommend [7–23].

5.18.5 Extraction of the Quantum Theory Out of Hilbert's Approach

The extraction of the main quantum equations out of our generalized Einstein–Hilbert action as given above has been—quite extensively—elaborated in [7, 8, 13]. Regarding a fairly compact presentation about how to obtain the Klein–Gordon [24], the Schrödinger [25, 26] and the Dirac equation [27, 28] from this concept, the reader is especially referred to [13]. Here, we will need only one derivation, which is the one of the metric Klein–Gordon equation directly out of the quantum Einstein field equations. We simply repeat from [19] (see there for more details and explanations).

We are looking for intrinsic solutions to the quantum Einstein field equations (405) as follows:

$$0 = \left(\begin{array}{c} -\frac{g^{ab}}{2F} \left(\begin{array}{c} 2F_{,ab}(n-2) - \Gamma_{ab}^{d} F_{,d}(n-2) \\ +F_{,d} g^{cd} \left(g_{ab,c} \frac{1}{2}(n-2) - 2g_{cb,a} \right) \end{array} \right) \\ -\frac{F_{,i} \cdot F_{,j}}{4F^{2}} g^{ij} \left((n-7)(n-2)\right) \end{array} \right)$$

$$0=R_{\alpha\beta}-\frac{g_{\alpha\beta}}{2}R+\left(\begin{array}{c}\frac{1}{2F}\left(\begin{array}{c}F_{,\alpha\beta}(n-2)+F_{,a}g^{ab}\left(g_{\beta b,\alpha}-g_{\beta\alpha,b}\right)\\ -F_{,\alpha}g^{ab}g_{\beta b,a}-F_{,\beta}g^{ab}g_{\alpha b,a}\\ +F_{,d}g^{cd}\frac{1}{2}n\left(\frac{2}{n}g_{\alpha c,\beta}-g_{\alpha c,\beta}-g_{\beta c,\alpha}+g_{\alpha\beta,c}\right)\\ -\frac{1}{2F}\left(F_{,\alpha}\cdot F_{,\beta}(3n-6)\right)\end{array}\right)\\ -\frac{\left(\partial_{\alpha}\Phi_{,\beta}-\Gamma^{*\gamma}_{\alpha\beta}\Phi_{,\gamma}-\frac{F\cdot g_{\alpha\beta}}{\sqrt{-gF^{n}}}\partial_{\mu}\left(\sqrt{-gF^{n}}\cdot\frac{g^{\mu\nu}}{F}\Phi_{,\nu}\right)\right)}{\Phi}\end{array}\right). \tag{439}$$

We start with the first equation in Eq. (439) and obtain:

$$0=\left(\begin{array}{c}-\frac{g^{ab}}{2F}\left(\begin{array}{c}2\left(F''f_{,a}\cdot f_{,b}+F'f_{,ab}\right)(n-2)-F'\Gamma^{d}_{ab}f_{,d}(n-2)\\ +F'f_{,d}g^{cd}\left(g_{ab,c}\frac{1}{2}(n-2)-2g_{cb,a}\right)\end{array}\right)\\ -(F')^{2}\frac{f_{,a}\cdot f_{,b}}{4F^{2}}g^{ab}\left((n-7)(n-2)\right)\end{array}\right)$$

$$=\left(\begin{array}{c}-\frac{g^{ab}F'}{2F}\left(\begin{array}{c}2f_{,ab}(n-2)-\Gamma^{d}_{ab}f_{,d}(n-2)\\ +f_{,d}g^{cd}\left(g_{ab,c}\frac{1}{2}(n-2)-2g_{cb,a}\right)\end{array}\right)\\ -\frac{f_{,a}\cdot f_{,b}}{2F}g^{ab}(n-2)\left(\frac{(F')^{2}}{2F}(n-7)+2F''\right)\end{array}\right). \tag{440}$$

As linearization condition, we demand:

$$0=\frac{(F')^{2}}{2F}(n-7)+2F'', \tag{441}$$

which can be satisfied via the function $F[f]$ as follows:

$$F[f]=\begin{cases}C_{F}\cdot\left(f+C_{f}\right)^{\frac{4}{n-3}} & n\neq 3\\ C_{F}\cdot e^{f\cdot C_{f}} & n=3.\end{cases} \tag{442}$$

Subsequently, Eq. (440) can be simplified and yields

$$0=g^{ab}\left(2f_{,ab}(n-2)-\Gamma_{ab}^{d}f_{,d}(n-2)+f_{,d}g^{cd}\left(g_{ab,c}\frac{1}{2}(n-2)-2g_{cb,a}\right)\right). \qquad (443)$$

Interestingly, for metrics without shear elements:

$$g_{ij}=\begin{pmatrix} g_{00} & \cdots & 0 \\ \vdots & \ddots & \vdots \\ 0 & \cdots & g_{n-1n-1} \end{pmatrix}, \qquad (444)$$

this converges to the ordinary Laplace operator, namely:

$$0=g^{ab}\left(2f_{,ab}(n-2)-\Gamma_{ab}^{d}f_{,d}(n-2)+f_{,d}g^{cd}\left(g_{ab,c}\frac{1}{2}(n-2)-2g_{cb,a}\right)\right)$$

$$\xrightarrow{g_{ij}=\begin{pmatrix} g_{00} & \cdots & 0 \\ \vdots & \ddots & \vdots \\ 0 & \cdots & g_{n-1n-1} \end{pmatrix}}$$

$$0=2(n-2)\Delta f. \qquad (445)$$

As said before, some critics may now argue that this equation is not of Klein–Gordon character as it does not contain any potential nor mass, but this author has already shown that this problem is easily solved by adding additional dimensions carrying the right properties to produce masses and potentials (e.g., [4, 7–10, 29, 30]).

5.18.6 Derivation of the Functional Derivative of the Scaled and Adjusted Einstein–Hilbert Action with Respect to the Function *f*

A third option to derive scalar quantum gravity field equations requires the use of the functional derivative:

$$\delta W=\frac{\delta W}{\delta f[\mathbf{y}]}=0=\frac{\delta}{\delta f}\int_V d^n x\left(\sqrt{g\cdot F^n}\cdot F\cdot R^{*}\right)$$

$$=\lim_{\in\to 0}\frac{1}{\in}\left(\begin{array}{c}\int_V d^n x\left(\sqrt{g\cdot F\left[f[\mathbf{x}]+\in\cdot\delta[\mathbf{x}-\mathbf{y}]\right]^n}\cdot R^{**}\left[f[\mathbf{x}]+\in\cdot\delta[\mathbf{x}-\mathbf{y}]\right]\right)\\ -\int_V d^n x\left(\sqrt{g\cdot F[f[\mathbf{x}]]^n}\cdot R^{**}[f[\mathbf{x}]]\right)\end{array}\right). \qquad (446)$$

With:

$$R^{**}=F\cdot R^{*}=R-\frac{g^{\alpha\beta}}{2F}\left(2F_{,\alpha\beta}(n-1)+F_{,d}g^{cd}\left(\begin{array}{c}\left(g_{\beta c,\alpha}+g_{\alpha c,\beta}-g_{\beta\alpha,c}\right)-2g_{c\beta,\alpha}\\+\frac{n}{2}\left(\left(-g_{\alpha c,\beta}-g_{\beta c,\alpha}+g_{\alpha\beta,c}\right)+g_{\alpha\beta,c}\right)\end{array}\right)\right)$$

$$+\frac{1}{4F^{2}}\left(F_{,\alpha}\cdot F_{,\beta}(3n-6)+g_{\alpha\beta}F_{,c}F_{,d}g^{cd}(4-n)\right)g^{\alpha\beta}$$

$$=R-\frac{g^{\alpha\beta}}{2F}\left(2F_{,\alpha\beta}(n-1)+F_{,d}g^{cd}\left(\begin{array}{c}\left(g_{\beta c,\alpha}+g_{\alpha c,\beta}-g_{\beta\alpha,c}\right)-2g_{c\beta,\alpha}\\+\frac{n}{2}\left(\left(-g_{\alpha c,\beta}-g_{\beta c,\alpha}+g_{\alpha\beta,c}\right)+g_{\alpha\beta,c}\right)\end{array}\right)\right)$$

$$+\frac{1}{4F^{2}}\left(F_{,\alpha}\cdot F_{,\beta}(3n-6)+g_{\alpha\beta}F_{,c}F_{,d}g^{cd}(4-n)\right)g^{\alpha\beta}$$

$$=R-\frac{g^{\alpha\beta}}{2F}F'\left(2f_{,\alpha\beta}(n-1)+f_{,d}g^{cd}\left(\begin{array}{c}g_{\alpha c,\beta}-g_{\beta\alpha,c}-g_{c\beta,\alpha}\\+\frac{n}{2}\left(2g_{\alpha\beta,c}-g_{\alpha c,\beta}-g_{\beta c,\alpha}\right)\end{array}\right)\right)$$

$$-(n-1)\frac{f_{,\alpha}\cdot f_{,\beta}}{4F^{2}}g^{\alpha\beta}\left(4FF''+F'\cdot F'(n-6)\right), \tag{447}$$

this gives:

$$\frac{\delta W}{\delta f}=0=\frac{\delta}{\delta f}\int_{V}d^{n}x\left(\sqrt{g\cdot F^{n}}\cdot R^{**}\right)$$

$$=-\left[\begin{array}{c}\left((n-1)\cdot\sqrt{g}\cdot F^{\frac{n}{2}-1}g^{\alpha\beta}\cdot F'\right)_{,\alpha\beta}-\sqrt{g}\cdot R\frac{dF^{\frac{n}{2}}}{df}\\-\frac{1}{2}\left(\sqrt{g}\cdot F^{\frac{n}{2}-1}g^{\alpha\beta}\cdot F'g^{cd}\left(\begin{array}{c}g_{\alpha c,\beta}-g_{\beta\alpha,c}-g_{c\beta,\alpha}\\+\frac{n}{2}\left(2g_{\alpha\beta,c}-g_{\alpha c,\beta}-g_{\beta c,\alpha}\right)\end{array}\right)\right)_{,d}\\+\frac{\sqrt{g}\cdot g^{\alpha\beta}}{2}\left(2f_{,\alpha\beta}(n-1)+f_{,d}g^{cd}\left(\begin{array}{c}g_{\alpha c,\beta}-g_{\beta\alpha,c}-g_{c\beta,\alpha}\\+\frac{n}{2}\left(2g_{\alpha\beta,c}-g_{\alpha c,\beta}-g_{\beta c,\alpha}\right)\end{array}\right)\right)\frac{d\left[F^{\frac{n}{2}-1}F'\right]}{df}\\+\frac{(n-1)}{2}\left(\begin{array}{c}-\left(f_{,\alpha}\cdot\sqrt{g}\cdot g^{\alpha\beta}F^{\frac{n}{2}-2}\left(4FF''+F'\cdot F'(n-6)\right)\right)_{,\beta}\\+\frac{f_{,\alpha}\cdot f_{,\beta}}{2}\sqrt{g}\cdot g^{\alpha\beta}\frac{d\left[F^{\frac{n}{2}-2}\left(4FF''+F'\cdot F'(n-6)\right)\right]}{df}\end{array}\right)\end{array}\right]_{\mathbf{y}}. \tag{448}$$

We have already presented the evaluation in appendix M and, thus, will only present the result here, which reads:

$$\frac{\delta W}{\delta f}=0$$

$$=-\left[F^{\frac{n}{2}-3}F'(n-1)\left(\begin{array}{c}\frac{\sqrt{g}\cdot g^{\alpha\beta}}{2}\frac{f_{,\alpha}f_{,\beta}}{2}\left(\frac{1}{2}(n-4)(n+2)(F')^{2}+2FF''\right)\\+\frac{(n+2)}{2}FF'\sqrt{g}\cdot\Delta f\\+F^{2}\left(\sqrt{g}\cdot g^{\alpha\beta}\right)_{,\alpha\beta}-\sqrt{g}\cdot\frac{R\cdot F^{2}}{(n-1)}\\-\frac{F^{2}}{2(n-1)}\left(\sqrt{g}\cdot g^{\alpha\beta}\cdot g^{ij}\left(\begin{array}{c}g_{i\beta,j}-g_{ij,\beta}-g_{\beta j,i}\\+\frac{n}{2}\left(2g_{ij,\beta}-g_{i\beta,j}-g_{j\beta,i}\right)\end{array}\right)\right)_{,\alpha}\end{array}\right)\right]_{y}. \quad (449)$$

It totally suffices to demand that the inner factor gives zero in order to fulfill the variational condition and, thus, we obtain:

$$\frac{\delta W}{\delta f}=0 \;\Rightarrow\; \Rightarrow$$

$$0=\left[\begin{array}{c}\frac{\sqrt{g}}{2}\left(\frac{f_{,\alpha}f_{,\beta}}{2}g^{\alpha\beta}\left(\frac{1}{2}(n-4)(n+2)(F')^{2}+2FF''\right)+(n+2)FF'\Delta f\right)\\+F^{2}\left(\begin{array}{c}\left(\sqrt{g}\cdot g^{\alpha\beta}\right)_{,\alpha\beta}-\frac{1}{2(n-1)}\left(\sqrt{g}\cdot g^{\alpha\beta}\cdot g^{ij}\left(\begin{array}{c}g_{i\beta,j}-g_{ij,\beta}-g_{\beta j,i}\\+\frac{n}{2}\left(2g_{ij,\beta}-g_{i\beta,j}-g_{j\beta,i}\right)\end{array}\right)\right)_{,\alpha}\\-\frac{\sqrt{g}\cdot R}{(n-1)}\end{array}\right)\end{array}\right]_{y}, \quad (450)$$

$$\xrightarrow{F=C+a\cdot f}$$

$$0=\left[\begin{array}{c}\frac{\sqrt{g}}{2}(n+2)\left(\frac{f_{,\alpha}f_{,\beta}}{4}g^{\alpha\beta}(n-4)a^2+(C+a\cdot f)\cdot a\cdot\Delta f\right)\\ +(C+a\cdot f)^2\left(\begin{array}{c}\left(\sqrt{g}\cdot g^{\alpha\beta}\right)_{,\alpha\beta}-\frac{\sqrt{g}\cdot R}{(n-1)}\\ -\frac{1}{2(n-1)}\left(\sqrt{g}\cdot g^{\alpha\beta}\cdot g^{ij}\left(\begin{array}{c}g_{i\beta,j}-g_{ij,\beta}-g_{\beta j,i}\\ +\frac{n}{2}\left(2g_{ij,\beta}-g_{i\beta,j}-g_{j\beta,i}\right)\end{array}\right)\right)_{,\alpha}\end{array}\right)\end{array}\right]_{\mathbf{y}}. \tag{451}$$

Instead of the setting of a linear function for $F[f]$ as used in Eq. (451), we aim for a linearization of Eq. (450) and demand the condition:

$$0=(n-4)(n+2)(F')^2+4FF'', \tag{452}$$

which can be satisfied by the function $F[f]$ as follows:

$$F[f]=C_F\cdot\left(f+C_f\right)^{\frac{4}{12+n\cdot(n-6)}}. \tag{453}$$

Subsequently, Eq. (450) can be simplified and yields

$$0=\left[\begin{array}{c}\frac{\sqrt{g}}{2}(n+2)F\frac{F'}{F}\Delta f-\frac{\sqrt{g}\cdot R}{(n-1)}\\ +\left(\sqrt{g}\cdot g^{\alpha\beta}\right)_{,\alpha\beta}-\frac{1}{2(n-1)}\left(\sqrt{g}\cdot g^{\alpha\beta}\cdot g^{ij}\left(\begin{array}{c}g_{i\beta,j}-g_{ij,\beta}-g_{\beta j,i}\\ +\frac{n}{2}\left(2g_{ij,\beta}-g_{i\beta,j}-g_{j\beta,i}\right)\end{array}\right)\right)_{,\alpha}\end{array}\right]$$

$$\Rightarrow$$

$$0=\left[\begin{array}{c}(n-1)(n+2)F\frac{F'}{2F}\Delta f-R\\ +\left(\frac{(n-1)}{\sqrt{g}}\left(\sqrt{g}\cdot g^{\alpha\beta}\right)_{,\alpha\beta}-\frac{1}{2\sqrt{g}}\left(\sqrt{g}\cdot g^{\alpha\beta}\cdot g^{ij}\left(\begin{array}{c}g_{i\beta,j}-g_{ij,\beta}-g_{\beta j,i}\\ +\frac{n}{2}\left(2g_{ij,\beta}-g_{i\beta,j}-g_{j\beta,i}\right)\end{array}\right)\right)_{,\alpha}\right)\end{array}\right]. \tag{454}$$

5.18.7 The Conventional and the Unconventional Physics Realm

5.18.7.1 The conventional physics realm

From the material presented in the previous sections and the appendices above, one might feel tempted to conclude that the scale factors to every metric may be interpreted as the soul, consciousness or mind of objects or—as this would be of special interest in connection with the so-called multiverse interpretation—as a mathematical tool for the switch to another world in the sense of Everett's approach [62–66]. That things are not so simple can be demonstrated by the fact that in our 4-dimensional space-time the corresponding scale factor provides the electromagnetic field as it was shown in section 2 of [2], has been repeated in chapter 2 of [4] and is represented here in appendix K. So, our question: how can we come to a gravity quantum description which embeds more dimensions with metric and quantum fields? The solution to this problem can be found via the following ansatz:

$$G_{\alpha\beta} = F_n\left[f_n\right]\begin{pmatrix} F_{n-1}\cdot\ldots\cdot F_2\cdot g_{00} & F_{n-1}\cdot\ldots\cdot F_2\cdot g_{10} & \cdots & \cdots & \cdots & g_{n0} \\ F_{n-1}\cdot\ldots\cdot F_2\cdot g_{10} & F_{n-1}\cdot\ldots\cdot F_2\cdot g_{11} & \cdots & \cdots & \cdots & \vdots \\ F_{n-1}\cdot\ldots\cdot F_3\cdot g_{20} & F_{n-1}\cdot\ldots\cdot F_3\cdot g_{21} & \ddots & \cdots & \cdots & \vdots \\ F_{n-1}\cdot\ldots\cdot F_4\cdot g_{30} & F_{n-1}\cdot\ldots\cdot F_4\cdot g_{31} & \cdots & \ddots & \cdots & \vdots \\ \cdots & \cdots & \cdots & \cdots & \ddots & \vdots \\ g_{n0} & \cdots & \cdots & \cdots & \cdots & g_{nn} \end{pmatrix}$$

$$F_i = F_i\left[f_i\right]. \tag{455}$$

Using the definition:

$$^{2}G_{\alpha\beta} = F_2\cdot{}^{2}g_{\alpha\beta}$$

$$^{3}G_{\alpha\beta} = F_3\cdot{}^{3}g_{\alpha\beta}$$

$$\vdots$$

$$^{n}G_{\alpha\beta} = F_n\cdot{}^{n}g_{\alpha\beta}, \tag{456}$$

we can write Eq. (455) as follows:

$$^{2}G_{\alpha\beta} = \left(F_2\cdot\left({}^{2}g_{\alpha\beta}\right)\right) = F_2\cdot\begin{pmatrix} {}^{2}g_{00} & {}^{2}g_{01} \\ {}^{2}g_{01} & {}^{2}g_{11} \end{pmatrix}$$

$$^{3}G_{\alpha\beta}=F_3\cdot{}^{3}g_{\alpha\beta}=F_3\cdot\begin{pmatrix}^{3}g_{00} & ^{3}g_{01} & ^{3}g_{02}\\ ^{3}g_{01} & ^{3}g_{11} & ^{3}g_{12}\\ ^{3}g_{02} & ^{3}g_{12} & ^{3}g_{22}\end{pmatrix}=F_3\cdot\begin{pmatrix}F_2\cdot\begin{pmatrix}^{2}g_{00} & ^{2}g_{01}\\ ^{2}g_{01} & ^{2}g_{11}\end{pmatrix} & \begin{matrix}^{3}g_{02}\\ ^{3}g_{12}\end{matrix}\\ \begin{matrix}^{3}g_{02} & ^{3}g_{12}\end{matrix} & ^{3}g_{22}\end{pmatrix}$$

$$\vdots$$

$$^{n}G_{\alpha\beta}=F_n\cdot\begin{pmatrix}\begin{pmatrix}\cdots F_3\cdot\begin{pmatrix}F_2\cdot\begin{pmatrix}^{2}g_{00} & ^{2}g_{01}\\ ^{2}g_{01} & ^{2}g_{11}\end{pmatrix} & \begin{matrix}^{3}g_{02}\\ ^{3}g_{12}\end{matrix}\\ \begin{matrix}^{3}g_{02} & ^{3}g_{12}\end{matrix} & ^{3}g_{22}\end{pmatrix}\cdots\\ \cdots\end{pmatrix} & \begin{matrix}^{n}g_{0m}\\ \vdots\\ \vdots\\ \vdots\end{matrix}\\ ^{n}g_{0m}\ \cdots\ \cdots\ \cdots\ \cdots & ^{n}g_{mm}\end{pmatrix};\quad m=n-1 \quad (457)$$

and obtain the corresponding multi-scaled Ricci scalar to be:

$$R^{*}=\sum_{m=1}^{n-1}\left(\sum_{\alpha=0}^{m}2\cdot{}^{m+1}R_{m\alpha}\ {}^{m+1}g^{m\alpha}-{}^{m+1}R_{mm}\ {}^{m+1}g^{mm}\right)+\sum_{m=2}^{n}\frac{F_m'}{F_m}\cdot\Delta_m f_m$$

$$\Delta_m f=\frac{1}{\sqrt{^{m}g}}\sum_{\alpha,\beta=0}^{m-1}\partial_\beta\left(\sqrt{^{m}g}\cdot{}^{m}g^{\alpha\beta}\cdot\partial_\alpha f\right). \quad (458)$$

Thereby, $^{j}R_{\alpha\beta}$ denote the Ricci tensor to the $^{j}g_{\alpha\beta}$ metric and all functions F_i have to fulfill the condition:

$$4\cdot F_iF_i''+F_i'F_i'(n-6)=0\ \Rightarrow\ F_i=\left(C_f\pm f^{a}\right)^{\frac{4}{n-2}};\quad n>2$$

$$4\cdot F_2F_2''+F_2'F_2'(n-6)=0\ \Rightarrow\ F_2=C_F\cdot e^{C_f\cdot f};\quad n=2. \quad (459)$$

Most interestingly, we obtain a sum of Laplace-operations in various dimensions on the scale-factors (wave functions) f plus curvature terms. Thus, what originally (on first sight) has been a typical General-Theory-of-Relativity task, namely the evaluation of the Ricci tensor and the subsequent Ricci scalar, evolves to a complex ensemble of quantum equations if only be brought into the right form. It should be explicitly pointed out that we can also construct string, brane, multi-verse and similar structures. Here we just give a few examples:

$$G_{\alpha\beta} = {}_nF\left[f_n\right]\begin{pmatrix} {}_2F_1\left[f_1\right]\cdot {}_1^2g_{\alpha\beta} & {}^2 0 & \cdots & \cdots & \cdots & {}^2 0 \\ {}^2 0 & {}_2F_2\left[f_2\right]\cdot {}_2^2g_{\alpha\beta} & \cdots & \cdots & \cdots & \vdots \\ {}^2 0 & {}^2 0 & \ddots & \cdots & \cdots & \vdots \\ {}^2 0 & {}^2 0 & \cdots & \ddots & \cdots & \vdots \\ \cdots & \cdots & \cdots & \cdots & \ddots & \vdots \\ {}^2 0 & \cdots & \cdots & \cdots & \cdots & {}_{n/2}^{2}g_{\alpha\beta} \end{pmatrix}$$

$${}_i^2g_{\alpha\beta} = \begin{pmatrix} {}_ig_{00} & {}_ig_{01} \\ {}_ig_{01} & {}_ig_{11} \end{pmatrix}; \quad {}_2F_i\left[f_i\right] = C_{f1i}\cdot e^{C_{fi}\cdot f_i} = C_{f1i}\cdot e^{C_{fi}\cdot f_i\left[x_i, y_i\right]}; \quad {}^2 0 = \begin{pmatrix} 0 & 0 \\ 0 & 0 \end{pmatrix}, \tag{460}$$

$$G_{\alpha\beta} = {}_nF\left[f_n\right]\begin{pmatrix} {}_3F_1\left[{}_3f_1\right]\cdot {}_1^3g_{\alpha\beta} & {}^2 0 & \cdots & \cdots & \cdots & {}^3 0 \\ {}^3 0 & {}_3F_2\left[{}_3f_2\right]\cdot {}_2^3g_{\alpha\beta} & \cdots & \cdots & \cdots & \vdots \\ {}^3 0 & {}^3 0 & \ddots & \cdots & \cdots & \vdots \\ {}^3 0 & {}^3 0 & \cdots & \ddots & \cdots & \vdots \\ \cdots & \cdots & \cdots & \cdots & \ddots & \vdots \\ {}^3 0 & \cdots & \cdots & \cdots & \cdots & {}_{n/3}^{3}g_{\alpha\beta} \end{pmatrix}$$

$${}_i^3g_{\alpha\beta} = \begin{pmatrix} {}_2F_i\left[{}_2f_i\right]\cdot\begin{pmatrix} {}_ig_{00} & {}_ig_{01} \\ {}_ig_{01} & {}_ig_{11} \end{pmatrix} & \begin{matrix} {}_ig_{02} \\ {}_ig_{11} \end{matrix} \\ \begin{matrix} {}_ig_{02} & \quad {}_ig_{21} \end{matrix} & {}_ig_{22} \end{pmatrix}; \quad {}_2F_i\left[{}_2f_i\right] = C_{f1i}\cdot e^{C_{fi}\cdot {}_2f_i} = C_{f1i}\cdot e^{C_{fi}\cdot {}_2f_i\left[x_i, y_i\right]}$$

$${}_3F_i\left[{}_3f_i\right] = C_{f1i}\cdot\left(C_{fi} + {}_3f_i\left[t_i, x_i, y_i, z_i\right]\right)^4; \quad {}^3 0 = \begin{pmatrix} 0 & 0 & 0 \\ 0 & 0 & 0 \\ 0 & 0 & 0 \end{pmatrix}, \tag{461}$$

$$G_{\alpha\beta} = {}_nF\left[f_n\right]\begin{pmatrix} {}_4F_1\left[f_1\right]\cdot {}_1^4g_{\alpha\beta} & {}^4 0 & \cdots & \cdots & \cdots & {}^4 0 \\ {}^4 0 & {}_4F_2\left[f_2\right]\cdot {}_2^4g_{\alpha\beta} & \cdots & \cdots & \cdots & \vdots \\ {}^4 0 & {}^4 0 & \ddots & \cdots & \cdots & \vdots \\ {}^4 0 & {}^4 0 & \cdots & \ddots & \cdots & \vdots \\ \cdots & \cdots & \cdots & \cdots & \ddots & \vdots \\ {}^4 0 & \cdots & \cdots & \cdots & \cdots & {}_{n/4}^{4}g_{\alpha\beta} \end{pmatrix}$$

$$
{}_{i}^{4}g_{\alpha\beta}=\begin{pmatrix} {}_{2}F_{i1}\left[{}_{2}f_{i1}\right]\cdot\begin{pmatrix} {}_{i}g_{00} & {}_{i}g_{01} \\ {}_{i}g_{01} & {}_{i}g_{11} \end{pmatrix} & {}^{2}0 \\ {}^{2}0 & {}_{2}F_{i2}\left[{}_{2}f_{i2}\right]\cdot\begin{pmatrix} {}_{i}g_{00} & {}_{i}g_{01} \\ {}_{i}g_{01} & {}_{i}g_{11} \end{pmatrix} \end{pmatrix};
$$

$$
{}_{4}F_{i}\left[f_{i}\right]=C_{f1i}\cdot\left(C_{fi}+f_{i}\left[t_{i},x_{i},y_{i},z_{i}\right]\right)^{2};\quad {}^{4}0=\begin{pmatrix} 0 & 0 & 0 & 0 \\ 0 & 0 & 0 & 0 \\ 0 & 0 & 0 & 0 \\ 0 & 0 & 0 & 0 \end{pmatrix}. \tag{462}
$$

When evaluating the Ricci scalar, we obtain—just as shown above with the Matryoshka structure from Eq. (155) or Eq. (457)—sums of Laplace operators applied on the various wave functions appearing in Eqs. (460) to (462). Almost arbitrary combinations of scaled sub-metrics in accordance with the examples given above are possible.

Examples and discussion will be presented elsewhere [67].

With this new multitude of options, however, the number of degrees of freedom does not end. As elaborated in [2] and [68], there is also the option of centers of gravity and centers of wave-functions, which simply means that we can seek for a satisfaction of Eq. (366) via variation with respect to the position of such centers. Seeing that, then the last line of Eq. (155) or Eq. (457) would become generalized as follows:

$$
{}^{n}G_{\alpha\beta}=F_{n}\left[X_{n}\right]\cdot\begin{pmatrix} \begin{pmatrix} \cdots F_{3}\left[X_{3}\right]\cdot\begin{pmatrix} F_{2}\left[X_{2}\right]\cdot\begin{pmatrix} {}^{2}g_{00} & {}^{2}g_{01} \\ {}^{2}g_{01} & {}^{2}g_{11} \end{pmatrix} & \begin{matrix} {}^{3}g_{02} \\ {}^{3}g_{12} \end{matrix} \\ \begin{matrix} {}^{3}g_{02} & {}^{3}g_{12} \end{matrix} & {}^{3}g_{22} \end{pmatrix} \cdots \\ \cdots \end{pmatrix} & \begin{matrix} {}^{n}g_{0m} \\ \vdots \\ \vdots \\ \vdots \end{matrix} \\ {}^{n}g_{0m}\quad \cdots\quad \cdots\quad \cdots\quad \cdots & {}^{n}g_{mm} \end{pmatrix}
$$

$$
X_{i}=x_{i}-\xi_{i};\quad {}^{i}g_{\alpha\beta}={}^{i}g_{\alpha\beta}\left[Y_{i}\right]={}^{i}g_{\alpha\beta}\left[y_{i}-\zeta_{i}\right], \tag{463}
$$

where we realize a huge number of possible combinations, all fulfilling Eq. (366) and many having the possibility to code not just matter, dark matter, energy, but also objects the common language might better express with the words “soul,” “mind” and “thought ensemble.”

For suitably chosen coordinates, we should be able to avoid terms like:

$$\sum_{m=1}^{n-1}\left(\sum_{\alpha=0}^{m}2\cdot{}^{m+1}R_{m\alpha}\;{}^{m+1}g^{m\alpha}-{}^{m+1}R_{mm}\;{}^{m+1}g^{mm}\right) \tag{464}$$

in Eq. (458) and subsequently obtain equations as follows:

$$R^{*}=\sum_{m=2}^{n}\frac{F_{m}{}'}{F_{m}}\cdot\Delta_{m}f_{m}$$

$$\Delta_{m}f=\frac{1}{\sqrt{{}^{m}g}}\sum_{\alpha,\beta=0}^{m-1}\partial_{\beta}\left(\sqrt{{}^{m}g}\cdot{}^{m}g^{\alpha\beta}\cdot\partial_{\alpha}f\right). \tag{465}$$

We realize that with a suitable eigenvalue solution for one function f_i, we might also obtain potentials, masses and other forms of inertia. Thereby, we also assume that we are able to restrict the coordinate choice and the variation to forms where we have Eq. (421) fulfilled and are just left with the scalar quantum Einstein field equation, such as the one given here as Eq. (375) or Eq. (376).

With the classical Dirac approach for the "factorization" of the Laplace operator, we should be able to force fermionic behavior into Eq. (465), but there is also another such possibility via a proper choice of the functions F_i. With the results from our previous publications [31–43], we can extract the option for the wrapper functions F_i as described in [69, 70]. Instead of a sum of various Laplace operators, we would then obtain the following equation:

$$R^{*}=\sum_{m=2}^{n}\frac{F_{m}{}'}{F_{m}}\cdot\left(f_{m}\Delta_{m}f_{m}-f_{m,\alpha}f_{m,\beta}\;{}^{m}g^{\alpha\beta}\right)=0$$

$$\Delta_{m}f=\frac{1}{\sqrt{{}^{m}g}}\sum_{\alpha,\beta=0}^{m-1}\partial_{\beta}\left(\sqrt{{}^{m}g}\cdot{}^{m}g^{\alpha\beta}\cdot\partial_{\alpha}f\right). \tag{466}$$

The solution can be found via the "Dirac" linearization as demonstrated in [69, 70]. Naturally, we could also have (via various choices of the functional wrapper F_i) mixed equations as follows:

$$R^{*}=\sum_{\forall i}\frac{F_{i}{}'}{F_{i}}(n_{i}-1)\left(f_{i}\Delta_{i}f_{i}-f_{i,\alpha}f_{i,\beta}\;{}^{i}g^{\alpha\beta}\right)+\sum_{\forall j}\frac{F_{j}{}'}{F_{j}}(n_{j}-1)\left(f_{j}\Delta_{j}f_{j}\right)=0$$

$$\Delta_{i}f=\frac{1}{\sqrt{{}^{i}g}}\sum_{\alpha,\beta=0}^{i-1}\partial_{\beta}\left(\sqrt{{}^{i}g}\cdot{}^{i}g^{\alpha\beta}\cdot\partial_{\alpha}f\right);\quad\Delta_{j}f=\frac{1}{\sqrt{{}^{j}g}}\sum_{\alpha,\beta=0}^{j-1}\partial_{\beta}\left(\sqrt{{}^{j}g}\cdot{}^{j}g^{\alpha\beta}\cdot\partial_{\alpha}f\right). \tag{467}$$

We see that in the realm of the conventional physics, which is to say, the physics following from Eq. (371), we already have a huge variety of possibilities to place (and potentially hide) something in there, which could be the residual of a formerly living entity, hence our previously made reference to the "soul" (also see our book [4]).

5.18.7.2 The unconventional physics realm

Now we simply propose that there are other solutions which are not covered by Eq. (371) but still follow from the Einstein–Hilbert action of ensembles of arbitrary properties and dimensionality. For reasons of classical perception, we name such solutions "chaos" and suggest that entities, consisting of ensembles of information and born in the conventional physics realm, could be transferred to the chaos and live on "in there." Thus, here we only need to derive the mathematical options for this chaos.

And why should this be of importance to the topic of "Mathematical Psychology"?

Well, because we think that a conscious entity (c.f. abstract of this book) has the ability to actually choose between the possibility to be "conventional" or "unconventional" and that therefore the latter better be rigorously derived, comprehensively observed and discussed, too. Above that, it does not seem to be too farfetched to assume that certain psychoses are just minds being trapped in the wrong solution to the Hamilton–Einstein–Hilbert extremal task (see Eq. (468) below), whether this solution is of "conventional" or "unconventional" style. The important thing to cure then is the knowledge about all such possibilities.

5.18.7.2.1 *What could the chaos be?*

Again, our starting point should be the Hamilton extremal principle and its manifestation for arbitrary Riemann attribute ensembles in the form of the Einstein–Hilbert action [1]:

$$\delta W = 0 = \delta \int_V d^n x \left(\sqrt{-g} \cdot \left(R - 2\Lambda + L_M \right) \right). \tag{468}$$

Generalization (scaling) of the metric as follows [2, 5]:

$$G_{\alpha\beta} = g_{\alpha\beta} \cdot F[f] \tag{469}$$

brings us to (here, for brevity, only the simplest metrics are considered, while general ones are discussed in the sections above; see also [4]):

$$\delta W = 0 = \delta \int_V d^n x \left(\sqrt{-g \cdot F^n} \times \left(\left(\begin{matrix} R + \frac{F'}{F}(1-n)\Delta f \\ + \frac{f_{,\alpha} f_{,\beta} g^{\alpha\beta}(1-n)}{4F^2}\left(4FF'' + (F')^2(n-6)\right) \end{matrix} \right) \frac{1}{F} \\ -2\Lambda + L_M \right) \right). \quad (470)$$

Performing the variation yields:

$$0 = \sqrt{-g \cdot F^n} \times \left(\begin{matrix} \boxed{R_{\alpha\beta} - \frac{g_{\alpha\beta}}{2} R + \kappa T_{\alpha\beta} + \Lambda \cdot g_{\alpha\beta}} \\ + \frac{g_{\alpha\beta}}{2} \left(\frac{1}{2F} \left(\begin{matrix} 2F_{,ij}(n-1) g^{ij} + 2\Gamma_{ij}^{a} F_{,a} g^{ij} \\ - F_{,i} g^{ab} g_{jb,a} g^{ij} - F_{,j} g^{ab} g_{ib,a} g^{ij} \\ - n\Gamma_{ij}^{d} F_{,d} g^{ij} + \frac{n}{2} F_{,d} g^{cd} g_{ab,c} g^{ab} \end{matrix} \right) \\ + \frac{F_{,i} \cdot F_{,j}}{4F^2} g^{ij} \left((n-6)(n-1)\right) \right) \\ - \frac{1}{2F} \left(\begin{matrix} F_{,\alpha\beta}(n-2) + F_{,ab} g_{\alpha\beta} g^{ab} \\ + F_{,a} g^{ab} \left(g_{\beta b,\alpha} - g_{\beta\alpha,b} \right) - F_{,\alpha} g^{ab} g_{\beta b,a} - F_{,\beta} g^{ab} g_{\alpha b,a} \\ + F_{,d} g^{cd} \frac{1}{2} n \left(\begin{matrix} \frac{2}{n} g_{\alpha c,\beta} - g_{\alpha c,\beta} - g_{\beta c,\alpha} \\ + g_{\alpha\beta,c} + \frac{1}{n} g_{\alpha\beta} g_{ab,c} g^{ab} \end{matrix} \right) \\ - \frac{1}{2F} \left(F_{,\alpha} \cdot F_{,\beta}(3n-6) + g_{\alpha\beta} F_{,c} F_{,d} g^{cd} (4-n) \right) \end{matrix} \right) \end{matrix} \right) \delta G^{\alpha\beta}, \quad (471)$$

where the red rectangle gives the classical Einstein field equations [3], while the rest describes what we know as Quantum Theory ... hence, we have a Quantum Gravity Theory.

Now, the usual path forward is to just make the term in parentheses zero, which—as shown in the chapters above—gives the classical field equations in the presence of quantum effects. This leads to all the physics we know, including General Theory of Relativity, Quantum Theory and even thermodynamics when considering ensembles of quantum and gravity centers [2, 4].

But this is only half of the truth ... in fact, it is only one third, because we also have the option of the terms G ($G = g * F^n$ ... determinant of the scaled metric) and the variation of the scaled metric:

$$\delta G^{\alpha\beta}, \tag{472}$$

being zero in order to fulfill the whole extremal principle. First, for us only the option of $g = 0$ is of interest, because in comparison to our classical physics, which is based on:

$$0=\left(\begin{array}{c} \boxed{R_{\alpha\beta}-\frac{g_{\alpha\beta}}{2}R+\kappa T_{\alpha\beta}+\Lambda\cdot g_{\alpha\beta}} \\ +\frac{g_{\alpha\beta}}{2}\left(\begin{array}{c}\frac{1}{2F}\left(\begin{array}{c}2F_{,ij}(n-1)g^{ij}+2\Gamma^{a}_{ij}F_{,a}g^{ij}\\ -F_{,i}g^{ab}g_{jb,a}g^{ij}-F_{,j}g^{ab}g_{ib,a}g^{ij}\\ -n\Gamma^{d}_{ij}F_{,d}g^{ij}+\frac{n}{2}F_{,d}g^{cd}g_{ab,c}g^{ab}\end{array}\right)\\ +\frac{F_{,i}\cdot F_{,j}}{4F^2}g^{ij}\left((n-6)(n-1)\right)\end{array}\right)\\ -\frac{1}{2F}\left(\begin{array}{c} F_{,\alpha\beta}(n-2)+F_{,ab}g_{\alpha\beta}g^{ab}\\ +F_{,a}g^{ab}\left(g_{\beta b,\alpha}-g_{\beta\alpha,b}\right)-F_{,\alpha}g^{ab}g_{\beta b,a}-F_{,\beta}g^{ab}g_{\alpha b,a}\\ +F_{,d}g^{cd}\frac{1}{2}n\left(\frac{2}{n}g_{\alpha c,\beta}-g_{\alpha c,\beta}-g_{\beta c,\alpha}+g_{\alpha\beta,c}+\frac{1}{n}g_{\alpha\beta}g_{ab,c}g^{ab}\right)\\ -\frac{1}{2F}\left(F_{,\alpha}\cdot F_{,\beta}(3n-6)+g_{\alpha\beta}F_{,c}F_{,d}g^{cd}(4-n)\right)\end{array}\right)\end{array}\right), \tag{473}$$

this is something completely different, a quasi paradigm shift and compared to what we know from our "daily" cognitive experience, "chaos" [65].

Thus, there may well have been a "time" before "time" in which this "g=0-chaos" reigned and which then somehow switched to the other, the ordinary physics option and created what we recognize as our universe.

The fact that the g=0-condition also demands the universe to have no volume and that—according to the great theoreticians—the big bang is considered to have started with a tiny (even singular) universe, kind of even supports this idea of a "g=0-chaos before time."

However, we should not forget the other option being:

$$\delta G^{\alpha\beta} = 0. \tag{474}$$

This also turns out to be a possibility, as we always can expand:

$$\delta G^{\alpha\beta} = g^{\alpha\beta}\delta\left(\frac{1}{F}\right) + \frac{\delta g^{\alpha\beta}}{F}. \tag{475}$$

Here we have an apparently lazy universe, with not much mood for any global variational dynamics. As the same holds for any system, including a conscious mind, the laziness aspect could be taken quite literally for some human beings (certain politicians, for instance) and their more or less inactive thinking apparatus usually "working" in the idle state ... if at all.

There would only be an intrinsic activity.

We cannot exclude the universe (or its creator) having awoken from such a state or even being an intermediate option in a timely and/or spatial aspect.

But as such a situation also allows for an extremely strange physics, we might just also consider this as a "chaos" state and thus "prove" the old genesis theory of the Hebrews with the two mathematical options:

$$\delta G^{\alpha\beta} = 0, \tag{476}$$

and:

$$g = 0, \tag{477}$$

right.

At this point, it should be added that the apparent simple chaos-condition $g = 0$ (c.f. Eq. (477)) is anything but simple, because the determinant of the metric tensor g_{ab} is given via:

$$g = \det\left[g_{ab}\right] = \frac{1}{(n-1)!}\cdot \varepsilon_{i_1 \dots i_{n-1}} \varepsilon^{j_1 \dots j_{n-1}} g_{i_1 j_1} \cdot \dots \cdot g_{i_{n-1} j_{n-1}}. \tag{478}$$

It should also be pointed out that so far, at least in the established physics, the variation with respect to the number of dimensions or the variation with respect to centers of gravity or quantum centers (c.f. corresponding derivation and discussion in [2, 4] and section "3.1 Brief Note about the Thermodynamic of Thoughts" in this book), has not been taken into account. Here, too, a non-classical-physics can arise and offer yet another explanation for the "chaos" before time. However, as this author has already discussed this variation in connection with the Bekenstein–Hawking problem of storing information in Black Holes (e.g., c.f. [4], pp. 342), we will not consider this approach here any further. The same holds for the variation with respect to gravity and quantum centers as demonstrated in [2, 4, 7, 71–75], which leads—among other things—to thermodynamics.

5.18.8 Towards the Metric Everett Mind

Before we can develop a multiverse or Everett theory of the mind, we have to derive a metric understanding about the whole multiverse concept. We start with a brief introduction to Everett's idea.

5.18.8.1 About Hugh Everett's many-world theory

5.18.8.1.1 *The idea*

We cite from [A]:

> "Everett's original work introduced the concept of a relative state. Two (or more) subsystems, after a general interaction, become entangled. Everett noted that such entangled systems can be expressed as the sum of products of states, where the two or more subsystems are each in a state relative to each other. After a measurement or observation one of the pair (or triple ...) is the measured, object or observed system, and one other member is the measuring apparatus (which may include an observer) having recorded the state of the measured system.
>
> In the example of Schrödinger's cat, after the box is opened, the entangled system is the cat, the poison vial and the observer. One relative triple of states would be the alive cat, the unbroken vial and the observer seeing an alive cat. Another relative triple of states would be the dead cat, the broken vial and the observer seeing a dead cat.
>
> The process of measurement or observation, or any correlation-inducing interaction, splits the system up into sets of relative states, where each set of relative states, forming a branch of the universal wavefunction, is consistent within itself, and all future measurements (including by multiple observers) will confirm this consistency.
>
> The many-worlds interpretation is DeWitt's popularization of Everett, who had referred to the combined observer–object system as split by an observation, each split corresponding to the different or multiple possible outcomes of an observation. These splits generate a branching tree, where each branch is a set of all the states relative to each other. DeWitt introduced the term "world" to describe a single branch of that tree, which is a consistent history. All observations or measurements in any branch are consistent with each other.
>
> Under the many-worlds interpretation, the Schrödinger equation, or its quantum field theory, relativistic analog, holds all the time, everywhere. An observation or measurement is modelled by applying the wave equation to the entire system, comprising the observer and the object being observed. One consequence is that

every observation can be thought of as causing the combined observer–object's wavefunction to change into a quantum superposition of two or more non-interacting branches, or split into many "worlds." Since many observation-like events have happened and are constantly happening, there are an enormous and growing number of simultaneously existing states.

If a system is composed of two or more subsystems, the system's state will be a superposition of products of the subsystems' states. Each product of subsystem states in the overall superposition evolves over time independently of other products. Once the subsystems interact, their states have become correlated or entangled and can no longer be considered independent. In Everett's terminology, each subsystem state was now correlated with its relative state, since each subsystem must now be considered relative to the other subsystems with which it has interacted."

5.18.8.1.2 *Roger Penrose's rejection of Everett's theory, or what does quantum gravity say?*

Again, citing from [A], we find that one essential point of criticism to Everett's theory arises from the fact that it is not based on a quantum gravity or Theory of Everything approach:

"Roger Penrose argues that the idea is flawed because it is based on an oversimplified version of quantum mechanics that does not account for gravity. In his view, applying conventional quantum mechanics to the universe implies the MWI, but the lack of a successful theory of quantum gravity negates the claimed universality of conventional quantum mechanics. According to Penrose, "the rules must change when gravity is involved." He further asserts that gravity helps anchor reality and "blurry" events have only one allowable outcome: "electrons, atoms, molecules, etc., are so minute that they require almost no amount of energy to maintain their gravity, and therefore their overlapping states. They can stay in that state forever, as described in standard Quantum Theory." On the other hand, "in the case of large objects, the duplicate states disappear in an instant due to the fact that these objects create a large gravitational field."

Our goal in this appendix is now to apply a Quantum Gravity Theory and see what Everett's suggestion would produce.

For this, we have already reconsidered the derivation of the Dirac equation [B] and thereby investigated the so-called Dirac spinor problem [C] and solved it in a completely new way. We also reconsidered the various states of the solutions of the Klein–Gordon and the Schrödinger equation

[D, E] and then especially concentrated on the multi-world hydrogen atom, where we reused some results from [F–H]. This was all done in our paper [I]. Some of the derivation will be repeated here.

Our approach is rigorously based on the Einstein–Hilbert action [J, K] in an arbitrary number of dimensions and a few generalizations [L–N].

5.18.8.1.3 *References for section 5.18.8.1*

A. https://en.wikipedia.org/wiki/Many-worlds_interpretation.

B. P. A. M. Dirac, *The Quantum Theory of the Electron*, Published 1 February 1928.DOI: 10.1098/rspa.1928.0023.

C. N. Schwarzer, *Dirac's Little Spinor Problem—Can Quantum Gravity Explain the Occurrence of This Postulated Object?*, self-published, ASIN: B0C3QWJSQS, Amazon Digital Services, 2023, Kindle.

D. E. Schrödinger, Quantisierung als Eigenwertproblem (erste Mitteilung), *Ann. Phys.*, 1926, 384(4), 361–376.

E. E. Schrödinger, Quantisierung als Eigenwertproblem (zweite Mitteilung), *Ann. Phys.*, 1926, 384(6), 489–527.

F. N. Schwarzer, *Derivation of the Schrödinger Hydrogen from the Einstein-Hilbert Action—The Metric Schrödinger Hydrogen*, self-published, ASIN: B0BT9Y54XP, Amazon Digital Services, 2023, Kindle.

G. N. Schwarzer, *Towards the Metric Quark—Derivation of More Spin-1/2-Objects from the Einstein-Hilbert Action*, self-published, Amazon Digital Services, B0BWDQDBFW, 2023, Kindle.

H. N. Schwarzer, *The Metric Electron—Derivation of Spin-1/2-Objects from the Einstein-Hilbert Action*, self-published, ASIN: B0BTPQM26X, Amazon Digital Services, 2023, Kindle.

I. N. Schwarzer, *Hugh Everett's Multiverse—How Quantum Gravity Revives an Old Idea and—Perhaps—Gives It a Second Life*, self-published, Amazon Digital Services, 2023, Kindle, ASIN: B0CCQWC6YX.

J. D. Hilbert, Die Grundlagen der Physik, Teil 1, *Göttinger Nachrichten*, 1915, 395–407.

K. N. Schwarzer, *The World Formula: A Late Recognition of David Hilbert's Stroke of Genius*, Jenny Stanford Publishing, ISBN: 9789814877206.

L. A. Einstein, Grundlage der allgemeinen Relativitätstheorie, *Annalen der Physik*, 49, ser. 4, 769–822.

M. N. Schwarzer, *The Math of Body, Soul, and the Universe*, Jenny Stanford Publishing, ISBN 9789814968249.

N. N. Schwarzer, *The Theory of Everything—Quantum and Relativity Is Everywhere—A Fermat Universe*, Jenny Stanford Publishing, 2020, ISBN-10: 9814774472.

5.18.8.2 The metric Everett mind

Can there be a metric path of construction leading to an Everett-like mindset?

Could it be that in a mind full of ideas the moment of decision making is just the selection (metric choice) of one of the simultaneously existing and possible free-floating "thought setups" belonging to one set of potentially infinite solutions to a problem?

Well, the recipe to such a metric mind description was just given above. Here now we want to develop it into a consistent apparatus. Thereby, for better understanding and illustration, we start with the classical quantum equations from Klein–Gordon and Schrödinger.

5.18.8.3 The metrically derived Klein–Gordon–Everett equation

Again we could just apply the metric setting Eq. (462) already including the condition (459) and obtain the following Ricci scalar when assuming ${}_nF[f_n] = 1$:

$$R^{*} = 0 = \sum_{i=1}^{4} \frac{F_i\,'}{F_i} \cdot \Delta_i f_i$$

$$\Delta_i f = \frac{1}{\sqrt{{}_i^5 g}} \sum_{\alpha,\beta=0}^{4} \partial_\beta \left(\sqrt{{}_i^5 g} \cdot {}_i^5 g^{\alpha\beta} \cdot \partial_\alpha f \right), \tag{479}$$

where we may just introduce the masses via one of the 5 dimensions (see section 6.11.2 in [4]) and solve Eq. (479) by demanding all resulting Klein–Gordon equations to vanish. Alternatively, we could leave ${}_nF[f_n]$ open and interpret all the various solutions to the f_i as solutions to different Everett worlds with metrics ${}_i^5 g^{\alpha\beta}$. The global volume or scaling function ${}_nF[f_n]$ then acts as a kind of selector, which decides which of the solutions is going to be realized in "our" metric ${}_{\mathrm{our}}^{\;\;5} g^{\alpha\beta}$. Unfortunately, with Eq. (462) only 4 different solutions would be coded. This problem, however, can easily be remediated via the following metric extension:

$$G_{\alpha\beta} = {}_nF\left[f_n\right]\cdot g_{\alpha\beta}$$
$$= {}_nF\left[f_n\right]\begin{pmatrix} {}_NF_1\left[f_1\right]\cdot {}^N_1g_{\alpha\beta} & {}^N0 & \cdots & \cdots & \cdots & {}^N0 \\ {}^N0 & {}_NF_2\left[f_2\right]\cdot {}^N_2g_{\alpha\beta} & \cdots & \cdots & \cdots & \vdots \\ {}^N0 & {}^N0 & \ddots & \cdots & \cdots & \vdots \\ {}^N0 & {}^N0 & \cdots & \ddots & \cdots & \vdots \\ \cdots & \cdots & \cdots & \cdots & \ddots & \vdots \\ {}^N0 & \cdots & \cdots & \cdots & \cdots & {}^N_{n/N}g_{\alpha\beta} \end{pmatrix}$$

$${}_NF_i\left[f_i\right] = C_{f1i}\cdot\left(C_{fi} + f_i\left[t_i, x_i, y_i, z_i, \ldots\right]\right)^{\frac{4}{N-2}}; \quad {}^N0 = \begin{pmatrix} 0 & \cdots 0 & 0 \\ \vdots & \ddots 0 & 0 \\ 0 & 0\ \ 0 & 0 \\ 0 & 0\ \ 0 & 0 \end{pmatrix}. \tag{480}$$

The subsequent Ricci scalar reads:

$$R^* = 0 = R + \frac{{}_nF'}{{}_nF}\cdot\Delta_n f_n + \sum_{i=1}^{n/N}\frac{F_i'}{F_i}\cdot\Delta_i f_i$$

$$\Delta_i f = \frac{1}{\sqrt{{}^N_ig}}\sum_{\alpha,\beta=0}^{N-1}\partial_\beta\left(\sqrt{{}^N_ig}\cdot {}^N_ig^{\alpha\beta}\cdot\partial_\alpha f\right);$$
$$\Delta_n f = \frac{1}{\sqrt{g}}\sum_{\alpha,\beta=0}^{n-1}\partial_\beta\left(\sqrt{g}\cdot g^{\alpha\beta}\cdot\partial_\alpha f\right). \tag{481}$$

Now, once again, we can either have the usual Klein–Gordon equations for the f_i via a Ricci curvature of the following kind:

$$R = \sum_{i=1}^{n/N}R_i = \sum_{i=1}^{n/N}\frac{F_i'}{F_i}\cdot m_i^2\cdot f_i$$

$$\Rightarrow \qquad 0 = \frac{{}_nF'}{{}_nF}\cdot\Delta_n f_n + \sum_{i=1}^{n/N}\frac{F_i'}{F_i}\cdot\left(\Delta_i f_i + m_i^2\cdot f_i\right), \tag{482}$$

or we obtain masses via certain coordinates of the sub-metrics as demonstrated in section 6.11.2 of [4] and set $R = 0$.

As N can be an arbitrary number, we can have functions f_i and sub-metrics ${}^N_ig^{\alpha\beta}$ for all possible solutions to the differential equation:

$$0 = \Delta_i f_i + m_i^2\cdot f_i \xrightarrow{\Delta_i f_i = \Delta_{i+1}f_i = \Delta f_i;\ m_i^2 = m_{i+1}^2 = m^2\ \ \forall i} 0 = \Delta f_i + m^2\cdot f_i. \tag{483}$$

With the global function ${}_nF[f_n]$ as Everett-selector (multi-world entanglement), we will see the realization of one of the many f_i in our metric ${}^{N}_{\text{our}}g^{\alpha\beta}$. All other options may be realized "elsewhere," which is to say in those other Everett worlds. However, the global function ${}_nF[f_n]$, seen as a selector, might just also be interpreted as the "divine decision maker," who decides the finally realized f_i for the observer. Renaming the "divine decision maker" as "the thinker" and considering him or her to be the owner of the mind (or just be the mind), we now also have the apparatus to mathematically describe the process of decision making for a consciously thinking individual.

Both interpretations, the multi-world and the "divine decision maker," solve the Schrödinger cat problem. With the existence of such a global function, it should be possible to actually measure the entanglement with the other worlds ... if they exist.

5.18.8.4 The metrically derived Schrödinger–Everett equation

The—or at least one—Schrödinger–Everett equation can be found by applying the transition from the usual Klein–Gordon to the Schrödinger equation and by just extending this to the various time-coordinates t_i for all sub-metrics ${}^{N}_{i}g^{\alpha\beta}$.

We start from our result (376) in vector form (index i) and form metrics of the type (380):

$$0 = R - \frac{F_i'}{F_i}(n-1)\Delta f_i + \frac{f_{i,\alpha} f_{i,\beta}(1-n)}{4F_i^2} g^{\alpha\beta}\left(4F_i F_i'' + F_i' F_i'(n-6)\right), \quad (484)$$

only that, for the reason of recognition, instead of the symbol f we apply the classical Greek symbol Ψ. We have applied condition (459), respectively (132). We also, in order to be close enough to the classical Schrödinger case, assume a Minkowski-like metric time component. We write Eq. (484) as follows:

$$0 = \begin{cases} \left(\Psi - C_{f0}\right)\cdot R + (1-n)\cdot\left(\Delta_{n-1}\Psi - \dfrac{\partial^2\Psi}{c^2\cdot\partial t^2}\right) & n>2 \\ R + C_{f0}\cdot(1-n)\cdot\left(\Delta_{n-1}\Psi - \dfrac{\partial^2\Psi}{c^2\cdot\partial t^2}\right) & n=2. \end{cases} \quad (485)$$

We ignore the $n = 2$-case and take it that R can be assumed being linear to f, which is to say that the curvature of space-time is caused by and proportional to the quantum effects residing in it. This gives us:

$$0 = \left(\Psi - C_{f0}\right) \cdot R_c \cdot \Psi + (1-n) \cdot \left(\Delta_{n-1}\Psi - \frac{\partial^2 \Psi}{c^2 \cdot \partial t^2} \right). \tag{486}$$

Assuming the quantum function to be small against the classical physics (an important aspect in Schrödinger's work anyway and also agrees with the practical observation that the space-time curvature is small as we experience our surrounding space as being mainly flat), we discard the quadratic term, resulting in:

$$\begin{aligned} 0 &= -C_{f0} \cdot R_c \cdot \Psi + (1-n) \cdot \left(\Delta_{n-1}\Psi - \frac{\partial^2 \Psi}{c^2 \cdot \partial t^2} \right) \\ &= \left[(1-n) \cdot \left(\Delta_{n-1} - \frac{\partial^2}{c^2 \cdot \partial t^2} \right) - C_R \right] \Psi. \end{aligned} \tag{487}$$

Then we can separate the time derivative and—for the reason of recognition in connection with Schrödinger's classical approach—we reformulate everything as follows:

$$\begin{aligned} 0 &= -C_{f0} \cdot R_c \cdot \Psi + (1-n) \cdot \left(\Delta_{n-1}\Psi - \frac{\partial^2 \Psi}{c^2 \cdot \partial t^2} \right) \\ &= \left[(1-n) \cdot \left(\Delta_{n-1} - \frac{\partial^2}{c^2 \cdot \partial t^2} \right) - C_R \right] \Psi \\ &\xrightarrow{\left(\Delta_{n-1} - \frac{\partial^2}{c^2 \cdot \partial t^2} \right) - \frac{C_R}{(1-n)} = C1 \cdot \frac{\partial_t^2}{c^2} + \Delta_{n-1} - M^2} \\ \Rightarrow \quad 0 &= \left[\Delta - M^2 \right] \Psi \\ &= \left[\left(C1 \cdot \frac{\partial_t^2}{c^2} + \underbrace{\frac{1}{\sqrt{g}} \partial_\alpha \sqrt{g} \cdot g^{\alpha\beta} \partial_\beta}_{\text{3D}-\Delta-\text{Operator}} \right) - M^2 \right] \Psi. \end{aligned} \tag{488}$$

Please note that our Greek indices α and β are running only from 1 to $n - 1$ and not—as usual—from 0 to $n - 1$, because the time or 0-component has been separated. Now we introduce a function $\Psi = \Phi + X$ and demand the following additional condition:

$$\partial_t \Psi = c^2 \cdot C2 \cdot (\Phi - \mathrm{X}). \tag{489}$$

Together with Eq. (488), we obtain:

$$0=-M^2(\Phi+\mathrm{X})+\begin{pmatrix} C1\cdot C2\cdot\partial_t(\Phi-\mathrm{X}) \\ +\frac{1}{\sqrt{g}}\partial_\alpha\sqrt{g}\cdot g^{\alpha\beta}\partial_\beta(\Phi+\mathrm{X}) \end{pmatrix}. \tag{490}$$

The following two equations summed up would result in Eq. (490):

$$0=-M^2\Phi+\left(C1\cdot C2\cdot\partial_t\Phi+\frac{1}{\sqrt{g}}\partial_\alpha\sqrt{g}\cdot g^{\alpha\beta}\partial_\beta\Phi\right)$$

$$0=-M^2\mathrm{X}+\left(\frac{1}{\sqrt{g}}\partial_\alpha\sqrt{g}\cdot g^{\alpha\beta}\partial_\beta\mathrm{X}-C1\cdot C2\cdot\partial_t\mathrm{X}\right). \tag{491}$$

We see that we have obtained a Schrödinger-like first derivative for the time-coordinate by the introduction of $\Psi=\Phi+X$. Along the way, we have found a matter and an anti-matter case, which leads us to the two Schrödinger equations (491).

According to the derivation above and our results from the section before, we can directly extract the Schrödinger–Everett equation as follows:

$$R^*=0=R+\frac{{}_nF'}{{}_nF}\cdot\Delta_n f_n+\sum_{i=1}^{n/N}\frac{F_i'}{F_i}\cdot\left(\Delta_i f_i\pm C1_i\cdot C2_i\cdot\partial_{ti}f_i\right)$$

$$\Delta_i f=\frac{1}{\sqrt{{}_i^Ng}}\sum_{\alpha,\beta=1}^{N-1}\partial_\beta\left(\sqrt{{}_i^Ng}\cdot{}_i^Ng^{\alpha\beta}\cdot\partial_\alpha f\right);$$

$$\Delta_n f=\frac{1}{\sqrt{g}}\sum_{\alpha,\beta=0}^{n-1}\partial_\beta\left(\sqrt{g}\cdot g^{\alpha\beta}\cdot\partial_\alpha f\right). \tag{492}$$

The usual Schrödinger equations for the f_i are now obtained via a Ricci curvature of the following kind:

$$R=\sum_{i=1}^{n/N}R_i=\sum_{i=1}^{n/N}\frac{F_i'}{F_i}\cdot m_i^2\cdot f_i$$

$$\Rightarrow\qquad 0=\frac{{}_nF'}{{}_nF}\cdot\Delta_n f_n+\sum_{i=1}^{n/N}\frac{F_i'}{F_i}\cdot\left(\Delta_i f_i\pm C1_i\cdot C2_i\cdot\partial_{ti}f_i+m_i^2\cdot f_i\right), \tag{493}$$

or we obtain masses via certain coordinates of the sub-metrics as demonstrated in section 6.11.2 of [4] and set $R=0$.

As in the Klein–Gordon case, N can be an arbitrary number, allowing us to have functions f_i and sub-metrics ${}_i^N g^{\alpha\beta}$ for all possible solutions to the differential equation:

$$0 = \Delta_i f_i \pm C1_i \cdot C2_i \cdot \partial_{ti} f_i + m_i^2 \cdot f_i$$

$$\xrightarrow{\Delta_i f_i = \Delta_{i+1} f_i = \Delta f_i;\ m_i^2 = m_{i+1}^2 = m^2\quad C1_i \cdot C2_i \cdot \partial_{ti} f_i = C1_{i+1} \cdot C2_{i+1} \cdot \partial_{ti} f_i = C1 \cdot C2 \cdot \partial_t f_i\quad \forall i}$$

$$0 = \Delta f_i \pm C1 \cdot C2 \cdot \partial_t f_i + m^2 \cdot f_i. \tag{494}$$

The global function ${}_nF[f_n]$ would again appear as multi-world entanglement or Everett-selector.

5.18.8.5 A demonstrator: The Everett–Schrödinger hydrogen atom

As we have already seen that the classical or quasi-classical hydrogen atom is such a nice illustrator for a "molecular dynamics of individuals" (c.f. section "2.2 The Mass Formation in the Quantum Gravity Model Picture" in this book and [76]), here we want to derive a completely metric Everett-version of the classical Schrödinger hydrogen atom.

Again, we apply the general global or multiverse metric (480) and split it up into many sub-metrics, which should all have the ability to provide the conditions for the Schrödinger hydrogen atom.

Thereby we allow for complex exponents for our metric appearances of the radius coordinate r and by moving to $N = 8$ dimensions and by changing the setting for the metrics ${}_k^N g^{\alpha\beta}$ according to:

$${}_k^N g_{\alpha\beta} = g_{\alpha\beta}^8 = \begin{pmatrix} c^2 & 0 & 0 & 0 & \cdots & 0 \\ 0 & 1 & 0 & 0 & \cdots & 0 \\ 0 & 0 & r_k^2 & 0 & \cdots & 0 \\ 0 & 0 & 0 & r_k^2 \cdot \sin^2 \varphi_{k1} & \cdots & 0 \\ \cdots & \cdots & \cdots & \cdots & \ddots & 0 \\ 0 & 0 & 0 & 0 & 0 & g_{77} \end{pmatrix}$$

$$\times F_k\left[f_k\left[t_k, r_k, \vartheta_k, \varphi_{k1}, \varphi_{k2}, \varphi_{k3}, \varphi_{k4}, \varphi_{k5}\right]\right];\quad F_k[f_k] = f_k^{\frac{2}{3}}$$

$$f_k\left[t_k, r_k, \vartheta_k, \varphi_{k1}, \varphi_{k2}, \varphi_{k3}, \varphi_{k4}, \varphi_{k5}\right] = f_{kt}[t] \cdot f_{kr}[r] \cdot f_{k\vartheta}[\vartheta] \cdot \prod_{i=1}^{5} f_{k\varphi_{ki}}\left[\varphi_{ki}\right];$$

$$g_{44} = \left(g_{\varphi_{k2}}\left[\varphi_{k2}\right]'\right)^2 \cdot r_k^{i \cdot s};$$

$$g_{55} = \left(g_{\varphi_{k3}}\left[\varphi_{k3}\right]'\right)^2 \cdot r_k^s;$$

$$g_{66} = \left(g_{\varphi_{k4}}\left[\varphi_{k4}\right]'\right)^2 \cdot r_k^{-i\cdot s};$$

$$g_{77} = -\left(g_{\varphi_{k5}}\left[\varphi_{k5}\right]'\right)^2 \cdot r_k^{-s}; \qquad (495)$$

we can get the partial differential equation for the Schrödinger hydrogen problem [25, 26]. It should be pointed out that we assumed the same speed of light in vacuum for all Everett "worlds," respectively all sub-metrics ${}_k^N g^{\alpha\beta}$.

For simplicity and brevity, we omit the index k from now on.

Our next step is to apply a separation approach and to partially "fix" some of the functions as follows (for i = 3, 4, 5[7]):

$$f_{\varphi_i}\left[\varphi_i\right] = f_{\varphi_i}\left[g_{\varphi_i}\left[\varphi_i\right]\right] \equiv f_{\varphi_i}\left[g_{\varphi_i}\right];$$

$$f_{\varphi_i}\left[g_{\varphi_i}\right] = C_{-i} \cdot e^{-i\cdot A_i \cdot g_{\varphi_i}} + C_{+i} \cdot e^{+i\cdot A_i \cdot g_{\varphi_i}} = C_{ci} \cdot \cos\left[A_i \cdot g_{\varphi_i}\right] + C_{si} \cdot \sin\left[A_i \cdot g_{\varphi_i}\right]$$

$$\text{especially} \quad (\text{classically}):$$

$$f_{\varphi 1}\left[g_{\varphi_1}\right] = C_{-1} \cdot e^{-i\cdot A_1 \cdot g_{\varphi_1}} + C_{+1} \cdot e^{+i\cdot A_1 \cdot g_{\varphi_1}} = C_{c1} \cdot \cos\left[\overbrace{A_1}^{\hat{=}m} \cdot g_{\varphi_1}\right] + C_{s1} \cdot \sin\left[\overbrace{A_1}^{\hat{=}m} \cdot g_{\varphi_1}\right], \qquad (496)$$

$$f_t\left[t\right] = C_{t1} \cdot \cos\left[c \cdot E \cdot t\right] + C_{t2} \cdot \sin\left[c \cdot E \cdot t\right], \qquad (497)$$

$$f_\vartheta\left[\vartheta = \varphi_0\right] = C_{P\vartheta} \cdot P_L^{A_1}\left[\cos\left[\vartheta = \varphi_0\right]\right] + C_{Q\vartheta} \cdot Q_L^{A_1}\left[\cos\left[\vartheta = \varphi_0\right]\right], \qquad (498)$$

with the associated Legendre polynomials $P_L^{A_1}, Q_L^{A_1}$. This gives us the following differential equation for the r-dependency of f[...]:

$$R^* = 0 = \left(\frac{L\cdot(L+1)}{r^2} + r^{i\cdot s} \cdot A_2^2 + r^s \cdot A_3^2 + r^{-i\cdot s} \cdot A_4^2 - r^{-s} \cdot A_5^2 - \overbrace{\Delta_{\text{3D-sphere}}\left[r\right]}^{2\frac{\partial}{r\cdot\partial r} + \frac{\partial^2}{\partial r^2}} + E^2\right)\Psi$$

[7]If not used as index "i" stands for the imaginary number $i = \sqrt{-1}$

$$\Psi=\Psi\left[r,\vartheta=\varphi_0,\varphi=\varphi_1\right]=f_r[r]\cdot f_\vartheta\left[\vartheta=\varphi_0\right]\cdot f_\varphi\left[\varphi=\varphi_1\right]$$

$$\Rightarrow \qquad 0=f_\vartheta[\vartheta]\cdot f[\varphi]\cdot\left(\begin{array}{c}\frac{L\cdot(L+1)}{r^2}+r^{i\cdot s}\cdot A_2^2+r^s\cdot A_3^2+r^{-i\cdot s}\cdot A_4^2\\ -r^{-s}\cdot A_5^2-\frac{2\partial}{r\cdot\partial r}-\frac{\partial^2}{\partial r^2}+E^2\end{array}\right)f_r[r]. \quad (499)$$

We see that for the choice of s = 1 and $f_{\varphi_i}\left[\varphi_i\right]=\text{const}$ for i = 2,3,4, we obtain:

$$0=f_\vartheta[\vartheta]\cdot f[\varphi]\cdot\left(\frac{L\cdot(L+1)}{r^2}-\frac{A_5^2}{r}-\frac{2\partial}{r\cdot\partial r}-\frac{\partial^2}{\partial r^2}+E^2\right)f_r[r], \quad (500)$$

which is just the classical Schrödinger hydrogen problem with the corresponding solution:

$$f_r[r]=\mathrm{e}^{-r\cdot E}\cdot r^L\cdot\left(C_U U\left[-nn,2\cdot(1+L),2\cdot r\cdot\mu\right]+C_L\cdot L_{nn}^{1+2\cdot L}\left[2\cdot r\cdot\mu\right]\right)$$

$$nn=\frac{A_5^2}{2\cdot E}-1-L. \quad (501)$$

Now we have a list of hydrogen solutions f_k with metrics ${}_k^N g^{\alpha\beta}$ and a global function ${}_nF[f_n]$, which would again appear as multi-world entanglement or Everett-selector.

5.18.8.6 The Everett mind

We realize that with the help of Eq. (480), we are able to construct the metric equivalent of a problem-solving machine (mind), which contains all possible outcomes to a certain problem in quantum mechanical entanglement as long as the governing global scaling function ${}_nF[f_n]$ is still undetermined, which is to say, "open" or "undecided." In this context, the function may be called the "decision-making function" or just "the selector."

In the simplest case, in difference to the approach (480), this selector is just a vector of constants V_i, which switches on or off the various solutions ${}_NF_i\left[f_i\right]\cdot{}_i^N g_{\alpha\beta}$ within the global ensemble:

$$G_{\alpha\beta} = V_n \cdot g_{\alpha\beta}$$

$$= \begin{pmatrix} V_1 \cdot {}_N F_1\left[f_1\right] \cdot {}_1^N g_{\alpha\beta} & {}^N 0 & \cdots & \cdots & \cdots & {}^N 0 \\ {}^N 0 & V_2 \cdot {}_N F_2\left[f_2\right] \cdot {}_2^N g_{\alpha\beta} & \cdots & \cdots & \cdots & \vdots \\ {}^N 0 & {}^N 0 & \ddots & \cdots & \cdots & \vdots \\ {}^N 0 & {}^N 0 & \cdots & \ddots & \cdots & \vdots \\ \cdots & \cdots & \cdots & \cdots & \ddots & \vdots \\ {}^N 0 & \cdots & \cdots & \cdots & \cdots & V_{n/N} \cdot {}_{n/N}^{N} g_{\alpha\beta} \end{pmatrix}$$

$${}_N F_i\left[f_i\right] = C_{f1i} \cdot \left(C_{fi} + f_i\left[t_i, x_i, y_i, z_i, \ldots\right]\right)^{\frac{4}{N-2}}; \quad {}^N 0 = \begin{pmatrix} 0 & \cdots 0 & 0 \\ \vdots & \ddots 0 & 0 \\ 0 & 0 \; 0 & 0 \\ 0 & 0 \; 0 & 0 \end{pmatrix}. \tag{502}$$

We will see that this approach also helps us to understand the occurrence of the function vector or spinor in connection with the Dirac equation [27]. We are going to work out this spinor or Dirac "problem" in the following sections of this appendix.

5.18.9 What Is the Dirac Problem?

In his original work [27], Dirac extracted his famous Dirac equation out of the Klein–Gordon equation via an operator factorization by means of quaternions (the Dirac matrices). At the end of the process, he realized that the wave function *f*, being a scalar in the Klein–Gordon equation could no longer remain a scalar but had to become some kind of vector in order to give the new Dirac equation any meaning (the reader can find a very good explanation in [24]). Thus, the vector *f* was postulated. Here, we are going to revisit the Dirac derivation, but instead of starting with the postulated Klein–Gordon equation, we want to use our results from the Einstein–Hilbert action. This way, we can be sure to have a true quantum gravity Dirac theory.

In the sections above, in quite a variety of ways, we ended up in scalar quantum gravity equations, which could all be brought into the Klein–Gordon form for the most commonly used metrics (especially the flat ones) and under certain conditions regarding the functional wrapper *F*[*f*]. So, here we have (at least) three fundamentally derived scalar quantum gravity equations, namely, Eqs. (376), (440) and (450). All these equations can be linearized via conditions (377), (441) and (452), respectively. After this linearization process, we obtain the standard quantum equations either directly or for most standard metrics. So, for instance, the metrics (369) and (444) for the cases (377) and (441), respectively, will almost directly bring us the Klein–Gordon equation. Thereby, the only problem left is the occurrence of mass, potential and energy, but as it was shown in our previous work (e.g., [2, 4, 7–23, 29, 30] that this only requires the introduction of entangled dimensions, we refrain from discussing this aspect here.

Here, our starting point shall be Eq. (371). Studying this equation closely, we realize that it contains a scalar and tensorial part. While we might consider the tensorial equation a generalized or quantum Einstein field equation, we are going to use the scalar one to derive the classical Dirac equation [27]. We already saw that by applying condition (374) and assuming metrics of the kind (369), our integrand in the Einstein–Hilbert action can dramatically be simplified. Setting this integrand equal to zero we obtain the following differential equation:

$$0 = R - \frac{F_i'}{F_i}(n-1)\Delta f_i + \frac{f_{i,\alpha} f_{i,\beta}(1-n)}{4F_i^2} g^{\alpha\beta}\left(4F_i F_i'' + F_i' F_i'(n-6)\right). \quad (503)$$

Please note that this is not the original Dirac starting point, but a general outcome of the Einstein–Hilbert action with just an assumption regarding the metric being of the type (369) and a restriction of the variation on the volume part of the latter (374). The only thing we—boldly—introduced (postulated) is the vector character of *f* and—consequently—*F*. We have set *f*→f_i and *F*→F_i. We know that this is a bold move and in principle, it should be considered downright wrong, because it contradicts the tensor conservation of Eq. (367), if—and only if—this vector would be a true one with respect to the metric base. In classical physics, the vector has been called a spinor and as such it is essential to the whole Dirac theory. However, within a metric theory its introduction suddenly renders arbitrary and even wrong. Nevertheless, we will see what we get and then try to sort out the problem later. After all, it may be considered a problem, scientists have lived with for almost a hundred years ... and, taking the success of the Dirac equation, they lived quite well with it.

Now we add in the Dirac starting point, who took for granted that:

(a) The space-time can be assumed Ricci flat and thus, $R = 0$.
(b) The Laplace operator term in Eq. (503) has an eigenvalue M^2.

Also setting $F[f] = f$ gives us:

$$0 = f_{i,\alpha} f_{i,\beta} g^{\alpha\beta} + \overbrace{M^2 \cdot \frac{4}{(n-6)}}^{\equiv \mu^2 = \boldsymbol{\mu}\cdot\boldsymbol{\mu}} \cdot f_i^2. \tag{504}$$

Now we apply the known relation between the Dirac matrices (note that all empty slots are standing for zeros; also note that here we only give the Dirac matrices in the Cartesian, respectively the Minkowski case for a 4-dimensional space-time):

$$\gamma^0 = \begin{pmatrix} 1 & & & \\ & 1 & & \\ & & -1 & \\ & & & -1 \end{pmatrix}; \quad \gamma^1 = \begin{pmatrix} & & & 1 \\ & & 1 & \\ & -1 & & \\ -1 & & & \end{pmatrix}$$

$$\gamma^2 = \begin{pmatrix} & & & -i \\ & & i & \\ & i & & \\ -i & & & \end{pmatrix}; \quad \gamma^3 = \begin{pmatrix} & & 1 & \\ & & & -1 \\ -1 & & & \\ & 1 & & \end{pmatrix}; \quad I = \begin{pmatrix} 1 & & & \\ & 1 & & \\ & & 1 & \\ & & & 1 \end{pmatrix}, \tag{505}$$

and the metric tensor, reading:

$$g^{\alpha\beta} \cdot I = \frac{\gamma^\alpha \gamma^\beta + \gamma^\beta \gamma^\alpha}{2}, \tag{506}$$

and put Eq. (504) into the following form:

$$
\begin{aligned}
& 0 = \left(f_{i,\alpha} f_{i,\beta} g^{\alpha\beta} + \mu^2 \cdot f_i^2\right) \cdot I \\
\Rightarrow \quad & = \Rightarrow f_{i,\alpha} f_{i,\beta} \left(\frac{\gamma^\alpha \gamma^\beta + \gamma^\beta \gamma^\alpha}{2}\right) + \frac{I+I}{2} \cdot m^2 \cdot f_i^2 = 0 \\
\Rightarrow \quad & f_{i,\alpha} f_{i,\beta} \gamma^\alpha \gamma^\beta + I \cdot m^2 \cdot f_i^2 \\
& = \left(f_{i,\alpha} \gamma^\alpha f_{i,\beta} \gamma^\beta + i \cdot I \cdot m \cdot f_i \overbrace{\left(f_{i,\beta}\gamma^\beta - f_{i,\alpha}\gamma^\alpha\right)}^{f_{i,\beta}\gamma^\beta - f_{i,\alpha}\gamma^\alpha = f_{i,\gamma}\gamma^\gamma - f_{i,\gamma}\gamma^\gamma = 0} + I \cdot m^2 \cdot f_i^2 \right) \\
& = \left(f_{i,\alpha}\gamma^\alpha + i \cdot I \cdot m \cdot f_i\right)\left(f_{i,\beta}\gamma^\beta - i \cdot I \cdot m \cdot f_i\right) = 0 \\
& \left(f_{i,\beta}\gamma^\beta - i \cdot I \cdot m \cdot f_i\right) = 0 \\
\Rightarrow \quad & f_{i,\alpha} f_{i,\beta} \gamma^\beta \gamma^\alpha + I \cdot m^2 \cdot f_i^2 = 0
\end{aligned}
$$

$$\Rightarrow \qquad \left(f_{i,\alpha}\gamma^{\alpha}+i\cdot I\cdot m\cdot f_i\right)\left(f_{i,\beta}\gamma^{\beta}-i\cdot I\cdot m\cdot f_i\right)=0. \tag{507}$$

This gives us the Dirac equation in its classical form [27]:

$$0=f_{i,\beta}\gamma^{\beta}-i\cdot I\cdot m\cdot f_i. \tag{508}$$

We should emphasize—again—that the Dirac matrices (505) are only valid for Cartesian coordinates, while in curvilinear coordinates quite some adjustments are needed [47].

Please note that in some cases the Dirac equation is also given via a separated time-derivative. This has been obtained by multiplication of Eq. (508) with $\beta = \gamma^0$. This leads to:

$$f_{i,t}=f_{i,\lambda}\alpha^{\lambda}-i\cdot\beta\cdot m\cdot f_i;\quad \lambda=1,2,3;\quad \alpha^{\lambda}=\begin{pmatrix}0 & \sigma_\lambda\\ \sigma_\lambda & 0\end{pmatrix}. \tag{509}$$

Thereby we have applied the Pauli matrices as we will discuss later and which are given in Eq. (538) and also used the β-matrix with

$$\beta=\begin{pmatrix}1 & & & \\ & 1 & & \\ & & -1 & \\ & & & -1\end{pmatrix}.$$

In order to make this consistent with the decomposition of the Klein–Gordon equation, we have to treat the f_i as scalars, which allows us to rewrite Eq. (507):

$$\begin{aligned}
&\left\{f_{i;\alpha},f_{i;\beta}\right\}=\left\{\nabla_\alpha f_i,\nabla_\beta f_i\right\}=\left\{f_{i;\alpha},f_{i;\beta}\right\}\Rightarrow 0=\left(f_{i;\alpha}f_{i;\beta}g^{\alpha\beta}+\mu^2\cdot f_i^{\,2}\right)\cdot I\\
\Rightarrow\quad &= \Rightarrow f_{i;\alpha}f_{i;\beta}\left(\frac{\gamma^{\alpha}\gamma^{\beta}+\gamma^{\beta}\gamma^{\alpha}}{2}\right)+\frac{I^2+I^2}{2}\cdot m^2\cdot f_i^2=0\\
\Rightarrow\quad & f_{i;\alpha}f_{i,\beta}\gamma^{\alpha}\gamma^{\beta}+I^2\cdot m^2\cdot f_i^2=0\\
\Rightarrow\quad & \left(f_{i;\alpha}\gamma^{\alpha}+i\cdot I\cdot m\cdot f_i\right)\left(f_{i;\beta}\gamma^{\beta}-i\cdot I\cdot m\cdot f_i\right)=0\\
\Rightarrow\quad & f_{i;\alpha}f_{i;\beta}\gamma^{\beta}\gamma^{\alpha}+I^2\cdot m^2\cdot f_i^2=0\\
\Rightarrow\quad & \left(f_{i;\alpha}\gamma^{\alpha}+i\cdot I\cdot m\cdot f_i\right)\left(f_{i;\beta}\gamma^{\beta}-i\cdot I\cdot m\cdot f_i\right)=0.
\end{aligned} \tag{510}$$

The advantage of the Dirac approach is clearly that here the structure of the equation with the quaternions already ensures the satisfaction of the eigenvalue equation for the Laplace operator term, which is to say the Klein–Gordon condition that we used above in Eq. (504):

$$\Delta f_i = M^2 \cdot f_i. \tag{511}$$

To our dismay, however, we find, that we actually would need:

$$\Delta f_i = -M^2 \cdot f_i \tag{512}$$

in order to make the whole derivation consistent with the classical Dirac operator factorization.

This can be "proven" as follows:

First, we write Eq. (512) in the following way:

$$\Delta f_i = g^{\alpha\beta} \nabla_\alpha \nabla_\beta f_i = -M^2 \cdot f_i. \tag{513}$$

Now we apply Eq. (506) again, thereby also use Eq. (505) and the β-matrix and obtain:

$$g^{\alpha\beta} \nabla_\alpha \nabla_\beta f_i + M^2 \cdot f_i = 0$$

$$0 = \left(g^{\alpha\beta} \nabla_\alpha \nabla_\beta f_i + M^2 \cdot f_i\right) \cdot I$$

$$\Rightarrow \qquad = \Rightarrow \left(\frac{\gamma^\alpha \gamma^\beta + \gamma^\beta \gamma^\alpha}{2}\right) \nabla_\alpha \nabla_\beta f_i + \frac{I^2 + I^2}{2} \cdot \tilde{m}^2 \cdot f_i = 0$$

$$\Rightarrow \qquad \gamma^\alpha \gamma^\beta \nabla_\alpha \nabla_\beta f_i + I^2 \cdot \tilde{m}^2 \cdot f_i = 0$$

$$\Rightarrow \qquad \left(\gamma^\alpha \nabla_\alpha + i \cdot I \cdot \tilde{m}\right)\left(\gamma^\beta \nabla_\beta - i \cdot I \cdot \tilde{m}\right) f_i = 0$$

$$\Rightarrow \qquad \gamma^\beta \gamma^\alpha \nabla_\alpha \nabla_\beta f_i + I^2 \cdot \tilde{m}^2 \cdot f_i = 0$$

$$\Rightarrow \qquad \left(\gamma^\alpha \nabla_\alpha + i \cdot I \cdot \tilde{m}\right)\left(\gamma^\beta \nabla_\beta - i \cdot I \cdot \tilde{m}\right) f_i = 0. \tag{514}$$

As it totally suffices for the functions f_i to satisfy the inner operator in the fifth and the last line of Eq. (514), we can demand:

$$\left(\gamma^\beta \nabla_\beta - i \cdot I \cdot \tilde{m}\right) f_i = 0 \;\Rightarrow\; \gamma^\beta \nabla_\beta f_i - i \cdot I \cdot \tilde{m} \cdot f_i = 0 \tag{515}$$

and thus, have "almost" reproduced our Dirac equation (508).

The only problem, apart from our "little" sign discrepancy in Eqs. (511) and (512) is that we note that in general we have $m \neq \tilde{m}$ and that therefore Eqs. (508) are not compatible with the essential Klein–Gordon condition (512). This condition, however, is needed in order to make the whole metric Dirac operation work at all.

What can be done?

Well, the problem is easily solved in either the right number of dimensions (this would be n = 10 (![8])) in the case of Eqs. (508) and (512) or a simple adjustment of the functional wrapper $F[f_i]$ in Eq. (503). We know from chapter 11 of [4] that with F given as:

$$4FF''+F'F'(n-6)=0 \;\;\Rightarrow\;\; F=\left(f_i\right)^{\frac{4}{n-2}};\;\; n>2, \tag{516}$$

and set into Eq. (503) this results in:

$$0=R-\frac{4\cdot(n-1)}{(n-2)\,f_i}\Delta f_i. \tag{517}$$

Now we simply assume R not to be zero as in the Dirac approach, but:

$$R=\frac{4\cdot(n-1)}{(n-2)}\cdot M^2 \tag{518}$$

and can directly finish the derivation via Eq. (514) without any incompatibility problem at all, because due to Eq. (516) the non-linear term in Eq. (503) has disappeared.

Even though this is a nice compact derivation, we should not ignore other possibilities, which, as it was shown in our previous paper [43] and has been represented in various ways in chapters 9 and 10 of [4], can be obtained via the degrees of freedom offering themselves in form of the wrapper function $F[f]$.

For instance: Knowing that instead of condition (516), we could also have any other condition in Eq. (503), such as (with an arbitrary function $H'[f]=\frac{\partial H[f]}{\partial f}$ = H'—for convenience we here omit the index i for the list of functions f_i):

$$\frac{4FF''+(F')^2(n-6)}{4F}=F'\cdot H'$$

$$\Rightarrow \qquad F=\begin{cases}\left(C_{f0}+\int_1^f e^{H[\phi]}\cdot\frac{n-2}{4}\cdot d\phi\right)^{\frac{4}{n-2}}\cdot C_{f1} & n>2\\ C_{f1}\cdot e^{C_{f0}\cdot\int_1^f e^{H[\phi]}d\phi} & n=2,\end{cases} \tag{519}$$

[8]We note that this is the number of dimensions the String Theory needs to describe fermions.

leading us to:

$$0 = R + \frac{F'}{F}(1-n)\Delta f + \frac{f_{,\alpha} f_{,\beta}(1-n)}{F} g^{\alpha\beta} H' F'$$

$$\Rightarrow \qquad -R = \frac{F'}{F}(1-n)\left(f \cdot \Delta f + f_{,\alpha} f_{,\beta} g^{\alpha\beta} H'\right). \qquad (520)$$

In order to get some orientation, we first apply a function $H = H[f] = C_{f0}$-f. The corresponding solution for $F[f]$ in the case of $n > 2$ would be:

$$\frac{4FF'' + (F')^2(n-6)}{4F} = F' \cdot H' = -F'$$

$$\Rightarrow \qquad F = C_{f1} \cdot e^{\frac{4}{n-2} \cdot \left(\ln\left[e^{f} - C_{f0}\right] - f\right)} \quad n > 2, \qquad (521)$$

where we detect an interesting minimum at f = ln $[C_{f0}]$ (Fig. R1). The finding merits quite some discussion, which will be presented elsewhere.

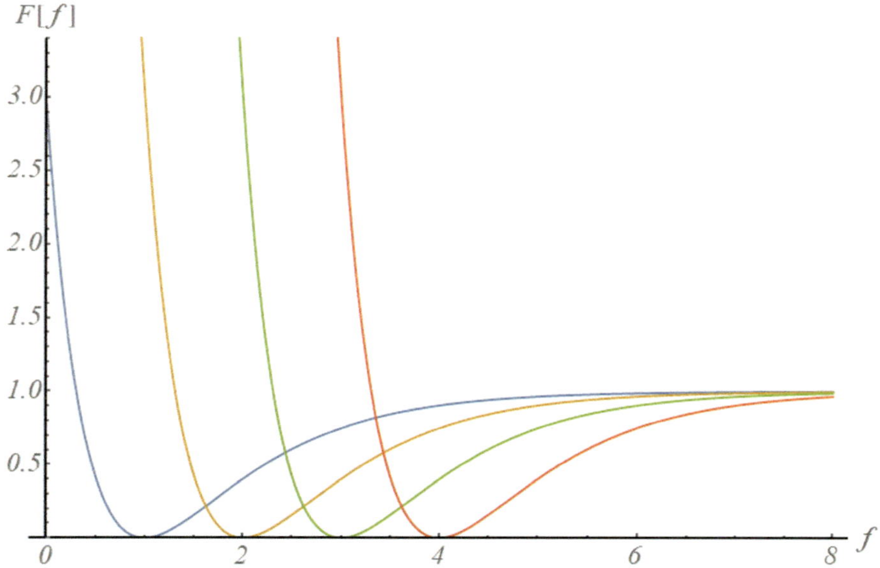

Figure R1 Illustration of the function $F[f]$ from Eq. (521) for a variety of settings regarding the constant C_{f0}, which is leading to minima at f = ln $[C_{f0}]$.

The simplest path forward in order to obtain a good starting point for the Dirac equation, however, is now to apply a function $H' = H'[f] = 1/F[f]/f$. The corresponding solution for $F[f]$ in the case of $n > 2$ would be:

$$\frac{4FF''+(F')^2(n-6)}{4F}=F'\cdot H'=-\frac{F'}{f}$$

$$\Rightarrow \qquad F=C_{f1}\cdot\left(C_{f0}+\ln[f]\right)^{\frac{4}{n-2}} \quad n>2. \tag{522}$$

With the Dirac condition for the Ricci-flat space-time $R = 0$, this gives us:

$$R^*=0=\frac{F'}{F}(1-n)\left(f\cdot\Delta f-f_{,\alpha}f_{,\beta}g^{\alpha\beta}\right)$$

$$\Rightarrow \qquad R^*=0=f\cdot\Delta f-f_{,\alpha}f_{,\beta}g^{\alpha\beta}, \tag{523}$$

from which we can derive the Dirac equation via the process given above with the Klein–Gordon condition (512). Introducing the index i again, namely, Eq. (523) becomes:

$$0=f_i\cdot\Delta f_i-f_{i,\alpha}f_{i,\beta}g^{\alpha\beta}\xrightarrow{\Delta f_i=-M^2 f_i}0=f_{i,\alpha}f_{i,\beta}g^{\alpha\beta}+M^2 f_i^2 \tag{524}$$

and we can directly move forward with Eqs. (507) and (510), only that this time instead of μ we have M:

$$0=\left(f_{i,\alpha}f_{i,\beta}g^{\alpha\beta}+M^2\cdot f_i^2\right)\cdot I$$

$$\Rightarrow \qquad =\Rightarrow f_{i,\alpha}f_{i,\beta}\left(\frac{\gamma^\alpha\gamma^\beta+\gamma^\beta\gamma^\alpha}{2}\right)+\frac{I^2+I^2}{2}\cdot m^2\cdot f_i^2=0$$

$$\Rightarrow \qquad f_{i,\alpha}f_{i,\beta}\gamma^\alpha\gamma^\beta+I^2\cdot m^2\cdot f_i^2=0$$

$$\Rightarrow \qquad \left(f_{i,\alpha}\gamma^\alpha+i\cdot I\cdot m\cdot f_i\right)\left(f_{i,\beta}\gamma^\beta-i\cdot I\cdot m\cdot f_i\right)=0$$

$$\Rightarrow \qquad f_{i,\alpha}f_{i,\beta}\gamma^\beta\gamma^\alpha+I^2\cdot m^2\cdot f_i^2=0$$

$$\Rightarrow \qquad \left(f_{i,\alpha}\gamma^\alpha+i\cdot I\cdot m\cdot f_i\right)\left(f_{i,\beta}\gamma^\beta-i\cdot I\cdot m\cdot f_i\right)=0. \tag{525}$$

This not only solves our little factor plus the sign problem from above but also gives us a clear metric understanding of the origin of the Dirac equation ... at least with respect to the scalar quantum gravity equation (376). With metric scaling functions of the kind (521) and Ricci flatness with respect to the unscaled metric tensor $g_{\alpha\beta}$, namely, we obtain Ricci scalars R^* of the type (523). Such equations can be brought into the Dirac form.

The general setting (519) also allows us another approach, which we will discuss next. Again, we leave the index i for the function list f_i aside. With the introduction of this function $H = H[f]$, we can develop the two differential operators in the first equation in Eq. (520) into one:

$$
\begin{aligned}
0 &= R-\frac{n-1}{F}\left(\frac{F'}{\sqrt{g}}\left(\sqrt{g}g^{\mu\nu}f_{,\mu}\right)_{,\nu}+\overbrace{\frac{1}{4F}\left(4FF''-(6-n)F'F'\right)}^{F'H'}f_{,\mu}f_{,\nu}g^{\mu\nu}\right)\\
&= R-(n-1)\frac{F'}{F}\left(\frac{1}{\sqrt{g}}\left(\sqrt{g}g^{\mu\nu}f_{,\mu}\right)_{,\nu}+H'f_{,\mu}f_{,\nu}g^{\mu\nu}\right)\\
&= R-(n-1)\frac{F'}{F}\left(\frac{1}{\sqrt{g}}\left(\sqrt{g}g^{\mu\nu}f_{,\mu}\right)_{,\nu}+\frac{\sqrt{g}}{\sqrt{g}}H'f_{,\mu}f_{,\nu}g^{\mu\nu}\right)\\
&= R-(n-1)\frac{F'}{F}\left(\frac{1}{\sqrt{g}}\left(\sqrt{g}\cdot g^{\mu\nu}H\cdot f_{,\mu}\right)_{,\nu}\right).
\end{aligned}
\tag{526}
$$

By defining a new metric $h^{\mu\nu}$ with the following properties:

$$\sqrt{g}\cdot g^{\mu\nu}H=\sqrt{h}\cdot h^{\mu\nu}, \tag{527}$$

we end up with an equation containing a simple Laplace operator as follows:

$$
\begin{aligned}
& 0 = R-(n-1)\frac{F'}{F}\left(\frac{1}{\sqrt{g}}\left(\sqrt{h}\cdot h^{\mu\nu}\cdot f_{,\mu}\right)_{,\nu}\right)\\
\Rightarrow\quad & 0 = \frac{\sqrt{g}}{\sqrt{h}}\cdot\frac{F}{F'}\cdot R_g-\frac{(n-1)}{\sqrt{h}}\left(\sqrt{h}\cdot h^{\mu\nu}\cdot f_{,\mu}\right)_{,\nu}\\
& \quad = \frac{\sqrt{g}}{\sqrt{h}}\cdot\frac{F}{F'}\cdot R_g-(n-1)\Delta_h f.
\end{aligned}
\tag{528}
$$

Thereby we need to distinguish between the metric for the Laplace operator applied on f, which is $h^{\mu\nu}$, and the one as the basis for the Ricci scalar R, which is $g^{\mu\nu}$. Now we set:

$$\frac{\sqrt{g}}{\sqrt{h}}\cdot\frac{F}{F'}\cdot R_g=(1-n)\cdot M^2\cdot f\xrightarrow{f\to f_i}=(1-n)\cdot M^2\cdot f_i \tag{529}$$

and obtain, once again, the perfect starting point for the classical Dirac approach (513), (514):

$$0=M^2\cdot f_i+\Delta_h f_i, \tag{530}$$

only that this time all differential operators are based on the metric $h^{\mu\nu}$ and not $g^{\mu\nu}$, which changes Eq. (513) to:

$$\Delta_h f_i = h^{\alpha\beta} \overset{h}{\nabla}_\alpha \overset{h}{\nabla}_\beta f_i = -M^2 \cdot f_i \tag{531}$$

and renders Eq. (514) as follows:

$$g^{\alpha\beta} \overset{h}{\nabla}_\alpha \overset{h}{\nabla}_\beta f_i + M^2 \cdot f_i = 0$$

$$0 = \left(g^{\alpha\beta} \overset{h}{\nabla}_\alpha \overset{h}{\nabla}_\beta f_i + M^2 \cdot f_i \right) \cdot I$$

$$\Rightarrow \qquad = \Rightarrow \left(\frac{\gamma^\alpha \gamma^\beta + \gamma^\beta \gamma^\alpha}{2} \right) \overset{h}{\nabla}_\alpha \overset{h}{\nabla}_\beta f_i + \frac{I^2 + I^2}{2} \cdot M^2 \cdot f_i = 0$$

$$\Rightarrow \qquad \gamma^\alpha \gamma^\beta \overset{h}{\nabla}_\alpha \overset{h}{\nabla}_\beta f_i + I^2 \cdot M^2 \cdot f_i = 0$$

$$\Rightarrow \qquad \left(\gamma^\alpha \overset{h}{\nabla}_\alpha + i \cdot I \cdot M \right) \left(\gamma^\beta \overset{h}{\nabla}_\beta - i \cdot I \cdot M \right) f_i = 0$$

$$\Rightarrow \qquad \gamma^\beta \gamma^\alpha \overset{h}{\nabla}_\alpha \overset{h}{\nabla}_\beta f_i + I^2 \cdot M^2 \cdot f_i = 0$$

$$\Rightarrow \qquad \left(\gamma^\alpha \overset{h}{\nabla}_\alpha + i \cdot I \cdot M \right) \left(\gamma^\beta \overset{h}{\nabla}_\beta - i \cdot I \cdot M \right) f_i = 0. \tag{532}$$

Before we want to go back to discuss the somehow problematic aspect of the postulation of the vector f_i, we should acknowledge Dirac's ingenious idea.

Seeing the simplicity of the derivation of the Dirac equation with the Einstein–Hilbert action as the starting point, as just performed here and comparing this with Dirac's path, we can only repeat [2, 4, 31–43] our amazement about "how on earth this genius came up with the idea of introducing the quaternions."

It is—we have no other word for it—a miracle [4, 31], which becomes more obvious the more one considers all the possible other options as we have investigated in our previous publications (especially [4]).

However, as we can see from the metric origin of the classical Dirac equation [27] and the possible alternatives being derived in the chapters 9 and 10 in [4], we conclude that Dirac's ingenious introduction of quaternions is not the only option, nor is it necessarily the best approach.

This becomes obvious in curvilinear coordinates [44, 45] or—even more so—in potentially non-Ricci-flat space-times.

5.18.9.1 What does the occurrence of spinors in quantum gravity mean?

In [4] we discussed metric scaling of the following kind:

$$G_{\alpha\beta} = \zeta^{c}_{\alpha b} f^{a} \cdot g_{a\beta} \tag{533}$$

and

$$G^{\alpha\beta} = \zeta^{\alpha b}_{c} f_{a} g^{a\beta}. \tag{534}$$

The necessary condition in order to obtain reasonable tensor equations would then be something like:

$$G^{\alpha\gamma} G_{\gamma\beta} = \zeta^{\alpha b}_{c} f_{a} g^{a\gamma} \zeta^{e}_{\beta f} f^{d} \cdot g_{\gamma d} = \zeta^{\alpha b}_{c} f_{\gamma} \zeta^{e}_{\beta f} f^{\gamma} = \delta^{\alpha}_{\beta}. \tag{535}$$

In addition, we would have to demand the terms $\zeta^{c}_{\alpha b} f^{a}$, $\zeta^{\alpha b}_{c} f_{a}$, to be somehow compatible with the usual tensor transformations, keeping the result a tensor. Finding solutions to this condition seems to be impossible, but when changing Eq. (535) to:

$$G^{\alpha\gamma} G_{\gamma\beta} + G^{\beta\gamma} G_{\gamma\alpha} = \delta^{\alpha}_{\beta} + \delta^{\beta}_{\alpha}, \tag{536}$$

which leads us to:

$$\zeta^{\alpha b}_{c} \zeta^{e}_{\beta f} + \zeta^{\beta b}_{c} \zeta^{e}_{\alpha f} = \delta^{\alpha}_{\beta} + \delta^{\beta}_{\alpha}; \quad f_{\gamma} f^{\gamma} = 1, \tag{537}$$

we can directly apply the Dirac Gamma matrices γ^{α}. Please note, however, that—as we are considering n-dimensional space-times—these are Dirac matrices in n dimensions.

But what does it mean that metric variations like Eqs. (533) and (534) are realized with the Dirac matrices and—consequently—also with spinors f_a?

This question shall be discussed in the next sub-section.

5.18.9.2 The meaning of the occurrence of spinors: A suggestion

Well, we already hinted it above: Seeing the need for the introduction of vector functions f_a or any other functional object of rank > 1 should be equivalent to an increase of the number of dimensions of the system in question. The example of the occurrence of spinors in connection with the Dirac equation will be treated here in order to find a truly geometrical meaning and extension.

Let us observe Fig. R2. We directly cite from [46] (p. 1135):

Take a cube (Fig. R2). Rotate it about one axis through 90°. Then pick another axis at right angles to the first. About it rotate the cube again through 90°. In this way the cube is carried from the orientation marked 'Initial' to that marked 'Final.' How can one make this net transformation in a single step, with a single rotation? In other words, what is the law for the combination of rotations?

...

To any rotation is associated a quantity (Hamilton's 'quaternions'; today's 'spin matrix' or 'spinor transformation' or 'rotation operator'):

$$Rot = \cos\left[\frac{\theta}{2}\right] - i\cdot\sin\left[\frac{\theta}{2}\right]\left(\sigma_x\cos[\alpha]+\sigma_y\cos[\beta]+\sigma_z\cos[\gamma]\right)$$

$$\text{with:}\ \ \sigma_x = \begin{Vmatrix}0 & 1\\ 1 & 0\end{Vmatrix};\ \ \sigma_y = \begin{Vmatrix}0 & -i\\ i & 0\end{Vmatrix};\ \ \sigma_z = \begin{Vmatrix}1 & 0\\ 0 & -1\end{Vmatrix}, \tag{538}$$

where θ is the angle of rotation and α, β, γ are the angles between the axis of rotation and the coordinate axes. A rotation described by Rot_1 followed by a rotation described by Rot_2 gives a net change in orientation described by the single rotation:

$$Rot_3 = Rot_1 Rot_2. \tag{539}$$

As we will not further discuss these spinors and their math here, the reader is referred to chapter 41 of [46]. The aspect we are interested in here lies in the fact that the object in Fig. R2 really needs to be an object with a certain volume, shape and so on, which is to say, a mathematical point would not need such a bother. It requires an object having a surface (just like the cube in Fig. R2) to show such a complexity when being rotated. We therefore conclude that the need to apply spinors and Dirac gamma matrices as scaling factors to metrics in order to obtain suitable descriptions of the physical reality arises from these objects not being mathematical points but having a surface. It directly leads us to non-point-like objects and thus, consequently, to systems of higher dimensions than the actual metric, which has originally spawned the need for a spinor, describes.

The same also holds, as already said above, in connection with the metric adaptations used in [4]:

$$G_{\alpha\beta} = \zeta^{a}_{\alpha\beta} f^{b}\cdot g_{ab}, \tag{540}$$

$$G^{\alpha\beta} = \zeta^{\alpha\beta}_{a} f_b g^{ab}, \tag{541}$$

$$G_{\alpha\beta} = \zeta^{a}_{\alpha b} f^{b} \cdot g_{a\beta}, \tag{542}$$

$$G^{\alpha\beta} = \zeta^{\alpha b}_{a} f_{b} g^{a\beta}, \tag{543}$$

where we have not explicitly used the Dirac gamma matrices but only saw the need for the introduction of wave functions *f* with ranks higher than 0.

Example: Let us consider a metric in 4 dimensions. We know that in order to obtain Dirac equations in this 4-dimensional space-time, we need to scale-factor the metric as given in Eqs. (533) and (534) (that is at least one possibility).

To give this operation a metric meaning, we may see two straightforward possibilities:

(A) We could conclude that this spinor scaling is necessary, because the points in our 4-dimensional space-time are in fact objects of a finite number of dimensions. In other words, the "points" in

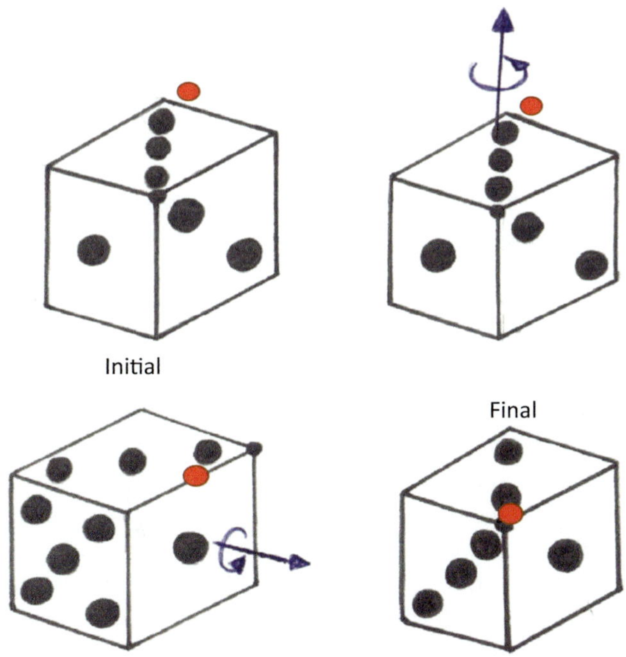

Figure R2 Cited from [46] (p. 1136) "Rotation about the vertical axis through 90°, followed by rotation about the horizontal axis through 90°, gives a net change in orientation that can be achieved by a single rotation through 120° about an axis emergent from the center through the corner" marked here with the red dot.

classical space-time cannot have zero dimensionality ... at least not when needing to be treated with the Dirac equation.

In other words: There are more dimensions than the number *n* indicates and these additional dimensions demand the need for a vector or spinor object behind the apparent scaling function *f*. This seems to justify the postulation of the f_i or f_a.

(B) The metric entanglement of dimensions via the function *F*[*f*] (being the scaling factor to all coordinates) may lead to the surface-coupling, which is to say the coupling of at least two coordinates, needed to allow for the non-point-behavior as shown in Fig. R2, regarding rotations.

On the other side, purely mathematically, it was our introduction of the gamma matrices which forced the vector function into existence. This looks like the introduction of the matrices, after all an artificial, if not to say even bold move in itself, brought the need for the vector function. There was no direct mathematical need for it, and thus the Dirac problem remains. We still have to answer the question:

Where does the wave-function vector in Eq. (503) metrically come from?

5.18.10 Where Does the Wave-Function Vector in the Dirac Equation Metrically Come from?

Considering Eq. (367), it becomes clear that in order to keep the tensorial character of the metric $G_{\alpha\beta}$, we have to demand the function *F* to be a scalar. So, from where does the "vector" of functions f_i for the Dirac approach come?

Can it be that the arbitrary or postulated character of the vector function introduction arises as an artefact, because of the Dirac equation only being an approximation of a more comprehensive Quantum Gravity Theory? This possibility will be discussed in section "5.18.17 The Spectral-Operator Decomposition" of this appendix.

The—or at least one—other answer can be found by applying the transition from the Klein–Gordon to the Schrödinger equation and extending this beyond the time-coordinate.

Partially repeating from the corresponding section in the appendices above, we start from our result (503) (index *i* omitted again), only that, for the reason of recognition, instead of the symbol *f* we apply the classical Greek symbol Ψ. We have applied condition (516), respectively (378). We also, in order to be close enough to the classical Schrödinger case, assume a Minkowski-like metric time component. We write Eq. (503) as follows:

$$0=\begin{cases}\left(\Psi-C_{f0}\right)\cdot R+(1-n)\cdot\left(\Delta_{n-1}\Psi-\frac{\partial^2\Psi}{c^2\cdot\partial t^2}\right) & n>2\\ R+C_{f0}\cdot(1-n)\cdot\left(\Delta_{n-1}\Psi-\frac{\partial^2\Psi}{c^2\cdot\partial t^2}\right) & n=2.\end{cases} \tag{544}$$

We ignore the n=2-case and take it that R can be assumed being linear to f, which is to say that the curvature of space-time is caused by and proportional to the quantum effects residing in it. This gives us:

$$0=\left(\Psi-C_{f0}\right)\cdot R_c\cdot\Psi+(1-n)\cdot\left(\Delta_{n-1}\Psi-\frac{\partial^2\Psi}{c^2\cdot\partial t^2}\right). \tag{545}$$

Assuming the quantum function to be small against the classical physics (an important aspect in Schrödinger's work anyway and also agrees with the practical observation that the space-time curvature is small as we experience our surrounding space as being mainly flat), we discard the quadratic term, resulting in:

$$\begin{aligned}0&=-C_{f0}\cdot R_c\cdot\Psi+(1-n)\cdot\left(\Delta_{n-1}\Psi-\frac{\partial^2\Psi}{c^2\cdot\partial t^2}\right)\\&=\left[(1-n)\cdot\left(\Delta_{n-1}-\frac{\partial^2}{c^2\cdot\partial t^2}\right)-C_R\right]\Psi.\end{aligned} \tag{546}$$

Then we can separate the time derivative and—for the reason of recognition in connection with Schrödinger's classical approach—we reformulate everything as follows:

$$\begin{aligned}0&=-C_{f0}\cdot R_c\cdot\Psi+(1-n)\cdot\left(\Delta_{n-1}\Psi-\frac{\partial^2\Psi}{c^2\cdot\partial t^2}\right)\\&=\left[(1-n)\cdot\left(\Delta_{n-1}-\frac{\partial^2}{c^2\cdot\partial t^2}\right)-C_R\right]\Psi\\&\xrightarrow{\left(\Delta_{n-1}-\frac{\partial^2}{c^2\cdot\partial t^2}\right)-\frac{C_R}{(1-n)}=C1\cdot\frac{\partial_t^2}{c^2}+\Delta_{n-1}-M^2}\end{aligned}$$

$$\Rightarrow\qquad 0=\left[\Delta-M^2\right]\Psi=\left[\left(C1\cdot\frac{\partial_t^2}{c^2}+\underbrace{\frac{1}{\sqrt{g}}\partial_\alpha\sqrt{g}\cdot g^{\alpha\beta}\partial_\beta}_{\text{3D}-\Delta-\text{Operator}}\right)-M^2\right]\Psi. \tag{547}$$

Please note that our Greek indices α and β are running only from 1 to n - 1 and not—as usual—from 0 to n - 1, because the time or 0-component has been separated. Now we introduce a function $\Psi = \Phi + X$ and demand the following additional condition:

$$\partial_t \Psi = c^2 \cdot C2 \cdot (\Phi - X). \tag{548}$$

Together with Eq. (547) we obtain:

$$0 = -M^2 (\Phi + X) + \left(C1 \cdot C2 \cdot \partial_t (\Phi - X) + \frac{1}{\sqrt{g}} \partial_\alpha \sqrt{g} \cdot g^{\alpha\beta} \partial_\beta (\Phi + X) \right). \tag{549}$$

The following two equations summed up would result in Eq. (549):

$$0 = -M^2 \Phi + \left(C1 \cdot C2 \cdot \partial_t \Phi + \frac{1}{\sqrt{g}} \partial_\alpha \sqrt{g} \cdot g^{\alpha\beta} \partial_\beta \Phi \right)$$

$$0 = -M^2 X + \left(\frac{1}{\sqrt{g}} \partial_\alpha \sqrt{g} \cdot g^{\alpha\beta} \partial_\beta X - C1 \cdot C2 \cdot \partial_t X \right). \tag{550}$$

We see that we have obtained a Schrödinger-like first derivative for the time-coordinate by the introduction of $\Psi = \Phi + X$. In order to proceed in this way, we might just assume another coordinate xx and introduce four new functions $\Phi = \Phi 1 + \Phi 2$; $X = X1 + X2$. With the assumption (here only for demonstration) of this coordinate being metrically Cartesian-like "organized," we may—just as with the time-coordinate above—take the following for granted:

$$\Phi[x] = C\Phi 1 * e^{+C\Phi \cdot xx} + C\Phi 2 * e^{-C\Phi \cdot xx};$$

$$X[x] = CX1 * e^{+CX \cdot xx} + CX2 * e^{-CX \cdot xx}. \tag{551}$$

Now we can demand:

$$\partial_x \Phi = C\Phi x \cdot (\Phi 1 - \Phi 2); \quad \partial_x \mathrm{X} = CXx \cdot (X1 - X2) \tag{552}$$

and obtain from Eq. (550):

$$0 = -M^2 (\Phi 1 + \Phi 2) + (C1 \cdot C2 \cdot \partial_t (\Phi 1 + \Phi 2) + \partial_x C\Phi x \cdot (\Phi 1 - \Phi 2))$$

$$\Rightarrow \qquad \begin{cases} 0 = -M^2 \Phi 1 + (C1 \cdot C2 \cdot \partial_t \Phi 1 + \partial_x C\Phi x \cdot \Phi 1) \\ 0 = -M^2 \Phi 2 + (C1 \cdot C2 \cdot \partial_t \Phi 2 - \partial_x C\Phi x \cdot \Phi 2) \end{cases}$$

$$0 = -M^2 (X1 - X2) + (\partial_x CXx \cdot (X1 - X2) - C1 \cdot C2 \cdot \partial_t (X1 + X2))$$

$$\Rightarrow \qquad \begin{cases} 0 = -M^2 X1 + (\partial_x CXx \cdot X1 - C1 \cdot C2 \cdot \partial_t X1) \\ 0 = -M^2 X2 + (-\partial_x CXx \cdot X2 - C1 \cdot C2 \cdot \partial_t X2). \end{cases} \tag{553}$$

As the more general curvilinear Dirac approach shows, this can also be done in non-Cartesian coordinates, but with the Klein–Gordon-to-Schrödinger transition, we see that the successive introduction of more functions with certain conditions regarding their first derivatives leads us to first-order differential equations instead of second-order ones.

We therefore conclude that the functions f_i in the Dirac approach are not really vectors in the sense of the metric $g_{\alpha\beta}$, but only lists of functional solutions to the fundamental Klein–Gordon equation (513) or even (503) with suitable settings for the functional wrapper $F[f_i]$. In the Dirac theory, they are therefore named spinors, which—of course—also has to do with the connection to rotation when using the Dirac matrices, which themselves consist of the Pauli matrices (538) (see section "5.18.9.2 The meaning of the occurrence of spinors: A suggestion" in this appendix above).

However, the question remains whether there could be a way to make f_i a true vector.

5.18.11 A True "Dirac"-Vector?

We go back to Eq. (503), set $R = 0$ and apply condition (378). As we have shown in [4] and quite some other publications that mass can always be obtained via suitable additional dimensions (entangled with the rest of the metric via the f_i), we end up with a Klein–Gordon equation. Alternatively, we could also just apply Eq. (516) with Eq. (518) or Eq. (531), which are also Klein–Gordon equations. Now we assume f to be a scalar with the following intrinsic structure:

$$f = \phi_\alpha g^{\alpha\beta} \phi_\beta. \tag{554}$$

It should be pointed out that such a setting would change the variation (adding an additional term), if the variation is performed with respect to the unscaled metric tensor g_{ab} and not G_{ab} as done above in order to avoid any surface terms, but this will be considered later.

The approach (554) allows us (just to pick one of our Klein–Gordon options) the following reconsideration of our derivation (514) (now with a scalar function f and not a list or spinor f_i):

$$g^{\alpha\beta}\nabla_\alpha\nabla_\beta\phi_a g^{ab}\phi_b + M^2\cdot\phi_a g^{ab}\phi_b = 0$$

$$\Rightarrow \quad \left(\frac{\gamma^\alpha\gamma^\beta + \gamma^\beta\gamma^\alpha}{2}\right)\nabla_\alpha\nabla_\beta\phi_a g^{ab}\phi_b + \frac{I^2+I^2}{2}\cdot\tilde{m}^2\cdot\phi_a g^{ab}\phi_b = 0$$

$$\Rightarrow \quad \gamma^\alpha\gamma^\beta\nabla_\alpha\nabla_\beta\phi_a g^{ab}\phi_b + I^2\cdot\tilde{m}^2\cdot\phi_a g^{ab}\phi_b = 0$$

$$\Rightarrow \qquad \left(\gamma^{\alpha}\nabla_{\alpha}+i\cdot I\cdot\tilde{m}\right)\left(\gamma^{\beta}\nabla_{\beta}-i\cdot I\cdot\tilde{m}\right)\phi_{a}g^{ab}\phi_{b}=0$$

$$\Rightarrow \qquad \gamma^{\beta}\gamma^{\alpha}\nabla_{\alpha}\nabla_{\beta}\phi_{a}g^{ab}\phi_{b}+I^{2}\cdot\tilde{m}^{2}\cdot\phi_{a}g^{ab}\phi_{b}=0$$

$$\Rightarrow \qquad \left(\gamma^{\alpha}\nabla_{\alpha}+i\cdot I\cdot\tilde{m}\right)\left(\gamma^{\beta}\nabla_{\beta}-i\cdot I\cdot\tilde{m}\right)\phi_{a}g^{ab}\phi_{b}=0. \qquad (555)$$

So, we have the typical Dirac equation:

$$\gamma^{\beta}\nabla_{\beta}f-i\cdot I\cdot\tilde{m}\cdot f=\gamma^{\beta}\nabla_{\beta}\phi_{a}g^{ab}\phi_{b}-i\cdot I\cdot\tilde{m}\cdot\phi_{a}g^{ab}\phi_{b}=0, \qquad (556)$$

only that something seems to be wrong, because with f being a scalar, the two addends in Eq. (556) do not really fit. Now we take it that not the scalar f is needed to satisfy this equation, but that the function ϕ_a would also do, meaning:

$$2\cdot\gamma^{\beta}\nabla_{\beta}\phi_{a}-i\cdot I\cdot\tilde{m}\cdot\phi_{a}=0. \qquad (557)$$

Our reasoning behind this bold claim is as follows:

$$g^{ab}\phi_{b}\gamma^{\beta}\nabla_{\beta}\phi_{a}+\phi_{a}\phi_{b}\overbrace{\gamma^{\beta}\nabla_{\beta}g^{ab}}^{=0}+\phi_{a}g^{ab}\gamma^{\beta}\nabla_{\beta}\phi_{b}-i\cdot I\cdot\tilde{m}\cdot\phi_{a}g^{ab}\phi_{b}=0$$

$$\xrightarrow{g^{ab}=g^{ba}}$$

$$2\cdot g^{ab}\phi_{b}\gamma^{\beta}\nabla_{\beta}\phi_{a}-i\cdot I\cdot\tilde{m}\cdot\phi_{a}g^{ab}\phi_{b}=0$$

$$\Rightarrow \qquad g^{ab}\phi_{b}\left(2\cdot\gamma^{\beta}\nabla_{\beta}-i\cdot I\cdot\tilde{m}\right)\phi_{a}=0. \qquad (558)$$

Please note that ϕ_a cannot be a solution to the Klein–Gordon equation as f is. This is in contrast to the classical Dirac spinor with f_i. We leave the investigation of the true-vector form and whether it gives any reasonable results for later.

Interested in avoiding the Dirac matrices, too, we reevaluate Eq. (555) as follows:

$$g^{\alpha\beta}\nabla_{\alpha}\nabla_{\beta}\phi_{a}g^{ab}\phi_{b}+M^{2}\cdot\phi_{a}g^{ab}\phi_{b}=0$$

$$\nabla_{\mu}\mathbf{E}^{\mu}\cdot\mathbf{E}^{\nu}\nabla_{\nu}\phi_{a}g^{ab}\phi_{b}+M^{2}\cdot\phi_{a}g^{ab}\phi_{b}=0$$

$$\mathbf{E}^{\mu}\cdot\mathbf{E}^{\nu}\nabla_{\mu}\nabla_{\nu}\phi_{a}g^{ab}\phi_{b}+M^{2}\cdot\phi_{a}g^{ab}\phi_{b}=0$$

$$\Rightarrow \qquad \left(\mathbf{E}^{\mu}\nabla_{\mu}+i\cdot M\right)\left(\mathbf{E}^{\nu}\nabla_{\nu}-i\cdot M\right)\phi_{a}g^{ab}\phi_{b}=0. \qquad (559)$$

So, we have the typical Dirac equation (this time with the base vectors):

$$\left(\mathbf{E}^{\nu}\nabla_{\nu}-i\cdot M\right)f=\left(\mathbf{E}^{\nu}\nabla_{\nu}-i\cdot M\right)\phi_{a}g^{ab}\phi_{b}=0 \qquad (560)$$

and the Dirac rank problem, because with f being a scalar the two addends in Eq. (556) do not really fit ... at least as long as M is also a scalar. Assuming once again that the scalar f is not necessarily needed to satisfy this equation, but that the function ϕ_a would do, meaning:

$$2\cdot\mathbf{E}^{\nu}\nabla_{\nu}\phi_a - i\cdot M\cdot\phi_a = 0. \tag{561}$$

Our reasoning behind this move is as follows:

$$g^{ab}\phi_b\mathbf{E}^{\nu}\nabla_{\nu}\phi_a + \phi_a\phi_b\overbrace{\mathbf{E}^{\nu}\nabla_{\nu}g^{ab}}^{=0} + \phi_a g^{ab}\mathbf{E}^{\nu}\nabla_{\nu}\phi_b - i\cdot M\cdot\phi_a g^{ab}\phi_b = 0$$

$$\xrightarrow{g^{ab}=g^{ba}}$$

$$2\cdot g^{ab}\phi_b\mathbf{E}^{\nu}\nabla_{\nu}\phi_a - i\cdot M\cdot\phi_a g^{ab}\phi_b = 0$$

$$\Rightarrow \qquad g^{ab}\phi_b\left(2\cdot\mathbf{E}^{\nu}\nabla_{\nu} - i\cdot M\right)\phi_a = 0. \tag{562}$$

Of course, things could dramatically be simplified with the assumption of a co-vector V_β with just constant components as follows:

$$f = \phi_\alpha g^{\alpha\beta}V_\beta. \tag{563}$$

In this case the derivations from Eqs. (558) and (562) read as follows:

$$g^{ab}V_b\gamma^{\beta}\nabla_{\beta}\phi_a + \phi_a V_b\overbrace{\gamma^{\beta}\nabla_{\beta}g^{ab}}^{=0} + \phi_a g^{ab}\overbrace{\gamma^{\beta}\nabla_{\beta}V_b}^{=0} - i\cdot I\cdot\tilde{m}\cdot\phi_a g^{ab}V_b = 0$$

$$g^{ab}V_b\gamma^{\beta}\nabla_{\beta}\phi_a - i\cdot I\cdot\tilde{m}\cdot\phi_a g^{ab}V_b = 0$$

$$\Rightarrow \qquad g^{ab}V_b\left(\gamma^{\beta}\nabla_{\beta} - i\cdot I\cdot\tilde{m}\right)\phi_a = 0, \tag{564}$$

$$g^{ab}V_b\mathbf{E}^{\nu}\nabla_{\nu}\phi_a + \phi_a V_b\overbrace{\mathbf{E}^{\nu}\nabla_{\nu}g^{ab}}^{=0} + \phi_a g^{ab}\overbrace{\mathbf{E}^{\nu}\nabla_{\nu}V_b}^{=0} - i\cdot M\cdot\phi_a g^{ab}V_b = 0$$

$$g^{ab}V_b\mathbf{E}^{\nu}\nabla_{\nu}\phi_a - i\cdot M\cdot\phi_a g^{ab}V_b = 0$$

$$\Rightarrow \qquad g^{ab}V_b\left(\mathbf{E}^{\nu}\nabla_{\nu} - i\cdot M\right)\phi_a = 0, \tag{565}$$

respectively, and as known from the classical Dirac theory, all solutions to the Dirac equations in the last lines in Eqs. (564) and (565) are also automatically solutions to the corresponding Klein–Gordon equation of f:

$$\left(g^{\alpha\beta}\nabla_{\alpha}\nabla_{\beta} + M^2\right)f = g^{\alpha\beta}\nabla_{\alpha}\nabla_{\beta}f + M^2\cdot f = g^{\alpha\beta}\nabla_{\alpha}\nabla_{\beta}\phi_a g^{ab}V_b + M^2\cdot\phi_a g^{ab}V_b$$

$$= g^{\alpha\beta}V_b\nabla_{\alpha}\nabla_{\beta}\phi_a g^{ab} + M^2\cdot\phi_a g^{ab}V_b$$

$$= V_b g^{\alpha\beta} \nabla_\alpha g^{ab} \nabla_\beta \phi_a + V_b g^{\alpha\beta} \nabla_\alpha \phi_a \overbrace{\nabla_\beta g^{ab}}^{=0} + M^2 \cdot \phi_a g^{ab} V_b$$

$$= V_b g^{\alpha\beta} \left(g^{ab} \nabla_\alpha \nabla_\beta \phi_a + \nabla_\beta \phi_a \overbrace{\nabla_\alpha g^{ab}}^{=0} \right) + g^{ab} V_b M^2 \cdot \phi_a$$

$$= V_b g^{\alpha\beta} g^{ab} \nabla_\alpha \nabla_\beta \phi_a + g^{ab} V_b M^2 \cdot \phi_a$$

$$= V_b g^{ab} \left(g^{\alpha\beta} \nabla_\alpha \nabla_\beta + M^2 \right) \phi_a = 0, \tag{566}$$

and hence, the metric scaling factor $F[f]$. However, as we can deduce from the last line of Eq. (566), it is the sum of $\sum_{a,b,\alpha,\beta}^{n-1} V_b g^{ab} \left(g^{\alpha\beta} \nabla_\alpha \nabla_\beta + M^2 \right) \phi_a$ which needs to end up zero and not necessarily each of the addends. Thus, our Dirac solutions need not fulfill the Klein–Gordon equation, as only their sum has to, but they can. But this is just the same situation as we have it with the corresponding Dirac equations (564) and (565). Nevertheless, there may be additional options which the classical approach does not cover.

5.18.11.1 Solving an old problem

With the structure for f as given in Eq. (563), which, by the way, we could also apply on the function Φ via:

$$\Phi = \phi_\alpha g^{\alpha\beta} V_\beta, \tag{567}$$

we automatically also solve the problem of the classical Dirac theory, where it never was properly understood how completely antagonistic solutions like matter and anti-matter or spin-up and spin-down co-exist in one and the same manifest context of reality. The new approach just solves this problem, because it may be that we require an ensemble of such antagonistic solutions for our functions vector ϕ_α within Eq. (563) or (567). In the end, it is the vector V_β, which selects one of such solutions for the manifestation in the real world.

This also—automatically—relativizes the need for Everett's multiverses [63] as we now have an alternative for the classical Dirac problem.

5.18.11.2 Getting a new problem

It has to be pointed out that in the case of the scaled metric function $F[f]$ and our approach (563), the Laplace or Klein–Gordon equation is only the

Einstein–Hilbert variational outcome in space-times with metrics of the following structure:

$$g_{ij} = \begin{pmatrix} g_{00} & \cdots & 0 \\ \vdots & \ddots & \vdots \\ 0 & \cdots & g_{n-1n-1} \end{pmatrix}; \quad g_{ii,i} = 0. \tag{568}$$

In the general cases, we have to deal with the following operator for the function *f*:

$$0 = R - \frac{F'}{2F} \begin{pmatrix} 2f_{,ij}(n-1)g^{ij} + 2\Gamma^a_{ij} f_{,a} g^{ij} \\ -f_{,i} g^{ab} g_{jb,a} g^{ij} - f_{,j} g^{ab} g_{ib,a} g^{ij} \\ -n\Gamma^d_{ij} f_{,d} g^{ij} + \frac{n}{2} f_{,d} g^{cd} g_{ab,c} g^{ab} \end{pmatrix}$$

$$= R \cdot \frac{8 \cdot \left(C_{f0} + f\right)}{n-2} - g^{ij} \begin{pmatrix} 2f_{,ij}(n-1) + (2-n)\Gamma^d_{ij} f_{,d} \\ -f_{,i} g^{ab} g_{jb,a} - f_{,j} g^{ab} g_{ib,a} + \frac{n}{2} f_{,j} g_{ab,i} g^{ab} \end{pmatrix}$$

$$= R \cdot \frac{8 \cdot \left(C_{f0} + f\right)}{n-2} - g^{ij} \begin{pmatrix} 2f_{,ij}(n-1) + (2-n)\Gamma^d_{ij} f_{,d} \\ +f_{,j} \frac{g^{ab}}{2}\left(n \cdot g_{ab,i} - 4 \cdot g_{ib,a}\right) \end{pmatrix}. \tag{569}$$

Consequently, defining:

$$R \cdot \frac{8 \cdot \left(C_{f0} + f\right)}{n-2} = m^2 \cdot f$$

$$\boxed{R} f = g^{ij} \lceil R_i \lceil R_j f = g^{ij} \left(2f_{,ij}(n-1) + (2-n)\Gamma^d_{ij} f_{,d} + f_{,j} \frac{g^{ab}}{2}\left(n \cdot g_{ab,i} - 4 \cdot g_{ib,a}\right) \right), \tag{570}$$

we have to find a solution to the following problem:

$$\left(m^2 - g^{ij} \lceil R_i \lceil R_j\right) f = 0$$

$$\Rightarrow \quad \left(m + B^{\mu} \lceil R_{\mu}\right)\left(m - B^{\nu} \lceil R_{\nu}\right) f = 0. \tag{571}$$

Thereby *f* is assumed to remain a scalar and the B^{μ} are—so far—unknown objects with the following properties:

$$\left(m + B^{\mu} \lceil R_{\mu}\right)\left(m - b^{\nu} \lceil R_{\nu}\right) f = 0$$

$$\left(m^2+m\overbrace{\left(-B^{\nu}\lceil R_{\nu}+B^{\mu}\lceil R_{\mu}\right)}^{0}+\overbrace{B^{\mu}\lceil R_{\mu}B^{\nu}\lceil R_{\nu}}^{=\lceil R_{\mu}\lceil R_{\nu}}\right)f=0. \tag{572}$$

We are going to investigate such objects elsewhere.

5.18.12 A Matrix-Free Dirac Equation?

We assume to have matrices with the—in difference to the identity (506)—following properties:

$$\frac{\gamma^{\alpha}\gamma^{\beta}+\gamma^{\beta}\gamma^{\alpha}}{2\cdot n}=\frac{\gamma_i^{j\alpha}\gamma_j^{i\beta}+\gamma_i^{j\beta}\gamma_j^{i\alpha}}{2\cdot n}=\frac{\delta_i^j\delta_j^i}{n}\cdot g^{\alpha\beta}=\frac{\delta_j^j}{n}\cdot g^{\alpha\beta}=g^{\alpha\beta}. \tag{573}$$

This allows us to put Eq. (504) into the form:

$$0=f_{i,\alpha}f_{i,\beta}g^{\alpha\beta}+\mu^2\cdot f_i^2$$

$$\Rightarrow \quad = \Rightarrow \lim_{q\to k}\lim_{p\to j}\left[f_{i,\alpha}f_{i,\beta}\left(\frac{\gamma_k^{j\alpha}\gamma_p^{q\beta}+\gamma_k^{j\beta}\gamma_p^{q\alpha}}{2\cdot n}\right)+\frac{m^2+m^2}{2}\cdot f_i^2\right]=0$$

$$\Rightarrow \quad \lim_{q\to k}\lim_{p\to j}f_{i,\alpha}f_{i,\beta}\gamma_k^{j\alpha}\gamma_p^{q\beta}+m^2\cdot f_i^2=0$$

$$\Rightarrow \quad \left(f_{i,\alpha}\gamma_k^{j\alpha}+i\cdot m\cdot f_i\right)\left(f_{i,\beta}\gamma_j^{k\beta}-i\cdot m\cdot f_i\right)=0$$

$$\lim_{k\to i}\left(f_{i,\beta}\gamma_j^{k\beta}-i\cdot m\cdot f_i\right)=0$$

$$\&$$

$$\Rightarrow \quad \lim_{q\to k}\lim_{p\to j}f_{i,\alpha}f_{i,\beta}\gamma_k^{j\beta}\gamma_p^{q\alpha}+m^2\cdot f_i^2=0$$

$$\Rightarrow \quad \left(f_{i,\alpha}\gamma_j^{k\alpha}+i\cdot m\cdot f_i\right)\left(f_{i,\beta}\gamma_k^{j\beta}-i\cdot m\cdot f_i\right)=0$$

$$\lim_{j\to i}\left(f_{i,\beta}\gamma_k^{j\beta}-i\cdot m\cdot f_i\right)=0. \tag{574}$$

This gives us the following Dirac equation:

$$\lim_{j\to i}\left(f_{i,\beta}\gamma_k^{j\beta}-i\cdot m\cdot f_i\right)=0. \tag{575}$$

5.18.13 Why Do We Need the Dirac Matrices at All?

Starting with first line in Eq. (574), we try a separation of the factors via base vectors as follows:

$$0 = f_{i,\alpha} f_{i,\beta} g^{\alpha\beta} + \mu^2 \cdot f_i^2$$

$$0 = f_{i,\alpha} f_{i,\beta} \mathbf{e}^{\alpha} \cdot \mathbf{e}^{\beta} + \mu^2 \cdot f_i^2$$

$$f_{i,\alpha} f_{i,\beta} \mathbf{e}^{\alpha} \cdot \mathbf{e}^{\beta} + \mu^2 \cdot f_i^2 + \overbrace{\left(f_{i,\beta} \mathbf{e}^{\beta} - f_{i,\alpha} \mathbf{e}^{\alpha}\right)}^{=0} \cdot i \cdot \mu \cdot f_i$$
$$= \left(f_{i,\alpha} \mathbf{e}^{\alpha} + i \cdot \mu \cdot f_i\right)\left(f_{i,\beta} \mathbf{e}^{\beta} - i \cdot \mu \cdot f_i\right). \tag{576}$$

We see that there is a problem with the different ranks of the differential and the non-differential terms in the two factors in the last line. The simplest way would be to make μ a vector $\boldsymbol{\mu}$ and rewrite the evaluation above as follows:

$$0 = f_{i,\alpha} f_{i,\beta} g^{\alpha\beta} + \mu^2 \cdot f_i^2$$

$$0 = f_{i,\alpha} f_{i,\beta} \mathbf{e}^{\alpha} \cdot \mathbf{e}^{\beta} + \overbrace{\boldsymbol{\mu} \cdot \boldsymbol{\mu}}^{=\mu^2} \cdot f_i^2$$

$$f_{i,\alpha} f_{i,\beta} \mathbf{e}^{\alpha} \cdot \mathbf{e}^{\beta} + \boldsymbol{\mu} \cdot \boldsymbol{\mu} \cdot f_i^2 + \overbrace{\left(f_{i,\beta} \mathbf{e}^{\beta} - f_{i,\alpha} \mathbf{e}^{\alpha}\right)}^{=0} \cdot i \cdot \boldsymbol{\mu} \cdot f_i$$
$$= \left(f_{i,\alpha} \mathbf{e}^{\alpha} + i \cdot \boldsymbol{\mu} \cdot f_i\right)\left(f_{i,\beta} \mathbf{e}^{\beta} - i \cdot \boldsymbol{\mu} \cdot f_i\right). \tag{577}$$

5.18.14 A Bit of Illustration of the Findings So Far

The reader may have noticed our lack of enthusiasm with respect to the evaluation of examples of our various Dirac equations. The reason for this laziness is nurtured from the fact that the original motivation for the search for the Dirac equation was the probability interpretation of the wave function (e.g., [24–28]). With the metric derivation of the wave function as the scale of the metric, it became clear that this function is related to the volume fluctuations of space-time ... in fact, it is such fluctuations [29]. Hence, the probability interpretation does not seem to be needed anymore and, thus, nor is the Dirac equation. As a justification for the need for the Dirac equation, we often hear the occurrence of the anti-matter solutions and the spin-1/2. Well, it has been already shown that spin-1/2. can also be obtained from our scalar quantum gravity equations (376), (440) and (450) either directly as special solutions to the angular part of the equation or for the right metric forcing two opposite spin states into existence [9, 10, 23]. Regarding the anti-matter solutions, we find that we also obtain them from the scalar quantum gravity equations (376), (440) and (450). For illustration with respect to the latter problem, we repeat the derivation of the wave function around spherical objects from [18] in appendix RA of this book. There, we find the typical time-dependency for such an object to be of the kind:

$$f[t]=e^{\pm i\cdot\frac{E}{h}\cdot t}\cdot C_f=e^{\pm i\cdot\frac{m_R\cdot c^2}{h}\cdot t}\cdot C_f, \tag{578}$$

where the oscillations in time are either driven by an energy E or a rest mass m_R. Of course, the interpretation of these oscillations should still be a matter of discussion, but as shown in appendix RA, the solution can also be combined with in- and outgoing waves of various kinds. The simplest ones are just pure radial waves (see appendix RA for derivation and explanation of the various symbols):

$$f[t,r]=\frac{i\cdot C_{ty}}{r}\cdot\begin{Bmatrix} e^{\pm i\cdot\frac{r_s}{2\cdot\ell_P^2}\cdot(r+c\cdot t)} \\ e^{\pm i\cdot\frac{r_s}{2\cdot\ell_P^2}\cdot(r-c\cdot t)} \end{Bmatrix}. \tag{579}$$

With sufficient compactness, such waves can appear as particles to the right measurement, which is to say with measurements not being designed to resolve the wave-like character of the waves. So, we have four wave-type solutions, namely:

$$f[t,r]=\frac{i\cdot C_{ty}}{r}\cdot\left\{e^{i\cdot\frac{r_s}{2\cdot\ell_P^2}\cdot(r+c\cdot t)},e^{-i\cdot\frac{r_s}{2\cdot\ell_P^2}\cdot(r+c\cdot t)},e^{i\cdot\frac{r_s}{2\cdot\ell_P^2}\cdot(r-c\cdot t)},e^{i\cdot\frac{r_s}{2\cdot\ell_P^2}\cdot(c\cdot t-r)}\right\}. \tag{580}$$

These are in- and outgoing waves with the oscillation going parallel and anti-parallel to the radius coordinate. Perhaps a somewhat more illustrative explanation for the difference between matter and anti-matter is just this parallel and anti-parallel behavior of the object's waves (moving or standing ones). But as we will see in the next section, there are also other possibilities to metrically explain matter and anti-matter.

5.18.15 What Is Anti-Matter?

5.18.15.1 The Dirac equation as an outcome of a scaled metric?

In summing up the above findings, one may come to the conclusion that the reason for the occurrence of the Dirac equation, which classically was always justified as a probability problem with the second-order time derivative residing in the Klein–Gordon equation, the half spin of the fermions and the existence of anti-matter, might be something completely different. In order to work out what this "other reason" could be, we consider the Laplace equation of a function F within a scaled metric with scale function F as given in Eq. (367):

$$\Delta_G F = 0. \tag{581}$$

This apparently simple equation can be expanded as follows:

$$\Delta_G F = \frac{1}{\sqrt{G}}\partial_\alpha\left(\sqrt{G}\cdot G^{\alpha\beta}\cdot\partial_\beta F\right) = \frac{1}{\sqrt{F^n\cdot g}}\partial_\alpha\left(\sqrt{F^n\cdot g}\cdot F^{-1}\cdot g^{\alpha\beta}\cdot\partial_\beta F\right)$$

$$= \frac{1}{F^{\frac{n}{2}}\cdot\sqrt{g}}\partial_\alpha\left(\sqrt{g}\cdot F^{\frac{n}{2}-1}\cdot g^{\alpha\beta}\cdot\partial_\beta F\right)$$

$$= \frac{1}{F^{\frac{n}{2}}\cdot\sqrt{g}}\left(\left(\frac{n}{2}-1\right)\cdot\sqrt{g}\cdot F^{\frac{n}{2}-2}\cdot F_{,\alpha}\cdot g^{\alpha\beta}\cdot\partial_\beta F + F^{\frac{n}{2}-1}\cdot\partial_\alpha\left(\sqrt{g}\cdot g^{\alpha\beta}\cdot\partial_\beta F\right)\right)$$

$$= \frac{n-2}{2\cdot F^2}\cdot F_{,\alpha}\cdot g^{\alpha\beta}\cdot\partial_\beta F + \frac{1}{F^1\cdot\sqrt{g}}\cdot\partial_\alpha\left(\sqrt{g}\cdot g^{\alpha\beta}\cdot\partial_\beta F\right)$$

$$= \frac{n-2}{2\cdot F^2}\cdot F_{,\alpha}\cdot g^{\alpha\beta}\cdot\partial_\beta F + \frac{\Delta_g F}{F^1} = 0$$

$$\Rightarrow$$

$$\frac{n-2}{2}\cdot\frac{F_{,\alpha}}{F}\cdot g^{\alpha\beta}\cdot\partial_\beta F + \Delta_g F = 0 \;\;\Rightarrow\;\; \frac{n-2}{2}\cdot F_{,\alpha}\cdot g^{\alpha\beta}\cdot F_{,\beta} + F\cdot\Delta_g F = 0. \tag{582}$$

This is—so it seems—again the perfect starting point for our "Dirac-derivation" from Eq. (504) onwards. For convenience and in order to show the "weak point," we will repeat the essential steps. First, we assume that the Laplace operator applied on F has an eigenvalue, making the last line in Eq. (582):

$$\Delta_G F = 0$$

$$\xrightarrow{\Delta_g F = \mu^2\cdot F}$$

$$\frac{n-2}{2}\cdot F_{,\alpha}\cdot g^{\alpha\beta}\cdot F_{,\beta} + F\cdot\Delta_g F = \frac{n-2}{2}\cdot F_{,\alpha}\cdot g^{\alpha\beta}\cdot F_{,\beta} + F\cdot\mu^2\cdot F = 0. \tag{583}$$

We further assume that the function F is in fact a list of functions F_i and obtain:

$$\frac{n-2}{2}\cdot F_{i,\alpha}\cdot g^{\alpha\beta}\cdot F_{i,\beta}+F_i^2\cdot\mu^2=0. \tag{584}$$

Applying the Dirac matrices and pulling the factor $\frac{n-2}{2}$ into the μ with the substitution $\mu\rightarrow\sqrt{\frac{n-2}{2}}\cdot\mu$, we can put Eq. (584) into the following form:

$$0=\left(F_{i,\alpha}F_{i,\beta}g^{\alpha\beta}+\mu^2\cdot F_i^2\right)\cdot I$$

$$\Rightarrow \quad F_{i,\alpha}F_{i,\beta}\left(\frac{\gamma^\alpha\gamma^\beta+\gamma^\beta\gamma^\alpha}{2}\right)+\frac{I+I}{2}\cdot\mu^2\cdot F_i^2=0$$

$$\Rightarrow \quad F_{i,\alpha}F_{i,\beta}\gamma^\alpha\gamma^\beta+I\cdot\mu^2\cdot F_i^2$$

$$=\left(F_{i,\alpha}\gamma^\alpha F_{i,\beta}\gamma^\beta+i\cdot I\cdot\mu\cdot F_i\overbrace{\left(F_{i,\beta}\gamma^\beta-F_{i,\alpha}\gamma^\alpha\right)}^{F_{i,\beta}\gamma^\beta-F_{i,\alpha}\gamma^\alpha=F_{i,\gamma}\gamma^\gamma-F_{i,\gamma}\gamma^\gamma=0}+I\cdot\mu^2\cdot F_i^2\right)$$

$$=\left(F_{i,\alpha}\gamma^\alpha+i\cdot I\cdot\mu\cdot F_i\right)\left(F_{i,\beta}\gamma^\beta-i\cdot I\cdot\mu\cdot F_i\right)=0$$

$$\left(F_{i,\beta}\gamma^\beta-i\cdot I\cdot\mu\cdot F_i\right)=0$$

$$F_{i,\alpha}F_{i,\beta}\gamma^\beta\gamma^\alpha+I\cdot\mu^2\cdot F_i^2=0$$

$$\Rightarrow \quad \left(F_{i,\alpha}\gamma^\alpha+i\cdot I\cdot\mu\cdot F_i\right)\left(F_{i,\beta}\gamma^\beta-i\cdot I\cdot\mu\cdot F_i\right)=0. \tag{585}$$

This gives us the Dirac equation in its classical form [12]:

$$0=F_{i,\beta}\gamma^\beta-i\cdot I\cdot\mu\cdot F_i. \tag{586}$$

Now we have to prove that our result also ensures the satisfaction of the eigenvalue equation for the Laplace operator term, which is to say the Klein–Gordon condition we used in Eq. (583):

$$\Delta F_i=\mu^2\cdot F_i. \tag{587}$$

First, we write the Laplace operator in Eq. (587) in the following way:

$$\Delta F_i=g^{\alpha\beta}\nabla_\alpha\nabla_\beta F_i=\mu^2\cdot F_i. \tag{588}$$

Now we apply the connection between the metric and the Dirac matrices again and obtain:

$$g^{\alpha\beta}\nabla_\alpha\nabla_\beta F_i - \mu^2 \cdot F_i = 0$$

$$0 = \left(g^{\alpha\beta}\nabla_\alpha\nabla_\beta F_i - \mu^2 \cdot F_i\right)\cdot I$$

$$\Rightarrow \qquad = \Rightarrow\left(\frac{\gamma^\alpha\gamma^\beta + \gamma^\beta\gamma^\alpha}{2}\right)\nabla_\alpha\nabla_\beta F_i - \frac{I^2 + I^2}{2}\cdot\mu^2\cdot F_i = 0$$

$$\Rightarrow \qquad \gamma^\alpha\gamma^\beta\nabla_\alpha\nabla_\beta F_i - I^2\cdot\mu^2\cdot F_i = 0$$

$$\Rightarrow \qquad \left(\gamma^\alpha\nabla_\alpha + I\cdot\mu\right)\left(\gamma^\beta\nabla_\beta - I\cdot\mu\right)F_i = 0$$

$$\Rightarrow \qquad \gamma^\beta\gamma^\alpha\nabla_\alpha\nabla_\beta F_i - I^2\cdot\mu^2\cdot F_i = 0$$

$$\Rightarrow \qquad \left(\gamma^\alpha\nabla_\alpha + I\cdot\mu\right)\left(\gamma^\beta\nabla_\beta - I\cdot\mu\right)F_i = 0. \tag{589}$$

As it totally suffices for the functions f_i to satisfy the inner operator in the fifth and last lines of Eq. (589), we can demand:

$$\left(\gamma^\beta\nabla_\beta - I\cdot\mu\right)f_i = 0 \;\Rightarrow\; \gamma^\beta\nabla_\beta f_i - I\cdot\mu\cdot f_i = 0. \tag{590}$$

We immediately realize that this does not agree with our result (586). We therefore conclude that the Dirac equation cannot "be forced into existence" via a simple scaled metric expansion approach of the likes (582). However, what about a small generalization as follows:

$$\Delta_G\varphi = 0. \tag{591}$$

The expanded version of this Laplace equation now reads:

$$\Delta_G\varphi = \frac{1}{\sqrt{G}}\partial_\alpha\left(\sqrt{G}\cdot G^{\alpha\beta}\cdot\partial_\beta\varphi\right) = \frac{1}{\sqrt{F^n\cdot g}}\partial_\alpha\left(\sqrt{F^n\cdot g}\cdot F^{-1}\cdot g^{\alpha\beta}\cdot\partial_\beta\varphi\right)$$

$$= \frac{1}{F^{\frac{n}{2}}\cdot\sqrt{g}}\partial_\alpha\left(\sqrt{g}\cdot F^{\frac{n}{2}-1}\cdot g^{\alpha\beta}\cdot\partial_\beta\varphi\right)$$

$$= \frac{1}{F^{\frac{n}{2}}\cdot\sqrt{g}}\left(\frac{n-2}{2}\cdot\sqrt{g}\cdot F^{\frac{n}{2}-2}\cdot F_{,\alpha}\cdot g^{\alpha\beta}\cdot\partial_\beta\varphi + F^{\frac{n}{2}-1}\cdot\partial_\alpha\left(\sqrt{g}\cdot g^{\alpha\beta}\cdot\partial_\beta\varphi\right)\right)$$

$$= \frac{n-2}{2}\cdot\frac{1}{F^2}\cdot F_{,\alpha}\cdot g^{\alpha\beta}\cdot\partial_\beta\varphi + \frac{1}{F^1\cdot\sqrt{g}}\cdot\partial_\alpha\left(\sqrt{g}\cdot g^{\alpha\beta}\cdot\partial_\beta\varphi\right)$$

$$= \frac{n-2}{2} \cdot \frac{1}{F^2} \cdot F_{,\alpha} \cdot g^{\alpha\beta} \cdot \partial_\beta \varphi + \frac{\Delta_g \varphi}{F^1} = 0$$

$$\Rightarrow \qquad \frac{n-2}{2} \cdot \frac{F_{,\alpha}}{F} \cdot g^{\alpha\beta} \cdot \partial_\beta \varphi + \Delta_g \varphi = 0$$

$$\Rightarrow \qquad \frac{n-2}{2} \cdot F_{,\alpha} \cdot g^{\alpha\beta} \cdot \varphi_{,\beta} + F \cdot \Delta_g \varphi = 0. \tag{592}$$

This time, we not only assume that the Laplace operator applied on φ has an eigenvalue but also demand:

$$\partial_\beta \varphi = -\partial_\beta F, \tag{593}$$

which leads the last two lines in Eq. (592) to read:

$$\Delta_G \varphi = 0$$

$$\xrightarrow{\Delta_g \varphi = \Delta_g F = -\mu^2 \cdot F}$$

$$\frac{n-2}{2} \cdot F_{,\alpha} \cdot g^{\alpha\beta} \cdot \varphi_{,\beta} + F \cdot \Delta_g \varphi = -\frac{n-2}{2} \cdot F_{,\alpha} \cdot g^{\alpha\beta} \cdot F_{,\beta} - F \cdot \mu^2 \cdot F = 0$$

$$\xrightarrow{\mu \to \sqrt{\frac{n-2}{2}} \cdot \mu}$$

$$-F_{,\alpha} \cdot g^{\alpha\beta} \cdot F_{,\beta} - F \cdot \mu^2 \cdot F = 0. \tag{594}$$

It has to be pointed out that instead of Eq. (593), we could also have a less rigid condition, reading:

$$\mathbf{e}^\beta \cdot \partial_\beta \varphi = -\mathbf{e}^\beta \cdot \partial_\beta F. \tag{595}$$

Again, we further assume that the function F is in fact a list of functions F_i and obtain:

$$F_{i,\alpha} \cdot g^{\alpha\beta} \cdot F_{i,\beta} + F_i^2 \cdot \mu^2 = 0. \tag{596}$$

Applying the Dirac matrices yields the evaluation (585) and gives the Dirac equation in its classical form [12]:

$$0 = F_{i,\beta} \gamma^\beta - i \cdot I \cdot \mu \cdot F_i. \tag{597}$$

Checking that our result also ensures the satisfaction of the eigenvalue equation for the Laplace operator term, which is to say the Klein–Gordon condition we used in Eq. (594):

$$\Delta F_i = -\mu^2 \cdot F_i, \tag{598}$$

can now be performed in the following way:

$$\Delta F_i = g^{\alpha\beta}\nabla_\alpha \nabla_\beta F_i = -\mu^2 \cdot F_i. \tag{599}$$

Now we apply the connection between the metric and the Dirac matrices again and—this time—obtain:

$$g^{\alpha\beta}\nabla_\alpha \nabla_\beta F_i + \mu^2 \cdot F_i = 0$$

$$0 = \left(g^{\alpha\beta}\nabla_\alpha \nabla_\beta F_i + \mu^2 \cdot F_i\right) \cdot I$$

$$\Rightarrow \qquad = \Rightarrow \left(\frac{\gamma^\alpha \gamma^\beta + \gamma^\beta \gamma^\alpha}{2}\right)\nabla_\alpha \nabla_\beta F_i + \frac{I^2 + I^2}{2} \cdot \mu^2 \cdot F_i = 0$$

$$\Rightarrow \qquad \gamma^\alpha \gamma^\beta \nabla_\alpha \nabla_\beta F_i + I^2 \cdot \mu^2 \cdot F_i = 0$$

$$\Rightarrow \qquad \left(\gamma^\alpha \nabla_\alpha + i \cdot I \cdot \mu\right)\left(\gamma^\beta \nabla_\beta - i \cdot I \cdot \mu\right) F_i = 0$$

$$\Rightarrow \qquad \gamma^\beta \gamma^\alpha \nabla_\alpha \nabla_\beta F_i + I^2 \cdot \mu^2 \cdot F_i = 0$$

$$\Rightarrow \qquad \left(\gamma^\alpha \nabla_\alpha + i \cdot I \cdot \mu\right)\left(\gamma^\beta \nabla_\beta - i \cdot I \cdot \mu\right) F_i = 0. \tag{600}$$

The result:

$$\left(\gamma^\beta \nabla_\beta - i \cdot I \cdot \mu\right) f_i = 0 \Rightarrow \gamma^\beta \nabla_\beta f_i - i \cdot I \cdot \mu \cdot f_i = 0 \tag{601}$$

is in perfect agreement with the outcome (597).

Thus, we have derived a purely mathematical "need" for a matrix equation coming out of Eq. (591), nurtured by the wish to linearize the expansion (592) with the assumed conditions (593) and (594), which was directly leading us to the Dirac equation. The only problem left would be to find a metric reason why this special Laplace equation (591) occurs in the first place. This will be investigated further below in the next section.

5.18.16 Is the Dirac Equation Only an Artefact of an Incomplete Approach?

In chapter 10 of [4] (especially section "10.10 The Gravity Dirac Equation"), we investigated the opportunity for the derivation of a "Gravity Dirac Field Equation." Using Eq. (405), we want to revisit this path here. Our reasoning starts with the following possibility regarding the adjustment of the kernel in the Einstein–Hilbert action:

$$0=\delta\int_V d^n x\left(\sqrt{-g}\cdot F^{\frac{n}{2}+q}\cdot\Phi_0\cdot R^*\right)$$

$$\Rightarrow\qquad 0=\delta\int_V d^n x\left(\sqrt{-g\cdot F^n}\cdot F^q\cdot\Phi_0\cdot R^*\right)$$

$$\xrightarrow{G_{\alpha\beta}=g_{\alpha\beta}\cdot F=g_{\alpha\beta}\cdot F[f];\quad F^q\cdot\Phi_0=\Phi}$$

$$\delta W=0=\delta\int_V d^n x\left(\sqrt{-G}\cdot\Phi\cdot R^*\right). \tag{602}$$

Per definition, we demand the vanishing surface term not only by assuming a clever choice of the variation but also due to a suitable selection of the ratio between metric scaling F and kernel scaling Φ. The problem to solve would be:

$$\left(R^*_{\mu\nu}-\frac{1}{2}R^*G_{\mu\nu}-\nabla_\mu\nabla_\nu+G_{\mu\nu}\Delta_G\right)\overset{\downarrow}{\Phi}\delta G^{\mu\nu}=0$$

$$\Rightarrow\qquad\left(R^*_{\mu\nu}-\frac{1}{2}R^*G_{\mu\nu}-\nabla_\mu\nabla_\nu+G_{\mu\nu}\Delta_G\right)\overset{\downarrow}{\Phi}\left(\frac{\delta g^{\mu\nu}}{F}+g^{\mu\nu}\delta\frac{1}{F}\right)$$

$$\left(R^*_{\mu\nu}-\frac{1}{2}R^*G_{\mu\nu}-\nabla_\mu\nabla_\nu+G_{\mu\nu}\Delta_G\right)\overset{\downarrow}{\Phi}\left(\frac{\delta g^{\mu\nu}}{F}-\frac{g^{\mu\nu}}{F^2}\delta F\right)=0$$

$$\left(R^*_{\mu\nu}-\frac{1}{2}R^*G_{\mu\nu}-\nabla_\mu\nabla_\nu+G_{\mu\nu}\Delta_G\right)\overset{\downarrow}{\Phi}\left(\frac{\delta g^{\mu\nu}}{F}-\frac{G^{\mu\nu}}{F}\delta F\right)=0. \tag{603}$$

Here the $\overset{\downarrow}{\Phi}$ symbol points out that the operators are only acting on the function Φ. We see that in order to find a solution for all variations, we have to have more than just a scalar degree of freedom. In other words, fixing just Φ will generally not suffice. This is no problem, because the equation above also has the metric tensor as a degree of freedom and thus enough components to adjust. It should be noted, however, that due to the separation of the variational term, the equation in the last line of Eq. (603) can be split into two equations, namely:

$$\left(R^*_{\mu\nu}-\frac{1}{2}R^*G_{\mu\nu}-\nabla_\mu\nabla_\nu+G_{\mu\nu}\Delta_G\right)\overset{\downarrow}{\Phi}\frac{\delta g^{\mu\nu}}{F}=0 \tag{604}$$

and:

$$-\left(R^{*}_{\mu\nu}-\frac{1}{2}R^{*}G_{\mu\nu}-\nabla_{\mu}\nabla_{\nu}+G_{\mu\nu}\Delta_{G}\right)\overset{\downarrow}{\mathbf{\Phi}}\frac{G^{\mu\nu}}{F}\delta F$$

$$=-\frac{G^{\mu\nu}}{F}\delta F\left(R^{*}_{\mu\nu}-\frac{1}{2}R^{*}G_{\mu\nu}-\nabla_{\mu}\nabla_{\nu}+G_{\mu\nu}\Delta_{G}\right)\Phi$$

$$=-\frac{\delta F}{F}\left(\left(1-\frac{n}{2}\right)\cdot R^{*}+(n-1)\cdot\Delta_{G}\right)\Phi=0$$

$$\Rightarrow\qquad\left(\frac{2-n}{n-1}\cdot R^{*}+\Delta_{G}\right)\Phi=0. \tag{605}$$

In Ricci-R^*-flat space-times, we would just obtain the Laplace equation with respect to the metric G_{ab}:

$$\Delta_G\Phi = 0 \tag{606}$$

and, as presented elsewhere, can obtain mass and potential via additional (suitably entangled) dimensions (e.g., [4, 19, 21, 30]), this way obtaining the usual quantum Klein–Gordon equation. This, as already shown above, can then be "decomposed" into the Dirac equation in the usual way.

We realize that Eq. (606) is exactly the starting point for our metrically derived Dirac equation from a scaled metric as discussed in the previous sub-section entitled "5.18.15.1 The Dirac equation as an outcome of a scaled metric?," where we needed an equation (591) in order to end up with the right conditions for the classical Dirac equation. Knowing now, however, where all the functions and conditions leading to the Dirac equation are coming from, we can easily explain—just to give an example—the difference between matter and anti-matter. We see that function Φ is part of the kernel of the Einstein–Hilbert action, while $F[f]$ is the scaling function for the metric. Taking the particle at rest, where we only have a dependency on the t-coordinate, the condition (593), now reading:

$$\partial_{\beta}\Phi=-\partial_{\beta}F, \tag{607}$$

respectively (weaker):

$$\mathbf{e}^{\beta}\cdot\partial_{\beta}\Phi=-\mathbf{e}^{\beta}\cdot\partial_{\beta}F, \tag{608}$$

forces the classical solution for the matter (minus sign in the exponent) and anti-matter (plus sign in the exponent) with:

$$\Phi=C_{0}\cdot e^{\pm i\cdot m\cdot t}, \tag{609}$$

to fix the corresponding function F as follows:

$$F = C_0 \cdot e^{\mp i \cdot m \cdot t}. \tag{610}$$

In other words, when taking the metric scaling as the function to determine the matter or anti-matter character of an object, the sign in Eq. (610) would be negative in the matter case and positive for anti-matter, while it is the other way round for the corresponding kernel scaling function Φ. Under the variational integral (602), we then have the following situation (only for the particle at rest):

$$\delta W = 0 = \delta \int_V d^n x \left(\sqrt{-G} \cdot \Phi \cdot R^* \right)$$

$$\xrightarrow{G_{\alpha\beta} = g_{\alpha\beta} \cdot F = g_{\alpha\beta} \cdot F[f]}$$

$$= \delta \int_V d^n x \left(\sqrt{-g} \cdot F^{\frac{n}{2}} \cdot \Phi \cdot R^* \right) = \delta \int_V d^n x \left(\sqrt{-g} \cdot \left(C_0 \cdot e^{\mp i \cdot m \cdot t} \right)^{\frac{n}{2}} \cdot C_0 \cdot e^{\pm i \cdot m \cdot t} \cdot R^* \right)$$

$$= \delta \int_V d^n x \left(\sqrt{-g} \cdot \left(C_0 \right)^{\frac{n}{2}+1} \cdot \left(e^{\mp i \cdot m \cdot t} \right)^{\frac{n}{2}-1} \cdot R^* \right)$$

$$= \delta \int_V d^n x \left(\sqrt{-g \cdot F} \cdot C_0 \cdot \left(e^{\mp i \cdot m \cdot t} \right)^{-1} \cdot R^* \right)$$

$$= \delta \int_V d^n x \left(\sqrt{-g \cdot F} \cdot \frac{C_0}{F} \cdot R^* \right) = \delta \int_V d^n x \left(\sqrt{-G} \cdot \frac{C_0}{F} \cdot R^* \right). \tag{611}$$

We realize that in Cartesian coordinates, solutions are rather simple and can be given in the form:

$$\Phi = C_0 \cdot e^{\pm i \cdot \left(m \cdot t + c_1 \cdot x_1 + c_2 \cdot x_2 + c_3 \cdot x_3 + \ldots \right)}, \tag{612}$$

$$F = C_0 \cdot e^{\mp i \cdot \left(m \cdot t + c_1 \cdot x_1 + c_2 \cdot x_2 + c_3 \cdot x_3 + \ldots \right)}. \tag{613}$$

5.18.17 The Spectral-Operator Decomposition?

But could there potentially be another motivation for the occurrence of the Dirac equation?

In order to investigate this question, we try to decompose the whole Eq. (603).

First, we define the following operator:

$$\widehat{N}_{\mu\nu}\Phi \equiv \left(R^{*}_{\mu\nu} - \frac{1}{2} R^{*} G_{\mu\nu} - \nabla_{\mu}\nabla_{\nu} + G_{\mu\nu}\Delta_{G} \right)\Phi = 0. \tag{614}$$

We immediately see that this operator is symmetric, because all its addends are and that therefore it should be possible to spectrally decompose it as follows:

$$\widehat{N}_{\mu\nu}\Phi = N\lambda N^{T}\Phi = N\eta\eta N^{T}\Phi = 0. \tag{615}$$

Thereby we have the eigenvectors $\mathbf{u}_i$ of the operator $\widehat{N}_{\mu\nu}$ forming our N and N^T as follows:

$$N = \begin{pmatrix} \leftarrow & \mathbf{u}_1 & \rightarrow \\ \leftarrow & \mathbf{u}_2 & \rightarrow \\ \leftarrow & \vdots & \rightarrow \end{pmatrix}; \quad N^{T} = \begin{pmatrix} \uparrow & \uparrow & \uparrow \\ \mathbf{u}_1 & \mathbf{u}_2 & \cdots \\ \downarrow & \downarrow & \downarrow \end{pmatrix} \tag{616}$$

and the eigenvalues λ_i constructing the matrices η via:

$$\eta = \begin{pmatrix} \sqrt{\lambda_1} & 0 & \cdots \\ 0 & \sqrt{\lambda_2} & \cdots \\ \vdots & \vdots & \ddots \end{pmatrix}. \tag{617}$$

As the matrices of the eigenvectors could also be defined as follows:

$$N^{T} = \begin{pmatrix} \leftarrow & \mathbf{u}_1 & \rightarrow \\ \leftarrow & \mathbf{u}_2 & \rightarrow \\ \leftarrow & \vdots & \rightarrow \end{pmatrix}; \quad N = \begin{pmatrix} \uparrow & \uparrow & \uparrow \\ \mathbf{u}_1 & \mathbf{u}_2 & \cdots \\ \downarrow & \downarrow & \downarrow \end{pmatrix}, \tag{618}$$

we conclude that to equation (405), (615) also intrinsic solutions of the type

$$\eta N^{T}\Phi = 0 \tag{619}$$

or

$$N\eta\Phi = 0 \tag{620}$$

should exist. Please note that in contrast to the derivation of the Dirac equation [27], our equations (619) and (620) do not demand the function Φ to become a function vector, because the rank and dimension of the operator are not changed when moving from Eq. (615) to Eqs. (619) and (620), but only the operational order is. Nor is a vector-, a matrix- or any other higher rank setting for the function Φ being anyhow prohibited as its origin in Eq. (404) shows. We will see the use of this additional degree of freedom further below, when trying to establish the connection to the classical Dirac equation.

Please also note that the evaluation above is always maintaining the rank and dimension of the operator and thus, has to be understood as follows:

$$\widehat{N}_{\mu\nu}\Phi=\left[N\lambda N^{T}\right]_{\mu\nu}\Phi=\left[N\eta\eta N^{T}\right]_{\mu\nu}\Phi=0, \tag{621}$$

$$\left[\eta N^{T}\right]_{\nu}\Phi=0, \tag{622}$$

$$\left[N\eta\right]_{\mu}\Phi=0. \tag{623}$$

So, why would we even bother to introduce the function vector then?

Well, we see that—as before—we require a vector or even a matrix Φ in order to obtain enough degrees of freedom to actually solve problem (623). The reason may be that we would not like to adjust the metric tensor instead of going the—potentially—simpler way of having to deal with something like (the maximum contraction option):

$$\left[\eta N^{T}\right]_{a\nu}^{b}\Phi^{\nu a}_{\ \ b}=0, \tag{624}$$

$$\left[N\eta\right]_{a\mu}^{b}\Phi^{\mu a}_{\ \ b}=0, \tag{625}$$

rather than:

$$\left[\eta N^{T}\right]\left[G_{\alpha\beta}=F\left[f\right]\cdot g_{\alpha\beta}\right]_{\nu}\Phi=0, \tag{626}$$

$$\left[N\eta\right]\left[G_{\alpha\beta}=F\left[f\right]\cdot g_{\alpha\beta}\right]_{\mu}\Phi=0. \tag{627}$$

In the latter case, we would have the task of finding the right metric tensor AND the function Φ, while the first option delegates the whole problem onto the multicomponent function object $\Phi^{\nu a}_{\ \ b}$ or $\Phi^{\mu a}_{\ \ b}$ with an assumed (or given) fixed metric tensor.

Is the whole Dirac approach just an approximated way to solve the metric-Φ-task in the form (624), (625) instead of the correct way (626), (627)?

It should be mentioned that we may even bring Eq. (604) into a form where we will be able to apply the Dirac matrices (505). For this we start with the mixed form of Eq. (614):

$$R_{\alpha}^{\beta}-\frac{\delta_{\alpha}^{\beta}}{2}R+\frac{1}{2F}\eth_{\alpha}^{\beta}\left[g_{\alpha\beta};F\right]-\frac{\left(G^{\beta\mu}\nabla_{\mu}\nabla_{\alpha}-\delta_{\alpha}^{\beta}\Delta_{G}\right)\Phi}{\Phi}=0, \tag{628}$$

with the abbreviation:

$$\frac{1}{2F}\eth_{\alpha}^{\beta}\left[g_{\alpha\beta};F\right]=\frac{1}{2F}\left(\begin{array}{c}\frac{\delta_{\alpha}^{\beta}}{2}\left(\left(\begin{array}{c}2F_{,ij}(n-1)g^{ij}+2\Gamma_{ij}^{a}F_{,a}g^{ij}\\-F_{,i}g^{ab}g_{jb,a}g^{ij}-F_{,j}g^{ab}g_{ib,a}g^{ij}\\-n\Gamma_{ij}^{d}F_{,d}g^{ij}+\frac{n}{2}F_{,d}g^{cd}g_{ab,c}g^{ab}\end{array}\right)+\frac{F_{,i}\cdot F_{,j}}{2F}g^{ij}\left((n-6)(n-1)\right)\right)\\-g^{\beta\gamma}\left(\begin{array}{c}F_{,\alpha\gamma}(n-2)+F_{,ab}g_{\alpha\gamma}g^{ab}\\+F_{,a}g^{ab}\left(g_{\gamma b,\alpha}-g_{\gamma\alpha,b}\right)-F_{,\alpha}g^{ab}g_{\gamma b,a}-F_{,\gamma}g^{ab}g_{\alpha b,a}\\+F_{,d}g^{cd}\frac{1}{2}n\left(\begin{array}{c}\frac{2}{n}g_{\alpha c,\gamma}-g_{\alpha c,\gamma}-g_{\gamma c,\alpha}\\+g_{\alpha\gamma,c}+\frac{1}{n}g_{\alpha\gamma}g_{ab,c}g^{ab}\end{array}\right)\\-\frac{1}{2F}\left(F_{,\alpha}\cdot F_{,\gamma}(3n-6)+g_{\alpha\gamma}F_{,c}F_{,d}g^{cd}(4-n)\right)\end{array}\right)\end{array}\right)\tag{629}$$

and using the spectral decomposition, and then bring it into the following form:

$$\begin{aligned}&R_{\alpha}^{\beta}-\frac{\delta_{\alpha}^{\beta}}{2}R+\frac{1}{2F}\eth_{\alpha}^{\beta}\left[g_{\alpha\beta};F\right]-\frac{\left(G^{\beta\mu}\nabla_{\mu}\nabla_{\alpha}-\delta_{\alpha}^{\beta}\Delta_{G}\right)\Phi}{\Phi}=0\\&\left(R_{\alpha}^{*\beta}-\frac{\delta_{\alpha}^{\beta}}{2}R^{*}-G^{\beta\mu}\nabla_{\mu}\nabla_{\alpha}+\delta_{\alpha}^{\beta}\nabla_{\mu}G^{\mu\nu}\nabla_{\nu}\right)\Phi=0\\&\left(R_{\alpha}^{*\beta}-\frac{\delta_{\alpha}^{\beta}}{2}R^{*}-\mathbf{E}^{\mu}\cdot\mathbf{E}^{\beta}\nabla_{\mu}\nabla_{\alpha}+\delta_{\alpha}^{\beta}\nabla_{\mu}\mathbf{E}^{\mu}\cdot\mathbf{E}^{\nu}\nabla_{\nu}\right)\Phi=0\\&\left(R_{\alpha}^{*\beta}-\frac{\delta_{\alpha}^{\beta}}{2}R^{*}-\mathbf{E}^{\mu}\nabla_{\mu}\cdot\left(\mathbf{E}^{\beta}\nabla_{\alpha}-\delta_{\alpha}^{\beta}\mathbf{E}^{\nu}\nabla_{\nu}\right)\right)\Phi=0.\end{aligned}\tag{630}$$

Now we take it that there are decompositions of the type:

$$\left(\mathbf{R}^{*\beta}\mathbf{R}^{*}{}_{\alpha}-\frac{\delta_{\alpha}^{\beta}}{2}\mathbf{R}^{*}\mathbf{R}^{*}-\mathbf{E}^{\mu}\cdot\mathbf{E}^{\beta}\nabla_{\mu}\nabla_{\alpha}+\delta_{\alpha}^{\beta}\nabla_{\mu}\mathbf{E}^{\mu}\cdot\mathbf{E}^{\nu}\nabla_{\nu}\right)\Phi=0$$

$$\xrightarrow{\mathbf{E}^{\alpha}\nabla_{\mu}=\mathbf{E}^{\mu}\nabla_{\alpha}\ \ \&\ \ \mathbf{R}^{*\alpha}=\mathbf{R}^{*}{}_{\alpha}}$$

$$\left(\begin{array}{c}\left(\mathbf{R}^{*\beta}\gamma^{0}-\frac{\delta_{\alpha}^{\sigma}}{\sqrt{2}}\mathbf{R}^{*}\gamma^{1}-\gamma^{2}\mathbf{E}^{\mu}\nabla_{\alpha}+\gamma^{3}\delta_{\alpha}^{\sigma}\nabla_{\mu}\mathbf{E}^{\mu}\right)\\ \times\left(\mathbf{R}^{*}{}_{\alpha}\gamma^{0}-\frac{\delta_{\sigma}^{\beta}}{\sqrt{2}}\mathbf{R}^{*}\gamma^{1}-\gamma^{2}\mathbf{E}^{\beta}\nabla_{\mu}+\gamma^{3}\delta_{\sigma}^{\beta}\nabla_{\mu}\mathbf{E}^{\mu}\right)\end{array}\right)\Phi=0. \qquad (631)$$

Naturally, in the general case, where we cannot fulfill the conditions in the second line above, we have to go back to Eq. (630) and apply general matrices similar to the ones used in Eq. (412), but with rank 4×4. More options are given in appendix G in our publication [7].

5.18.18 The Four-Dimension Question

Why do we have a space-time of 4 dimensions ... at least on bigger scales?

We try to answer this question with the help of our general quantum gravity equation (603), only that—this time—we assume the scalar function Φ to actually be a scalar product of the following kind:

$$\Phi = \phi_{\alpha}G^{\alpha\beta}\phi_{\beta}. \qquad (632)$$

This changes the variational task (602) and renders the result (603) as follows:

$$\left[\left(R^{*}{}_{\mu\nu}-\frac{1}{2}R^{*}G_{\mu\nu}-\nabla_{\mu}\nabla_{\nu}+G_{\mu\nu}\Delta_{G}\right)\left[\phi_{\alpha}\overset{\downarrow}{G^{\alpha\beta}}\phi_{\beta}\right]+\phi_{\mu}\phi_{\nu}R^{*}\right]\delta G^{\mu\nu}=0$$

$$\Rightarrow$$

$$\left(\frac{\delta g^{\mu\nu}}{F}+g^{\mu\nu}\delta\frac{1}{F}\right)\left[\left(R^{*}{}_{\mu\nu}-\frac{1}{2}R^{*}G_{\mu\nu}-\nabla_{\mu}\nabla_{\nu}+G_{\mu\nu}\Delta_{G}\right)\left[\phi_{\alpha}G^{\alpha\beta}\phi_{\beta}\right]+\phi_{\mu}\phi_{\nu}\cdot R^{*}\right]$$

$$\left(\frac{\delta g^{\mu\nu}}{F}-\frac{g^{\mu\nu}}{F^{2}}\delta F\right)\left[\left(R^{*}{}_{\mu\nu}-\frac{1}{2}R^{*}G_{\mu\nu}-\nabla_{\mu}\nabla_{\nu}+G_{\mu\nu}\Delta_{G}\right)\left[\phi_{\alpha}G^{\alpha\beta}\phi_{\beta}\right]+\phi_{\mu}\phi_{\nu}\cdot R^{*}\right]=0$$

$$\left(\frac{\delta g^{\mu\nu}}{F}-\frac{G^{\mu\nu}}{F}\delta F\right)\left[\left(R^{*}{}_{\mu\nu}-\frac{1}{2}R^{*}G_{\mu\nu}-\nabla_{\mu}\nabla_{\nu}+G_{\mu\nu}\Delta_{G}\right)\overbrace{\left[\phi_{\alpha}G^{\alpha\beta}\phi_{\beta}\right]}^{=\Phi}+\phi_{\mu}\phi_{\nu}\cdot R^{*}\right]=0.$$

(633)

As demonstrated above, the expression in the last line of Eq. (633) can be split into two equations, namely:

$$\frac{\delta g^{\mu\nu}}{F}\left[\left(R^{*}{}_{\mu\nu}-\frac{1}{2}R^{*}G_{\mu\nu}-\nabla_{\mu}\nabla_{\nu}+G_{\mu\nu}\Delta_{G}\right)\overbrace{\left[\phi_{\mu}G^{\mu\nu}\phi_{\nu}\right]}^{=\Phi}+\phi_{\mu}\phi_{\nu}R^{*}\right]=0 \quad (634)$$

and:

$$-\frac{G^{\mu\nu}}{F}\delta F\left[\left(R^{*}{}_{\mu\nu}-\frac{1}{2}R^{*}G_{\mu\nu}-\nabla_{\mu}\nabla_{\nu}+G_{\mu\nu}\Delta_{G}\right)\overbrace{\left[\phi_{\alpha}G^{\alpha\beta}\phi_{\beta}\right]}^{=\Phi}+\phi_{\mu}\phi_{\nu}\cdot R^{*}\right]$$

$$=-\frac{G^{\mu\nu}}{F}\delta F\left[\left(R^{*}{}_{\mu\nu}-\frac{1}{2}R^{*}G_{\mu\nu}-\nabla_{\mu}\nabla_{\nu}+G_{\mu\nu}\Delta_{G}\right)\Phi+\phi_{\mu}\phi_{\nu}\cdot R^{*}\right]$$

$$=-\frac{\delta F}{F}\left[\left(\left(1-\frac{n}{2}\right)\cdot R^{*}+(n-1)\cdot\Delta_{G}\right)\Phi+\phi_{\mu}G^{\mu\nu}\phi_{\nu}\cdot R^{*}\right]$$

$$=-\frac{\delta F}{F}\left[\left(\left(1-\frac{n}{2}\right)\cdot R^{*}+(n-1)\cdot\Delta_{G}\right)\Phi+\Phi\cdot R^{*}\right]=0$$

$$\Rightarrow \qquad \left(\frac{4-n}{n-1}\cdot R^{*}+\Delta_{G}\right)\Phi=0. \quad (635)$$

As before, in Ricci-R^*-flat space-times we would just obtain the Laplace equation with respect to the metric G_{ab}:

$$\Delta_G\Phi = 0 \quad (636)$$

and, as presented elsewhere, can obtain mass and potential via additional (suitably entangled) dimensions (e.g., [4, 19, 21, 30]). In this way, we also obtain the usual quantum Klein–Gordon equation. This, as already shown above, can then be "decomposed" into the Dirac equation in the usual way.

However, we would also obtain the pure Laplace equation when just setting $n = 4$, which, surprisingly, is the number of dimensions of our ordinary space-time.

Of course, at first sight, one might be tempted to say that this whole consideration has some kind of arbitrariness about it, because our setting for the function Φ could also be:

$$\Phi=\Xi\left(\phi_{\alpha}G^{\alpha\beta}\phi_{\beta}\right) \quad (637)$$

and then assume that one simply has to adjust the function $\Xi\left(\phi_{\alpha}G^{\alpha\beta}\phi_{\beta}\right)$ such that it makes the Ricci scalar disappear in any number of dimensions. But as one can easily show, this is not possible. In fact, as demonstrated above, it only works in 4 dimensions. Using Eq. (637), namely, changes Eq. (633) to:

$$\left[\begin{array}{c}\left(R^{*}_{\mu\nu}-\frac{1}{2}R^{*}G_{\mu\nu}-\nabla_{\mu}\nabla_{\nu}+G_{\mu\nu}\Delta_{G}\right)\left[\overset{\downarrow}{\Xi\left(\phi_{\alpha}G_{\alpha\beta}\phi_{\beta}\right)}\right]\\ +\phi_{\mu}\phi_{\nu}\cdot\frac{\partial\Xi\left(\phi_{\alpha}G^{\alpha\beta}\phi_{\beta}\right)}{\partial\phi_{\alpha}G^{\alpha\beta}\phi_{\beta}}R^{*}\end{array}\right]\delta G^{\mu\nu}=0$$

$$\Rightarrow$$

$$\left(\frac{\delta g^{\mu\nu}}{F}-\frac{G^{\mu\nu}}{F}\delta F\right)\left[\begin{array}{c}\left(R^{*}_{\mu\nu}-\frac{1}{2}R^{*}G_{\mu\nu}-\nabla_{\mu}\nabla_{\nu}+G_{\mu\nu}\Delta_{G}\right)\overbrace{\Xi\left(\phi_{\alpha}G^{\alpha\beta}\phi_{\beta}\right)}^{=\Phi}\\ +\phi_{\mu}\phi_{\nu}\cdot\frac{\partial\Xi\left(\phi_{\alpha}G^{\alpha\beta}\phi_{\beta}\right)}{\partial\phi_{\alpha}G^{\alpha\beta}\phi_{\beta}}R^{*}\end{array}\right]=0. \quad (638)$$

Subsequently we obtain the following tensor:

$$\frac{\delta g^{\mu\nu}}{F}\left[\begin{array}{c}\left(R^{*}_{\mu\nu}-\frac{1}{2}R^{*}G_{\mu\nu}-\nabla_{\mu}\nabla_{\nu}+G_{\mu\nu}\Delta_{G}\right)\left[\overbrace{\Xi\left(\phi_{\alpha}G^{\alpha\beta}\phi_{\beta}\right)}^{=\Phi}\right]\\ +\phi_{\mu}\phi_{\nu}\cdot\frac{\partial\Xi\left(\phi_{\alpha}G^{\alpha\beta}\phi_{\beta}\right)}{\partial\phi_{\alpha}G^{\alpha\beta}\phi_{\beta}}R^{*}\end{array}\right]=0 \quad (639)$$

and the scalar equation:

$$-\frac{G^{\mu\nu}}{F}\delta F\left[\begin{array}{c}\left(R^{*}_{\mu\nu}-\frac{1}{2}R^{*}G_{\mu\nu}-\nabla_{\mu}\nabla_{\nu}+G_{\mu\nu}\Delta_{G}\right)\overbrace{\Xi\left(\phi_{\alpha}G^{\alpha\beta}\phi_{\beta}\right)}^{=\Phi}\\ +\phi_{\mu}\phi_{\nu}\cdot\frac{\partial\Xi\left(\phi_{\alpha}G^{\alpha\beta}\phi_{\beta}\right)}{\partial\phi_{\alpha}G^{\alpha\beta}\phi_{\beta}}R^{*}\end{array}\right]$$

$$=-\frac{G^{\mu\nu}}{F}\delta F\left[\left(R^{*}_{\mu\nu}-\frac{1}{2}R^{*}G_{\mu\nu}-\nabla_{\mu}\nabla_{\nu}+G_{\mu\nu}\Delta_{G}\right)\Phi+\phi_{\mu}\phi_{\nu}\cdot\frac{\partial\Xi\left(\phi_{\alpha}G^{\alpha\beta}\phi_{\beta}\right)}{\partial\phi_{\alpha}G^{\alpha\beta}\phi_{\beta}}R^{*}\right]$$

$$=-\frac{\delta F}{F}\left[\left(\left(1-\frac{n}{2}\right)\cdot R^{*}+(n-1)\cdot\Delta_{G}\right)\Phi+\phi_{\mu}G^{\mu\nu}\phi_{\nu}\cdot\frac{\partial\Xi\left(\phi_{\alpha}G^{\alpha\beta}\phi_{\beta}\right)}{\partial\phi_{\alpha}G^{\alpha\beta}\phi_{\beta}}\cdot R^{*}\right]$$

$$=-\frac{\delta F}{F}\left[\left(\left(1-\frac{n}{2}\right)\cdot R^{*}+(n-1)\cdot\Delta_{G}\right)\Phi+\Phi\cdot\frac{\partial\Xi\left(\phi_{\alpha}G^{\alpha\beta}\phi_{\beta}\right)}{\partial\phi_{\alpha}G^{\alpha\beta}\phi_{\beta}}\cdot R^{*}\right]=0$$

$$\Rightarrow\qquad\left(\frac{2\cdot\left(1+\frac{\partial\Xi\left(\phi_{\alpha}G^{\alpha\beta}\phi_{\beta}\right)}{\partial\phi_{\alpha}G^{\alpha\beta}\phi_{\beta}}\right)-n}{n-1}\cdot R^{*}+\Delta_{G}\right)\Phi=0. \tag{640}$$

We realize that only in 4-dimensional space-times, the Ricci-curvature would vanish from the scalar equation with a simple scalar-product approach (632).

5.18.19 Vanishing Ricci Scalar Also in Space-Times with $n \neq 4$?

5.18.19.1 6-Dimensional space-times

While functional approaches of the kind (637) are of no use when being interested in letting the Ricci scalar vanish for $n \neq 4$, we could still assume higher-order dependencies as follows:

$$\Phi=\phi_{\alpha\gamma}G^{\alpha\beta}G^{\gamma\eta}\phi_{\beta\eta} \tag{641}$$

or:

$$\Phi=\phi_{\alpha}\phi_{\gamma}G^{\alpha\beta}G^{\gamma\eta}\phi_{\eta}\phi_{\beta}. \tag{642}$$

Picking Eq. (642) changes the variational task (602) and renders the result (603) as follows:

$$\left[\begin{array}{c}\left(R^{*}_{\mu\nu}-\frac{1}{2}R^{*}G_{\mu\nu}-\nabla_{\mu}\nabla_{\nu}+G_{\mu\nu}\Delta_{G}\right)\left[\phi_{\alpha}\phi_{\gamma}\overset{\downarrow}{G^{\alpha\beta}}G^{\gamma\eta}\phi_{\eta}\phi_{\beta}\right]\\+\left(\phi_{\alpha}G^{\alpha\beta}\phi_{\beta}+\phi_{\gamma}G^{\gamma\eta}\phi_{\eta}\right)\phi_{\mu}\phi_{\nu}\cdot R^{*}\end{array}\right]\delta G^{\mu\nu}$$

$$\left[\begin{array}{c}\left(R^{*}_{\mu\nu}-\frac{1}{2}R^{*}G_{\mu\nu}-\nabla_{\mu}\nabla_{\nu}+G_{\mu\nu}\Delta_{G}\right)\left[\phi_{\alpha}\phi_{\gamma}\overset{\downarrow}{G^{\alpha\beta}}G^{\gamma\eta}\phi_{\eta}\phi_{\beta}\right]\\+2\left(\phi_{\alpha}G^{\alpha\beta}\phi_{\beta}\right)\phi_{\mu}\phi_{\nu}\cdot R^{*}\end{array}\right]\delta G^{\mu\nu}=0$$

$$\Rightarrow$$

$$\left(\frac{\delta g^{\mu\nu}}{F}-\frac{G^{\mu\nu}}{F}\delta F\right)\left[\begin{array}{c}\left(R^{*}{}_{\mu\nu}-\frac{1}{2}R^{*}G_{\mu\nu}-\nabla_{\mu}\nabla_{\nu}+G_{\mu\nu}\Delta_{G}\right)\\ \times\left[\phi_{\alpha}\phi_{\gamma}G^{\alpha\overset{\downarrow}{\beta}}G^{\gamma\eta}\phi_{\eta}\phi_{\beta}\right]+2\left(\phi_{\alpha}G^{\alpha\beta}\phi_{\beta}\right)\phi_{\mu}\phi_{\nu}\cdot R^{*}\end{array}\right]=0. \tag{643}$$

Again, we split the expression in the last line of Eq. (643) into two equations:

$$\frac{\delta g^{\mu\nu}}{F}\left[\begin{array}{c}\left(R^{*}{}_{\mu\nu}-\frac{1}{2}R^{*}G_{\mu\nu}-\nabla_{\mu}\nabla_{\nu}+G_{\mu\nu}\Delta_{G}\right)\left[\phi_{\alpha}\phi_{\gamma}G^{\alpha\overset{\downarrow}{\beta}}G^{\gamma\eta}\phi_{\eta}\phi_{\beta}\right]\\ +2\left(\phi_{\alpha}G^{\alpha\beta}\phi_{\beta}\right)\phi_{\mu}\phi_{\nu}\cdot R^{*}\end{array}\right]=0 \tag{644}$$

and:

$$-\frac{G^{\mu\nu}}{F}\delta F\left[\begin{array}{c}\left(R^{*}{}_{\mu\nu}-\frac{1}{2}R^{*}G_{\mu\nu}-\nabla_{\mu}\nabla_{\nu}+G_{\mu\nu}\Delta_{G}\right)\left[\phi_{\alpha}\phi_{\gamma}G^{\alpha\overset{\downarrow}{\beta}}G^{\gamma\eta}\phi_{\eta}\phi_{\beta}\right]\\ +2\left(\phi_{\alpha}G^{\alpha\beta}\phi_{\beta}\right)\phi_{\mu}\phi_{\nu}\cdot R^{*}\end{array}\right]$$

$$=-\frac{G^{\mu\nu}}{F}\delta F\left[\left(R^{*}{}_{\mu\nu}-\frac{1}{2}R^{*}G_{\mu\nu}-\nabla_{\mu}\nabla_{\nu}+G_{\mu\nu}\Delta_{G}\right)\Phi+2\left(\phi_{\alpha}G^{\alpha\beta}\phi_{\beta}\right)\phi_{\mu}\phi_{\nu}\cdot R^{*}\right]$$

$$=-\frac{\delta F}{F}\left[\left(\left(1-\frac{n}{2}\right)\cdot R^{*}+(n-1)\cdot\Delta_{G}\right)\Phi+\phi_{\mu}G^{\mu\nu}\phi_{\nu}2\left(\phi_{\alpha}G^{\alpha\beta}\phi_{\beta}\right)\cdot R^{*}\right]$$

$$=-\frac{\delta F}{F}\left[\left(\left(1-\frac{n}{2}\right)\cdot R^{*}+(n-1)\cdot\Delta_{G}\right)\Phi+2\Phi\cdot R^{*}\right]=0$$

$$\Rightarrow\qquad\left(\frac{6-n}{n-1}\cdot R^{*}+\Delta_{G}\right)\Phi=0. \tag{645}$$

This time we have the term with the Ricci-scalar to vanish in 6-dimensional space-times.

Instead of Eq. (642), applying Eq. (641) results in:

$$\left[\begin{array}{c}\left(R^{*}{}_{\mu\nu}-\frac{1}{2}R^{*}G_{\mu\nu}-\nabla_{\mu}\nabla_{\nu}+G_{\mu\nu}\Delta_{G}\right)\left[\phi_{\alpha\gamma}G^{\alpha\overset{\downarrow}{\beta}}G^{\gamma\eta}\phi_{\beta\eta}\right]\\ +\left(\phi_{\alpha\mu}G^{\alpha\beta}\phi_{\beta\nu}+\phi_{\mu\gamma}G^{\gamma\eta}\phi_{\nu\eta}\right)\cdot R^{*}\end{array}\right]\delta G^{\mu\nu}$$

$$\left[\begin{array}{c}\left(R^{*}{}_{\mu\nu}-\frac{1}{2}R^{*}G_{\mu\nu}-\nabla_{\mu}\nabla_{\nu}+G_{\mu\nu}\Delta_{G}\right)\left[\overset{\downarrow}{\Phi}\right]\\ +2\left(\phi_{\alpha\mu}G^{\alpha\beta}\phi_{\beta\nu}\right)\cdot R^{*}\end{array}\right]\delta G^{\mu\nu}=0$$

$$\Rightarrow$$

$$\left(\frac{\delta g^{\mu\nu}}{F}-\frac{G^{\mu\nu}}{F}\delta F\right)\left[\left(R^{*}{}_{\mu\nu}-\frac{1}{2}R^{*}G_{\mu\nu}-\nabla_{\mu}\nabla_{\nu}+G_{\mu\nu}\Delta_{G}\right)\Phi+2\left(\phi_{\alpha\mu}G^{\alpha\beta}\phi_{\beta\nu}\right)\cdot R^{*}\right]=0.$$

(646)

Splitting the expression in the last line of Eq. (646) into two equations gives us:

$$\frac{\delta g^{\mu\nu}}{F}\left[\left(R^{*}{}_{\mu\nu}-\frac{1}{2}R^{*}G_{\mu\nu}-\nabla_{\mu}\nabla_{\nu}+G_{\mu\nu}\Delta_{G}\right)\Phi+2\left(\phi_{\alpha\mu}G^{\alpha\beta}\phi_{\beta\nu}\right)\cdot R^{*}\right]=0 \qquad (647)$$

and:

$$-\frac{G^{\mu\nu}}{F}\delta F\left[\left(R^{*}{}_{\mu\nu}-\frac{1}{2}R^{*}G_{\mu\nu}-\nabla_{\mu}\nabla_{\nu}+G_{\mu\nu}\Delta_{G}\right)\Phi+2\left(\phi_{\alpha\mu}G^{\alpha\beta}\phi_{\beta\nu}\right)\cdot R^{*}\right]$$

$$=-\frac{\delta F}{F}\left[\left(\left(1-\frac{n}{2}\right)\cdot R^{*}+(n-1)\cdot\Delta_{G}\right)\Phi+2\left(\phi_{\alpha\mu}G^{\mu\nu}G^{\alpha\beta}\phi_{\beta\nu}\right)\cdot R^{*}\right]$$

$$=-\frac{\delta F}{F}\left[\left(\left(1-\frac{n}{2}\right)\cdot R^{*}+(n-1)\cdot\Delta_{G}\right)\Phi+2\Phi\cdot R^{*}\right]=0$$

$$\Rightarrow \qquad \left(\frac{6-n}{n-1}\cdot R^{*}+\Delta_{G}\right)\Phi=0. \qquad (648)$$

Similar results can be obtained with:

$$\Phi=\phi_{\alpha\beta}G^{\alpha\beta}G^{\gamma\eta}\phi_{\gamma\eta}. \qquad (649)$$

5.18.19.1.1 *Dimensional mass creation in 6 dimensions*

Taking Eq. (648), assuming $n = 6$ and introducing a metric for the coordinates $t, r, \vartheta, \varphi, u, v$ of the form:

$$G_{\alpha\beta}^{6}=\begin{pmatrix}-c^{2}&0&0&0&0&0\\0&1&0&0&0&0\\0&0&r^{2}&0&0&0\\0&0&0&r^{2}\cdot\sin[\vartheta]&0&0\\0&0&0&0&1&0\\0&0&0&0&0&1\end{pmatrix}\cdot f[t,r,\vartheta,\varphi], \qquad (650)$$

just requires the setting of Φ in the following way:

$$\Phi[t,r,\vartheta,\varphi,u,v]=\begin{pmatrix}(C_{uc}\cdot\cos(m\cdot u)+C_{us}\cdot\sin(m\cdot u))\\ \times\left(C_{vc}\cdot\cos\left(\sqrt{E}\cdot v\right)+C_{vs}\cdot\sin\left(\sqrt{E}\cdot v\right)\right)\end{pmatrix}\cdot\Phi_S[t,r,\vartheta,\varphi] \quad (651)$$

in order to obtain the Klein–Gordon equation with mass m and energy E:

$$\left(\overbrace{\frac{6-n}{n-1}}^{=0}\cdot R^* + \Delta_G\right)\Phi=\left(\Delta_{[t,r,\vartheta,\varphi]}-E-m^2\right)\Phi_S[t,r,\vartheta,\varphi]=0. \quad (652)$$

5.18.19.1.2 *Alternative example of dimensional mass creation*

The consideration of the situation in 6 dimensions has some importance with respect to the simplest cases of dimensional mass creation and spin. We will therefore repeat the corresponding derivations.

In [48–50], we have demonstrated that a classical Klein–Gordon equation can also directly be obtained from the Ricci scalar of a scaled metric (391). Thereby we introduced a special de Sitter vacuum metric, which gave mass terms in the corresponding Ricci scalar-based Klein–Gordon equation in various numbers of dimensions. For completeness, we repeat the main results here.

Introducing a metric for the coordinates t, r, ϑ, φ, u, v of the form:

$$g^6_{\alpha\beta}=\begin{pmatrix}-c^2 & 0 & 0 & 0 & 0 & 0\\ 0 & 1 & 0 & 0 & 0 & 0\\ 0 & 0 & r^2 & 0 & 0 & 0\\ 0 & 0 & 0 & r^2\cdot\sin[\vartheta] & 0 & 0\\ 0 & 0 & 0 & 0 & g[v] & 0\\ 0 & 0 & 0 & 0 & 0 & g[v]\end{pmatrix}\cdot f[t,r,\vartheta,\varphi], \quad (653)$$

with the entanglement solution for u and v as follows:

$$g[v]=\frac{C_{v1}^2}{C_{v0}^2\cdot\left(1+\cosh\left[C_{v1}\cdot\left(v+C_{v2}\right)\right]\right)}, \quad (654)$$

gives us the following Ricci scalar from Eq. (391) with condition (377):

$$R^*=\frac{1}{F[f]^3}\cdot\left(-C_{N2}\cdot F[f]\cdot\frac{\partial F[f]}{\partial f}\cdot\Delta_g f\right);\quad F[f]=C_f+f[t,r,\vartheta,\varphi]$$

$$\xrightarrow{6D} R^* = \frac{5}{f^2} \cdot \left(\frac{C_{v0}^2}{5} \cdot f + \frac{\partial^2 f}{c^2 \cdot \partial t^2} - \Delta_{\text{3D-sphere}} f \right); \quad f = f[t,r,\vartheta,\varphi]; \quad C_f = 0.$$

(655)

Here $\Delta_{\text{3D-sphere}} f$ denotes the Laplace operator in spherical coordinates.

As for $R^* = 0$ and $\frac{C_{v0}^2}{5} = \frac{m^2 \cdot c^2}{\hbar^2}$ (m = mass, c = speed of light in vacuum, $\hbar$ = reduced Planck constant), we clearly have obtained the quantum Klein–Gordon equation and we can easily conclude that the Higgs field can obviously be constructed by a suitable set of additional, entangled coordinates (c.f. section "The Metric Origin of Scalar Fields like the Higgs Field and Symmetry Breaking" in [13]).

In connection with a non-vanishing cosmological constant, we have also shown in [49] that with:

$$g_{\alpha\beta}^{6} = \begin{pmatrix} -c^2 \cdot h[r] & 0 & 0 & 0 & 0 & 0 \\ 0 & \frac{1}{h[r]} & 0 & 0 & 0 & 0 \\ 0 & 0 & r^2 & 0 & 0 & 0 \\ 0 & 0 & 0 & r^2 \cdot \sin[\vartheta] & 0 & 0 \\ 0 & 0 & 0 & 0 & g[v] & 0 \\ 0 & 0 & 0 & 0 & 0 & g[v] \end{pmatrix}$$

$$\text{with:} \quad h[r] = 1 + \frac{C_v^2 \cdot r^2}{6} - r^2 \cdot \frac{\Lambda}{3} - \frac{C_r}{r}; \quad C_v^2 = \Lambda, \tag{656}$$

we could have a complete vacuum solution, where we have:

$$G^{\alpha\beta} + \Lambda g^{\alpha\beta} = R^{\alpha\beta} - \frac{1}{2} R g^{\alpha\beta} + \Lambda g^{\alpha\beta} = 0. \tag{657}$$

In this case, the condition $C_v^2 = \Lambda$ provides the necessary balance between matter and cosmological constant. As the constant C_r would be arbitrary, it could be set to zero in order to avoid any singularities.

We see that the metric (144) would be a vacuum solution of the Einstein field equations, but still the structure (universe) it describes contains matter. Assuming that the condition (145) should be a boundary condition for the object described by Eq. (144), we obtain a balance rule for the mass parameter C_v and the cosmological constant. The object would have a resulting non-zero Ricci curvature of:

$$R = 3 \cdot \Lambda. \tag{658}$$

Similarly (see [48]), we could construct a metric as follows:

$$g^{6}_{\alpha\beta} = \begin{pmatrix} -c^2 \cdot h[r] & 0 & 0 & 0 & 0 & 0 \\ 0 & \frac{1}{h[r]} & 0 & 0 & 0 & 0 \\ 0 & 0 & r^2 & 0 & 0 & 0 \\ 0 & 0 & 0 & r^2 \cdot \sin[\vartheta] & 0 & 0 \\ 0 & 0 & 0 & 0 & a^2 \cdot g[v] & 0 \\ 0 & 0 & 0 & 0 & 0 & a^2 \end{pmatrix}$$

$$\text{with:}\ \ h[r] = 1 + \frac{S^2 \cdot r^2}{3 \cdot a^2} - r^2 \cdot \frac{\Lambda}{3} - \frac{C_r}{r};\ \ S^2 = \frac{a^2 \cdot \Lambda}{2}. \tag{659}$$

Now we shall consider S a spin-like matter option in addition to our masses as given above.

This time we have:

$$g_{1,2}[v] = \begin{Bmatrix} C_{S1} \sin\left[S\left(v - C_{S2}\right)\right]^2 \\ C_{S1} \cos\left[S\left(v - C_{S2}\right)\right]^2 \end{Bmatrix}. \tag{660}$$

For more information and discussion (especially with respect to the extension to arbitrary coordinates and the flatness problem of our universe), the reader is referred to the original papers [48–50].

5.18.19.2 8-Dimensional space-times

Instead of the functional approach (637), we assume a higher-order dependence as follows:

$$\Phi = \phi_{\alpha\gamma} G^{\alpha\beta} G^{\gamma\eta} G^{\rho\lambda} \phi_{\beta\eta} \phi_{\rho\lambda} \tag{661}$$

or all possible interchanges of indices of this kind or:

$$\Phi = \phi_{\alpha} \phi_{\gamma} \phi_{\rho} G^{\alpha\beta} G^{\gamma\eta} G^{\rho\lambda} \phi_{\lambda} \phi_{\eta} \phi_{\beta} \tag{662}$$

or even:

$$\Phi = \phi_{\alpha\gamma\rho} G^{\alpha\beta} G^{\gamma\eta} G^{\rho\lambda} \phi_{\beta\eta\lambda}. \tag{663}$$

Picking Eq. (662) changes the variational task (602) and renders the result (603) as follows:

$$\delta G^{\mu\nu}\left[\begin{array}{c}\left(R^{*}_{\mu\nu}-\frac{1}{2}R^{*}G_{\mu\nu}-\nabla_{\mu}\nabla_{\nu}+G_{\mu\nu}\Delta_{G}\right)\phi_{\alpha}\phi_{\gamma}\phi_{\rho}G^{\alpha\beta}G^{\gamma\eta}G^{\rho\lambda}\phi_{\lambda}\phi_{\eta}\phi_{\beta}\\+\left(\phi_{\alpha}\phi_{\gamma}G^{\alpha\beta}G^{\gamma\eta}\phi_{\eta}\phi_{\beta}+\phi_{\alpha}\phi_{\rho}G^{\alpha\beta}G^{\rho\lambda}\phi_{\lambda}\phi_{\beta}+\phi_{\gamma}\phi_{\rho}G^{\gamma\eta}G^{\rho\lambda}\phi_{\lambda}\phi_{\eta}\right)\phi_{\mu}\phi_{\nu}\cdot R^{*}\end{array}\right]=$$

$$\delta G^{\mu\nu}\left[\begin{array}{c}\left(R^{*}_{\mu\nu}-\frac{1}{2}R^{*}G_{\mu\nu}-\nabla_{\mu}\nabla_{\nu}+G_{\mu\nu}\Delta_{G}\right)\Phi\\+3\left(\phi_{\alpha}\phi_{\gamma}G^{\alpha\beta}G^{\gamma\eta}\phi_{\eta}\phi_{\beta}\right)\phi_{\mu}\phi_{\nu}\cdot R^{*}\end{array}\right]=0$$

$$\Rightarrow$$

$$\left(\frac{\delta g^{\mu\nu}}{F}-\frac{G^{\mu\nu}}{F}\delta F\right)\left[\begin{array}{c}\left(R^{*}_{\mu\nu}-\frac{1}{2}R^{*}G_{\mu\nu}-\nabla_{\mu}\nabla_{\nu}+G_{\mu\nu}\Delta_{G}\right)\Phi\\+3\left(\phi_{\alpha}\phi_{\gamma}G^{\alpha\beta}G^{\gamma\eta}\phi_{\eta}\phi_{\beta}\right)\phi_{\mu}\phi_{\nu}\cdot R^{*}\end{array}\right]=0. \quad (664)$$

Again, we split the expression in the last line of Eq. (664) into two equations:

$$\frac{\delta g^{\mu\nu}}{F}\left[\begin{array}{c}\left(R^{*}_{\mu\nu}-\frac{1}{2}R^{*}G_{\mu\nu}-\nabla_{\mu}\nabla_{\nu}+G_{\mu\nu}\Delta_{G}\right)\Phi\\+3\left(\phi_{\alpha}\phi_{\gamma}G^{\alpha\beta}G^{\gamma\eta}\phi_{\eta}\phi_{\beta}\right)\phi_{\mu}\phi_{\nu}\cdot R^{*}\end{array}\right]=0 \quad (665)$$

and:

$$-\frac{G^{\mu\nu}}{F}\delta F\left[\begin{array}{c}\left(R^{*}_{\mu\nu}-\frac{1}{2}R^{*}G_{\mu\nu}-\nabla_{\mu}\nabla_{\nu}+G_{\mu\nu}\Delta_{G}\right)\Phi\\+3\left(\phi_{\alpha}\phi_{\gamma}G^{\alpha\beta}G^{\gamma\eta}\phi_{\eta}\phi_{\beta}\right)\phi_{\mu}\phi_{\nu}\cdot R^{*}\end{array}\right]$$

$$=-\frac{\delta F}{F}\left[\left(\left(1-\frac{n}{2}\right)\cdot R^{*}+(n-1)\cdot\Delta_{G}\right)\Phi+\phi_{\mu}G^{\mu\nu}\phi_{\nu}3\left(\phi_{\alpha}\phi_{\gamma}G^{\alpha\beta}G^{\gamma\eta}\phi_{\eta}\phi_{\beta}\right)\cdot R^{*}\right]$$

$$=-\frac{\delta F}{F}\left[\left(\left(1-\frac{n}{2}\right)\cdot R^{*}+(n-1)\cdot\Delta_{G}\right)\Phi+3\Phi\cdot R^{*}\right]=0$$

$$\Rightarrow \quad \left(\frac{8-n}{n-1}\cdot R^{*}+\Delta_{G}\right)\Phi=0. \quad (666)$$

The term with the Ricci-scalar now vanishes in 8-dimensional space-times.

Applying Eq. (663) results in:

$$\left[\begin{array}{c}\left(R^*_{\mu\nu}-\frac{1}{2}R^*G_{\mu\nu}-\nabla_\mu\nabla_\nu+G_{\mu\nu}\Delta_G\right)\Phi=\phi_{\alpha\gamma\rho}G^{\alpha\beta}G^{\gamma\eta}G^{\rho\lambda}\phi_{\beta\eta\lambda}\\ +\left(\phi_{\mu\gamma\rho}G^{\gamma\eta}G^{\rho\lambda}\phi_{\nu\eta\lambda}+\phi_{\alpha\mu\rho}G^{\alpha\beta}G^{\rho\lambda}\phi_{\beta\nu\lambda}+\phi_{\alpha\gamma\mu}G^{\alpha\beta}G^{\gamma\eta}\phi_{\beta\eta\nu}\right)\cdot R^*\end{array}\right]\delta G^{\mu\nu}$$

$$\left[\begin{array}{c}\left(R^*_{\mu\nu}-\frac{1}{2}R^*G_{\mu\nu}-\nabla_\mu\nabla_\nu+G_{\mu\nu}\Delta_G\right)\left[\overset{\downarrow}{\Phi}\right]\\ +3\left(\phi_{\mu\gamma\rho}G^{\gamma\eta}G^{\rho\lambda}\phi_{\nu\eta\lambda}\right)\cdot R^*\end{array}\right]\delta G^{\mu\nu}=0$$

$$\Rightarrow\quad\left(\frac{\delta g^{\mu\nu}}{F}-\frac{G^{\mu\nu}}{F}\delta F\right)\left[\begin{array}{c}\left(R^*_{\mu\nu}-\frac{1}{2}R^*G_{\mu\nu}-\nabla_\mu\nabla_\nu+G_{\mu\nu}\Delta_G\right)\Phi\\ +3\left(\phi_{\mu\gamma\rho}G^{\gamma\eta}G^{\rho\lambda}\phi_{\nu\eta\lambda}\right)\cdot R^*\end{array}\right]=0. \quad (667)$$

Splitting the expression in the last line of Eq. (646) into two equations gives us:

$$\frac{\delta g^{\mu\nu}}{F}\left[\left(R^*_{\mu\nu}-\frac{1}{2}R^*G_{\mu\nu}-\nabla_\mu\nabla_\nu+G_{\mu\nu}\Delta_G\right)\Phi+3\left(\phi_{\mu\gamma\rho}G^{\gamma\eta}G^{\rho\lambda}\phi_{\nu\eta\lambda}\right)\cdot R^*\right]=0 \quad (668)$$

and:

$$-\frac{G^{\mu\nu}}{F}\delta F\left[\begin{array}{c}\left(R^*_{\mu\nu}-\frac{1}{2}R^*G_{\mu\nu}-\nabla_\mu\nabla_\nu+G_{\mu\nu}\Delta_G\right)\Phi\\ +3\left(\phi_{\mu\gamma\rho}G^{\gamma\eta}G^{\rho\lambda}\phi_{\nu\eta\lambda}\right)\cdot R^*\end{array}\right]$$

$$=-\frac{\delta F}{F}\left[\left(\left(1-\frac{n}{2}\right)\cdot R^*+(n-1)\cdot\Delta_G\right)\Phi+3\Phi\cdot R^*\right]=0$$

$$\Rightarrow\quad\left(\frac{8-n}{n-1}\cdot R^*+\Delta_G\right)\Phi=0. \quad (669)$$

Similar results can be obtained with all the options (661).

5.18.19.2.1 *An example*

Taking the last line from Eq. (669) and assuming $n = 8$, we obtain a pure Laplace equation:

$$\Delta_G\Phi = 0. \quad (670)$$

Introducing a metric of the type as derived in appendix RB and applying it in Eq. (670) shows us that the apparently simple equation expands as follows:

$$\Delta_G \Phi = \frac{1}{\sqrt{G}} \partial_\alpha \left(\sqrt{G} \cdot G^{\alpha\beta} \cdot \partial_\beta \Phi \right) = \frac{1}{\sqrt{F^n \cdot g}} \partial_\alpha \left(\sqrt{F^n \cdot g} \cdot F^{-1} \cdot g^{\alpha\beta} \cdot \partial_\beta \Phi \right)$$

$$= \frac{1}{F^{\frac{n}{2}} \cdot \sqrt{g}} \partial_\alpha \left(\sqrt{g} \cdot F^{\frac{n}{2}-1} \cdot g^{\alpha\beta} \cdot \partial_\beta \Phi \right)$$

$$= \frac{1}{F^{\frac{n}{2}} \cdot \sqrt{g}} \left(\sqrt{g} \cdot F^{\frac{n}{2}-2} \cdot F_{,\alpha} \cdot g^{\alpha\beta} \cdot \partial_\beta \Phi + F^{\frac{n}{2}-1} \cdot \partial_\alpha \left(\sqrt{g} \cdot g^{\alpha\beta} \cdot \partial_\beta \Phi \right) \right)$$

$$= \frac{1}{F^2} \cdot F_{,\alpha} \cdot g^{\alpha\beta} \cdot \partial_\beta \Phi + \frac{1}{F^1 \cdot \sqrt{g}} \cdot \partial_\alpha \left(\sqrt{g} \cdot g^{\alpha\beta} \cdot \partial_\beta \Phi \right)$$

$$= \frac{1}{F^2} \cdot F_{,\alpha} \cdot g^{\alpha\beta} \cdot \partial_\beta \Phi + \frac{\Delta_g \Phi}{F^1} = 0$$

$$\Rightarrow \qquad \frac{F_{,\alpha}}{F} \cdot g^{\alpha\beta} \cdot \partial_\beta \Phi + \Delta_g \Phi = 0$$

$$\Rightarrow \qquad F_{,\alpha} \cdot g^{\alpha\beta} \cdot \Phi_{,\beta} + F \cdot \Delta_g \Phi = 0. \tag{671}$$

Thereby, we kept—for the time being and for the sake of generality—the number of dimensions n and did not fix it to $n = 8$. In order to find a simple solution to the condition $R^* = 0$, we applied condition (377) for the function $F[f]$ (see the evaluation in appendix RB). This leads to the following equation for the functions f and Φ:

$$F' \cdot f_{,\alpha} \cdot g^{\alpha\beta} \cdot \Phi_{,\beta} + F \cdot \Delta_g \Phi = \frac{4}{n-2} \cdot \left(f + C_f \right)^{\frac{4}{n-2}-1} \cdot f_{,\alpha} \cdot g^{\alpha\beta} \cdot \Phi_{,\beta}$$
$$+ \left(f + C_f \right)^{\frac{4}{n-2}} \cdot \Delta_g \Phi = 0$$

$$\Rightarrow \qquad \frac{4}{n-2} \cdot f_{,\alpha} \cdot g^{\alpha\beta} \cdot \Phi_{,\beta} + \left(f + C_f \right) \cdot \Delta_g \Phi$$
$$= \xrightarrow{C_f = 0} \frac{4}{n-2} \cdot f_{,\alpha} \cdot g^{\alpha\beta} \cdot \Phi_{,\beta} + f \cdot \Delta_g \Phi = 0. \tag{672}$$

Starting with the assumption:

$$\Phi_{,\beta} = q \cdot f_{,\beta}; \quad \Delta_g \Phi = m^2 \cdot f \tag{673}$$

gives us:

$$\frac{4\cdot q}{n-2}\cdot f_{,\alpha}\cdot g^{\alpha\beta}\cdot f_{,\beta}+f^2\cdot m^2=0\xrightarrow{\mu^2=m^2\cdot\frac{n-2}{4\cdot q}}f_{,\alpha}\cdot g^{\alpha\beta}\cdot f_{,\beta}+f^2\cdot\mu^2=0.\tag{674}$$

Introducing the functional vector f_i and the generalized gamma matrices (in n dimensions [54]) again, plus using the property:

$$g^{\alpha\beta}\cdot I=\frac{\gamma^\alpha\gamma^\beta+\gamma^\beta\gamma^\alpha}{2},\tag{675}$$

we can put Eq. (674) into the following form:

$$\begin{aligned}
&0=\left(f_{i,\alpha}f_{i,\beta}g^{\alpha\beta}+\mu^2\cdot f_i^2\right)\cdot I\\
\Rightarrow\quad&=\Rightarrow f_{i,\alpha}f_{i,\beta}\left(\frac{\gamma^\alpha\gamma^\beta+\gamma^\beta\gamma^\alpha}{2}\right)+\frac{I+I}{2}\cdot\mu^2\cdot f_i^2=0\\
\Rightarrow\quad&f_{i,\alpha}f_{i,\beta}\gamma^\alpha\gamma^\beta+I\cdot\mu^2\cdot f_i^2\\
&=\left(f_{i,\alpha}\gamma^\alpha f_{i,\beta}\gamma^\beta+i\cdot I\cdot\mu\cdot f_i\overbrace{\left(f_{i,\beta}\gamma^\beta-f_{i,\alpha}\gamma^\alpha\right)}^{f_{i,\beta}\gamma^\beta-f_{i,\alpha}\gamma^\alpha=f_{i,\gamma}\gamma^\gamma-f_{i,\gamma}\gamma^\gamma=0}+I\cdot\mu^2\cdot f_i^2\right)\\
&=\left(f_{i,\alpha}\gamma^\alpha+i\cdot I\cdot\mu\cdot f_i\right)\left(f_{i,\beta}\gamma^\beta-i\cdot I\cdot\mu\cdot f_i\right)=0\\
&\left(f_{i,\beta}\gamma^\beta-i\cdot\beta\cdot\mu\cdot f_i\right)=0\\
\Rightarrow\quad&f_{i,\alpha}f_{i,\beta}\gamma^\beta\gamma^\alpha+I\cdot\mu^2\cdot f_i^2=0\\
\Rightarrow\quad&\left(f_{i,\alpha}\gamma^\alpha+i\cdot I\cdot\mu\cdot f_i\right)\left(f_{i,\beta}\gamma^\beta-i\cdot I\cdot\mu\cdot f_i\right)=0.
\end{aligned}\tag{676}$$

This gives us the Dirac equation in its classical form [27], only that we have an n-dimensional space-time and general curvilinear coordinates:

$$0=f_{i,\beta}\gamma^\beta-i\cdot I\cdot\mu\cdot f_i.\tag{677}$$

Unfortunately, just as shown above in section "5.18.15.1 The Dirac equation as an outcome of a scaled metric?," this will not allow us to also give the same Dirac equation for the second condition in Eq. (673). Again, we have an inconsistency.

While our choice for $F[f]$ and the corresponding general solution for f (see appendix RB) ensures a vanishing Ricci scalar R^*, we also realize quite some complication for the construction of the Dirac equation. Thus, for simplicity, we assume:

$$\Phi_{,\beta} = -F_{,\beta}; \quad \Delta_g \Phi = -m^2 \cdot F. \tag{678}$$

It should be pointed out that such an assumption could also be generalized as follows:

$$\Phi_{,\beta} = q_\beta^\lambda \cdot F_{,\lambda}; \quad \Delta_g \Phi = \overbrace{\mathbf{m} \cdot \mathbf{m}}^{=-m^2} \cdot F, \tag{679}$$

which we will not consider here.

For completeness only, we point out that the following metric changes would also help us to avoid the necessity to fix the function F. The simplest "correction" would be to omit the potential term in g_{44}:

$$G_{\alpha\beta}^8 = \begin{pmatrix} -c^2 & 0 & 0 & 0 & \cdots & 0 \\ 0 & 1 & 0 & 0 & \cdots & 0 \\ 0 & 0 & r^2 & 0 & \cdots & 0 \\ 0 & 0 & 0 & r^2 \cdot \sin^2 \varphi_1 & \cdots & 0 \\ \cdots & \cdots & \cdots & \cdots & \ddots & 0 \\ 0 & 0 & 0 & 0 & 0 & g_{77} \end{pmatrix}$$

$$\times F\left[f\left[t,r,\vartheta,\varphi,\tau,\rho,\theta,\phi\right]\right]; \quad F[f] = f^{\frac{2}{3}}$$

$$f\left[t,r,\vartheta,\varphi,\tau,\rho,\theta,\phi\right] = f_t[t] \cdot f_r[r] \cdot f_\vartheta[\vartheta] \cdot f_\varphi[\varphi] \cdot f_\tau[\tau] \cdot f_\rho[\rho] \cdot f_\theta[\theta] \cdot f_\phi[\phi];$$

$$g_{44} = c_4^2; \quad g_{55} = 1; \quad g_{66} = \rho^2; \quad g_{77} = \rho^2 \cdot \sin[\theta]^2. \tag{680}$$

In this case, we can apply the solutions (741) to (746) ($c_4 = 1$) and obtain the following differential equation for the scaled Ricci scalar R^*:

$$R^* = 0 = \left(\frac{\ell \cdot (\ell+1)}{\rho^2} - \frac{2 \cdot \partial}{\rho \cdot \partial \rho} - \frac{\partial^2}{\partial \rho^2} + \frac{L \cdot (L+1)}{r^2} + A_2^2 - \frac{\partial}{r \cdot \partial r} - \frac{\partial^2}{\partial r^2} - E_t^2 \right) f. \tag{681}$$

As we can directly separate the two radii, we obtain the two differential equations:

$$\left(\frac{\ell \cdot (\ell+1)}{\rho^2} - \frac{2 \cdot \partial}{\rho \cdot \partial \rho} - \frac{\partial^2}{\partial \rho^2} \right) f_\rho = E_\rho^2 \cdot f_\rho, \tag{682}$$

$$0 = \left(\frac{L \cdot (L+1)}{r^2} + A_2^2 - \frac{\partial}{r \cdot \partial r} - \frac{\partial^2}{\partial r^2} - E_t^2 + E_\rho^2 \right) f_r. \tag{683}$$

Both equations give us the spherical Bessel functions as solutions via:

$$f_\rho[\rho]=C_j\cdot j_\ell[E_\rho\cdot\rho]+C_y\cdot y_\ell[E_\rho\cdot\rho], \tag{684}$$

$$f_r[r]=C_{rj}\cdot j_L\left[\sqrt{E_t^2+E_\rho^2-A_2^2}\cdot r\right]+C_{ry}\cdot y_L\left[\sqrt{E_t^2+E_\rho^2-A_2^2}\cdot r\right]. \tag{685}$$

When insisting on potential dependencies in r as given in appendix RD and still wanting to end up with $R^* = 0$, we have to move to a higher number of dimensions as follows:

$$G_{\alpha\beta}^{12}=\begin{pmatrix} -c^2 & 0 & 0 & 0 & \cdots & 0 \\ 0 & 1 & 0 & 0 & \cdots & 0 \\ 0 & 0 & r^2 & 0 & \cdots & 0 \\ 0 & 0 & 0 & r^2\cdot\sin^2\varphi_1 & \cdots & 0 \\ \cdots & \cdots & \cdots & \cdots & \ddots & 0 \\ 0 & 0 & 0 & 0 & 0 & g_{1111} \end{pmatrix}$$

$$\times F\left[f\left[t,r,\vartheta,\varphi,\tau,\rho,\theta,\phi,x_1,x_2,x_3,x_4\right]\right];\quad F[f]=f^{\frac{2}{5}}$$

$$f\left[t,r,\vartheta,\varphi,\tau,\rho,\theta,\phi,x_1,\ldots\right]$$

$$=f_t[t]\cdot f_r[r]\cdot f_\vartheta[\vartheta]\cdot f_\varphi[\varphi]\cdot f_\tau[\tau]\cdot f_\rho[\rho]\cdot f_\theta[\theta]\cdot f_\phi[\phi]\cdot\ldots;$$

$$g_{44}=c_4^2;\quad g_{55}=1;\quad g_{66}=\rho^2;\quad g_{77}=\rho^2\cdot\sin[\theta]^2;$$

$$g_{88}=r^2;\quad g_{99}=r^{-2};\quad g_{1010}=r^{i\cdot 2};\quad g_{1111}=r^{-i\cdot 2}. \tag{686}$$

The interested reader can find the solution to this metric for the function f satisfying the condition $R^* = 0$ in [9, 10, 30]. For convenience, we have placed some repetition into appendix RD and only give the rest here. We start with the partial solutions (741) to (746) ($c_4 = 1$) and:

$$f_{x1}[x_1]=C_{x_1 1}\cdot\cos[A_6\cdot x_1]+C_{x_1 2}\cdot\sin[A_6\cdot x_1]$$

$$f_{x2}[x_2]=C_{x_2 1}\cdot\cos[A_7\cdot x_2]+C_{x_2 2}\cdot\sin[A_7\cdot x_2]$$

$$f_{x3}[x_3]=f_{x4}[x_4]=\text{const.} \tag{687}$$

We end up with the following differential equation:

$$0=\left(\begin{array}{c}\left(\frac{\ell\cdot(\ell+1)}{\rho^2}-\frac{2\cdot\partial}{\rho\cdot\partial\rho}-\frac{\partial^2}{\partial\rho^2}+L\cdot(L+1)+A_6^2\right)\cdot\frac{1}{r^2}\\ +A_7^2\cdot r^2+A_2^2-\frac{2\partial}{r\cdot\partial r}-\frac{\partial^2}{\partial r^2}-E_t^2\end{array}\right)f. \tag{688}$$

Again, we can directly separate the two radii and obtain two ordinary differential equations:

$$\left(\frac{\ell\cdot(\ell+1)}{\rho^2}-\frac{2\cdot\partial}{\rho\cdot\partial\rho}-\frac{\partial^2}{\partial\rho^2}+A_2^2\right)f_\rho=E_\rho^2\cdot f_\rho, \tag{689}$$

$$0=\left(\left(L\cdot(L+1)+A_6^2\right)\cdot\frac{1}{r^2}+E_\rho^2+A_7^2\cdot r^2+A_2^2-\frac{2\partial}{r\cdot\partial r}-\frac{\partial^2}{\partial r^2}-E_t^2\right)f. \tag{690}$$

The first equation (689) gives:

$$f_\rho[\rho]=C_j\cdot j_\ell\left[\sqrt{E_\rho^2-A_2^2}\cdot\rho\right]+C_y\cdot y_\ell\left[\sqrt{E_\rho^2-A_2^2}\cdot\rho\right], \tag{691}$$

again, while the second one (690) requires the hypergeometric function $U[a,b,z]$ and the Laguerre polynomials $L_{nn}^{\sqrt{L\cdot(1+L)}}$:

$$f_r[r]=\frac{2^{\frac{P}{4}}e^{-\frac{1}{2}\cdot A_7\cdot r^2}\cdot r^{\frac{P}{2}}}{r^{3/2}}\times\left(\begin{array}{c}C_U\cdot U\left[-nn,1+Q,A_7\cdot r^2\right]\\ +C_L\cdot L_{nn}^{Q}\left[A_7\cdot r^2\right]\end{array}\right)$$

$$nn=\frac{E_t^2-E_\rho^2-2\cdot A_7\cdot P}{4\cdot A_7};$$

$$P=2+\sqrt{4\cdot A_6^2+(1+2\cdot L)^2};$$

$$Q=\frac{1}{2}\sqrt{4\cdot A_6^2+(1+2\cdot L)^2}. \tag{692}$$

The corresponding interpretation of the solution, especially for the momentum numbers, is quite interesting (c.f. appendix RB), but we will leave this task for later or to the interested reader (whatever comes first).

We see that our derivation for the function F and f automatically brought us the Ricci curvature R^* in Eq. (605). However, when not being interested in fixing the function F, we could also just apply:

$$\Phi = G^{\alpha_1\beta_1}\phi_{\alpha_1}\phi_{\beta_1}G^{\alpha_2\beta_2}\phi_{\alpha_2}\phi_{\beta_2}G^{\alpha_3\beta_3}\phi_{\alpha_3}\phi_{\beta_3}G^{\alpha_4\beta_4}\phi_{\alpha_4}\phi_{\beta_4}G^{\alpha_5\beta_5}\phi_{\alpha_5}\phi_{\beta_5}$$

$$= \prod_{i=1}^{5} G^{\alpha_i\beta_i}\phi_{\alpha_i}\phi_{\beta_i}, \tag{693}$$

which automatically brings the R^*-term in Eq. (603) to vanish, because we obtain:

$$\left(\frac{12-n}{n-1}\cdot R^* + \Delta_G\right)\Phi = \xrightarrow{n=12} \Delta_G\Phi = 0. \tag{694}$$

5.18.19.3 Generalization to 2*k Dimensions with k = 1,2,3, ...

With the recipes elaborated above, it is now quite simple to generalize the concept. With approaches for the function Φ as follows:

$$\Phi = \phi_{\alpha_i...\alpha_k}\phi_{\beta_i...\beta_k}\prod_{i=1}^{k} G^{\alpha_i\beta_i}, \tag{695}$$

$$\Phi = \prod_{i=1}^{k} G^{\alpha_i\beta_i}\phi_{\alpha_i}\phi_{\beta_i} \tag{696}$$

and all possible options in between these two forms, we are always able to produce scalar quantum gravity equations of the from:

$$\left(\frac{2\cdot(k+1)-n}{n-1}\cdot R^* + \Delta_G\right)\Phi = 0. \tag{697}$$

Thus, the term with the Ricci scalar vanishes for $2(k+1) = n$ and we are left with a simple Laplace equation for the function Φ.

5.18.19.4 Important note regarding the transition to the classical theory

In order to obtain the classical Dirac equations, please note that instead of Eq. (554), Eqs. (563) and (567) have to be applied. Consequently, the generalized case (696) should read:

$$\Phi = \prod_{i=1}^{k} G^{\alpha_i\beta_i}\phi_{\alpha_i}V_{\beta_i} \tag{698}$$

and:

$$f = \prod_{i=1}^{k} G^{\alpha_i\beta_i}f_{\alpha_i}V_{\beta_i}, \tag{699}$$

with V_{β_i} being a vector of constants.

5.18.20 Another Way to Avoid the Dirac Hassle with Φ

In this section, we will show that for metrics of the type (380), the complicated Dirac approach can completely be avoided while only the functional wrapper $F[f]$ and the Laplace equation for f play a role.

Our starting point shall be the scalar quantum gravity equation with Φ as resulting from Eqs. (603) and (605):

$$\delta W = 0 = \delta \int_V d^n x \left(\sqrt{-G} \cdot \Phi \cdot R^* \right)$$

$$\xrightarrow{\frac{\delta g^{\mu\nu}}{F} \ll g^{\mu\nu} \delta \frac{1}{F}}$$

$$\left(\frac{2-n}{n-1} \cdot R^* + \Delta_G \right) \Phi = 0. \tag{700}$$

Now we add the operator expansion from Eq. (671), where we saw that:

$$\Delta_G \Phi = \frac{1}{\sqrt{G}} \partial_\alpha \left(\sqrt{G} \cdot G^{\alpha\beta} \cdot \partial_\beta \Phi \right) = \frac{1}{\sqrt{F^n \cdot g}} \partial_\alpha \left(\sqrt{F^n \cdot g} \cdot F^{-1} \cdot g^{\alpha\beta} \cdot \partial_\beta \Phi \right)$$

$$= \frac{1}{F^{\frac{n}{2}} \cdot \sqrt{g}} \partial_\alpha \left(\sqrt{g} \cdot F^{\frac{n}{2}-1} \cdot g^{\alpha\beta} \cdot \partial_\beta \Phi \right)$$

$$= \frac{1}{F^{\frac{n}{2}} \cdot \sqrt{g}} \left(\sqrt{g} \cdot F^{\frac{n}{2}-2} \cdot F_{,\alpha} \cdot g^{\alpha\beta} \cdot \partial_\beta \Phi + F^{\frac{n}{2}-1} \cdot \partial_\alpha \left(\sqrt{g} \cdot g^{\alpha\beta} \cdot \partial_\beta \Phi \right) \right)$$

$$= \frac{1}{F^2} \cdot F_{,\alpha} \cdot g^{\alpha\beta} \cdot \partial_\beta \Phi + \frac{1}{F^1 \cdot \sqrt{g}} \cdot \partial_\alpha \left(\sqrt{g} \cdot g^{\alpha\beta} \cdot \partial_\beta \Phi \right)$$

$$= \frac{1}{F^2} \cdot F_{,\alpha} \cdot g^{\alpha\beta} \cdot \partial_\beta \Phi + \frac{\Delta_g \Phi}{F^1} = 0. \tag{701}$$

Bringing Eqs. (700) and (701) together yields:

$$\left(\frac{2-n}{n-1} \cdot R^* + \frac{1}{F^2} \cdot F_{,\alpha} \cdot g^{\alpha\beta} \cdot \partial_\beta + \frac{\Delta_g}{F} \right) \Phi$$

$$\left(\frac{2-n}{n-1} \cdot F^2 \cdot R^* + F_{,\alpha} \cdot g^{\alpha\beta} \cdot \partial_\beta + F \cdot \Delta_g \right) \Phi$$

$$\frac{2-n}{n-1} \cdot F^2 \cdot R^* \cdot \Phi + F_{,\alpha} \cdot g^{\alpha\beta} \cdot \Phi_{,\beta} + F \cdot \Delta_g \Phi = 0. \tag{702}$$

Now we set $\Phi = \Phi[f]$, which converts the Laplace operator in the last line of (702) to:

$$\Delta_g \Phi = \Delta_g \Phi[f] = g^{\alpha\beta} \cdot \Phi'' \cdot f_{,\alpha} \cdot f_{,\beta} + \Phi' \cdot \Delta_g f. \tag{703}$$

Assuming:

$$\Delta_g f = 0 \tag{704}$$

and:

$$R^* = 0 \tag{705}$$

simplifies the last line in Eq. (702) as follows:

$$F_{,\alpha} \cdot g^{\alpha\beta} \cdot \Phi_{,\beta} + F \cdot g^{\alpha\beta} \cdot \Phi'' \cdot f_{,\alpha} \cdot f_{,\beta} = 0. \tag{706}$$

Also assuming $F = F[f]$ with the setting from Eq. (377) not only linearizes the differential equation for R^*, but—for many important metrics (c.f. Eq. (380))—also satisfies condition (705) with property (704) AND automatically gives us Eq. (706) in the form:

$$F' \cdot g^{\alpha\beta} \cdot \Phi' \cdot f_{,\alpha} \cdot f_{,\beta} + F \cdot g^{\alpha\beta} \cdot \Phi'' \cdot f_{,\alpha} \cdot f_{,\beta} = 0$$

$$Z' = \frac{\partial Z}{\partial f}; \quad Z'' = \frac{\partial^2 Z}{\partial f^2}$$

$\Rightarrow$

$$F' \cdot \Phi' + F \cdot \Phi'' = 0. \tag{707}$$

While function $F[f]$ would be fixed due to Eq. (377), which leads to Eq. (378), the solution for the function Φ reads:

$$\Phi[f] = \begin{cases} \left(\left(f + C_f\right)^{\frac{n-6}{n-2}} - \left(1 + C_f\right)^{\frac{n-6}{n-2}} \right) C_{\Phi 1} + C_{\Phi 2} & n \neq 2, n \neq 6 \\ C_{\Phi 1} \cdot \ln\left(f + C_f\right) + C_{\Phi 2} & n = 6 \\ C_{\Phi 1} \cdot \left(e^{-C_f} - e^{-C_f \cdot f}\right) + C_{\Phi 2} & n = 2. \end{cases} \tag{708}$$

5.18.21 The Collapse of the Wave Function [56–58]

We have (at least) three fundamentally derived scalar quantum gravity equations, namely, Eqs. (376), (440) and (450). All of these equations can be linearized via conditions (377), (441) and (452), respectively. After this linearization process, we obtain the standard quantum equations either directly or for most standard metrics (e.g., Eqs. (369) and (444) for the cases

(377) and (441), respectively). We also saw that there is a variety of options to obtain Dirac equations from the non-linear scalar quantum gravity equation. Thereby it was shown that at least some of these equations can either be brought into Klein–Gordon form or made suitable for the application of the Dirac approach (as demonstrated with the technique in Eqs. (519) to (529)). Switching between the two options changes the metric tensor but still allows a solution of the problem as a whole.

This brings us to the following conjectures:

(A) A "universal selector" or "global decision maker" as introduced and discussed in connection with the Everett "multiverse model" in section "5.18.8 Towards the Metric Everett Mind" of this appendix is responsible for the collapse (which in this case isn't really a collapse, but a decision).

(B) The collapse of the wave functions process is just a transition from either the Dirac option or the Klein–Gordon outcome AND/OR from either linearity to non-linearity, thereby forcing the wave function to leave its state of superposition of eigenfunctions and to take on the state of only one "observable" solution. It is the fragility of the settings for Klein–Gordon or Dirac and linearity, only possible via a delicate setting for the metric "wrapper" $F[f]$ and—as shown in this appendix—also the wrapper for the variational kernel of the Einstein–Hilbert action $\Phi[f]$ of the wave function via conditions (377), (441), (452) and (707), which, when the system is nudged (e.g., by the measurement) and pushed out of its comfortable stupor in a metrically restricted state allowing for the linear equations (379), (443) and (451) into the much higher-dimensional reality with the non-linear equations (376), (440) and (450). Similarly, we can have a switch from Dirac to Klein–Gordon linearity, which is also decided via delicate settings for the functional metric or variational-integral-kernel wrappers $F[f]$ and $\Phi[f]$. The "nudging" destroys the frail setting, which allowed for the occurrence of linearity in the first place and forces the system onto a non-linear partial differential apparatus or—if staying linear—onto either a Dirac or a Klein–Gordon path. This causes the collapse and the destruction of the coherence of the eigenfunction ensemble. Actually, one might just put it like this: The nudging wakes the system up and forces it to make a decision.

In this section, we are going to discuss option B.

And what about Schrödinger's cat?

Well, in order to achieve the coherence of a system as big as a cat, including, Schrödinger's box and the complicated killing mechanism, one had to cool down everything so close to the absolute zero that the cat would be dead anyway. Thus, the question does not arise at all.

5.18.22 Bearing on "Quantum Psychology"

Regarding the connection of all the Dirac quantum derivations to our "Mathematical Psychology," we refer to the main part of the book and section "4.2 More Quantum Gravity Topics and Their Bearing on Mathematical Psychology." However, as we consider it our most important result, we want to repeat the following insight here:

Our results come with a warning against those mind-controllers and totalitarians out there:

With thoughts being of fundamentally quantum gravity character,

there is ALWAYS the fundamental possibility of a mental escape from any thought prison [18].

5.18.23 Conclusions from the Findings in Appendix R

What in the classical Quantum Theory appears as wave function is just the volume or scale fluctuation of the metric of the space or space-time in question and the kernel-jitter of the Einstein–Hilbert action in a holistic Quantum Gravity Theory. Thereby the classical wave function determines the metric volume perturbation and vanishes in the far distance of the quantum object under consideration. As the metric, respectively its volume, usually is much bigger than the fluctuations of the same and as the quantum effects tend to wear off rather quickly, Quantum Theory only becomes important at smaller scales (usually, which is to say under ordinary energy, temperature and gravity conditions).

What classically appears as potential is just the entanglement of dimensions.

Mass and energy are the oscillation of dimensions, thus, of the same source and that's why identical. Hence, the Einstein mass-energy equivalence.

We also conclude:

(a) Any space or space-time of properties, attributes, degrees of freedom or—YES—dimensions, apparently no matter what it is, can produce coherent states due to possible settings for the volume part of the metric, allowing for linear partial differential equations and thus, additivity of eigen-solutions for the wave functions. These eigen-solutions, respectively, the ensembles of eigen-solutions, are just perturbations existing on these very properties, attributes, degrees of freedom or—YES—dimensions in accordance with our **linearized** quantum gravity field equations.

(b) Thereby, in order to reach the state of additivity and linearity, the system has an option for a second-order differential outcome (Klein–Gordon character for simpler, which is to say the usual metrics) or a first order, which is to say Dirac-like evolution. The difference decides the Fermionic and the Bosonic behavior.

(c) When the system is "nudged" (perturbed), the frail setting, which allowed for the linearity of the quantum gravity field equations, is destroyed and the additivity is lost. The system "has to make a decision" about which eigenstate to exist in and the wave function collapses, because—after the nudging—it has to satisfy a non-linear field equation.

The important finding, when deriving the Klein-Gordon–Dirac bifurcation or the wave function collapse via our metric approach, is the generality of the effect. It is by no means restricted to small particles and quantum fields. But by adjusting the parameters (Planck constant and mass) to the experimental findings, we see that indeed only "things" with small masses and a limited number of degrees of freedom have a chance of showing significant coherence in order to achieve the delicate states allowing for quantum gravity linearity. Schrödinger's cat can therefore never be realized, because coherence of a system as big as a cat requires cooling the "thing" down to absolute zero, thereby killing the cat for sure and rendering the question about the combined dead-alive-state of the Schrödinger cat completely meaningless. However, with the right setting, systems of thought can be constructed in a way where quantum gravity becomes dominant and our usual classical mechanics is worthless.

5.18.24 Appendix RA: Hawking Radiation around a Black Hole

In his original work [59], where Hawking derived the effect of particle creation and subsequent radiation of Black Holes, Hawking's starting point was the free Klein–Gordon equation of a massless scalar field. Most interestingly, the Klein–Gordon equation is exactly the outcome of the Einstein–Hilbert action in its generalized forms:

$$\delta W = 0 = \delta \int_V d^n x \left(\sqrt{-g \cdot F^n} \times \left(\left(R - \frac{1}{2F} \begin{pmatrix} 2F_{,\alpha\beta}(n-1)g^{\alpha\beta} + 2\Gamma^{a}_{\alpha\beta}F_{,a}g^{\alpha\beta} \\ -F_{,\alpha}g^{ab}g_{\beta b,a}g^{\alpha\beta} - F_{,\beta}g^{ab}g_{\alpha b,a}g^{\alpha\beta} \\ -n\Gamma^{d}_{\alpha\beta}F_{,d}g^{\alpha\beta} + \frac{n}{2}F_{,d}g^{cd}g_{ab,c}g^{ab} \end{pmatrix} - \frac{F_{,\alpha} \cdot F_{,\beta}}{4F^2} g^{\alpha\beta}\left((n-6)(n-1)\right) \right) \frac{1}{F} - 2\Lambda + L_M \right) \right) \tag{709}$$

or

$$0 = \left[\begin{array}{l} \frac{\sqrt{-g}}{2}\left(\frac{f_{,\alpha}f_{,\beta}}{2} g^{\alpha\beta}\left(\frac{1}{2}(n-4)(n+2)(F')^2 + 2FF''\right) + \boxed{(n+2)FF'\Delta f} \right) \\ +F^2 \left(\begin{array}{c} \left(\sqrt{-g} \cdot g^{\alpha\beta}\right)_{,\alpha\beta} - \frac{\sqrt{-g} \cdot R}{(n-1)} \\ -\frac{1}{2(n-1)} \left(\sqrt{-g} \cdot g^{\alpha\beta} \cdot g^{ij} \begin{pmatrix} g_{i\beta,j} - g_{ij,\beta} - g_{\beta j,i} \\ +\frac{n}{2}\left(2g_{ij,\beta} - g_{i\beta,j} - g_{j\beta,i}\right) \end{pmatrix} \right)_{,\alpha} \end{array} \right) \end{array} \right], \tag{710}$$

as derived in the main part of this appendix above. One should point out that the following derivation also works for the even more general field equations (404) (see also [19]), only that things are lengthier. When assuming the "weak gravity" condition:

$$\delta G^{\alpha\beta} = G^{\alpha\beta} \cdot \delta_0 + \overbrace{G^{ab}\delta^{\alpha\beta}_{ab}}^{\text{"Gravity"}} \xrightarrow{\forall\, \delta^{\alpha\beta}_{ab} \ll \delta_0} = G^{\alpha\beta} \cdot \delta_0, \tag{711}$$

and further assuming that the observer (the one who sees the Black Hole radiation) has some distance to the Black Hole, we are allowed to apply much simpler spherical symmetry rather than the Schwarzschild one. Hawking, namely, did not apply the Klein–Gordon equation for the Black Hole metric, but the one of a flat space-time.

And in fact, by using the metric of a Schwarzschild object in an n-dimensional space-time:

$$g_{\alpha\beta}^{N}=\begin{pmatrix} -c^2\cdot f[r] & 0 & 0 & 0 & \cdots & 0 \\ 0 & \dfrac{1}{f[r]} & 0 & 0 & \cdots & 0 \\ 0 & 0 & r^2 & 0 & \cdots & 0 \\ 0 & 0 & 0 & r^2\cdot\sin^2\varphi_1 & \cdots & 0 \\ \cdots & \cdots & \cdots & \cdots & \ddots & 0 \\ 0 & 0 & 0 & 0 & 0 & g_{aa} \end{pmatrix}$$

$$g_{44}=r^2\cdot\sin^2\varphi_1\cdot\sin^2\varphi_2;\quad g_{aa}=r^2\cdot\prod_{j=1}^{a-2}\sin^2\varphi_j;\quad a=N-1=n;$$

$$f[r]=1-\frac{r_S^{N-3}}{r^{N-3}}=1-\frac{r_S^{n-2}}{r^{n-2}},\tag{712}$$

we see that for big distances to the Black Hole $r \gg r_S$, the Schwarzschild metric goes over to the metric of a flat space-time of an n-sphere, respectively, a flat space-time given in n-sphere coordinates.

5.18.24.1 The static field around a Schwarzschild object

Here, we want to derive the corresponding solution of a slightly generalized time-3-sphere, given via the following metric (from [12, 52, 60]):

$$g_{\alpha\beta}=\begin{pmatrix} -H^2\cdot c^2 & 0 & 0 & 0 \\ 0 & A^2 & 0 & 0 \\ 0 & 0 & B^2\cdot r^2 & 0 \\ 0 & 0 & 0 & D^2\cdot\sin[\vartheta]^2\cdot r^2 \end{pmatrix}.\tag{713}$$

Demanding a vanishing cosmological constant

$$\Lambda = 0,\tag{714}$$

we have shown above that the resulting Einstein field equations:

$$0=\begin{pmatrix} R_{\alpha\beta}-\frac{g_{\alpha\beta}}{2}R+\frac{g_{\alpha\beta}}{2}\begin{pmatrix} \frac{1}{2F}\begin{pmatrix} 2F_{,ij}(n-1)g^{ij}+2\Gamma_{ij}^{a}F_{,a}g^{ij} \\ -F_{,i}g^{ab}g_{jb,a}g^{ij}-F_{,j}g^{ab}g_{ib,a}g^{ij} \\ -n\Gamma_{ij}^{d}F_{,d}g^{ij}+\frac{n}{2}F_{,d}g^{cd}g_{ab,c}g^{ab} \end{pmatrix} \\ +\frac{F_{,i}\cdot F_{,j}}{4F^{2}}g^{ij}\left((n-6)(n-1)\right) \end{pmatrix} \\ -\frac{1}{2F}\begin{pmatrix} F_{,\alpha\beta}(n-2)+F_{,ab}g_{\alpha\beta}g^{ab} \\ +F_{,a}g^{ab}\left(g_{\beta b,\alpha}-g_{\beta\alpha,b}\right)-F_{,\alpha}g^{ab}g_{\beta b,a}-F_{,\beta}g^{ab}g_{\alpha b,a} \\ +F_{,d}g^{cd}\frac{1}{2}n\left(\frac{2}{n}g_{\alpha c,\beta}-g_{\alpha c,\beta}-g_{\beta c,\alpha}+g_{\alpha\beta,c}+\frac{1}{n}g_{\alpha\beta}g_{ab,c}g^{ab}\right) \\ -\frac{1}{2F}\left(F_{,\alpha}\cdot F_{,\beta}(3n-6)+g_{\alpha\beta}F_{,c}F_{,d}g^{cd}(4-n)\right) \end{pmatrix} \\ +\kappa T_{\alpha\beta}+\Lambda\cdot F\cdot g_{\alpha\beta} \end{pmatrix} \tag{715}$$

give a perfect Klein–Gordon equation when demanding condition (711), we expand here as follows:

$$\delta G^{\alpha\beta}=\frac{1}{F}\left(g^{\alpha\beta}\cdot\delta_{0}+\overbrace{g^{ab}\delta_{ab}^{\alpha\beta}}^{\text{Gravity}}\right)\xrightarrow{\forall\,\delta_{ab}^{\alpha\beta}\ll\delta_{0}}=\frac{g^{\alpha\beta}}{F}\cdot\delta_{0}. \tag{716}$$

In [18], it was shown that this condition is not even needed to derive the Klein–Gordon equation directly from the Einstein–Hilbert action as done here, but for brevity, we will stick with the slightly simpler path. This may be justified by the fact that gravity is very weak in comparison to the other fundamental forces.

Also setting the function $F[f]$ via $F[f=f[t,r,\vartheta,\varphi]]=(C_1+f)^2$ yields:

$$R^{*}=-C_{N2}\cdot\frac{\frac{\partial F[f]}{\partial f}\cdot\Delta_{g}f}{F[f]^{2}}\xrightarrow{4D}=-3\cdot\frac{(C_1+2\cdot f)}{\frac{1}{4}\cdot(C_1+2\cdot f)^{2}}\cdot\Delta_{g}f=-\frac{12\cdot\Delta_{g}f}{C_1+2\cdot f}, \tag{717}$$

$$\Rightarrow\qquad -\frac{C_1+2\cdot f}{12}\cdot R^{*}=\Delta_{g}f=\Delta_{3D-g}f-\frac{f^{(2,0,0,0)}}{H^{2}\cdot c^{2}}. \tag{718}$$

Now we substitute as follows in Eq. (718):

$$-\frac{C_1+2\cdot\tilde{f}}{12}\cdot R^{*}=\Delta_g\tilde{f}=\Delta_{3D-g}\tilde{f}-\frac{\tilde{f}^{(2,0,0,0)}}{H^2\cdot c^2}$$

$$f\equiv\frac{C_1+2\cdot\tilde{f}}{12}\quad\Rightarrow\quad -f\cdot R^{*}=\Delta_g\left(6\cdot f-\frac{C_1}{2}\right)$$

$$=6\cdot\Delta_g f=6\cdot\left(\Delta_{3D-g}f-\frac{f^{(2,0,0,0)}}{H^2\cdot c^2}\right). \tag{719}$$

Applying the separation approach $f[t,r,\vartheta,\varphi] = g[t] * h[r] * Y[\vartheta] * Z[\varphi]$ with the separation parameters C_t and m in (718) gives us the following immediate solution for $g[t]$ and $Z[\varphi]$:

$$g[t]=C_{t1}\cdot\cos\left[H\cdot c\cdot C_t\cdot t\right]+C_{t2}\cdot\sin\left[H\cdot c\cdot C_t\cdot t\right], \tag{720}$$

$$Z_m[\varphi]=C_{\varphi 1}\cdot\cos[D\cdot m\cdot\varphi]+C_{\varphi 2}\cdot\sin[D\cdot m\cdot\varphi]. \tag{721}$$

The two remaining differential equations can be constructed as follows:

$$-\frac{m^2}{\sin[\vartheta]^2}+\frac{\cot[\vartheta]Y'[\vartheta]+Y''[\vartheta]}{B^2Y[\vartheta]}=l(1+l), \tag{722}$$

$$\frac{l(1+l)}{r^2}-\frac{\left(2\cdot h'[r]+r\cdot h''[r]\right)}{B^2\cdot r\cdot h[r]}=\frac{B^2\cdot C_t^2}{A^2}-\frac{R^{*}}{6}. \tag{723}$$

While the first equation of the two gives the usual solution with the Legendre Polynomials, as also known from the Schrödinger hydrogen solution [25, 26], only in a slightly generalized manner (watch the asymmetry-parameter B):

$$Y_k^m[\vartheta]=C_{P\nu}\cdot P_k^{B\cdot m}[\cos[\vartheta]]+C_{Q\nu}\cdot Q_k^{B\cdot m}[\cos[\vartheta]];\quad k=\frac{1}{2}\left(\sqrt{1+4B^2l(1+l)}-1\right)$$

$$B^2l(1+l)\equiv L(1+L), \tag{724}$$

we have a rather non-Schrödinger-like solution with respect to the radius coordinate in the case of $R^{*}=0$:

$$h[r]=C_j\cdot j_k\left[\frac{B^2}{A}\cdot C_t\cdot r\right]+C_y\cdot y_k\left[\frac{B^2}{A}\cdot C_t\cdot r\right];\quad k=\frac{1}{2}\left(\sqrt{1+4B^2l(1+l)}-1\right). \tag{725}$$

The reader can find an extensive discussion on this solution in [52, 60]. Formally speaking, the Einstein–Hilbert action and its generalized field equation (125) directly give us metric and a quantum field for a Schwarzschild object with Schwarzschild radius r_S. Thereby, the quantum halo or wave function surrounding the object can be evaluated for the far distance to be:

$$f[t,r,\vartheta,\varphi]=\begin{pmatrix} \left(C_{t1}\cdot\cos\left[H\cdot c\cdot C_t\cdot t\right]+C_{t2}\cdot\sin\left[H\cdot c\cdot C_t\cdot t\right]\right) \\ \times\left(C_{\varphi1}\cdot\cos\left[D\cdot m\cdot\varphi\right]+C_{\varphi2}\cdot\sin\left[D\cdot m\cdot\varphi\right]\right) \\ \times\left(C_{Pv}\cdot P_k^{B\cdot m}\left[\cos[\vartheta]\right]+C_{Qv}\cdot Q_k^{B\cdot m}\left[\cos[\vartheta]\right]\right) \\ \times\left(C_j\cdot j_k\left[\frac{B^2}{A}\cdot C_t\cdot r\right]+C_y\cdot y_k\left[\frac{B^2}{A}\cdot C_t\cdot r\right]\right) \end{pmatrix}$$

$$k=\frac{1}{2}\left(\sqrt{1+4B^2l(1+l)}-1\right). \tag{726}$$

In the case of a perfectly symmetric "time+3-sphere" with $A = B = C = 1$ and also $H = 1$ this solution simplifies to:

$$f[t,r,\vartheta,\varphi]=\begin{pmatrix} \left(C_{t1}\cdot\cos\left[c\cdot C_t\cdot t\right]+C_{t2}\cdot\sin\left[c\cdot C_t\cdot t\right]\right) \\ \times\left(C_{\varphi1}\cdot\cos\left[m\cdot\varphi\right]+C_{\varphi2}\cdot\sin\left[m\cdot\varphi\right]\right) \\ \times\left(C_{Pv}\cdot P_l^m\left[\cos[\vartheta]\right]+C_{Qv}\cdot Q_l^m\left[\cos[\vartheta]\right]\right) \\ \times\left(C_j\cdot j_l\left[C_t\cdot r\right]+C_y\cdot y_l\left[C_t\cdot r\right]\right) \end{pmatrix}. \tag{727}$$

We realize that the solution is even quantized as only certain parameters and quantum numbers lead to reasonable (infinity free) results. Excluding the obvious infinities and not considering half spins either (c.f. appendix H in [12], for instance), reduces our wave solution to:

$$f[t,r,\vartheta,\varphi]=\begin{pmatrix} \left(C_{t1}\cdot\cos\left[H\cdot c\cdot C_t\cdot t\right]+C_{t2}\cdot\sin\left[H\cdot c\cdot C_t\cdot t\right]\right) \\ \times\left(C_{\varphi1}\cdot\cos\left[D\cdot m\cdot\varphi\right]+C_{\varphi2}\cdot\sin\left[D\cdot m\cdot\varphi\right]\right) \\ \times C_j\cdot C_{Pv}\cdot P_k^{B\cdot m}\left[\cos[\vartheta]\right]\cdot j_k\left[\frac{B^2}{A}\cdot C_t\cdot r\right] \end{pmatrix}. \tag{728}$$

Now we relate this wave function to the Dirac particle and therefore set $C_{t2} = \pm i^*C_{t1}$, which results in:

$$f[t,r,\vartheta,\varphi]=e^{\pm i\cdot H\cdot c\cdot C_t\cdot t}\begin{pmatrix} \times\left(C_{\varphi1}\cdot\cos\left[D\cdot m\cdot\varphi\right]+C_{\varphi2}\cdot\sin\left[D\cdot m\cdot\varphi\right]\right) \\ \times\underbrace{C_{t1}\cdot C_j\cdot C_{Pv}}_{\equiv C_0}\cdot P_k^{B\cdot m}\left[\cos[\vartheta]\right]\cdot j_k\left[\frac{B^2}{A}\cdot C_t\cdot r\right] \end{pmatrix}. \tag{729}$$

From basic Quantum Theory, we know that a particle at rest has the time dependency [12]:

$$f[t]=e^{\pm i\cdot\frac{m_R\cdot c^2}{\hbar}\cdot t}\cdot C_f, \tag{730}$$

with m_R giving the rest mass of the particle and $\hbar$ denoting the reduced Planck constant and the time factor in Eq. (729) we then extract the following relation:

$$\pm i\cdot\frac{m_R\cdot c^2}{\hbar}=\pm i\cdot H\cdot c\cdot C_t \;\Rightarrow\; \frac{m_R\cdot c}{\hbar}=H\cdot C_t. \tag{731}$$

Inserting the Schwarzschild radius $r_s=\frac{2\cdot m_R\cdot G}{c^2}$ (G is Newton's constant and c is the speed of light in vacuum), thereby substituting the rest mass, leaves us with:

$$\frac{r_s\cdot c^3}{2\cdot\hbar\cdot G}=H\cdot C_t \;\Rightarrow\; C_t=\frac{r_s\cdot c^3}{2\cdot H\cdot\hbar\cdot G}=\frac{r_s}{2\cdot H}\cdot\left(\frac{c^3}{\hbar\cdot G}\right)=\frac{r_s}{2\cdot H\cdot\ell_P^2}. \tag{732}$$

Here ℓ_P denotes the Planck length. This changes Eq. (729) to:

$$f[t,r,\vartheta,\varphi]=e^{\pm i\cdot\frac{c\cdot r_s}{2\cdot\ell_P^2}\cdot t}\begin{pmatrix}\times\left(C_{\varphi1}\cdot\cos[D\cdot m\cdot\varphi]+C_{\varphi2}\cdot\sin[D\cdot m\cdot\varphi]\right)\\ \times C_0\cdot P_k^{B\cdot m}[\cos[\vartheta]]\cdot j_k\left[\frac{B^2}{A}\cdot\frac{r_s}{2\cdot H\cdot\ell_P^2}\cdot r\right]\end{pmatrix}. \tag{733}$$

Assuming a perfectly symmetric sphere with $A = B = C = 1$ and also $H = 1$ plus setting $m = k = 0$, we would end up with the following "ground state":

$$f[t,r,\vartheta,\varphi]=f[t,r]=C_0\cdot C_{\varphi1}\cdot e^{\pm i\cdot\frac{c\cdot r_s}{2\cdot\ell_P^2}\cdot t}\cdot j_0\left[\frac{r_s}{2\cdot\ell_P^2}\cdot r\right]$$

$$=C_0\cdot C_{\varphi1}\cdot e^{\pm i\cdot\frac{c\cdot r_s}{2\cdot\ell_P^2}\cdot t}\cdot\frac{\sin\left[\frac{r_s}{2\cdot\ell_P^2}\cdot r\right]}{r}. \tag{734}$$

Thus, the simplest wave function field we would observe around a Black Hole (from a sufficiently far distance) would be oscillating in both time and radius. With the Planck length being so extremely small, namely, $\ell_P = 1.616255(18)\times 10^{-35}$ meters, these oscillations are of an extremely high frequency even for small Schwarzschild objects. But high frequencies usually stand for high energies. In other words, extreme amounts of energy could be stored in such a field, and as it is of static character, it might just be this field which contains all the mass of the object. There would be no need for a singularity anymore. Even a relatively small object like the earth—as a Black Hole—would have an incredible number of 1.7578×10^{67} oscillations on just 1 meter and 5.273×10^{75} oscillations per second.

Taking the ground state (734) and remembering our functional setting for the scaling function $F[f] = (C_1 + f)^2$, the whole metric would asymptotically tend towards the flat space with increasing r in the following way:

$$G_{\alpha\beta} = \begin{pmatrix} C_1^2 + 2 \cdot C_1 \cdot \overbrace{C_0 \cdot C_{\varphi 1}}^{C_{0\varphi}} \cdot e^{\pm i \cdot \frac{c \cdot r_s}{2 \cdot \ell_P^2} \cdot t} \cdot \dfrac{\sin\left[\frac{r_s}{2 \cdot \ell_P^2} \cdot r\right]}{r} \\ + C_{0\varphi}^2 \cdot e^{\pm i \cdot \frac{c \cdot r_s}{\ell_P^2} \cdot t} \cdot \dfrac{\sin^2\left[\frac{r_s}{2 \cdot \ell_P^2} \cdot r\right]}{r^2} \end{pmatrix} g_{\alpha\beta}^{\text{flat}}. \tag{735}$$

It should be pointed out that the ground state field reduction with increasing radius r, going with $1/r$, is the smallest and also the dominant field decrease. All other terms, resulting from higher-order solutions of the spherical Bessel functions, namely, also sport $1/r$-character plus $1/r^q$ (with $q > 1$). In other words, the corresponding higher-order fields disappear much faster with the increase in the radial distance r.

5.18.24.2 And how could the static field lead to radiation?

Instead of Eq. (734), we consider the more general t-r-dependent solution:

$$f[t,r] = \begin{pmatrix} \left(C_{t1} \cdot \cos\left[\frac{c \cdot r_s}{2 \cdot \ell_P^2} \cdot t\right] + C_{t2} \cdot \sin\left[\frac{c \cdot r_s}{2 \cdot \ell_P^2} \cdot t\right] \right) \\ \times \left(C_j \cdot j_0\left[\frac{r_s}{2 \cdot \ell_P^2} \cdot r\right] + C_y \cdot y_0\left[\frac{r_s}{2 \cdot \ell_P^2} \cdot r\right] \right) \end{pmatrix}$$

$$= \begin{pmatrix} \left(C_{t1} \cdot \cos\left[\frac{c \cdot r_s}{2 \cdot \ell_P^2} \cdot t\right] + C_{t2} \cdot \sin\left[\frac{c \cdot r_s}{2 \cdot \ell_P^2} \cdot t\right] \right) \\ \times \dfrac{\left(c_j \cdot \sin\left[\frac{r_s}{2 \cdot \ell_P^2} \cdot r\right] + c_y \cdot \cos\left[\frac{r_s}{2 \cdot \ell_P^2} \cdot r\right] \right)}{r} \end{pmatrix}. \tag{736}$$

When demanding:

$$c_j \cdot C_{t2} = \mp c_y \cdot C_{t1}; \quad c_y \cdot C_{t2} = \pm c_j \cdot C_{t1}, \tag{737}$$

we can apply the addition theorems of the trigonometric functions and obtain the following solution:

$$f[t,r]$$

$$= \frac{1}{r} \begin{pmatrix} C_{t1} \cdot c_y \cdot \cos\left[\frac{r_s}{2 \cdot \ell_P^2} \cdot r\right] \cos\left[\frac{c \cdot r_s}{2 \cdot \ell_P^2} \cdot t\right] + C_{t2} \cdot c_y \cdot \cos\left[\frac{r_s}{2 \cdot \ell_P^2} \cdot r\right] \sin\left[\frac{c \cdot r_s}{2 \cdot \ell_P^2} \cdot t\right] \\ + c_j \cdot C_{t1} \cdot \cos\left[\frac{c \cdot r_s}{2 \cdot \ell_P^2} \cdot t\right] \sin\left[\frac{r_s}{2 \cdot \ell_P^2} \cdot r\right] + c_j \cdot C_{t2} \cdot \sin\left[\frac{c \cdot r_s}{2 \cdot \ell_P^2} \cdot t\right] \sin\left[\frac{r_s}{2 \cdot \ell_P^2} \cdot r\right] \end{pmatrix}$$

$$= \frac{1}{r} \begin{pmatrix} C_{t1} \cdot c_y \cdot \cos\left[\frac{r_s}{2 \cdot \ell_P^2} \cdot r\right] \cos\left[\frac{c \cdot r_s}{2 \cdot \ell_P^2} \cdot t\right] \mp c_y \cdot C_{t1} \cdot \sin\left[\frac{r_s}{2 \cdot \ell_P^2} \cdot r\right] \sin\left[\frac{c \cdot r_s}{2 \cdot \ell_P^2} \cdot t\right] \\ + c_j \cdot C_{t1} \cdot \sin\left[\frac{r_s}{2 \cdot \ell_P^2} \cdot r\right] \cos\left[\frac{c \cdot r_s}{2 \cdot \ell_P^2} \cdot t\right] \pm c_j \cdot C_{t1} \cdot \cos\left[\frac{r_s}{2 \cdot \ell_P^2} \cdot r\right] \sin\left[\frac{c \cdot r_s}{2 \cdot \ell_P^2} \cdot t\right] \end{pmatrix}$$

$$= \frac{1}{r} \left(C_{t1} \cdot c_y \cdot \cos\left[\frac{r_s}{2 \cdot \ell_P^2} \cdot r \pm \frac{c \cdot r_s}{2 \cdot \ell_P^2} \cdot t\right] + C_{t1} \cdot c_j \cdot \sin\left[\frac{r_s}{2 \cdot \ell_P^2} \cdot r \pm \frac{c \cdot r_s}{2 \cdot \ell_P^2} \cdot t\right] \right)$$

$$= \frac{C_{t1}}{r} \left(c_y \cdot \cos\left[\frac{r_s}{2 \cdot \ell_P^2} \cdot (r \pm c \cdot t)\right] + c_j \cdot \sin\left[\frac{r_s}{2 \cdot \ell_P^2} \cdot (r \pm c \cdot t)\right] \right). \tag{738}$$

These are clearly in- and outgoing waves with the following dependencies:

$$f[t,r] = \frac{C_{t1}}{r} \left(c_y \cdot \cos\left[\frac{r_s}{2 \cdot \ell_P^2} \cdot (r \pm c \cdot t)\right] + c_j \cdot \sin\left[\frac{r_s}{2 \cdot \ell_P^2} \cdot (r \pm c \cdot t)\right] \right)$$

$$(r + c \cdot t): \quad C_{t1} = \mp i \cdot C_{t2}; \quad c_j = \pm i \cdot c_y$$

$$(r - c \cdot t): \quad C_{t1} = \pm i \cdot C_{t2}; \quad c_j = \pm i \cdot c_y$$

$$\Rightarrow$$

$$f[t,r] = \begin{Bmatrix} \frac{\mp i \cdot C_{t2}}{r} c_y \cdot \left(\cos\left[\frac{r_s}{2 \cdot \ell_P^2} \cdot (r + c \cdot t)\right] \pm i \cdot \sin\left[\frac{r_s}{2 \cdot \ell_P^2} \cdot (r + c \cdot t)\right] \right) \\ \frac{\pm i \cdot C_{t2}}{r} c_y \cdot \left(\cos\left[\frac{r_s}{2 \cdot \ell_P^2} \cdot (r - c \cdot t)\right] \pm i \cdot \sin\left[\frac{r_s}{2 \cdot \ell_P^2} \cdot (r - c \cdot t)\right] \right) \end{Bmatrix}$$

$$=\left\{\begin{matrix}\frac{\mp i\cdot C_{t2}}{r}c_y\cdot e^{\pm i\cdot\frac{r_s}{2\cdot\ell_P^2}\cdot(r+c\cdot t)}\\ \frac{\pm i\cdot C_{t2}}{r}c_y\cdot e^{\pm i\cdot\frac{r_s}{2\cdot\ell_P^2}\cdot(r-c\cdot t)}\end{matrix}\right\}. \tag{739}$$

For convenience, we introduce a new constant C_c, reading:

$$c_y\cdot C_{t2}=i\cdot C_c, \tag{740}$$

which gives us:

$$f[t,r]=\left\{\begin{matrix}\pm\frac{C_c}{r}\cdot e^{\pm i\cdot\frac{r_s}{2\cdot\ell_P^2}\cdot(r+c\cdot t)}\\ \mp\frac{C_c}{r}\cdot e^{\pm i\cdot\frac{r_s}{2\cdot\ell_P^2}\cdot(r-c\cdot t)}\end{matrix}\right\}. \tag{741}$$

Of course, more such waves could be constructed from higher-order spherical Bessel functions in our general solution (726), potentially even with angular dependencies ϑ and φ. These waves can be of arbitrary complexity and, thus, can store large amounts of information. Here, however, we will not investigate these higher-order options, as it only was our goal to show how radiation can be obtained for a Schwarzschild object without artificially bringing in a Klein–Gordon field. We saw that such a field was already inside the generalized Einstein–Hilbert action for a scaled metric, respectively the subsequent quantum Einstein field equations (125).

5.18.25 Appendix RB: A Metric Object with "Hidden" Spin-1/2

We apply the following metric on our scalar field equation (376):

$$G_{\alpha\beta}^{8}=\begin{pmatrix}-c^2 & 0 & 0 & 0 & \cdots & 0\\ 0 & 1 & 0 & 0 & \cdots & 0\\ 0 & 0 & r^2 & 0 & \cdots & 0\\ 0 & 0 & 0 & r^2\cdot\sin^2\varphi_1 & \cdots & 0\\ \cdots & \cdots & \cdots & \cdots & \ddots & 0\\ 0 & 0 & 0 & 0 & 0 & g_{77}\end{pmatrix}$$

$$\times F\left[f\left[t,r,\vartheta,\varphi,\tau,\rho,\theta,\phi\right]\right];\quad F[f]=f^{\frac{2}{3}}$$

$$f[t,r,\vartheta,\varphi,\tau,\rho,\theta,\phi]=f_t[t]\cdot f_r[r]\cdot f_\vartheta[\vartheta]\cdot f_\varphi[\varphi]\cdot f_\tau[\tau]\cdot f_\rho[\rho]\cdot f_\theta[\theta]\cdot f_\phi[\phi];$$

$$g_{44}=c_4^2\cdot r^{-2};\quad g_{55}=1;\quad g_{66}=\rho^2;\quad g_{77}=\rho^2\cdot\sin[\theta]^2. \tag{742}$$

Thereby we have used two spherical coordinate sets and when using the following partial solutions discussed elsewhere [9, 10] as follows:

$$f_\varphi[\varphi]=C_{\varphi-}\cdot e^{-i\cdot A\cdot\varphi}+C_{\varphi+}\cdot e^{+i\cdot A\cdot\varphi}, \tag{743}$$

$$f_\tau[\tau]=C_{\tau 1}\cdot\cos[c_4\cdot A_2\cdot\tau]+C_{\tau 2}\cdot\sin[c_4\cdot A_2\cdot\tau], \tag{744}$$

$$f_\vartheta[\vartheta]=C_{P\vartheta}\cdot P_L^A[\cos[\vartheta]]+C_{Q\vartheta}\cdot Q_L^A[\cos[\vartheta]], \tag{745}$$

$$f_\phi[\phi]=C_{\phi-}\cdot e^{-i\cdot B\cdot\phi}+C_{\phi+}\cdot e^{+i\cdot B\cdot\phi}, \tag{746}$$

$$f_t[t]=C_{t1}\cdot\cos[c\cdot E_t\cdot t]+C_{t2}\cdot\sin[c\cdot E_t\cdot t], \tag{747}$$

$$f_\theta[\theta]=C_{P\theta}\cdot P_\ell^B[\cos[\theta]]+C_{Q\theta}\cdot Q_\ell^B[\cos[\theta]], \tag{748}$$

with the associated Legendre polynomials $P_L^{A_1}, Q_L^{A_1}$. This gives us the following Ricci scalar R^* and the subsequent differential equation for the *r*-*ρ*-dependency of *f*[...]:

$$R^*=0=\left(\frac{\ell\cdot(\ell+1)}{\rho^2}-\frac{2\cdot\partial}{\rho\cdot\partial\rho}-\frac{\partial^2}{\partial\rho^2}+\frac{L\cdot(L+1)}{r^2}+A_2^2\cdot r^2-\frac{\partial}{r\cdot\partial r}-\frac{\partial^2}{\partial r^2}-E_t^2\right)f. \tag{749}$$

As we can directly separate the two radii with the approach from Eq. (742), we obtain the two differential equations:

$$\left(\frac{\ell\cdot(\ell+1)}{\rho^2}-\frac{2\cdot\partial}{\rho\cdot\partial\rho}-\frac{\partial^2}{\partial\rho^2}\right)f_\rho=E_\rho^2\cdot f_\rho, \tag{750}$$

$$0=\left(\frac{L\cdot(L+1)}{r^2}+A_2^2\cdot r^2-\frac{\partial}{r\cdot\partial r}-\frac{\partial^2}{\partial r^2}-E_t^2+E_\rho^2\right)f_r. \tag{751}$$

While Eq. (750) gives us the spherical Bessel functions again via:

$$f_\rho[\rho]=C_j\cdot j_\ell[E_\rho\cdot\rho]+C_y\cdot y_\ell[E_\rho\cdot\rho], \tag{752}$$

we obtain the following solution for f_r:

$$f_r[r]=\frac{2^{\frac{1}{2}\left(1+\sqrt{L\cdot(1+L)}\right)}\mathrm{e}^{-\frac{1}{2}\cdot A_2\cdot r^2}\cdot\left(r^2\right)^{\frac{1}{2}\left(1+\sqrt{L\cdot(1+L)}\right)}}{r}$$

$$\times\left(C_U\cdot U\left[-nn,1+\sqrt{L\cdot(1+L)},A_2\cdot r^2\right]+C_L\cdot L_{nn}^{\sqrt{L\cdot(1+L)}}\left[A_2\cdot r^2\right]\right)$$

$$nn=\frac{E_t^2-E_\rho^2-2\cdot A_2\cdot\left(1+\sqrt{L\cdot(1+L)}\right)}{4\cdot A_2}. \qquad (753)$$

We recognize the hypergeometric function $U[a,b,z]$ and the Laguerre polynomials $L_{nn}^{\sqrt{L\cdot(1+L)}}$ in the solutions above, which we know from the Schrödinger hydrogen solution (e.g., [9]). The only reasonable solution (singularity free) can be found for the case of L = 0 and Eq. (753) simplifies to:

$$f_r[r]=\sqrt{2}\cdot\mathrm{e}^{-\frac{1}{2}\cdot A_2\cdot r^2}\cdot\left(C_U\cdot U\left[-nn,1,A_2\cdot r^2\right]+C_L\cdot L_{nn}^{0}\left[A_2\cdot r^2\right]\right)$$

$$nn=\frac{E_t^2-E_\rho^2-2\cdot A_2}{4\cdot A_2}=\frac{E_t^2-E_\rho^2}{4\cdot A_2}-\frac{1}{2}. \qquad (754)$$

Figure RB1 shows the corresponding f_r-distribution for three settings of E_t.

In order to also allow for $L\neq 0$-states, we change the metric (742) as follows:

$$G_{\alpha\beta}^{8}=\begin{pmatrix}-c^2 & 0 & 0 & 0 & \cdots & 0\\ 0 & 1 & 0 & 0 & \cdots & 0\\ 0 & 0 & r^2 & 0 & \cdots & 0\\ 0 & 0 & 0 & r^2\cdot\sin^2\varphi_1 & \cdots & 0\\ \cdots & \cdots & \cdots & \cdots & \ddots & 0\\ 0 & 0 & 0 & 0 & 0 & g_{77}\end{pmatrix}$$

$$\times F\left[f\left[t,r,\vartheta,\varphi,\tau,\rho,\theta,\phi\right]\right];\quad F[f]=f^{\frac{2}{3}}$$

$$f\left[t,r,\vartheta,\varphi,\tau,\rho,\theta,\phi\right]=f_t[t]\cdot f_r[r]\cdot f_\vartheta[\vartheta]\cdot f_\varphi[\varphi]\cdot f_\tau[\tau]\cdot f_\rho[\rho]\cdot f_\theta[\theta]\cdot f_\phi[\phi];$$

$$g_{44}=c_4^2\cdot r^{-2};\quad g_{55}=r^2;\quad g_{66}=r^2\cdot\rho^2;\quad g_{77}=r^2\cdot\rho^2\cdot\sin[\theta]^2, \qquad (755)$$

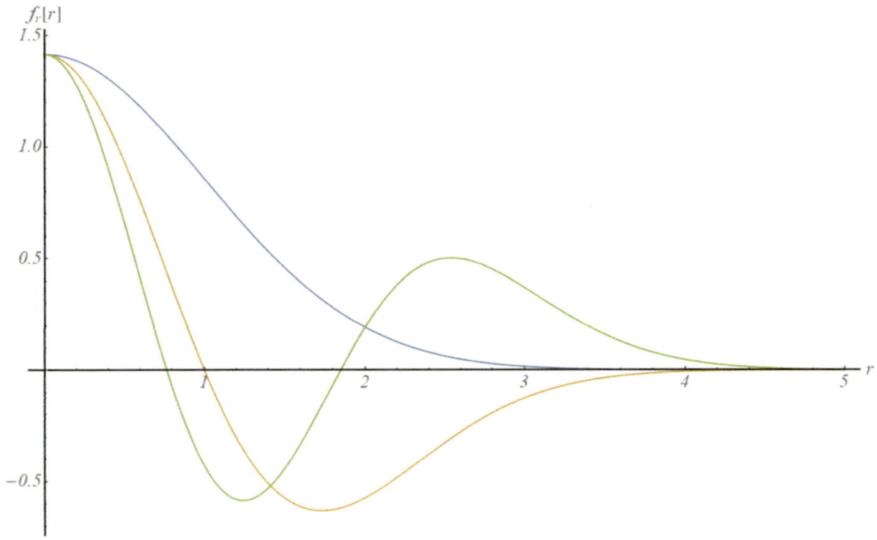

Figure RB1 f_r-distribution for solution (754) with setting $A_2 = 1$, $E_\rho = \pm 1$, $L = 0$ and $E_t = \sqrt{3}, \sqrt{7}, \sqrt{11}$, resulting in nn = 0 (blue), 1 (yellow), 2 (green).

and keeping the partial solutions from Eqs. (743) to (748), we end up with the following differential equation:

$$0 = \left(\begin{array}{l} \left(\frac{\ell \cdot (\ell+1)}{\rho^2} - \frac{2 \cdot \partial}{\rho \cdot \partial \rho} - \frac{\partial^2}{\partial \rho^2} - \frac{18}{7} + L \cdot (L+1) \right) \cdot \frac{1}{r^2} \\ + A_2^2 \cdot r^2 - \frac{4\partial}{r \cdot \partial r} - \frac{\partial^2}{\partial r^2} - E_t^2 \end{array} \right) f. \tag{756}$$

Again, we can directly separate the two radii and obtain two differential equations:

$$\left(\frac{\ell \cdot (\ell+1)}{\rho^2} - \frac{2 \cdot \partial}{\rho \cdot \partial \rho} - \frac{\partial^2}{\partial \rho^2} \right) f_\rho = E_\rho^2 \cdot f_\rho, \tag{757}$$

$$0 = \left(\left(L \cdot (L+1) - \frac{18}{7} + E_\rho^2 \right) \cdot \frac{1}{r^2} + A_2^2 \cdot r^2 - \frac{4\partial}{r \cdot \partial r} - \frac{\partial^2}{\partial r^2} - E_t^2 \right) f_r. \tag{758}$$

Thereby, we have already applied solution (752) in Eq. (758). The solution for f_r reads:

$$f_r[r] = \frac{2^{\frac{1}{2}(1+L_x)} e^{-\frac{1}{2}\cdot A_2 \cdot r^2} \cdot \left(r^2\right)^{\frac{1}{2}(1+L_x)}}{r^{5/2}}$$
$$\times\left(C_U \cdot U\left[-nn, 1+L_x, A_2\cdot r^2\right] + C_L \cdot L_{nn}^{L_x}\left[A_2 \cdot r^2\right]\right)$$

$$nn = \frac{E_t^2 - 2\cdot A_2 \cdot (1+L_x)}{4\cdot A_2}; \quad L_x = \frac{1}{2}\cdot\sqrt{4\cdot\left(E_\rho^2 + L\cdot(1+L)\right) - \frac{9}{7}}. \qquad (759)$$

Setting $L = 0$, we find no singularity-free solution unless $L_x \geq 2$. Subsequently, we have to demand nn ≥ 0. Figure RB2 shows the corresponding r-distribution for three settings of E_t.

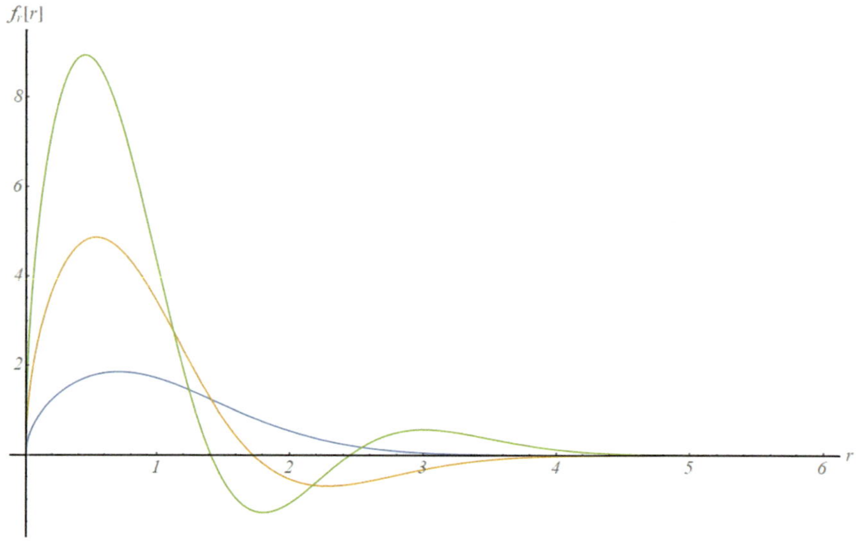

Figure RB2 f_r-distribution for solution (754) with setting $A_2 = 1$, $E_\rho = \pm\frac{11}{2\sqrt{7}}$, $L = 0$ and $E_t = \sqrt{6}, \sqrt{10}, \sqrt{14}$, resulting in nn = 0 (blue), 1 (yellow), 2 (green).

Choosing $L = 1/2$, we can obtain reasonable (free of singularities) solutions, for $L_x \geq 2$. Subsequently, we have to demand nn ≥ 2. Figure RB3 shows the corresponding r-distribution for three settings of E_t.

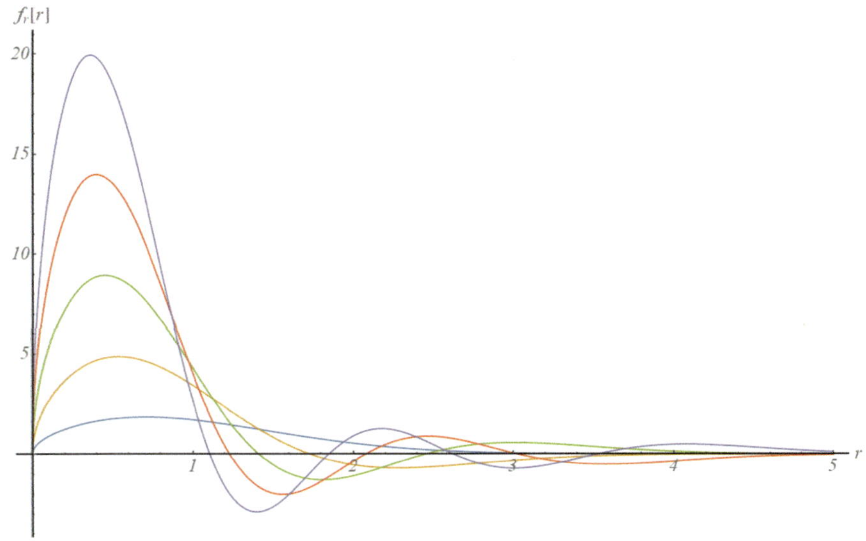

Figure RB3 f_r-distribution for solution (754) with setting $A_2 = 1$, $E_\rho = \pm\frac{5}{\sqrt{7}}$, $L = 1/2$ and $(E_t)^2 = 6, 10, 14, 18, 22$, resulting in nn = 0 (blue), 1 (yellow), 2 (green), 3 (red), 4 (violet).

5.18.25.1 Does spin-1/2 really exist?

It is quite illustrative to compare our finding with the corresponding half spin solution for the classical Schrödinger hydrogen case. Here $L = 1/2$ would also lead to $L_x = 2$ (c.f. Eq. (759)). This is quite interesting, because it leads to the question:

Is the spin-1/2 only an illusion?

After all, the r-behavior we obtain is, thanks to $L_x = 2 = L * (L + 1)$ for the case $L = 1/2$, definitively the one of a half spin object. In addition, we have the two states for E_ρ, being the states of two "spin-like energies" rather than spin orientations, but in connection with our metric (755) they contribute to the spin.

However, if spin-1/2 was only an illusion, which is to say an object pretending to have spin-1/2 by containing some kind of intrinsic sphere with always two states for the momentum-connected parameter E_ρ, should this approach not also—automatically—provide a metric understanding of the Pauli exclusion principle [51]?

Well, whatever the ominous additional radius ρ might be, we know that for odd integers L_ρ, the spherical Bessel functions from Eq. (752) result in opposite signed outcomes for the two parameters $E_\rho = \pm X$. Thus, the total space-time curvature along ρ of two objects possessing opposite signed E_ρ leads to zero space-time and thus, probably a lower total energy. On the other hand, in the case of equally signed E_ρ, the total curvature increases, leading to higher energy states. Nature definitively prefers the state with lower total curvature and hence, the Pauli exclusion.

5.18.25.2 Conclusions to our new spin-1/2 option

It was shown that a scaled metric automatically provides Schrödinger and Schrödinger-like equations [24, 25, 26] directly out of the Einstein–Hilbert action when adding additional dimensions with certain radius dependencies. Thereby, we also found metric dependencies producing confining potentials and bringing about apparent half spin objects ... but without the problematic 720-degree rotation abnormity of the classical spinors.

Our metric scale thereby takes on the classical quantum mechanical wave function. This just means that our space-time is distorted due to quantum processes via this wave function.

5.18.26 Appendix RC: A Few Words about Spin-1/2, -3/2, -5/2 and so on

In the following, we want to investigate potential half spin solutions for the associated Legendre polynomials, which we needed in various forms within this appendix.

Here, as an example, we intend to consider the simple form:

$$Y_l^m[v] = C_{Pv} \cdot P_l^m[\cos[v]] + C_{Qv} \cdot Q_l^m[\cos[v]]. \tag{760}$$

Observing the solution $Y[v]$ more closely, we find that there exist singularity-free spin $l=n/2$-solutions for $m = -n/2$ with $n = 1,3,5,7, \ldots$ in the case of $C_{Pv} \neq 0$, $C_{Qv} = 0$ and for $m = +n/2$ in the case of $C_{Pv} = 0$, $C_{Qv} \neq 0$. Figures RC1 and RC2 illustrate the corresponding distribution within the domain of definition for the angle v from 0 to π.

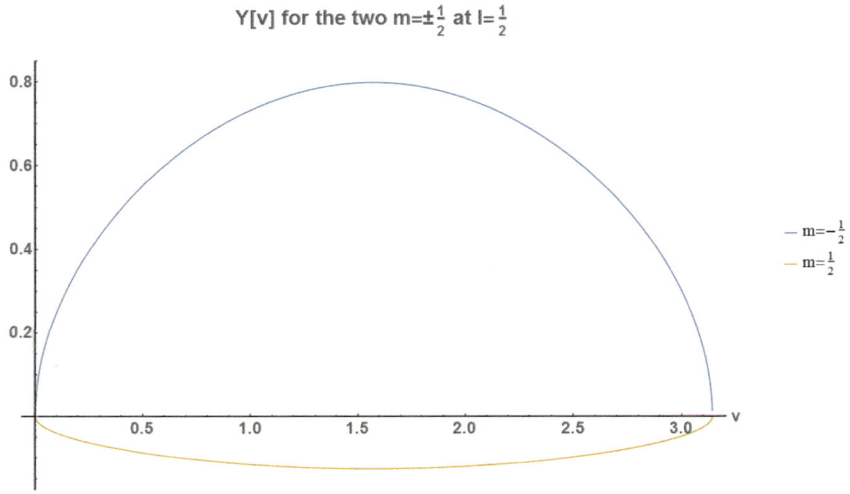

Figure RC1 Spin l=1/2 situation with the two possible spin states $m = \pm 1/2$ according to our angular quantum gravity solution (724). For better illustration and comparability, we divided the "Q-Legendres" by 10.

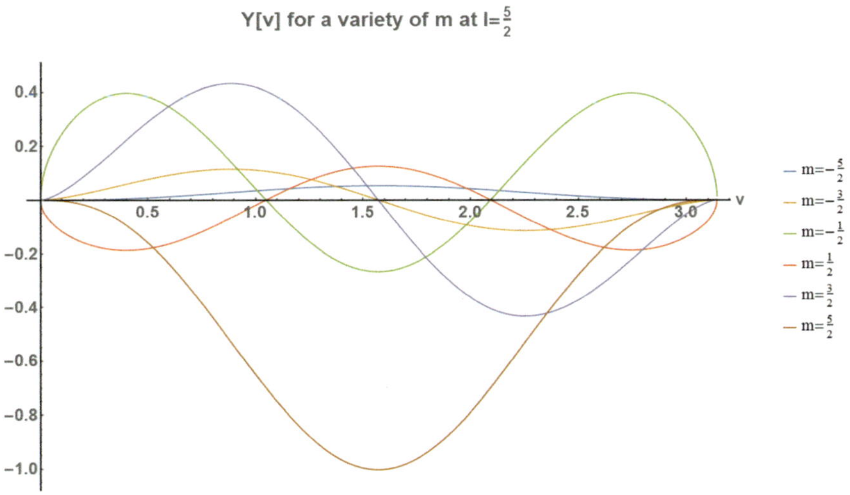

Figure RC2 Spin $l = 5/2$ situation with the six possible spin states $m = \pm\{1/2, 3/2, 5/2\}$ according to our angular quantum gravity solution (724). Regarding the evaluation, see text. For better illustration, we divided the "Q-Legendres" by 10.

The general solution to the partial differential equation resulting from the Schrödinger hydrogen problem [24, 25, 26] can be given as follows:

$$f_{n,l,m}[t,r,\vartheta,\varphi]=g[t]\cdot\Psi_{n,l,m}[r,\vartheta,\varphi]=g[t]\cdot e^{i\cdot m\cdot\varphi}\cdot P_l^m[\cos\vartheta]\cdot R_{n,l}[r]$$

$$=\left(C_1\cdot\cos\left[c\cdot C_{tt}\cdot t\right]+C_2\cdot\sin\left[c\cdot C_{tt}\cdot t\right]\right)\cdot N\cdot e^{-\rho/2}\cdot\rho^l\cdot L_{n-l-1}^{2l+1}[\rho]\cdot Y_l^m[\vartheta,\varphi]$$

$$\rho=\frac{2\cdot r}{n\cdot a_0};\quad N=\sqrt{\left(\frac{2}{n\cdot a_0}\right)^3\frac{(n-l-1)!}{2\cdot n\cdot(n+l)!}}. \tag{761}$$

It totally suffices here to consider the classical problems in order to also discuss the angular aspects of the metric (755) as the results for these coordinates in both cases are the same functional dependencies.

For nostalgic reasons, we used the symbol "n" (here not the dimension) as the so-called main quantum number. All quantum numbers n, l and the parameter a_0 depend on the constants E, A_i in connection with our solutions to metric (755) above and have to satisfy certain quantum conditions in order to result in singularity-free solutions for $f[...]$.

Regarding the conditions for the quantum numbers n, l and m, we not only have the usual:

$$\{n,\mathrm{l},\mathrm{m}\}\in\mathbb{Z};\quad n\geq 0;\quad l<n;\quad -l\leq m\leq +l, \tag{762}$$

but also found the suitable solutions for the half spin forms as discussed above and derived in Figs. RC1 and RC2. The corresponding main quantum numbers for half spin l-numbers with $l = 1/2, 3/2, ...$ are simply (just as before with the integers) $n = l + 1 = 3/2,5/2,7/2, ...$

It should explicitly be noted, however, that the usual spherical harmonics are inapplicable in cases of half spin. For $\{n,l,m\}=\{1/2,3/2,5/2,7/2, ...\}$ the wave function (761) has to be adapted as follows:

$$f_{n,l,m}[t,r,\vartheta,\varphi]=e^{\pm i\cdot c\cdot C_t\cdot t}\cdot N\cdot e^{-\rho/2}\cdot\rho^l\cdot L_{n-l-1}^{2l+1}[\rho]\cdot Z_m[\varphi]\cdot\begin{Bmatrix}P_l^{m<0}[\cos\vartheta]\\ Q_l^{m>0}[\cos\vartheta]\end{Bmatrix}$$

$$=e^{\pm i\cdot c\cdot C_t\cdot t}\cdot N\cdot e^{-\rho/2}\cdot\rho^l\cdot L_{n-l-1}^{2l+1}[\rho]\cdot\begin{Bmatrix}\cos[m\cdot\varphi]\\ \sin[m\cdot\varphi]\end{Bmatrix}\cdot\begin{Bmatrix}P_l^{m<0}[\cos\vartheta]\\ Q_l^{m>0}[\cos\vartheta]\end{Bmatrix}. \tag{763}$$

Thereby it was elaborated elsewhere [52] that in fact the sin- and the cos-functions seem to make the Pauli exclusion [51] and not the "+" and "–" of the m. However, in order to have the usual Fermionic statistic, we can simply define as follows:

$$f_{n,l,m}\left[t,r,\vartheta,\varphi\right]=e^{\pm i\cdot c\cdot C_t\cdot t}\cdot N\cdot e^{-\rho/2}\cdot\rho^l\cdot L_{n-l-1}^{2l+1}\left[\rho\right]\cdot\begin{Bmatrix}\sin\left[m\cdot\varphi\right]_{m<0}\\ \cos\left[m\cdot\varphi\right]_{m>0}\end{Bmatrix}\cdot\begin{Bmatrix}P_l^{m<0}\left[\cos\vartheta\right]\\ Q_l^{m>0}\left[\cos\vartheta\right]\end{Bmatrix}. \tag{764}$$

As derived in [52], the resolution of the degeneration with respect to half spin requires a break of the symmetry, which we achieved by introducing elliptical geometry instead of the spherical one in [52].

Thus, we have metrically derived a fairly general "hydrogen atom." In addition to the Schrödinger structure, our form also sports a time-dependent factor clearly showing the options for matter and anti-matter via the ±-sign in the $g[t]$-function. We also found the half spin states and were able to resolve the spin-degeneration via a simple symmetry break by switching from spherical to elliptical coordinates (see [52] regarding the evaluation). A few illustrations of the half spin solutions are given in section "Illustration of Spin-1/2-Solutions" of [24].

5.18.27 Appendix RD: Obtaining Purely Metric Potentials

It was shown in [9, 10, 30] that in connection with condition (374) and the subsequent scalarized quantum Einstein field equations (379), a potential can be constructed via four additional dimensions. We demonstrated that we can obtain any desired shape of potential in r, via a set of additional dimensions (coordinates) with always four such coordinates for each exponent in r. So, being interested in a potential of the type: const, $1/r$, $1/r^2$, for instance, our additional metric components should look as follows:

$$g_{\alpha\beta}^{12}=\begin{pmatrix}c^2 & 0 & 0 & 0 & \cdots & 0\\ 0 & 1 & 0 & 0 & \cdots & 0\\ 0 & 0 & r^2 & 0 & \cdots & 0\\ 0 & 0 & 0 & r^2\cdot\sin^2\varphi_1 & \cdots & 0\\ \cdots & \cdots & \cdots & \cdots & \ddots & 0\\ 0 & 0 & 0 & 0 & 0 & g_{n-1n-1}\end{pmatrix}$$

$$\times F\left[f\left[t,r,\vartheta,\varphi_1,\varphi_2,\varphi_3,\varphi_4,\ldots,\varphi_9\right]\right];\quad n=12;\quad F[f]=f^{\frac{4}{12-2}}=f^{\frac{2}{5}}$$

$$f\left[t,r,\vartheta,\varphi_1,\varphi_2,\varphi_3,\varphi_4,\ldots,\varphi_9\right]=f_t\left[t\right]\cdot f_r\left[r\right]\cdot f_\vartheta\left[\vartheta\right]\cdot\prod_{i=1}^{9}f_{\varphi_i}\left[\varphi_i\right];$$

$$s=1\begin{cases} g_{44}=\left(g_{\varphi_2}\left[\varphi_2\right]'\right)^2\cdot r^{i\cdot s};\quad g_{55}=\left(g_{\varphi_3}\left[\varphi_3\right]'\right)^2\cdot r^{s}; \\ g_{66}=\left(g_{\varphi_4}\left[\varphi_4\right]'\right)^2\cdot r^{-i\cdot s};\quad g_{77}=-\left(g_{\varphi_5}\left[\varphi_5\right]'\right)^2\cdot r^{-s}; \end{cases}$$

$$s=2\begin{cases} g_{88}=\left(g_{\varphi_2}\left[\varphi_2\right]'\right)^2\cdot r^{i\cdot s};\quad g_{99}=\left(g_{\varphi_3}\left[\varphi_3\right]'\right)^2\cdot r^{s}; \\ g_{1010}=\left(g_{\varphi_4}\left[\varphi_4\right]'\right)^2\cdot r^{-i\cdot s};\quad g_{1111}=-\left(g_{\varphi_5}\left[\varphi_5\right]'\right)^2\cdot r^{-s}. \end{cases} \tag{765}$$

Introducing the potential function $k[r]$ again, this approach can be simplified as follows:

$$g_{\alpha\beta}^{n}=\begin{pmatrix} -c^2 & 0 & 0 & 0 & \cdots & 0 \\ 0 & 1 & 0 & 0 & \cdots & 0 \\ 0 & 0 & r^2 & 0 & \cdots & 0 \\ 0 & 0 & 0 & r^2\cdot\sin^2\varphi_1 & \cdots & 0 \\ \cdots & \cdots & \cdots & \cdots & \ddots & 0 \\ 0 & 0 & 0 & 0 & 0 & g_{n-1n-1} \end{pmatrix}$$

$$\times F\left[f\left[t,r,\vartheta,\varphi_1,\varphi_2,\varphi_3,\varphi_4,\varphi_5\right]\right];\quad n=8;\quad F[f]=f^{\frac{4}{n-2}}=f^{\frac{2}{3}}$$

$$f\left[t,r,\vartheta,\varphi_1,\varphi_2,\varphi_3,\varphi_4,\varphi_5\right]=f_t[t]\cdot f_r[r]\cdot f_\vartheta[\vartheta]\cdot\prod_{i=1}^{5}f_{\varphi_i}\left[\varphi_i\right];$$

$$g_{44}=\left(g_{\varphi_2}\left[\varphi_2\right]'\right)^2\cdot k[r]^{i\cdot s};\quad g_{55}=\left(g_{\varphi_3}\left[\varphi_3\right]'\right)^2\cdot k[r]^{s};$$

$$g_{66}=\left(g_{\varphi_4}\left[\varphi_4\right]'\right)^2\cdot k[r]^{-i\cdot s};\quad g_{77}=\left(g_{\varphi_5}\left[\varphi_5\right]'\right)^2\cdot k[r]^{-s}. \tag{766}$$

Using the separation solution from above with:

$$f_{\varphi_i}\left[\varphi_i\right]=f_{\varphi_i}\left[g_{\varphi_i}\left[\varphi_i\right]\right]\equiv f_{\varphi_i}\left[g_{\varphi_i}\right];$$

$$f_{\varphi_i}\left[g_{\varphi_i}\right]=C_{-i}\cdot e^{-i\cdot A_i\cdot g_{\varphi_i}}+C_{+i}\cdot e^{+i\cdot A_i\cdot g_{\varphi_i}}=C_{ci}\cdot\cos\left[A_i\cdot g_{\varphi_i}\right]+C_{si}\cdot\sin\left[A_i\cdot g_{\varphi_i}\right]$$

$$\text{especially}\quad(\text{classically}):$$

$$f_{\varphi 1}\left[g_{\varphi_1}\right]=C_{-1}\cdot e^{-i\cdot A_1\cdot g_{\varphi_1}}+C_{+1}\cdot e^{+i\cdot A_1\cdot g_{\varphi_1}}=C_{c1}\cdot\cos\left[\overbrace{A_1}^{\hat{=}m}\cdot g_{\varphi_1}\right]+C_{s1}\cdot\sin\left[\overbrace{A_1}^{\hat{=}m}\cdot g_{\varphi_1}\right], \tag{767}$$

$$f_t[t] = C_{t1} \cdot \cos[c \cdot E \cdot t] + C_{t2} \cdot \sin[c \cdot E \cdot t], \tag{768}$$

$$f_\vartheta\left[\vartheta = \varphi_0\right] = C_{P\vartheta} \cdot P_L^{A_1}\left[\cos\left[\vartheta = \varphi_0\right]\right] + C_{Q\vartheta} \cdot Q_L^{A_1}\left[\cos\left[\vartheta = \varphi_0\right]\right], \tag{769}$$

and inserting everything into Eq. (379), we obtain the following differential equation in r:

$$R^* = 0 = \begin{pmatrix} \frac{L \cdot (L+1)}{r^2} + k[r]^{i \cdot s} \cdot A_2^2 + k[r]^s \cdot A_3^2 \\ +k[r]^{-i \cdot s} \cdot A_4^2 + k[r]^{-s} \cdot A_5^2 - \overbrace{\Delta_{\text{3D-sphere}}}^{2\frac{\partial}{r \cdot \partial r} + \frac{\partial^2}{\partial r^2}}[r] - E^2 \end{pmatrix} \Psi$$

$$\Psi = \Psi\left[r, \vartheta = \varphi_0, \varphi = \varphi_1\right] = f_r[r] \cdot f_\vartheta\left[\vartheta = \varphi_0\right] \cdot f_\varphi\left[\varphi = \varphi_1\right]$$

$$\Rightarrow$$

$$0 = f_\vartheta[\vartheta] \cdot f[\varphi] \cdot \begin{pmatrix} \frac{L \cdot (L+1)}{r^2} + k[r]^{i \cdot s} \cdot A_2^2 + k[r]^s \cdot A_3^2 + k[r]^{-i \cdot s} \cdot A_4^2 \\ +k[r]^{-s} \cdot A_5^2 - \frac{2\partial}{r \cdot \partial r} - \frac{\partial^2}{\partial r^2} - E^2 \end{pmatrix} f_r[r]. \tag{770}$$

We see that for the choice of $k[r]$ as:

$$k[r] = -\frac{1}{A_3^2}\left(\frac{V}{r} - E_P^2\right), \tag{771}$$

$s = 1$ and $f_{\varphi_i}\left[\varphi_i\right] = \text{const}$ for i = 2,4,5, we obtain:

$$0 = f_\vartheta[\vartheta] \cdot f[\varphi] \cdot \left(\frac{L \cdot (L+1)}{r^2} - \frac{V}{r} + E_P^2 - \frac{2\partial}{r \cdot \partial r} - \frac{\partial^2}{\partial r^2} - E^2\right) f_r[r]$$

$$\Rightarrow \qquad \left(-\frac{L \cdot (L+1)}{r^2} + \frac{V}{r} + 2\frac{\partial}{r \cdot \partial r} + \frac{\partial^2}{\partial r^2}\right) f_r[r] = \left(E_P^2 - E^2\right) f_r[r] \tag{772}$$

and, thus, the classical Schrödinger equation again (regarding its solution see sub-section above).

5.18.28 References for Appendix R

1. D. Hilbert, Die Grundlagen der Physik, Teil 1, *Göttinger Nachrichten*, 1915, 395–407.
2. N. Schwarzer, *The World Formula: A Late Recognition of David Hilbert's Stroke of Genius*, Jenny Stanford Publishing, ISBN: 9789814877206.

3. A. Einstein, Grundlage der allgemeinen Relativitätstheorie, *Annalen der Physik*, 49, ser. 4, 769–822.
4. N. Schwarzer, *The Math of Body, Soul, and the Universe*, Jenny Stanford Publishing, ISBN 9789814968249.
5. N. Schwarzer, *The Theory of Everything—Quantum and Relativity Is Everywhere—A Fermat Universe*, Jenny Stanford Publishing, 2020, ISBN-10: 9814774472.
6. C. A. Sporea, *Notes on f(R) Theories of Gravity*, 2014, arxiv.org/pdf/1403.3852.pdf.
7. N. Schwarzer, *How Systems Evolve—A Red Pill Course on Evolution*, self-published, ASIN: B0B6RGBKQ4, Amazon Digital Services, 2022, Kindle.
8. N. Schwarzer, *Mathematical Psychology—Tools for the Understanding, Simulation, Mitigation and Potential Prevention of Mass Formation Psychoses*, self-published, ASIN: B0BC5T1HB4, Amazon Digital Services, 2022, Kindle.
9. N. Schwarzer, *Towards the Metric Quark—Derivation of More Spin-1/2-Objects from the Einstein-Hilbert Action*, self-published, Amazon Digital Services, B0BWDQDBFW, 2023, Kindle.
10. N. Schwarzer, *The Metric Electron—Derivation of Spin-1/2-Objects from the Einstein-Hilbert Action*, self-published, ASIN: B0BTPQM26X, Amazon Digital Services, 2023, Kindle.
11. N. Schwarzer, *The Foreseeable Quantum Enigma Disaster, Part 10b of Medical Socio-Economic Quantum Gravity*, self-published, Amazon Digital Services, 2021, Kindle, ASIN: B096PJRNTK.
12. N. Schwarzer, *The Quantum Gravity War—How Will the Nearby Unification of Physics Change Future Warfare?*, Jenny Stanford Publishing, ISBN: 9789814968584.
13. N. Schwarzer, *Quantum Gravity Systems Science*, self-published, ASIN: B0BRL8SKSB, Amazon Digital Services, 2023, Kindle.
14. N. Schwarzer, *Why Tucker Carlson Is Perfectly Right about the WEF-People—A Mathematical Proof*, self-published, ASIN: B0BSTGKC6F, Amazon Digital Services, 2023, Kindle.
15. N. Schwarzer, *Derivation of the Harmonic Oscillator from the Einstein-Hilbert Action—The Metric Harmonic Oscillator*, self-published, ASIN: B0BSXJCJVB, Amazon Digital Services, 2023, Kindle.
16. N. Schwarzer, *Derivation of the Schrödinger Hydrogen from the Einstein-Hilbert Action—The Metric Schrödinger Hydrogen*, self-published, ASIN: B0BT9Y54XP, Amazon Digital Services, 2023, Kindle.

17. N. Schwarzer, *Derivation Newton's Gravity Law from Quantum Einstein Field Equations*, self-published, Amazon Digital Services, ASIN: B0BXPS2ZJH, 2023, Kindle.
18. N. Schwarzer, *Derivation Hawking Radiation of Black Holes from Quantum Einstein Field Equations*, self-published, Amazon Digital Services, ASIN: B0BXQ5ZG4Z, 2023, Kindle.
19. N. Schwarzer, *Quantum Gravity—Brief Derivation of Quantum Einstein Field Equations*, self-published, Amazon Digital Services, ASIN: B0BWTWZHP5, 2023, Kindle.
20. N. Schwarzer, *Quantum Gravity and Einstein's Famous Moon Problem—The Theory of Perspectivity*, self-published, Amazon Digital Services, ASIN: B0BXPQQWZN, 2023, Kindle.
21. N. Schwarzer, *Shouldn't Quantum Gravity Solve the 3-Generation Problem*?, self-published, Amazon Digital Services, ASIN: B0BXPXP33L, 2023, Kindle.
22. N. Schwarzer, *About Apparent Asymmetries in the 3-Generations*, self-published, Amazon Digital Services, ASIN: B0BXPYGMNF, 2023, Kindle.
23. N. Schwarzer, *The Universe's Half-Spin Deception Game*, self-published, Amazon Digital Services, ASIN: B0BXPKWPQZ, 2023, Kindle.
24. H. Haken, H. Chr. Wolf, *Atom- und Quantenphysik*, 4th ed., in German, Springer Heidelberg, 1990, ISBN 0-387-52198-4.
25. E. Schrödinger, Quantisierung als Eigenwertproblem (erste Mitteilung), *Ann. Phys.*, 1926, 384(4), 361–376.
26. E. Schrödinger, Quantisierung als Eigenwertproblem (zweite Mitteilung), *Ann. Phys.*, 1926, 384(6), 489–527.
27. P. A. M. Dirac, *The Quantum Theory of the Electron*, Published 1 February 1928. DOI: 10.1098/rspa.1928.0023.
28. C. Cohen-Tannoudji, B. Diu, F. Laloë, Franck, *Quantenmechanik 1&2*, 2. Auflage, Walter de Gruyter, Berlin—New York, 1999, ISBN-10: 3110626004.
29. N. Schwarzer, *Metric Interpretation of Quantum Theory—How Can Quantum Gravity Help Us to Understand Quantum Strangeness*, self-published, Amazon Digital Services, ASIN: B0BXQRKPV7, 2023, Kindle.
30. N. Schwarzer, *The Metric Potential—How Can Quantum Gravity Help Us to Understand the Origin of Forces*, self-published, Amazon Digital Services, ASIN: B0BXSHQZX1, 2023, Kindle.
31. N. Schwarzer, *The Dirac Miracle, Part 8a of Medical Socio-Economic Quantum Gravity*, self-published, Amazon Digital Services, ASIN: B0963Z1Z74, May 2021, Kindle.

32. N. Schwarzer, *How Einstein gives Dirac, Klein-Gordon and Schrödinger*, self-published, Amazon Digital Services, ASIN: B071K2Y4V2, 2017, Kindle.
33. N. Schwarzer, *Einstein Had It, But He Did Not See It—Part XXIV: A Variety of Solutions and the Dirac-Schwarzschild-Particle*, self-published, Amazon Digital Services, 2018, Kindle, ASIN: B0796GZX8R.
34. N. Schwarzer, *Einstein Had It, But He Did Not See It—Part XLIII: A Selection of Einstein-Dirac-Particles*, self-published, Amazon Digital Services, 2018, Kindle, ASIN: B07DPVRZ8B.
35. N. Schwarzer, *Einstein Already Had It, But He Did Not See It—Part LXIII: Einstein-Field-Equations = Dirac² (+) Klein-Gordon*, self-published, Amazon Digital Services, 2018, Kindle, ASIN: B07JVFV4HP.
36. N. Schwarzer, *Einstein Had It, But He Did Not See It—Part LXXX: Short Note on the Killing of Dirac*, self-published, Amazon Digital Services, 2018, Kindle, ASIN: B07NKZVF61.
37. N. Schwarzer, *Science Riddles—Riddle No. 16: How to Understand the Dirac Equation*?, self-published, Amazon Digital Services, 2019, Kindle, ASIN: B07VFW2Z3F.
38. N. Schwarzer, *Science Riddles—Riddle No. 17: How Einstein Becomes First Order and Goes Dirac—Can We Factorize the Einstein-Field-Equations*?, self-published, Amazon Digital Services, 2019, Kindle, ASIN: B07VV9FG7K.
39. N. Schwarzer, *Einstein Had It, But He Did Not See It—Part LXXXI: More Dirac Killing*, self-published, Amazon Digital Services, 2019, Kindle, ASIN: B07WW1G6N7.
40. N. Schwarzer, *Science Riddles—Riddle No. 23: Why Does the Dirac Equation Work So Well*?, self-published, Amazon Digital Services, 2019, Kindle, ASIN: B0818V32JY.
41. N. Schwarzer, *Brief Proof of Hilbert's World Formula—Dirac, Klein-Gordon, Schrödinger, Einstein, Evolution and the 2nd Law of Thermodynamics All from One Origin*, self-published, Amazon Digital Services, 2020, Kindle, ASIN: B08585TRB8.
42. N. Schwarzer, *The New Dirac Equation: Originally Postulated, Now Geometrically Derived and—Along the Way—Generalized to Arbitrary Dimensions and Arbitrary Metrics Space-Times*, self-published, Amazon Digital Services, 2020, Kindle, ASIN: B085636Q88.
43. N. Schwarzer, *The Metric Dirac Equation Revisited and the Geometry of Spinors, Part 8 of Medical Socio-Economic Quantum Gravity*, self-

published, Amazon Digital Services, April 2021, Kindle, ASIN: B08Y96TL3D.

44. E. Ley-Koo, R. E. Wang, Dirac equation in orthogonal curvilinear coordinates, *Revista Mexicana de Fisica*, 1988, 34(2), 206–919, https://rmf.smf.mx/pdf/rmf/34/2/34_2_296.pdf.
45. T. Brennan, *General Relativity and the Dirac Equation*, 2020, https://quantumthom.com/GenRelAndDiracEqH2.pdf.
46. Ch. W. Misner, K. S. Thorne, J. A. Wheeler, *Gravitation*, 20th ed., W. H. Freeman and Company, New York, 1997, ISBN 0-7167-0344.
47. https://en.wikipedia.org/wiki/Dirac_equation_in_curved_spacetime.
48. N. Schwarzer, *Science Riddles—Riddle No. 11: What Is Mass*?, self-published, Amazon Digital Services, 2019, Kindle.
49. N. Schwarzer, *Science Riddles—Riddle No. 12: Is There a Cosmological Spin*?, self-published, Amazon Digital Services, 2019, Kindle.
50. N. Schwarzer, *Science Riddles—Riddle No. 13: How to Solve the Flatness Problem*?, self-published, Amazon Digital Services, 2019, Kindle.
51. W. Pauli, Über den Zusammenhang des Abschlusses der Elektronengruppen im Atom mit der Komplexstruktur der Spektren 1925, *Zeitschrift für Physik*, 31, 765–783, Bibcode:1925ZPhy...31..765P. doi:10.1007/BF02980631.
52. N. Schwarzer, *Einstein Had It, But He Did Not See It—Part LXXIX: Dark Matter Options*, self-published, Amazon Digital Services, 2016, Kindle.
53. https://de.wikipedia.org/wiki/Relativistisches_Wasserstoffproblem.
54. https://en.wikipedia.org/wiki/Higher-dimensional_gamma_matrices.
55. N. Schwarzer, *The Death of Schrödinger's Cat—How Quantum Gravity Solves the Problem of the Wave Function Collapse*, self-published, Amazon Digital Services, 2016, Kindle, ASIN: B0C2444JZX.
56. W. Greiner, Quantenmechanik—Teil 1 Einführung, in: *Theoretische Physik*, 5th ed., in German, Verlag Harri Deutsch, Frankfurt am Main, vol. 4, 1992, ISBN 3-8171-1206-8.
57. F. Schwabl, *Quantenmechanik*, 2nd ed., in German, Springer-Verlag Berlin Heidelberg New York, 1990, ISBN 3-540-52359-6.
58. https://en.wikipedia.org/wiki/Wave_function_collapse.
59. S. W. Hawking, Particle creation by Black Holes, *Commun. Math. Phys.*, 1975, 43, 199–220 (projecteuclid.org, PDF; 2,8 MB).
60. N. Schwarzer, *Einstein Had It, But He Did Not See It—Part LXXXII: Half Spin Hydrogen*, self-published, Amazon Digital Services, 2019, Kindle.

61. N. Schwarzer, *Understanding Quantum Tunneling—How Can Quantum Gravity Help Us to Get the Gist about this Strange Phenomenon*, Amazon Digital Services, 2023, Kindle, ASIN: B0C241FQH3.
62. N. Schwarzer, *Life After Death—What Does Quantum Gravity Say*, Amazon Digital Services, 2023, Kindle, ASIN: B0CB5WS4DD.
63. N. Schwarzer, *Is Mass Really Only a Scalar? How a Quantum Gravity Derived Dirac Equation Might Say Otherwise*, self-published, Amazon Digital Services, 2023, Kindle, ASIN: B0CCQQ4B7G.
64. N. Schwarzer, *Hugh Everett's Multiverse—How Quantum Gravity Revives an Old Idea and—Perhaps—Gives It a Second Life*, self-published, Amazon Digital Services, 2023, Kindle, ASIN: B0CCQWC6YX.
65. N. Schwarzer, *What Was Before Time?—How Quantum Gravity Might Give an Answer*, Amazon Digital Services, 2023, Kindle, ASIN: B0CB3ZKKNL.
66. N. Schwarzer, *Cracking the Black Hole Code—Can the Fall of the Covariance Principle Solve the Information and the Singularity Problem*, Amazon Digital Services, 2023, Kindle, ASIN: B0CB41SR1C.
67. N. Schwarzer, *Strings, Branes, Friedmanns and the Matryoshka Universe, Part 7D of Medical Socio-Economic Quantum Gravity*, self-published, Amazon Digital Services, December 2022, Kindle.
68. N. Schwarzer, *Towards Quantum Einstein Field Equations, Part 7 of Medical Socio-Economic Quantum Gravity*, self-published, Amazon Digital Services, December 2020, Kindle, ASIN: B08NG6H22X.
69. N. Schwarzer, *Dirac's Little Spinor Problem—Can Quantum Gravity Explain the Occurrence of this Postulated Object*?, self-published, ASIN: B0C3QWJSQS, Amazon Digital Services, 2023, Kindle.
70. N. Schwarzer, *How Fermions Get Massy—Can Quantum Gravity Answer this Question*?, self-published, Amazon Digital Services, ASIN: B0C5TM9ZCG, 2023, Kindle.
71. N. Schwarzer, *Quantum Gravity Thermodynamics—And It May Get Hotter*, www.amazon.com, ASIN: B07XC2JW7F.
72. N. Schwarzer, *Science Riddles—Riddle No. 20: Second Law of Thermodynamics—Where Is Its Fundamental Origin*?, www.amazon.com, ASIN: B07Y79BTT9s.
73. N. Schwarzer, *Science Riddles—Riddle No. 21: Evolution—Where Is Its Fundamental Origin*?, www.amazon.com, ASIN: B07YKL37DL.

74. N. Schwarzer, *Quantum Gravity Thermodynamics II—Derivation of the Second Law of Thermodynamics and the Metric Driving Force of Evolution*, www.amazon.com, ASIN: B07XWPXF3G.
75. N. Schwarzer, *Brief Proof of Hilbert's World Formula—Dirac, Klein-Gordon, Schrödinger, Einstein, Evolution and the 2nd Law of Thermodynamics All from One Origin*, self-published, Amazon Digital Services, 2020, Kindle, ASIN: B08585TRB8.
76. S. Noiré (T. Bodan, co-author), *The Womanizer and His Image of Diseases—About Love Fields and Isolation Traps—A Victim Reveals, Illustrated Version*, self-published, BoD Classic, ISBN 9783756841370.

5.19 Appendix S: Hate or Love, One and the Same Source for Both?

It appears—depending on the personal ability to see humor as a skill to "still laugh"—either as a bad joke or as a cruel connection at least, however, as a funny irony that it is the math of a potentially all-destructive vacuum bomb (Fig. S1) that this author has "accidentally" derived for completely different reasons in a previous publication [62], which gives us the suitable structures also for the individual love and emotion fields. The irony here lies in the fact that the same emotional structure, which can produce field orbitals of love, can also be applied for the manifestation of animosity and hate and with this, the mathematical connection to the "Big Bang Bomb" [61] becomes almost bone-chilling.

Of course, as we already elaborated in [62], using such a bomb makes no sense in combat with a participant who intends to live after, but as such a bomb could be the ultimate threat (terroristic or just as fundamental equalizer or doomsday weapon), one should better know about it. So, it seems that such a thing is extremely powerful—YES indeed—but also quite useless if one does not intend to blow up the whole universe or use it as a "doomsday bomb" or ultimate threat (like the Israelis with the a-bomb they—so they claim officially—don't have). However, with the mathematical connection between love and animosity fields and such a device, as it will be worked out in this appendix, established, one cannot ignore the strange realization about hate and love sometimes also being of quite destructive character should always better be treated with quite some respect. It does probably not need the example of Helene of Troy to point this out. The simple "everyday" observation of the destructiveness of certain love fields [63] and the often inevitable decomposition of families with the subsequent suffering and often significant chance reduction for the children reveals the love bomb connection to everybody quite obviously.

Figure S1 Cover from [61].

But isn't love also able to produce something good?

Of course, and for this we think it is worthwhile to reconsider the operated and steered version of the Big Bang Bomb, which then, obviously, is not a bomb anymore.

So, how about a well-controlled bomb, which is not a bomb but just a trigger? ... a trigger for the energy residing at each and every point in space-time (no mathematical but Planck points are being meant here) and which can be set free on demand whenever one wishes to? ... and what could we potentially learn with respect to our love-field bomb there?

Yes indeed, this sounds too phantastic to be true, but shouldn't we better make sure that we know about such an option in case there is such a possibility? After all, there must have been a trigger which had set loose the whole Big Bang story of our universe. So why should this not be possible a second, a third or an X time? And why—when considering the love or hate fields—should this not also be possible "in the small," respectively, in the world of thoughts or the realm of the mind?

"Because there is no suitable metric solution allowing for such concentrated energy states," critics might answer.

Well, this may have been true until the discovery of the scaled metric and the subsequent quantum gravity field equations. Apart from the Big Bang, namely, we meanwhile have another piece of circumstantial evidence for the possible existence of such confined "space-time energy" within the Planck volumina of our universe. We simply need to observe the so-called ρ-sphere-metrics we have discussed in the appendix of chapter "The Quantum Gravity Turing Machine" in [62] (see also section "2.7 The Size of a Thought" and appendix Q). There we found the possibility for the existence of metric objects of spherical symmetry with increasing mass for decreasing size.

Thus, at least theoretically, such objects can exist and then ...

Space itself is the bomb (or contains it), and tickled[9] with the right frequencies, parts of it can be made to blow up in a controlled manner.

The theory seems to predict the existence of such a bomb when

(a) there is a Higgs field;
(b) there is a photonic background.

And there is also the additional circumstantial evidence of a space-time, which seems to expand when being compressed (see [64] and appendix F in the book [62]).

---

[9]With respect to the bomb device, one simply imagines it like a laser being pointed at the spot where you would like to have the epi-center of the explosion. Then the character of this "laser" could also decide upon the amount of energy being set free (possibly reaching from just enough to make a human head to burst, via incoming warheads to blow up in a great enough distance from your position, to a complete army being wiped out in just one "tickling" stroke). With respect to the love and hate field problem, one probably assumes the steering mind to have some means to keep the "thing" under control.

Of course, the concept, if being controllable, could also lead to an inexhaustible source of energy, whereby the latter is just harvested out of space-time. Regarding "the world of thoughts and feelings" it appears quite appropriate to remind oneself of "the power of love," which historically turned so often into a quite destructive force that probably no further elaboration is needed in order to illustrate the connection.

Thus, we are going to rederive the mathematical principles of the Big Bang Bomb and more in the next sub-sections of this appendix. We will see that—most interestingly—also quite some options for the so-called dark matter or dark energy can be obtained. With respect to the latter, we immediately associate many other aspects, like the involuntary thoughts or the unconscious mind, for our "mathematical psychology" and—again—it is almost frightening to see the cosmic scale in the individual universe of a single mind.

5.19.1 The Big Bang Bomb as a Simple Quantum Gravity Vacuum Solution

Starting with the following metric:

$$g_{\alpha\beta}=\begin{pmatrix} -H^2\cdot c^2 & 0 & 0 & 0 \\ 0 & A^2 & 0 & 0 \\ 0 & 0 & B^2\cdot r^2 & 0 \\ 0 & 0 & 0 & D^2\cdot\sin[\vartheta]^2\cdot r^2 \end{pmatrix} \tag{773}$$

and demanding a vanishing cosmological constant

$$\Lambda = 0, \tag{774}$$

we have shown in [65] and the appendices above that the resulting Ricci scalar R^* of the transformed metric:

$$R^* = R^*_{\alpha\beta}G^{\alpha\beta} = \begin{pmatrix} \Gamma^{\sigma}_{\alpha\beta,\sigma}g^{\alpha\beta} - \Gamma^{\sigma}_{\beta\sigma,\alpha}g^{\alpha\beta} - \Gamma^{\mu}_{\sigma\alpha}\Gamma^{\sigma}_{\beta\mu}g^{\alpha\beta} + \Gamma^{\sigma}_{\alpha\beta}\Gamma^{\mu}_{\sigma\mu}g^{\alpha\beta} \\ +\frac{F'}{F}(1-n)\Delta f + \frac{f_{,\alpha}f_{,\beta}g^{\alpha\beta}(1-n)}{4F^2}\left(4FF''+(F')^2(n-6)\right) \end{pmatrix}\frac{1}{F}$$

$$= \left(R + \frac{F'}{F}(1-n)\Delta f + \frac{f_{,\alpha}f_{,\beta}g^{\alpha\beta}(1-n)}{4F^2}\left(4FF''+(F')^2(n-6)\right)\right)\frac{1}{F}$$

$$\text{with:}\quad F=F[f];\quad F'=\frac{\partial F[f]}{\partial f};\quad F''=\frac{\partial^2 F[f]}{\partial f^2} \tag{775}$$

and the corresponding 4D wrapper-function $F[f=f[t,r,\vartheta,\varphi]]=(C_1+f)^2$ reads:

$$R^* = \frac{1}{F[f]^3}\cdot\left(\begin{array}{c}\left(C_{N1}\cdot\left(\frac{\partial F[f]}{\partial f}\right)^2 - C_{N2}\cdot F[f]\cdot\frac{\partial^2 F[f]}{\partial f^2}\right)\cdot\left(\tilde{\nabla}_g f\right)^2 \\ -C_{N2}\cdot F[f]\cdot\frac{\partial F[f]}{\partial f}\cdot\Delta_g f\end{array}\right)$$

$$=-C_{N2}\cdot\frac{\frac{\partial F[f]}{\partial f}\cdot\Delta_g f}{F[f]^2}\xrightarrow{4D}=-3\cdot\frac{(C_1+2\cdot f)}{\frac{1}{4}\cdot(C_1+2\cdot f)^2}\cdot\Delta_g f=-\frac{12\cdot\Delta_g f}{C_1+2\cdot f}, \tag{776}$$

$$\Rightarrow\qquad -\frac{C_1+2\cdot f}{12}\cdot R^* = \Delta_g f = \Delta_{3D-g} f - \frac{f^{(2,0,0,0)}}{H^2\cdot c^2}. \tag{777}$$

Now we substitute in Eq. (777) as follows:

$$-\frac{C_1+2\cdot\tilde{f}}{12}\cdot R^* = \Delta_g \tilde{f} = \Delta_{3D-g}\tilde{f} - \frac{\tilde{f}^{(2,0,0,0)}}{H^2\cdot c^2}$$

$$f\equiv\frac{C_1+2\cdot\tilde{f}}{12}\quad\Rightarrow\quad -f\cdot R^* = \Delta_g\left(6\cdot f-\frac{C_1}{2}\right)$$

$$=6\cdot\Delta_g f = 6\cdot\left(\Delta_{3D-g} f - \frac{f^{(2,0,0,0)}}{H^2\cdot c^2}\right). \tag{778}$$

Applying the separation approach $f[t,r,\vartheta,\varphi] = g[t] * h[r] * Y[\vartheta] * Z[\varphi]$ with the separation parameters C_t and m (c.f. sub-section "Flat Space Situation" in [66]) in Eq. (777) gives us the following immediate solution for $g[t]$ and $Z[\varphi]$:

$$g[t]=C_{t1}\cdot\cos[H\cdot c\cdot C_t\cdot t]+C_{t2}\cdot\sin[H\cdot c\cdot C_t\cdot t], \tag{779}$$

$$Z_m[\varphi]=C_{\varphi1}\cdot\cos[D\cdot m\cdot\varphi]+C_{\varphi2}\cdot\sin[D\cdot m\cdot\varphi]. \tag{780}$$

The two remaining differential equations can be constructed as follows:

$$-\frac{m^2}{\sin[\vartheta]^2}+\frac{\cot[\vartheta]Y'[\vartheta]+Y''[\vartheta]}{B^2 Y[\vartheta]}=l(1+l), \tag{781}$$

$$\frac{l(1+l)}{r^2}-\frac{(2\cdot h'[r]+r\cdot h''[r])}{B^2\cdot r\cdot h[r]}=\frac{B^2\cdot C_t^2}{A^2}-\frac{R^*}{6}. \tag{782}$$

While the first equation of the two gives the usual solution with the Legendre Polynomials, as also known from the Schrödinger hydrogen solution [67], only in slightly generalized manner (watch the asymmetry-parameter B):

$$Y_k^m[\vartheta]=C_{Pv}\cdot P_k^{B\cdot m}[\cos[\vartheta]]+C_{Qv}\cdot Q_k^{B\cdot m}[\cos[\vartheta]];\quad k=\frac{1}{2}\left(\sqrt{1+4B^2 l(1+l)}-1\right)$$

$$B^2 l(1+l)\equiv L(1+L), \tag{783}$$

we have a rather non-Schrödinger-like solution with respect to the radius coordinate in the case of $R^* = 0$:

$$h[r]=C_j\cdot j_k\left[\frac{B^2}{A}\cdot C_t\cdot r\right]+C_y\cdot y_k\left[\frac{B^2}{A}\cdot C_t\cdot r\right];\quad k=\frac{1}{2}\left(\sqrt{1+4B^2 l(1+l)}-1\right). \tag{784}$$

The reader can find an extensive discussion on this solution in [65, 66]. In the case $l = 0$, we can simplify the last two results to:

$$Y[\vartheta]=C_{\vartheta 1}\cdot\cosh\left[m\left(\ln\left[\cot\left[\frac{\vartheta}{2}\right]\right]\right)\right]-i\cdot C_{\vartheta 2}\cdot\sinh\left[m\left(\ln\left[\cot\left[\frac{\vartheta}{2}\right]\right]\right)\right], \tag{785}$$

$$h[r]=\frac{C_{r1}}{r}\cdot e^{-i\cdot r\cdot\sqrt{C_t^2}}+\frac{C_{r2}}{r}\cdot e^{i\cdot r\cdot\sqrt{C_t^2}} \tag{786}$$

where we have set $H = A = B = D = 1$.

In [65], we answered the question whether we might apply inner-outer approaches in connection with solution (786). A setting of the kind $C_{r2} = -C_{r1}$ allows us to avoid any $r = 0$-singularity within the function $h[r]$. The rest is just oscillations along the radius coordinate and, if connected with the corresponding time function $g[t]$ (779), thereby choosing $C_{t2} = \pm i^* C_{t1}$, allows us to obtain radial photonic waves:

$$\begin{aligned} g[t]\cdot h[r]&=\frac{C_{r1}\cdot C_{t1}}{r}\cdot e^{\pm i\cdot c\cdot t\cdot C_t}\left(e^{-i\cdot r\cdot C_t}-e^{i\cdot r\cdot C_t}\right)\\ &=\frac{C_{r1}\cdot C_{t1}}{r}\cdot\left(e^{-i\cdot C_t\cdot(r\pm c\cdot t)}-e^{i\cdot C_t\cdot(r\pm c\cdot t)}\right). \end{aligned} \tag{787}$$

But the number of options does not end here, because we could simply demand to have a boundary $r = b$ at which our radial photonic structure should end its oscillations. For this, we simply require the evaluation of the zeros of the following equation:

$$\frac{\left(e^{-i\cdot b\cdot C_t}-e^{i\cdot b\cdot C_t}\right)}{b}=0 \tag{788}$$

with respect to C_t. In result, we will obtain a series of quantized $C_t \to C_j$, with j being a quantum number. Please note that we could choose the boundary condition (788) in such a way that we see Eq. (787) as the inner solution and the Schwarzschild metric function for the time component as the outer one (c.f. corresponding section "Inner and Outer Approaches" in [65], where we also introduced a generalized Schwarzschild solution). One might try a function $g_r[r]$ within the outer solution. The boundary conditions should then read:

$$C_r \cdot \frac{\left(\mathrm{e}^{-i \cdot b \cdot C_t} - \mathrm{e}^{i \cdot b \cdot C_t}\right)}{b} = 1 - \frac{r_s}{g_r[b]}$$

$$C_r \cdot \left[\frac{i \cdot \left(\mathrm{e}^{-i \cdot b \cdot C_t} - \mathrm{e}^{i \cdot b \cdot C_t}\right)}{b} - \frac{\left(\mathrm{e}^{-i \cdot b \cdot C_t} - \mathrm{e}^{i \cdot b \cdot C_t}\right)}{b^2}\right] = \frac{r_s}{g_r[b]^2} \cdot g'_r[b]. \tag{789}$$

Direct solution with $g_r[r]$, however, does not provide us with a suitable radius function for the outer region:

$$r_s \cdot \left[1 - C_r \cdot \frac{\left(\mathrm{e}^{-i \cdot r \cdot C_t} - \mathrm{e}^{i \cdot r \cdot C_t}\right)}{r}\right]^{-1} = g_r[r]. \tag{790}$$

In [65], we therefore only searched for asymptotic solutions, which most surprisingly still led to forms perfectly mirroring the experimental findings in connection with the Schwarzschild metric.

Now we want to get a feeling about the possible values for these C_j, and by taking the Dirac solution for a particle at rest (e.g., [7]) we find that:

$$C_j = \frac{M \cdot c}{\hbar}. \tag{791}$$

In order to stay in the metric picture, we substitute the mass M by the Schwarzschild radius $r_s = \frac{2 \cdot M \cdot G}{c^2}$ and obtain:

$$C_j = r_s \cdot \frac{c^3}{2 \cdot \hbar \cdot G} = \frac{r_s}{2 \cdot \ell_P^2}. \tag{792}$$

Remember that ℓ_P denotes the Planck length of about 1.616×10^{-35} meters.

Now we assume a space full of objects, being made just out of spherical photons of the kind (787), which all end at a certain maximum radius $r = b$. The idea for such a construction has been already discussed in a series of partially fictive, partially non-fictive stories (e.g., see the introduction part of [4]). The spherical objects were named "Friedmanns" as they resemble minuscule universes. It was also shown within the series of "Einstein had it..." (see the reference lists of this book) that such a structure quite easily explains the quantum properties observable in this universe (alternatively [4] and references therein).

Additionally, assuming that these little objects have a boundary which itself is just as small as the Planck length times a certain number q, giving the objects a diameter of $2 \cdot \ell_P \cdot q$, we end up with the condition $b = \ell_P \cdot q$. Setting $C_{r1} = i$ gives us the boundary condition and its solutions as follows:

$$\frac{2 \cdot \sin\left[C_j \cdot b\right]}{b} = 0 \Rightarrow C_{j,k} = \left\{\frac{2 \cdot \pi \cdot j}{b}, \frac{\pi + 2 \cdot \pi \cdot k}{b}\right\}; \quad j,k = 0,1,2,3,.... \tag{793}$$

This yields the subsequent lowest possible wave number ($j = k = 0$) to be:

$$C_{0,0} = \frac{\pi}{\ell_P \cdot q}, \tag{794}$$

which, according to the Dirac-exponent expressed in metric units (792), corresponds to a Schwarzschild radius as:

$$C_{0,0} = \frac{\pi}{\ell_P \cdot q} = \frac{r_s}{2 \cdot \ell_P^2} \quad \Rightarrow \quad r_s = \frac{2 \cdot \pi \cdot \ell_P}{q}. \tag{795}$$

The next three wave numbers would be:

$$C_{1,0} = \frac{2 \cdot \pi}{\ell_P \cdot q} = \frac{r_s}{2 \cdot \ell_P^2} \quad \Rightarrow \quad r_s = \frac{4 \cdot \pi \cdot \ell_P}{q}$$

$$C_{0,1} = \frac{3 \cdot \pi}{\ell_P \cdot q} = \frac{r_s}{2 \cdot \ell_P^2} \quad \Rightarrow \quad r_s = \frac{6 \cdot \pi \cdot \ell_P}{q}$$

$$C_{2,0} = \frac{4 \cdot \pi}{\ell_P \cdot q} = \frac{r_s}{2 \cdot \ell_P^2} \quad \Rightarrow \quad r_s = \frac{8 \cdot \pi \cdot \ell_P}{q}. \tag{796}$$

We immediately see that in order to avoid the formation of Black Holes, which would appear the moment we would have the condition $r_s \geq \ell_P$, we need to have certain minimum object sizes (q-numbers) in order to allow the object to hold a certain photon of a given wave number (or consisting of such). Thus, the simplest object only containing one wave with $C_{0,0}$ would require a $q > 2 * \pi$ in order to avoid to turn into a Black Hole. This corresponds to a minimum size of the object as follows:

$$C_{0,0}: \; b > r_s \quad \Rightarrow \quad b > 2 \cdot \pi \cdot \ell_P. \tag{797}$$

In conclusion, a universe being considered as grain-like with small photonic spheres to make up its fundamental building blocks restricts the smallest possible size of these structures to a minimum of at least two times π times the Planck length. With the size of these structures close to this limit[10], the vacuum energy density of the universe would be almost equal to the Planck energy density of 4.633×10^{113} J/m^3. Increasing the object-sizes b leads to a decrease of the corresponding vacuum energy density proportional to b^{-3}.

Thus, spherical photonic objects with quantum solutions of the type (787) would contain significant amounts of energy even at sizes of many thousand times the Planck length (please also see footnote below). On the other hand, it is known that by using the upper limit of the cosmological constant, the vacuum energy of free space has been estimated to be 10^{-9} J/m^3.

Please note therefore that the assumption of small structures as fundamental building blocks should not be restricted to such photonic ensembles. However, there would also be another, perhaps more interesting, if not to say elegant, explanation for the obvious discrepancy of the observed cosmological vacuum energy density and the quantum theoretically necessary one. One might simply assume that the spherical photons are standing waves being trapped inside the boundary $r = b$. With the parameter settings already introduced in order to avoid r=0-singularities, we automatically obtain standing waves of the types:

$$g[t]\cdot h[r]=\frac{C_{r1}\cdot C_{t1}}{r}\cdot \mathrm{e}^{\pm i\cdot c\cdot t\cdot C_t}\left(\mathrm{e}^{-i\cdot r\cdot C_t}\text{-}\mathrm{e}^{i\cdot r\cdot C_t}\right)$$

$$=\frac{C_{r1}\cdot C_{t1}}{r}\cdot \mathrm{e}^{\pm i\cdot c\cdot t\cdot C_j}\cdot 2\cdot \sin\left[r\cdot C_j\right]. \tag{798}$$

With C_j chosen in accordance with condition (793), we could easily have ensured an inner-outer constellation of the kind:

$$g[t]\cdot h[r]=\begin{cases}\dfrac{C_{r1}\cdot C_{t1}}{r}\cdot \mathrm{e}^{\pm i\cdot c\cdot t\cdot C_j}\cdot 2\cdot \sin\left[r\cdot C_j\right] & \text{for} \quad r\le b\\ 0 \quad \text{for} \quad r>b. \end{cases} \tag{799}$$

As a by-product even higher wave numbers would now be possible as long as they are following the boundary condition (793) and sum up to standing waves (798) being completely reflected at the boundary $r = b$. Then it might well be that the true vacuum energy density would be even higher than just

[10]Please note that here we do not consider the possibility of solutions to the quantum Einstein field equations with completely different scale and size behavior as derived in [5, 41, 42, 68, 69].

the Planck energy density, but as this energy is trapped inside our photonic spheres, it has no effect on anything bigger than $r = b$... this does at least hold for the physical means we have at hand so far. As a consequence, we might also see Black Holes just as such spherical photons with all their energy being trapped inside a sphere of the size $r \leq b$ and consisting of an ensemble of spherical photons all forming standing waves of the kind as given in Eq. (798).

And why can the waves not come out of the spherical photon-like objects?

Because—so the possibly simplest explanation—outside there is no substrate which could transport them. With the ordinary space being subdivided into just many photon-spheres of sizes $r = b$ (and potentially bigger), waves with wave numbers bigger than $C_t = C_j = 1/b$ will find no spatial ground on which they could propagate. Only waves of lengths of several Planck lengths find enough spatial (substrate) material to be supported and can exist within our usual space-time. Between the two scales, there is a gap ensuring the absence of certain ultra-energetic photons in our visible part of the universe. With such a constellation we, which is to say our visible universe(s), are sitting on an energy bomb millinillion times more powerful in each piece of Planck volume than our whole universe could bring together. One way to understand the nature of the "substrate" might simply be to assume additional dimensions which are curled up at very low scales (Planck scale). So, for instance, we find spherical photonic solutions in principle also higher-dimensional spaces and they are all sporting a radial dependency of the kind:

$$h[r] = \frac{C_J \cdot J_k\left[C_t \cdot r\right] + C_Y \cdot Y_k\left[C_t \cdot r\right]}{r^{\frac{n-1}{2}}}; \quad k = \frac{1}{2}\sqrt{1 + 4l(1+l) - 2n + n^2}. \qquad (800)$$

Here J, Y are the Bessel functions of the first and second kind, l is the momentum quantum number as before and n gives the number of angular coordinates forming the n+1-sphere. We see that with $C_Y = 0$ we can always construct a free-of singularities solution and the zeros of the J-Bessel functions ensure that $h[r = b] = 0$ can always be satisfied by a proper choice of C_t. Simply assuming that one of the angular coordinates/dimensions is compactified and does not allow the hyper-n+1-photon to propagate outside the n+1-sphere leads to the desired confinement of the photonic energy.

What happens if space has been compressed as in Black Holes?

In Black Holes, on the other hand, these photonic spheres are broken up to much smaller sub-structures (perhaps even something close to a continuum), allowing objects with wavelengths following condition $C_j > 1/b$ to exist and to freely propagate. In connection with the compactified dimensions, a Black Hole could easily be understood as a sphere of higher dimensions than the surrounding 3+1-dimensional space-time.

What does this mean in connection with the overall universe?

The same restriction probably also holds for the outside of our universe, where we would expect to find no wavelength smaller than $1/B$, with B now being the current radial size of our universe. This should at least hold as long as there is not another universe like ours (and existing outside ours) also holding a set of higher energy photons compared to its own surrounding vast—almost—nothingness.

The Universe as a Fractal Pattern

The attentive reader might already guess where this is leading us, namely to a self-similar structure with patterns of photon-spheres, potentially existing in higher-dimensional bubbles, actually being universes, inside bigger photon-spheres, inside bigger… and so on. As we already saw in [70] that the center of a Black Hole could be very well the perfect birthplace for a universe with the properties as those we find in our cosmos, we might find it less peculiar to understand the world as fractal. As in addition we have many other examples in many different fields pointing towards fractal structures and repetitive patterns [71, 72, 73, 74], we should not be surprised at all by our illustrative and—at least in the opinion of this author—rather obvious interpretation of the mathematical findings above.

And how to set this trapped energy free?

We better not even try to answer this question here… even if we knew it, which does not mean that we actually do know it.

And what has this to do with "Mathematical Psychology"?

Well, here we simply "answer" with just another question, respectively two questions to our dear readers, namely:

(1) Have you never experienced the perception or sensation of trapped intense feelings, thoughts or strong emotions "somehow enclosed inside the back of your mind"?

(2) And have you never come to the conclusion that some of these feelings, thoughts or emotions remain hidden, simply because the energy or impulse they might set free if being let loose would be disastrous?

Similarly, we could argue with respect to dark matter and dark energy, which we will therefore consider in the next sub-section of this appendix.

5.19.2 Ordinary and Potentially Dark Matter in Spherical Coordinates

In this section of appendix S, we intend to just bring together a few metric solutions, potentially fit to present certain particles or at least aspects of particles, which—in the intention of this book—would allow us to mathematically describe compactified—which is to say particle like—thoughts or emotions as orbitals... either to ensure a certain kind of individuality due to emotion field boundaries or in order to allow for the interaction with other individual's fields via orbitals with bonding capacity. Based on the similarity in chemistry, one might call the latter valence fields. Later on, we want to discuss these solutions under the pretext of finding those which are most prone to present either matter or dark matter or—perhaps—something else. Thereby, in order to avoid too much referencing through the whole book, we are going to repeat also some basics even though they might already have been brought up earlier in this book.

5.19.2.1 Flat space situation

The case with spatial spherical coordinates was partially investigated in connection with the Schwarzschild metric in [75] and with the Robertson–Walker metric in [45, 46]. Here we shall directly start with the flat space vacuum case:

$$g_{\alpha\beta}^{\text{flat}} = \begin{pmatrix} -c^2 & 0 & 0 & 0 \\ 0 & 1 & 0 & 0 \\ 0 & 0 & r^2 & 0 \\ 0 & 0 & 0 & r^2 \cdot \sin^2 \vartheta \end{pmatrix}. \tag{801}$$

For convenience, here we want to repeat the case $R^* = 0$, which was investigated in [75] and above (however, in a generalized form and a slightly different context). Here now, we also want to investigate a spherical symmetry with hyperbolic angular coordinates.

From Eq. (775) we obtain the simple 4D-Laplacian:

$$R^{*}=-3\cdot\frac{\left(C_{1}+\frac{C_{1}^{2}\cdot f}{2\cdot C_{2}}\right)}{\left(C_{1}\cdot f+\frac{C_{1}^{2}\cdot f^{2}}{4\cdot C_{2}}+C_{2}\right)^{2}}\cdot\Delta_{g}f=0$$

$$=\Delta_{g}f=\Delta_{\text{3D-sphere}}f-\frac{f^{(2,0,0,0)}}{c^{2}}. \tag{802}$$

Applying the separation approach $f[t,r,\vartheta,\varphi]$ = $g[t]$ * $h[r]$ * $Y[\vartheta,\varphi]$ in Eq. (802) gives us the solution for $g[t]$:

$$g[t]=C_{t1}\cdot\cos[c\cdot C_{t}\cdot t]+C_{t2}\cdot\sin[c\cdot C_{t}\cdot t], \tag{803}$$

and the spherical Bessel functions $j_n[z]$ and $y_n[z]$ for $h[r]$ as already derived in [75], only that here the solution reads as follows:

$$h[r]=C_{j}\cdot j_{k}[C_{t}\cdot r]+C_{y}\cdot y_{k}[C_{t}\cdot r]$$

$$k=\frac{\sqrt{1+4\cdot\omega^{2}}-1}{2}=\frac{\sqrt{1+4\cdot l\cdot(l+1)}-1}{2}=0,1,2,3,\ldots\text{ for }l=0,1,2,3,\ldots. \tag{804}$$

Solutions which do not produce singularities or do grow infinitely with respect to r are only found for the spherical Bessel function of the first kind $j_k[z]$ with the usual $\omega^2 = l * (l + 1)$ and l = 1,2,3, ... Also, as discussed in sub-section "The r_s = 0-Case Now without Feeding from Classical Quantum Theory" of [75] and very different from the Schrödinger hydrogen solution, we have no main quantum number. We are going to discuss this aspect further below, but before we can come to that, we need to cover a few other things first.

Please note that in addition to the usual spherical harmonics $Y_{l}^{m}[\vartheta,\varphi]$ for the usual spherical metric, we could also have a hyperbolical option, which can be given as follows (simply set $r_s = \Lambda = 0$ in order to have the Einstein–Hilbert vacuum situation there, too):

$$g_{\alpha\beta}^{hyp}=\begin{pmatrix} h\cdot g_{t}'[t]^{2}\cdot\alpha[r] & 0 & 0 & 0\\ 0 & \mathrm{a}\cdot G_{r}'[r]^{2}\alpha[r]^{-1} & 0 & 0\\ 0 & 0 & g_{22} & 0\\ 0 & 0 & 0 & g_{33}\end{pmatrix}$$

$$g_{22}=b\cdot \mathrm{g}_{\vartheta}'\left[c_{\vartheta}\vartheta\right]^{2} G_{r}\left[r\right]^{2}\cdot\begin{Bmatrix}\sinh\left[\mathrm{g}_{\vartheta}\left[c_{\vartheta}\vartheta\right]\right]^{-2}\\ \cosh\left[\mathrm{g}_{\vartheta}\left[c_{\vartheta}\vartheta\right]\right]^{-2}\end{Bmatrix}$$

$$g_{33}=d\cdot \mathrm{g}_{\varphi}'\left[c_{\varphi}\varphi\right]^{2}\cdot G_{r}\left[r\right]^{2}\cdot\begin{Bmatrix}\sinh\left[\mathrm{g}_{\vartheta}\left[c_{\vartheta}\vartheta\right]\right]^{-2}\\ \cosh\left[\mathrm{g}_{\vartheta}\left[c_{\vartheta}\vartheta\right]\right]^{-2}\end{Bmatrix}$$

$$a[r]=\left(C_1+\frac{r_s}{G_r[r]}+\frac{\Lambda}{3\cdot c^2}\cdot G_r[r]^2\right). \tag{805}$$

We also note that in contrast to the Schwarzschild metric that we have considered in [75], it does not matter which spherical harmonics we choose. Both forms:

$$Y_l^m\left[\vartheta,\varphi\right]=e^{i\cdot m\cdot\varphi}\cdot\left(C_{jP}\cdot P_{l_j}^{m_j}\left[\cos\vartheta\right]+C_{jQ}\cdot Q_{l_j}^{m_j}\left[\cos\vartheta\right]\right), \tag{806}$$

$$Y_l^m\left[\vartheta,\varphi\right]=e^{i\cdot m\cdot\varphi}\cdot\left(C_{jP}\cdot P_{l_j}^{m_j}\left[\tanh\vartheta\right]+C_{jQ}\cdot Q_{l_j}^{m_j}\left[\tanh\vartheta\right]\right), \tag{807}$$

being the spherical and the spherical-hyperbolical one, respectively, give reasonable distributions with the conditions for the quantum numbers m and l as demanded in the paragraph above.

Quite some discussion can be found in [75, 76]. There we also elaborated that the quantum solutions of our transformation are principally additive in the case of $R^* = 0$. It was demonstrated in [45, 46] that while the Einstein field equations are not additive with respect to the metric, we find that our differential equations for gravity quantum fields (802) are perfectly linear operations on f (if obtained with wrapper functions $F[f]$ fulfilling condition:

$$C_{N1}\cdot\left(\frac{\partial F[f]}{\partial f}\right)^2-C_{N2}\cdot F[f]\cdot\frac{\partial^2 F[f]}{\partial f^2}=0). \tag{808}$$

This gives us the option of superposition. Thus, applying the separation approach in the form of a sum of functions $f=\sum_{j=1}^{\Omega} f_j$, with all $f_j[t,r,\vartheta,\varphi] = g_j[t] * h_j[r] * Y_j[\vartheta,\varphi]$, and solving the angular part in the usual way (c.f. [67]) converts Eq. (802) to:

$$\Delta_g f=\Delta_g\sum_{j=1}^{\Omega} f_j=0. \tag{809}$$

Now we simply demand that each of the summands gives zero and have:

$$\Delta_g f = \Delta_g \sum_{j=1}^{\Omega} f_j = \sum_{j=1}^{\Omega} \Delta_g f_j = 0 \Rightarrow \Delta_g f_j = 0. \tag{810}$$

The separation calculation with our form $f = \sum_{j=1}^{\Omega} f_j$ from above has now simply to be performed for each function f_j separately. Subsequently we obtain a set of solutions $Y_{jl}^{m}[\vartheta,\varphi]$, $h_j[r]$ and $g_j[t]$ with parameters C_{jt} and, ω_j respectively, l_j and m_j as follows:

$$Y_{jl}^{m}[\vartheta,\varphi] = e^{i\cdot m_j\cdot\varphi}\cdot\left(C_{jP}\cdot P_{l_j}^{m_j}[\cos\vartheta] + C_{jQ}\cdot Q_{l_j}^{m_j}[\cos\vartheta]\right), \tag{811}$$

$$Y_{jl}^{m}[\vartheta,\varphi] = e^{i\cdot m_j\cdot\varphi}\cdot\left(C_{jP}\cdot P_{l_j}^{m_j}[\tanh\vartheta] + C_{jQ}\cdot Q_{l_j}^{m_j}[\tanh\vartheta]\right), \tag{812}$$

for the usual spherical metric and the spherical hyperbolical metric (805), respectively plus:

$$g_j[t] = C_{jt1}\cdot\cos\left[c\cdot C_{jt}\cdot t\right] + C_{jt2}\cdot\sin\left[c\cdot C_{jt}\cdot t\right] \tag{813}$$

for the time functions and:

$$h_{jl}[r] = C_{jj}\cdot j_{jk}\left[C_{jt}\cdot r\right] + C_{jy}\cdot y_{jk}\left[C_{jt}\cdot r\right];\quad k_j = \frac{\sqrt{1+4\omega_j^{\,2}}-1}{2} \tag{814}$$

for the radius functions. Of special interest are superpositions with respect to the quantum number *l*, which allow us to construct many radius distributions without compromising the time-dependency of the corresponding object (see [45, 46]). The outcome would be a series approach via the *l*-values of our spherical Bessel functions $h_l[r]$. This could still be combined with many frequency values C_{jt} as given in Eq. (814):

$$f_j^{m}[t,r,\vartheta,\varphi] = g_j[t]\cdot\sum_{l=0}^{\infty} C_{jj}[l]\cdot j_{jk}\left[C_{jt}\cdot r\right]\cdot Y_{jl}^{m}[\vartheta,\varphi];$$

$$k = k_l = \frac{\sqrt{1+4\cdot l\cdot(l+1)}-1}{2} = 0,1,2,3,\ldots \text{ for } l = 0,1,2,3,\ldots. \tag{815}$$

This time, however, they come as clearly discrete values C_{jt} with respect to the frequencies in order to guarantee timely stable outcomes. Even plane waves can be constructed. So, it was shown by H. Murayama (http://hitoshi.berkeley.edu/221b/bessel.pdf) that an infinite sum of our spatial solutions can give a plane wave in the *z*-direction as follows:

$$e^{i\cdot C_{jt}\cdot z}=\sum_{l=0}^{\infty}(2\cdot l+1)\cdot i^{l}\cdot j_{jk}\left[C_{jt}\cdot r\right]\cdot P_{l}\left[\cos[\vartheta]\right];$$

$$k=k_{l}=\frac{\sqrt{1+4\cdot l\cdot(l+1)}-1}{2}=0,1,2,3,\ldots \text{ for } l=0,1,2,3,\ldots. \quad (816)$$

The separation parameter C_{jt} here acts as a wave number and we can have many such waves simultaneously. The interesting aspect here would be that we have obtained a pure z-axis-dependency for a wave function out of a series of spherical solutions. The complete wave function including the time-coordinate would be:

$$f[t,z]=e^{i\cdot C_{jt}\cdot z}\cdot\left(C_{jt1}\cdot\cos\left[c\cdot C_{jt}\cdot t\right]+C_{jt2}\cdot\sin\left[c\cdot C_{jt}\cdot t\right]\right), \quad (817)$$

which clearly gives an electromagnetic wave (e.g., c.f. [3], chapter "The Photon"). With respect to our "Mathematical Psychology," the finding shows us how just one single individual with a strong enough mindset and determination and without the help of other individuals (idea-spreaders) can make its own ideas appear as quasi-homogeneous and plane wave distributed as if just being there and not coming from this one source.

Now we want to consider situations where a radial boundary at $r = b$ should be present demanding the Sturm–Liouville boundary value problem of the kind [77]:

$$h[b]\cdot\cos[\beta]-b^{2}\cdot h'[b]\sin[\beta]=0$$

$$\text{with: } h[r=b]=h[b]=j_{l}\left[C_{t}\cdot(r=b)\right]$$

$$\Rightarrow \quad j_{l}\left[C_{t}\cdot b\right]\cdot\cos[\beta]-C_{t}\cdot b^{2}\cdot j'_{l}\left[C_{t}\cdot b\right]\cdot\sin[\beta]=0. \quad (818)$$

Thereby the parameter β decides upon the character of the boundary condition. For example, a pure zero-value boundary would clearly have $\beta = 0$. Automatically, with the condition (818), we obtain a restriction or quantization condition for the parameter C_t as now only some of the infinitely many "wave numbers" C_t would allow to satisfy it. We may therefore name the suitable C_t fulfilling condition (818) C_{tj}. As these parameters are themselves only valid for a certain l-number, we shall leave the separation index t aside and have the quantum parameter C_{jl}.

Thus, restriction of the size of the object to a certain radial size of $r = b$ or simply defining certain boundary conditions there automatically gives us a radial quantization.

The evaluation of the solutions to Eq. (818) would give us an infinite number of nonnegative values $z_{jl} = C_{jl} * b$. The n-th eigenfunction is then of the kind $j_l[C_{jl} * b]$ and it was presented in [78] that the following normalizations:

$$N_j^l = \frac{b^3}{2}\left(j_l'\left[z_{jl}\right]^2 + \frac{j_l'\left[z_{jl}\right]\cdot j_l\left[z_{jl}\right]}{z_{jl}} + \left(1 - \frac{l\cdot(l+1)}{z_{jl}^2}\right) j_l\left[z_{jl}\right]^2 \right) \quad (819)$$

allow the expansion of any function $f[r,\vartheta,\varphi]$ defined on $r \le b$ in the following way:

$$f[r,\vartheta,\varphi] = \sum_{j=1}^{\infty}\sum_{l=0}^{\infty}\sum_{m=-l}^{l} S_{jlm}\cdot\frac{j_l\left[C_{jl}\cdot r\right]}{\sqrt{N_j^l}}\cdot Y_{jl}^m\left[\vartheta,\varphi\right] \equiv \sum_{j=1}^{\infty}\sum_{l=0}^{\infty}\sum_{m=-l}^{l}\Psi_{jlm}\left[r,\vartheta,\varphi\right]$$

$$S_{jlm} = \int_0^b\int_0^\pi\int_0^{2\pi} f\left[r,\vartheta,\varphi\right]\cdot\Psi^*_{jlm}\left[r,\vartheta,\varphi\right]\cdot r^2\cdot\sin[\vartheta]\cdot d\varphi\cdot d\vartheta\cdot dr. \quad (820)$$

The normalizations (819) simplify significantly in the case of zero-value boundary conditions:

$$N_j^l = \frac{b^3}{2}\cdot j_{l+1}\left[z_{jl}\right]^2 \quad (821)$$

and derivative boundary conditions:

$$N_1^0 = \frac{b^3}{2};\quad N_j^l = \frac{b^3}{2}\left(1 - \frac{l\cdot(l+1)}{z_{jl}^2}\right) j_l\left[z_{jl}\right]^2. \quad (822)$$

Now within this universe we assume to have a limit for the highest possible wave number (which automatically correlates with the highest possible energy density). This restricts our options for the choice of C_{jl} to a certain maximum number. Thus, knowing the boundary $r = b$ and its conditions would allow us to evaluate the highest possible C_{jl}. It is also clear that with the main quantum number being restricted to a certain maximum, not every function can be presented by our set of eigenfunctions as shown in Eq. (820). With j being restricted—due to limited energy and thus, values for C_{jl}—the $f[r,\vartheta,\varphi]$ cannot be arbitrary but have to be finite sums of $\Psi_{jlm}\left[r,\vartheta,\varphi\right]$. As before, in order to get a feeling about the possible values for C_{jl}, we recapitulate the Dirac solution for a particle at rest (e.g., [7]) and find that:

$$C_{jl} = \frac{M\cdot c}{\hbar}. \quad (823)$$

In order to stay in the metric picture, we substitute the mass m by the Schwarzschild radius $r_s = \frac{2 \cdot M \cdot G}{c^2}$ and obtain:

$$C_{jl} = r_s \cdot \frac{c^3}{2 \cdot \hbar \cdot G} = \frac{r_s}{2 \cdot \ell_P^2}. \tag{824}$$

Here ℓ_P denotes the Planck length of about 1.616×10^{-35} meters.

With this finding, we should have all the ingredients we need to reconsider the Bekenstein thought experiment [49, 50], which we already covered in this book. For generalization, it should be pointed out that by taking into account a special scaling factor "a" of the kind:

$$g_{\alpha\beta}^{\text{flat}} = \begin{pmatrix} -c^2 & 0 & 0 & 0 \\ 0 & a^2 & 0 & 0 \\ 0 & 0 & a^2 \cdot r^2 & 0 \\ 0 & 0 & 0 & a^2 \cdot r^2 \cdot \sin^2 \vartheta \end{pmatrix}, \tag{825}$$

the quantum gravity equation (802) changes to:

$$R^* = 0 = \Delta_g f = \Delta_{\text{3D-sphere}} f - a^2 \cdot \frac{f^{(2,0,0,0)}}{c^2}. \tag{826}$$

Keeping the separation approach and the solution (803) for $g[t]$ changes our result for $h[r]$ to:

$$h[r] = C_j \cdot j_l \left[a \cdot C_t \cdot r \right] + C_y \cdot y_l \left[a \cdot C_t \cdot r \right]. \tag{827}$$

Subsequently the evaluation for Eq. (824) gives:

$$C_{jl} = \frac{M \cdot c}{a \cdot \hbar} \Rightarrow C_{jl} = \frac{r_s}{a} \cdot \frac{c^3}{2 \cdot \hbar \cdot G} = \frac{r_s}{2 \cdot a \cdot \ell_P^2}. \tag{828}$$

We are going to see the importance of the scale factor in connection with the so-called Regge-trajectories [79] in the next sub-section.

5.19.2.2 (Potentially spontaneous or field induced) symmetry breaking and subsequent energy quantization

Instead of the perfect spherical symmetry, we now assume an elliptical one and introduce the following flat space metric with coordinates t, u, v, φ:

$$g_{\alpha\beta}^{\text{flat}}=\begin{pmatrix} -c^2 & 0 & 0 & 0 \\ 0 & a^2\cdot\left(H^2=\sinh[u]^2+\cos[v]^2\right) & 0 & 0 \\ 0 & 0 & a^2\cdot H^2 & 0 \\ 0 & 0 & 0 & a^2\cdot\cosh[u]^2\cdot\sin^2[v] \end{pmatrix}.$$

(829)

Here "a" shall stand for a simple scaling factor. For convenience, we apply $F[f] = (f + C_f)^2$ to the case $R^* = 0$. From Eq. (775), we obtain the following 4D-Laplacian:

$$R^*=0=\Delta_g f=\Delta_{3D-g} f-\frac{f^{(2,0,0,0)}}{c^2}\cdot a^2\cdot\left(\sinh[u]^2+\cos[v]^2\right). \quad (830)$$

Applying the separation approach $f[t,u,v,\varphi] = g[t] * h[u] * Y[v] * Z[\varphi]$ with the previously used separation parameters C_t and m (c.f. subsection "5.19.2.1 Flat space situation") in Eq. (802) gives us the following immediate solution for $g[t]$ and $Z[\varphi]$:

$$g[t]=C_{t1}\cdot\cos[c\cdot C_t\cdot t]+C_{t2}\cdot\sin[c\cdot C_t\cdot t], \quad (831)$$

$$Z[\varphi]=C_{\varphi1}\cdot\cos[m\cdot\varphi]+C_{\varphi2}\cdot\sin[m\cdot\varphi]. \quad (832)$$

The two remaining differential equations can be constructed as follows:

$$\frac{1}{\sin[v]}\cdot\frac{\partial\left(\sin[v]\cdot Y'[v]\right)}{\partial v}+\left(l\cdot(l+1)-\frac{m^2}{\sin[v]^2}\right)\cdot Y[v]=0, \quad (833)$$

$$\frac{1}{\cosh[u]}\cdot\frac{\partial\left(\cosh[u]\cdot h'[u]\right)}{\partial u}-\left(l\cdot(l+1)-\frac{m^2}{\cosh[u]^2}\right)\cdot h[u]=-C_t^2\cdot a^2\cdot H^2. \quad (834)$$

While the first equation of the two gives the usual solution:

$$Y_l^m[v]=C_{Pv}\cdot P_l^m[\cos[v]]+C_{Qv}\cdot Q_l^m[\cos[v]], \quad (835)$$

we see no way to also solve Eq. (834) in closed form. However, near the equator $v = 0$ and close to the origin $u = 0$, we have $H = 1$ and the subsequent approximated solution would read:

$$h_l^m[u]=\left(C_{Pu}\cdot P_k^q[\tanh[u]]+C_{Qu}\cdot Q_k^q[\tanh[u]]\right)\cdot\left(\cosh[u]^{-2}\right)^{1/4}$$

$$k=\frac{1}{2}(2\cdot m-1);\quad q=\frac{1}{2}\sqrt{(1+2\cdot l)^2-2\cdot a^2\cdot C_t^2}. \quad (836)$$

As known from the Schrödinger hydrogen problem, we have the conditions $C_{Qv} = 0$, $l = 0, 1, 2, 3, \ldots$ and $-l \leq m \leq l$. Only allowing for solutions free of singularities also in Eq. (836), we need to ensure $q = 0$ for any pairing of m, l. This, most interestingly, automatically quantizes C_t just as we had found before with the perfect spherical symmetry constrained by a radial boundary $r = b$ in the sub-section above.

Thus, breaking the symmetry automatically quantizes the wave numbers C_t for the time function $g[t]$. In contrast to our spherical situation as considered above, however, where we needed to add a boundary to get some C_t-quantization at all, but having this quantization completely independent from the angular one, this time the quantization of the wave numbers C_t is clearly coupled to the quantum number l via:

$$(1+2\cdot l)^2 - 2\cdot a^2\cdot C_t^2 = 0 \quad\Rightarrow\quad C_{tl} = \pm\frac{1+2\cdot l}{a\cdot\sqrt{2}}. \tag{837}$$

Just as a side note, it should be pointed out that what in the classical Schrödinger hydrogen problem gives us the main quantum number, namely the radial potential $1/r$, here now is produced by either the asymmetry or an additional boundary b.

Now back to our solution (837). Obviously, there is a non-zero positive solution for C_t, which is to be found for $l = m = 0$ and gives:

$$C_{t0} = \pm\frac{1}{a\cdot\sqrt{2}}. \tag{838}$$

Comparison with the Dirac exponent (c.f. Eqs. (823) and (824)) directly connects the quantum number l, also called the "angular momentum quantum number," with the rest mass of the object or its effective Schwarzschild radius r_s as follows:

$$C_{tl} = r_s\cdot\frac{c^3}{2\cdot\hbar\cdot G} = \frac{r_s}{2\cdot\ell_P^2} \quad\Rightarrow\quad r_s = (\pm)\frac{\ell_P^2}{a}\cdot\sqrt{2}\cdot(1+2\cdot l). \tag{839}$$

Thereby we have the inverse of the scaling factor "a" times the Planck length squared as a proportionality constant. Thus, the lowest possible—potentially effective—mass of an elliptic object with scaling factor a corresponds to a Schwarzschild radius of:

$$r_{s0} = (\pm)\sqrt{2}\cdot\frac{\ell_P^2}{a}. \tag{840}$$

An important additional aspect of the broken symmetry is the creation of more degrees of freedom. So, for our ellipsoid coordinates, we would have the three main radii instead of just one for the perfect sphere. In this case, the covariance principle comes under quite some pressure [88], and this author thinks that the application of such an option is one of the tricks of the conscious mind to distinguish itself from a computer (see the abstract of this book).

5.19.2.3 Short note about Regge trajectories

It is interesting to note that we could reproduce the Regge trajectories, giving a linear connection between angular momentum l and the square of the mass of a variety of particles (e.g., [79]). For this we have to assume the scaling factor a to be dependent on the number l as $a \sim C_a + \sqrt{l}$. This results in an l-dependency for C_{tl} and r_s as follows:

$$C_{tl} = \pm\frac{1+2\cdot l}{a\cdot\sqrt{2}} \sim \pm\frac{1+2\cdot l}{\left(C_a+\sqrt{l}\right)\cdot\sqrt{2}} \sim C_? + \sqrt{l} \quad \Rightarrow \quad C_{tl}^2 \sim l$$

$$\text{and with}: C_{tl} = r_s \cdot \frac{c^3}{2\cdot\hbar\cdot G} = \frac{r_s}{2\cdot\ell_P^2} \quad \Rightarrow \quad r_s^2 \sim l. \qquad (841)$$

Now it is not illogical to assume that after a particle scattering experiment, not just the angular momentum changes but also the radial scale of the object. In order to obtain a linear Regge-like behavior between the wave number C_t squared (and subsequently also the mass squared M^2) and the angular momentum, we saw that we would also need the radial scale of the object to increase with the square root of l. Here, however, classical physics would immediately suggest a radius increase caused by the centrifugal forces coming from the growing angular momentum. As our objects are of metric character, this would mean that space has been stretched due to the angular momentum and we might even like to apply some elastic theory in order to obtain the right "material behavior" allowing for this kind of $a^2{\sim}1$-dependency in connection with a given geometry of the object. From classical mechanics, we know that the moment of inertia I_z for ordinary bodies is proportional to their radius squared (measured from the axis of revolution, here the z-axis). Thus, any object satisfying something like $I_z{\sim}l$ would give the necessary behavior. The observant reader must have already seen that this is just another form of Einstein's mass energy relation $E{\sim}m$ in the case of rotation.

5.19.2.4 A few results

For entertainment, we evaluate the $h[u]$-behavior for a variety of numbers m (see Figs. S1 and S2). In order to make the degenerate states $-m$ and $+m$ distinguishable, we subtracted a small constant from the curves with $-m$.

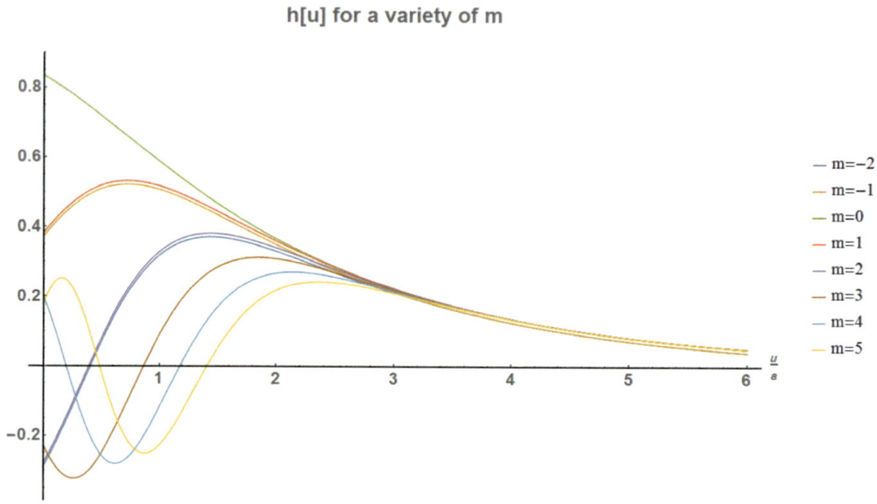

Figure S1 Distribution of function $h[u]$ from Eq. (836) for $m = -2$ to $m = 5$ with $q = 0$ (see condition (837)) and $C_{Pu} = 1$ and $C_{Qu} = 0$. The quantum states $m = +k$ and $m = -k$ appear degenerate. To make them still distinguishable, we subtracted a small constant from the curves with $-m$.

Just as add-on information, it should be mentioned that there are quite some options of additional metric solutions to the elliptic case also with non-vanishing cosmological constant (de Sitter situations). Here we only want to give one example via the approach:

$$g_{\alpha\beta}^{\text{flat}} = \begin{pmatrix} -c^2 \cdot \left(g = g[t]\right)^{-1} & 0 & 0 & 0 \\ 0 & a^2 \cdot H^2 \cdot g & 0 & 0 \\ 0 & 0 & a^2 \cdot H^2 \cdot g & 0 \\ 0 & 0 & 0 & a^2 \cdot \cosh[u]^2 \cdot \sin^2[v] \cdot g \end{pmatrix}. \tag{842}$$

The solution for g would be:

$$g[t] = \frac{1}{3} \cdot c^2 \cdot t^2 \cdot \Lambda \pm 2 \cdot c \cdot t \cdot C_{\Lambda t} \cdot \sqrt{\frac{\Lambda}{3}} + C_{\Lambda t}^2. \tag{843}$$

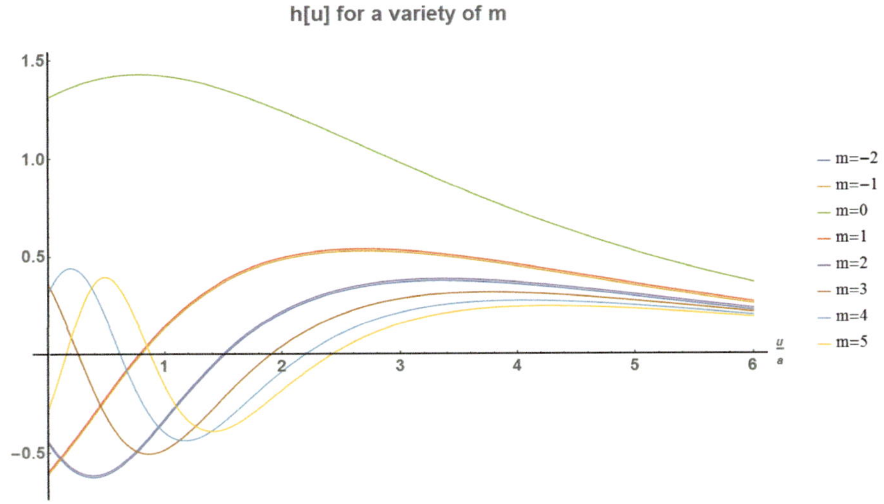

Figure S2 Distribution of function $h[u]$ from (836) for $m = -2$ to $m = 5$ with $q = 0$ (see condition (837)) and $C_{Pu} = 0$ and $C_{Qu} = 1$. The quantum states $m = +k$ and $m = -k$ appear degenerate. To make them still distinguishable, we subtracted a small constant from the curves with $-m$.

This has an interesting aspect regarding the transformed Ricci scalar as it changes the Laplace equation in the R^*=0-case to a Poisson-like equation as follows (again we have $F[f] = (f + C_f)^2$):

$$R^* = \frac{3}{2c^2\left(\cos[2v]+\cosh[2u]\right)\left(C_f+f\right)^3 g[t]} \times \left(\begin{array}{l} -8c^2 \left(\begin{array}{l} \left(\csc[y]^2 - sech[x]^2\right) f^{(0,0,0,2)} + \cot[y] f^{(0,0,1,0)} \\ + f^{(0,0,2,0)} + \tanh[x] f^{(0,1,0,0)} + f^{(0,2,0,0)} \end{array} \right) \\ + a^2 \cdot \left(\cosh[2u]+\cos[2v]\right) \cdot \left(\begin{array}{l} \left(C_f+f\right) g'[t]^2 + 8 g[t] g'[t] f^{(1,0,0,0)} \\ + 2g[t]\left(\left(C_f+f\right) g''[t] + 2g[t] f^{(2,0,0,0)}\right) \end{array} \right) \end{array} \right). \tag{844}$$

Changing the approach for $F[f]$ to $F[f] = f$ gives squares of first derivatives and requires the consideration of Dirac-like approaches as discussed in appendix R, section “5.18.9 What Is the Dirac Problem?” and the following. In the case of elliptic coordinates, however, this would lead to very complicated expressions. Therefore, we leave the task to the mathematically skilled and interested reader.

Global References

1. D. Hilbert, Die Grundlagen der Physik, Teil 1, *Göttinger Nachrichten*, 1915, 395–407.
2. A. Einstein, Grundlage der allgemeinen Relativitätstheorie, *Annalen der Physik*, 49, ser. 4, 769–822.
3. N. Schwarzer, *The Theory of Everything—Quantum and Relativity Is Everywhere—A Fermat Universe*, Jenny Stanford Publishing, 2020, ISBN-10: 9814774472.
4. N. Schwarzer, *The World Formula: A Late Recognition of David Hilbert 's Stroke of Genius*, Jenny Stanford Publishing, ISBN: 9789814877206.
5. N. Schwarzer, *The Math of Body, Soul and the Universe*, Jenny Stanford Publishing, ISBN 9789814968249.
6. M. Desmet, *The Psychology of Totalitarianism*, Chelsea Green Publishing, 2022, ISBN 978-1645021728.
7. P. A. M. Dirac, *The Quantum Theory of the Electron*, Published 1 February 1928, Proceedings of the Royal Society A. DOI: 10.1098/rspa.1928.0023.
8. E. Schrödinger, Quantisierung als Eigenwertproblem (erste Mitteilung), *Ann. Phys.*, 1926, 384(4), 361–376.
9. E. Schrödinger, Quantisierung als Eigenwertproblem (zweite Mitteilung), *Ann. Phys.*, 1926, 384(6), 489–527.
10. N. Schwarzer, *Science Riddles—Riddle No. 11: What Is Mass*?, self-published, Amazon Digital Services, 2019, Kindle.
11. N. Schwarzer, *Science Riddles—Riddle No. 12: Is There a Cosmological Spin*?, self-published, Amazon Digital Services, 2019, Kindle.
12. N. Schwarzer, *Science Riddles—Riddle No. 13: How to Solve the Flatness Problem*?, self-published, Amazon Digital Services, 2019, Kindle.
13. W. Greiner, J. Reinhardt, *Feldquantisierung, Theoretische Physik, Band 7A*, Harri Deutsch & Thun, Frankfurt am Main, 1993, ISBN 387144975X.

14. P. W. *Higgs,* Broken Symmetries and the Masses of Gauge Bosons, *Phys. Rev. Lett.*, Published 19 October 1964, 13, 508.
15. C. A. Sporea, *Notes on f(R) Theories of Gravity*, 2014, arxiv.org/pdf/1403.3852.pdf.
16. M. Desmet, *Lacan's Logic of Subjectivity—A Walk on the Graph of Desire*, QWL Press, 2019, Ghent, ISBN 9789089318695.
17. Solomon, Robert C. *Subjectivity*, ed., Honderich, Ted. Oxford Companion to Philosophy, Oxford University Press, 2005, p. 900, as cited on https://en.wikipedia.org/wiki/Subjectivity.
18. W. Greiner, J. Reinhardt, *Theoretische Physik, Band 7—Quanten Elektro Dynamik*, Verlag Harri Deutsch & Thun, Frankfurt am Main, 1995, ISBN 3-8171-1426-5.
19. A. Einstein, B. Podolsky, N. Rosen, Can quantum mechanical description of physical reality be considered complete?, *Phys. Rev.*, 1935, 47, 777.
20. A. Einstein, N. Rosen, The Particle Problem in the General Theory of Relativity, *Phys. Rev.*, 1935, 48, 73.
21. J. S. Bell, On the Einstein-Podolsky-Rosen paradox, *Physics*, 1(3), 1964, 195–200.
22. J. Maldacena, L. Susskind, *Cool Horizons for Entangled Black Holes*, 2013, arXiv:1306.0533v2.
23. L. Susskind, *Copenhagen vs Everett, Teleportation, and ER=EPR*, 2016 arXiv:1604.02589v2.
24. G. N. Remmen, N. Bao, J. Pollack, *Entanglement Conservation, ER=EPR, and a New Classical Area Theorem for Wormholes*, 2016, arXiv:1604.08217v1.
25. N. Schwarzer, *Science Riddles—Riddle No. 6: What Is Time*?, self-published, Amazon Digital Services, 2019, Kindle.
26. N. Schwarzer, *Science Riddles—Riddle No. 10: What, Where and Why Is the Higgs Field*?, self-published, Amazon Digital Services, 2019, Kindle.
27. www.astro.umd.edu/~richard/ASTRO340/class23_RM_2015.pdf.
28. en.wikipedia.org/wiki/Flatness_problem.
29. N. Schwarzer, *Einstein Had It, But He Did Not See It—Part LXXXIII: Quantum Relativity—The Two Sides of the Same Medal*, self-published, Amazon Digital Services, 2019, Kindle.
30. N. Schwarzer, *Is There an Ultimate, Truly Fundamental and Universal Computer Machine*?, self-published, Amazon Digital Services, 2019, Kindle.
31. N. Schwarzer, *Einstein Had It, But He Did Not See It—Part V: Amendment to the Heisenberg Uncertainty Principle*, self-published, Amazon Digital Services, 2017, Kindle.
32. N. Schwarzer, *Strings, Branes, Friedmanns and the Matryoshka Universe, Part 7D of Medical Socio-Economic Quantum Gravity*, self-published, Amazon Digital Services, December 2022, Kindle.

33. T. Bodan, *7 Days—How to Explain the World to My Dying Child, Illustrated Version*, self-published, BoD Classic, ISBN 9783752639728.
34. T. Bodan, *The Eighth Day—Two Jews against The Third Reich, Illustrated Version*, self-published, BoD Classic, ISBN 9783753417257.
35. T. Bodan, *7 Days—How to Explain the World to My Dying Child, 7 Tage—Wie erkläre ich meinem sterbenden Kind die Welt*, in German, self-published, Amazon, 2016, ASIN: 1520917562.
36. T. Bodan, *The Eighth Day—Holocaust and the World's Biggest Mysteries: The Other Final Solution—With Mathematical Elaborations (English Edition)*, self-published, Amazon Digital Services, 2015, Kindle.
37. T. Bodan, *The Eighth Day*, self-published, Amazon Digital Services, 2015, Kindle.
38. N. Schwarzer, T. Bodan, *Sherlock, Watson, Einstein—Part 1: The Mystery of Entanglement and the Spooky Action at a Distance*, self-published, Amazon Digital Services, 2018, Kindle.
39. N. Schwarzer, T. Bodan, *Sherlock, Watson, Stalin Part 2: The Hell of Gender, Merkel, Communism and the Dictatorship of Parasites*, self-published, Amazon Digital Services, 2018, Kindle.
40. T. Bodan, *EU vs. Britain—The Other Monkey Trial*, self-published, Amazon, 2017, ASIN: 1521304793.
41. N. Schwarzer, *The Quantum Black Hole, Part 7B of Medical Socio-Economic Quantum Gravity*, self-published, Amazon Digital Services, March 2021, Kindle.
42. N. Schwarzer, *Quantum Cosmology—A Simple Answer to the Flatness Riddle, Part 7C of Medical Socio-Economic Quantum Gravity*, self-published, Amazon Digital Services, April 2021, Kindle.
43. G. Saad, *The Parasitic Mind: How Infectious Ideas Are Killing Common Sense*, 2020, Regnery Publishing, ISBN-10 162157959X.
44. en.wikipedia.org/wiki/Higgs_boson.
45. N. Schwarzer, *Einstein Had It, But He Did Not See It—Part LXXVI: Quantum Universes—We Don't Need No... an Inflation*, self-published, Amazon Digital Services, 2019, Kindle.
46. N. Schwarzer, *Einstein Had It, But He Did Not See It—Part LXXVII: Matter Is Nothing and So Nothing Matters*, self-published, Amazon Digital Services, 2019, Kindle.
47. N. Schwarzer, *The Photon*, self-published, Amazon Digital Services, 2016, Kindle.
48. K. Schwarzschild, Über das Gravitationsfeld einer Kugel aus inkompressibler Flüssigkeit nach der Einsteinschen Theorie [On the gravitational field of a ball of incompressible fluid following Einstein's theory], *Sitzungsberichte der Königlich-Preussischen Akademie der Wissenschaften*, in German, 1916, Berlin, pp. 424–434.

49. J. D. Bekenstein, Black holes and entropy, *Phys. Rev. D*, 1973, 7, 2333–2346.

50. J. D. Bekenstein, Information in the holographic universe, *Scientific American*, August 2003, 289(2), p. 61.

51. P. Frank, Propagation of error and reliability of global air temperature projection, *Front. Earth Sci.*, September 6, 2019, https://doi.org/10.3389/feart.2019.00223 (or: www.frontiersin.org/articles/10.3389/feart.2019.00223/full).

52. www.quora.com/What-does-Michael-Mann-s-court-battle-loss-mean-to-the-notion-of-climate-change.

53. N. Schwarzer, *Science Riddles—Riddle No. 8: Could the Schwarzschild Metric Contain Its Own Quantum Solution*?, self-published, Amazon Digital Services, 2018, Kindle.

54. L. Susskind, *The Black Hole War: My battle with Stephen Hawking to Make the World Safe for Quantum Mechanics*, Little, Brown, 2008.

55. N. Schwarzer, *How Systems Evolve—A Red Pill Course on Evolution*, self-published, Amazon Digital Services, 2022, Kindle.

56. M. M. Vopsen, The information catastrophe, *AIP Adv.* 2020, 10, 085014; https://doi.org/10.1063/5.0019941.

57. N. Schwarzer, *Science Riddles—Riddle No. 20: Second Law of Thermodynamics—Where Is Its Fundamental Origin*?, www.amazon.com, ASIN: B07Y79BTT9s.

58. N. Schwarzer, *Science Riddles—Riddle No. 21: Evolution—Where Is Its Fundamental Origin*?, www.amazon.com, ASIN: B07YKL37DL.

59. N. Schwarzer, *Quantum Gravity Thermodynamics II—Derivation of the Second Law of Thermodynamics and the Metric Driving Force of Evolution*, www.amazon.com, ASIN: B07XWPXF3G.

60. C. Domb, M. S. *Green, Phase Transitions and Critical Phenomena*, Academic Press, London, New York, San Francisco, vol. 6, 1976, ISBN 0-12-220306-2.

61. N. Schwarzer, *Science Riddles—Riddle No. 3: Was (IS) There a Big Bang Bomb*?, self-published, Amazon Digital Services, 2019, Kindle.

62. N. Schwarzer, *The Quantum Gravity War—How Will the Nearby Unification of Physics Change Future Warfare*?, Jenny Stanford Publishing, ISBN: 9789814968584.

63. S. Noiré (T. Bodan Co-authoring), *The Womanizer and His Image of Diseases—About Love Fields and Isolation Traps—A Victim Reveals, Illustrated Version*, self-published, BoD Classic, ISBN 9783756841370.

64. N. Schwarzer, *Towards Quantum Einstein Field Equations, Part 7 of Medical Socio-Economic Quantum Gravity*, self-published, Amazon Digital Services, December 2020, Kindle.

65. N. Schwarzer, *Einstein Had It, But He Did Not See It—Part LXXXII: Half Spin Hydrogen*, self-published, Amazon Digital Services, 2019, Kindle.

66. N. Schwarzer, *Einstein Had It, But He Did Not See It—Part LXXIX: Dark Matter Options*, self-published, Amazon Digital Services, 2019, Kindle.
67. H. Haken, H. Chr. Wolf, *Atom- und Quantenphysik*, 4th ed., in German, Springer Heidelberg, 1990, ISBN 0-387-52198-4.
68. N. Schwarzer, *The Dirac Miracle, Part 8a of Medical Socio-Economic Quantum Gravity*, self-published, Amazon Digital Services, May 2021, Kindle.
69. N. Schwarzer, *My Horcruxes—A Curvy Math to Salvation, Part 9 of Medical Socio-Economic Quantum Gravity*, self-published, Amazon Digital Services, 2021, Kindle.
70. N. Schwarzer, *Einstein Had It, But He Did Not See It—Part LII: Why We Might Live Inside a Black Hole*, self-published, Amazon Digital Services, 2018, Kindle.
71. N. Schwarzer, *Einstein Had It, But He Did Not See It—Part LIII: They Are Everywhere! Why There Are So Many Sigmoid-Dependencies in This World*, self-published, Amazon Digital Services, 2018, Kindle.
72. N. Schwarzer, *Einstein Had It, But He Did Not See It—Part LX: The Hitch Hiker's Guide to the Creation of a Universe*, self-published, Amazon Digital Services, 2018, Kindle.
73. N. Schwarzer, *Einstein Had It, But He Did Not See It—Part LXI: Don't Panic in a Universe with Only One Dimension*, self-published, Amazon Digital Services, 2018, Kindle.
74. St. Wolfram, The Foundations of Mathematics and Mathematica, online at www.stephenwolfram.com/publications/academic/foundations-mathematics-mathematica.pdf.
75. N. Schwarzer, *Einstein Had It, But He Did Not See It—Part LXXV: The Metric Creation of Matter*, www.amazon.com, ASIN: B07ND3LWZJ.
76. N. Schwarzer, *Einstein Had It, But He Did Not See It—Part LXXVIII: About a Possible Quantum Gravity Alternative to Schwarzschild and Co.*, www.amazon.com, ASIN: B07NXRL2BH.
77. N. N. Lebedev, *Special Functions and Their Applications*, Translated from Russian by R. A. Silverman, Chapter 5, Dover, 1972, pp. 98–142.
78. Q. Wang, O. Ronneberger, H. Burkhardt, *Fourier Analysis in Polar and Spherical Coordinates*—Internal Report 1/08, https://lmb.informatik.uni-freiburg.de/papers/download/wa_report01_08.pdf.
79. D. Ebert, R. N. Faustov, V. O. Galkin, Mass Spectra and Regge Trajectories of Light Mesons in the Relativistic Quark Model, *Phys. Rev. D*, 2009, 79, 114029, arXiv:0903.5183.
80. N. Schwarzer, *Love Fields—The Application of Mathematical Psychology for the Understanding and Use of the Supposedly Greatest Force in the Universe,*

upcoming book-project with Jenny Stanford Publishing (e.g. see: https://www.jennystanford.com/9789814968249/the-math-of-body-soul-and-the-universe/#tab-4).

81. W. Greiner, Quantenmechanik—Teil 1 Einführung, in: *Theoretische Physik*, 5th ed., in German, Verlag Harri Deutsch, Frankfurt am Main, vol. 4, 1992, ISBN 3-8171-1206-8.

82. F. Schwabl, *Quantenmechanik*, 2nd ed., in German, Springer-Verlag Berlin Heidelberg New York, 1990, ISBN 3-540-52359-6.

83. N. Schwarzer, *The Metric Potential—How Can Quantum Gravity Help Us to Under-Stand the Origin of Forces*, self-published, Amazon Digital Services, ASIN: B0BXSHQZX1, 2023, Kindle.

84. N. Schwarzer, *The Quantum Gravity Well—From Particle Physics to Psychology—A True Theory of Everything?*, self-published, Amazon Digital Services, ASIN: B0C3QV9S1Y, 2023, Kindle.

85. N. Schwarzer, *Understanding Quantum Tunneling—How Can Quantum Gravity Help Us to Get the Gist about This Strange Phenomenon*, self-published, Amazon Digital Services, 2023, Kindle, ASIN: B0C241FQH3.

86. N. Schwarzer, *Metric Interpretation of Quantum Theory—How Can Quantum Gravity Help Us to Understand Quantum Strangeness*, self-published, Amazon Digital Services, ASIN: B0BXQRKPV7, 2023, Kindle.

87. N. Schwarzer, *The Wave Particle Dualism—How Does Quantum Gravity Explain this Apparent Peculiarity*, self-published, Amazon Digital Services, ASIN: B0C5TMS8W5, 2023, Kindle.

88. N. Schwarzer, *The Covariance Principle Is Dead, Long Live the Covariance Principle—How Quantum Gravity Kills a Cornerstone of Physics*, self-published, Amazon Digital Services, 2023, Kindle, ASIN: B0C96L41LS.

Index

For Product Safety Concerns and Information please contact our EU representative GPSR@taylorandfrancis.com Taylor & Francis Verlag GmbH, Kaufingerstraße 24, 80331 München, Germany